Computational Statistics

Computational Statistics

Geof H. Givens
Jennifer A. Hoeting

A JOHN WILEY & SONS, INC., PUBLICATION

Published by John Wiley & Sons, Inc., Hoboken, New Jersey.
Published simultaneously in Canada.

For general information on our other products and services please contact our Customer Care Department within the U.S. at 877-762-2974, outside the U.S. at 317-572-3993 or fax 317-572-4002.

Wiley also publishes its books in a variety of electronic formats. Some content that appears in print, however, may not be available in electronic format.

Library of Congress Cataloging-in-Publication Data is available.

ISBN 0-471-46124-5

Printed in the United States of America

10 9 8 7 6 5 4

To our parents,

Jim, Eileen, Alan, and Arlene

Preface

This book covers most topics needed to develop a broad and thorough working knowledge of modern statistical computing and computational statistics. We seek to develop a practical understanding of how and why existing methods work, enabling readers to use modern statistical methods effectively. Since many new methods are built from components of existing techniques, our ultimate goal is to provide scientists with the tools they need to contribute new ideas to the field.

Achieving these goals requires familiarity with diverse topics in statistical computing, computational statistics, computer science, and numerical analysis. Our choice of topics reflects our view of what is central to this evolving field, and what will be interesting and useful for our readers. We pragmatically assigned priority to topics that can be of the most benefit to students and researchers most quickly.

Some topics we omitted represent important areas of past and present research in the field, but their priority here is lowered by the availability of high-quality software. For example, the generation of pseudo-random numbers is a classic topic, but one that we prefer to address by giving students reliable software. Some topics, such as numerical linear algebra, are on the borderline. Such topics are critical for many applications, yet good routines are generally available. In our judgment, the frequency with which one must shelve the routines and dig into the details of numerical linear algebra falls (barely) below the threshold we set for inclusion in this book. Among the classic topics we have chosen to cover are optimization and numerical integration. We include these because (i) they are cornerstones of frequentist and Bayesian inference; (ii) routine application of available software often fails for hard problems; and (iii) the methods themselves are often secondary components of other statistical computing algorithms.

Our use of the adjective *modern* is potentially troublesome: there is no way that this book can cover all the latest, greatest techniques. We have not even tried. Some topics, such as heuristic search and Markov chain Monte Carlo, simply move too quickly. We have instead tried to offer a reasonably up-to-date survey of a broad portion of the field, while leaving room for diversions and esoterica. Some topics (e.g., principal curves and tabu search) are included simply because they are interesting and provide very different perspectives on familiar problems. Perhaps a future researcher may draw ideas from such topics to design a creative and effective new algorithm.

Our target audience includes graduate students in statistics and related fields, working statisticians, and quantitative empirical scientists in other fields. We hope such readers may use the book when applying standard methods and developing new methods.

The level of mathematics expected of the reader does not extend much beyond Taylor series and linear algebra. Breadth of mathematical training is more helpful than depth. Essential review is provided in Chapter 1. More advanced readers will find greater mathematical detail in the wide variety of high-quality books available on specific topics, many of which are referenced in the text. Other readers caring less about analytical details may prefer to focus on our descriptions of algorithms and examples.

The expected level of statistics is equivalent to that obtained by a graduate student in his or her first year of study of the theory of statistics and probability. An understanding of maximum likelihood methods, Bayesian methods, elementary asymptotic theory, Markov chains, and linear models is most important. Many of these topics are reviewed in Chapter 1.

With respect to computer programming, we find that good students can learn as they go. However, a working knowledge of a suitable language allows implementation of the ideas covered in this book to progress much more quickly. We have chosen to forgo any language-specific examples, algorithms, or coding. For those wishing to learn a language while they study this book, we recommend you choose a high-level, interactive package that permits the flexible design of graphical displays and includes supporting statistics and probability functions. At the time of writing, we recommend S-Plus, R, and MATLAB.[1] These are the sort of languages often used by researchers during the development of new statistical computing techniques, and are suitable for implementing all the methods we describe, except in some cases for problems of vast scope or complexity. Of course, lower-level languages such as C++ can also be used, and are favored for professional grade implementation of algorithms after researchers have refined the methodology.

Even adept computer programmers may have little understanding of how mathematics is carried out in the binary world of a computer. Mysterious problems with

---

[1]Websites for these software packages are www.insightful.com, www.r-project.org, and www.mathworks.com, respectively. R is free software reimplementing portions of S-Plus; the others are commercial.

full-rank matrices that appear noninvertible, integrals and likelihoods that vanish, numerical approximations that appear more precise than they really are, and other oddities are not unusual. While not dismissing the importance of computer arithmetic and numerically stable computation, we prefer to focus on the big picture of how algorithms work and to sweep under the rug some of the nitty-gritty numerical computation details.

The book is organized into three major parts: optimization (Chapters 2, 3, and 4), integration (Chapters 5, 6, 7, and 8), and smoothing (Chapters 10, 11, and 12). Chapter 9 adds another essential topic, the bootstrap. The chapters are written to stand independently, so a course can be built by selecting the topics one wishes to teach. For a one-semester course, our selection typically weights most heavily topics from Chapters 2, 5, 6, 7, 9, 10, and 11. With a leisurely pace or more thorough coverage, a shorter list of topics could still easily fill a semester course. There is sufficient material here to provide a thorough one-year course of study, notwithstanding any supplemental topics one might wish to teach.

A variety of homework problems are included at the end of each chapter. Some are straightforward, while others require the student to develop a thorough understanding of the model/method being used, to carefully (and perhaps cleverly) code a suitable technique, and to devote considerable attention to the interpretation of results.

The datasets discussed in the text and problems are available from the book website, *www.stat.colostate.edu/computationalstatistics*. You are holding a fourth printing of the book, which corrects and improves issues identified in earlier printings. Errata for each printing can be found on the book website. Responsibility for all errors lies with us.

Geof H. Givens and Jennifer A. Hoeting

Fort Collins, Colorado

Acknowledgments

We owe a great deal of intellectual debt to Adrian Raftery, who deserves special thanks not only for his teaching and advising, but also for his unwavering support and his seemingly inexhaustible supply of good ideas. In addition, we thank our influential advisors and teachers at the University of Washington Statistics Department, including David Madigan, Werner Stuetzle, and Judy Zeh. Of course, each of our chapters could be expanded into a full-length book, and great scholars have already done so. We owe much to their efforts, upon which we relied when developing our course and our manuscript.

The course upon which this book is based was developed and taught by us at Colorado State University from 1994 onwards. We thank our colleagues in the Statistics Department for their continued support. The late Richard Tweedie merits particular acknowledgment for his mentoring during the early years of our careers. Thanks also are due to our long-suffering students who have been semi-willing guinea pigs over the years. Portions of this book were written at the Department of Mathematics and Statistics, University of Otago, New Zealand, whose faculty graciously hosted us during sabbatical.

Our manuscript was greatly improved through the constructive reviews of John Bickham, Kate Cowles, Jan Hannig, Alan Herlihy, David Hunter, Devin Johnson, Michael Newton, Doug Nychka, Steve Sain, David W. Scott, N. Scott Urquhart, Haonan Wang, Darrell Whitley, and eight anonymous referees. Our editor Steve Quigley and the folks at Wiley were supportive and helpful during the publication process. We thank Nélida Pohl for permission to adapt her photograph in the cover design. We also owe special note of thanks to Zube (a.k.a. John Dzubera), who kept our own computers running despite our best efforts.

Funding from National Science Foundation (NSF) CAREER grant #SBR-9875508 was a significant source of support for the first author during the preparation of this book. He also thanks his colleagues and friends in the North Slope Borough, Alaska, Department of Wildlife Management for their longtime research support. The second author gratefully acknowledges the support of STAR Research Assistance Agreement CR-829095 awarded to Colorado State University by the U.S. Environmental Protection Agency (EPA). The views expressed here are solely those of the authors. NSF and EPA do not endorse any products or commercial services mentioned herein.

Finally, we thank our parents, to whom the book is dedicated, for enabling and supporting our educations and for providing us with the "stubborness gene" necessary for graduate school, the tenure track, or book publication—take your pick!

G. H. G. and J. A. H.

Contents

1

Review

This chapter reviews notation and background material in mathematics, probability, and statistics. Readers may wish to skip this chapter and turn directly to Chapter 2, returning here only as needed.

1.1 MATHEMATICAL NOTATION

We use boldface to distinguish a vector $\mathbf{x} = (x_1, \ldots, x_p)$ or a matrix $\mathbf{M}$ from a scalar variable x or a constant M. A vector-valued function $\mathbf{f}$ evaluated at $\mathbf{x}$ is also boldfaced, as in $\mathbf{f}(\mathbf{x}) = (f_1(\mathbf{x}), \ldots, f_p(\mathbf{x}))$. The transpose of $\mathbf{M}$ is denoted $\mathbf{M}^T$.

Unless otherwise specified, all vectors are considered to be column vectors, so, for example, a $n \times p$ matrix can be written as $\mathbf{M} = (\mathbf{x}_1 \ \ldots \ \mathbf{x}_n)^T$. Let $\mathbf{I}$ denote an identity matrix, and $\mathbf{1}$ and $\mathbf{0}$ denote vectors of ones and zeros, respectively.

A symmetric square matrix $\mathbf{M}$ is *positive definite* if $\mathbf{x}^T \mathbf{M} \mathbf{x} > 0$ for all nonzero vectors $\mathbf{x}$. Positive definiteness is equivalent to the condition that all eigenvalues of $\mathbf{M}$ are positive. $\mathbf{M}$ is *nonnegative definite* or *positive semidefinite* if $\mathbf{x}^T \mathbf{M} \mathbf{x} \geq 0$ for all nonzero vectors $\mathbf{x}$.

The derivative of a function f, evaluated at x, is denoted $f'(x)$. When $\mathbf{x} = (x_1, \ldots, x_p)$, the *gradient* of f at $\mathbf{x}$ is $\mathbf{f}'(\mathbf{x}) = \left(\frac{df(\mathbf{x})}{dx_1}, \ldots, \frac{df(\mathbf{x})}{dx_p} \right)$. The *Hessian matrix* for f at $\mathbf{x}$ is $\mathbf{f}''(\mathbf{x})$ having (i, j)th element equal to $\frac{d^2 f(\mathbf{x})}{dx_i \, dx_j}$. The negative Hessian has important uses in statistical inference.

Let $\mathbf{J}(\mathbf{x})$ denote the *Jacobian matrix* evaluated at $\mathbf{x}$ for the one-to-one mapping $\mathbf{y} = \mathbf{f}(\mathbf{x})$. The (i, j)th element of $\mathbf{J}(\mathbf{x})$ is equal to $\frac{df_i(\mathbf{x})}{dx_j}$.

A *functional* is a real-valued function on a space of functions. For example, if $T(f) = \int_0^1 f(x)\,dx$, then the functional T maps suitably integrable functions onto the real line.

The indicator function $1_{\{A\}}$ equals 1 if A is true and 0 otherwise. The real line is denoted $\Re$, and p-dimensional real space is $\Re^p$.

1.2   TAYLOR'S THEOREM AND MATHEMATICAL LIMIT THEORY

First, we define standard "big oh" and "little oh" notation for describing the relative orders of convergence of functions. Let the functions f and g be defined on a common, possibly infinite interval. Let z_0 be a point in this interval or a boundary point of it (i.e., $-\infty$ or ∞). We require $g(z) \neq 0$ for all $z \neq z_0$ in a neighborhood of z_0. Then we say

$$f(z) = \mathcal{O}(g(z)) \tag{1.1}$$

if there exists a constant M such that $|f(z)| \leq M|g(z)|$ as $z \to z_0$. For example, $\frac{n+1}{3n^2} = \mathcal{O}(n^{-1})$, and it is understood that we are considering $n \to \infty$. If $\lim_{z \to z_0} f(z)/g(z) = 0$, then we say

$$f(z) = o(g(z)). \tag{1.2}$$

For example, $f(x_0 + h) - f(x_0) = hf'(x_0) + o(h)$ as $h \to 0$ if f is differentiable at x_0. The same notation can be used for describing the convergence of a sequence $\{x_n\}$ as $n \to \infty$, by letting $f(n) = x_n$.

Taylor's theorem provides a polynomial approximation to a function f. Suppose f has finite $(n+1)$th derivative on (a, b) and continuous nth derivative on $[a, b]$. Then for any $x_0 \in [a, b]$ distinct from x, the Taylor series expansion of f about x_0 is

$$f(x) = \sum_{i=0}^{n} \frac{1}{i!} f^{(i)}(x_0)(x - x_0)^i + R_n, \tag{1.3}$$

where $f^{(i)}(x_0)$ is the ith derivative of f evaluated at x_0, and

$$R_n = \frac{1}{(n+1)!} f^{(n+1)}(\xi)(x - x_0)^{n+1} \tag{1.4}$$

for some point ξ in the interval between x and x_0. As $|x - x_0| \to 0$, note that $R_n = \mathcal{O}(|x - x_0|^{n+1})$.

The multivariate version of Taylor's theorem is analogous. Suppose f is a real-valued function of a p-dimensional variable $\mathbf{x}$, possessing continuous partial derivatives of all orders up to and including $n+1$ with respect to all coordinates, in an open convex set containing $\mathbf{x}$ and $\mathbf{x}_0 \neq \mathbf{x}$. Then

$$f(\mathbf{x}) = f(\mathbf{x}_0) + \sum_{i=1}^{n} \frac{1}{i!} D^{(i)}(f; \mathbf{x}_0, \mathbf{x} - \mathbf{x}_0) + R_n, \tag{1.5}$$

where

$$D^{(i)}(f; \mathbf{x}, \mathbf{y}) = \sum_{j_1=1}^{p} \cdots \sum_{j_i=1}^{p} \left\{ \left(\frac{d^i}{dt_{j_1} \cdots dt_{j_i}} f(\mathbf{t}) \Big|_{\mathbf{t}=\mathbf{x}} \right) \prod_{k=1}^{i} y_{j_k} \right\} \tag{1.6}$$

and

$$R_n = \frac{1}{(n+1)!} D^{(n+1)}(f; \boldsymbol{\xi}, \mathbf{x} - \mathbf{x}_0) \tag{1.7}$$

for some $\boldsymbol{\xi}$ on the line segment joining $\mathbf{x}$ and $\mathbf{x}_0$. As $|\mathbf{x} - \mathbf{x}_0| \to 0$, note that $R_n = \mathcal{O}(|\mathbf{x} - \mathbf{x}_0|^{n+1})$.

The *Euler–Maclaurin formula* is useful in many asymptotic analyses. If f has $2n$ continuous derivatives in $[0, 1]$, then

$$\int_0^1 f(x)\, dx = \frac{f(0) + f(1)}{2}$$

$$-\sum_{i=0}^{n-1} \frac{b_{2i}(f^{(2i-1)}(1) - f^{(2i-1)}(0))}{(2i)!} - \frac{b_{2n} f^{(2n)}(\xi)}{(2n)!}, \tag{1.8}$$

where $0 \leq \xi \leq 1$, $f^{(j)}$ is the jth derivative of f, and $b_j = B_j(0)$ can be determined using the recursion relation

$$\sum_{j=0}^{m} \binom{m+1}{j} B_j(z) = (m+1)z^m \tag{1.9}$$

initialized with $B_0(z) = 1$. The proof of this result is based on repeated integrations by parts [328].

Finally, we note that it is sometimes desirable to approximate the derivative of a function numerically, using finite differences. For example, the ith component of the gradient of f at $\mathbf{x}$ can be approximated by

$$\frac{df(\mathbf{x})}{dx_i} \approx \frac{f(\mathbf{x} + \epsilon_i \mathbf{e}_i) - f(\mathbf{x} - \epsilon_i \mathbf{e}_i)}{2\epsilon_i}, \tag{1.10}$$

where ϵ_i is a small number and $\mathbf{e}_i$ is the unit vector in the ith coordinate direction. Typically, one might start with, say, $\epsilon_i = 0.01$ or 0.001 and approximate the desired derivative for a sequence of progressively smaller ϵ_i. The approximation will generally improve until ϵ_i becomes small enough that the calculation is degraded and eventually dominated by computer roundoff error introduced by subtractive cancellation. Introductory discussion of this approach and a more sophisticated Richardson extrapolation strategy for obtaining greater precision are provided in [328]. Finite differences can also be used to approximate the second derivative of f at $\mathbf{x}$ via

$$\frac{df(\mathbf{x})}{dx_i\, dx_j} \approx \frac{1}{4\epsilon_i \epsilon_j} \bigg(f(\mathbf{x} + \epsilon_i \mathbf{e}_i + \epsilon_j \mathbf{e}_j) - f(\mathbf{x} + \epsilon_i \mathbf{e}_i - \epsilon_j \mathbf{e}_j)$$

$$- f(\mathbf{x} - \epsilon_i \mathbf{e}_i + \epsilon_j \mathbf{e}_j) + f(\mathbf{x} - \epsilon_i \mathbf{e}_i - \epsilon_j \mathbf{e}_j) \bigg) \tag{1.11}$$

with similar sequential precision improvements.

1.3 STATISTICAL NOTATION AND PROBABILITY DISTRIBUTIONS

We use capital letters to denote random variables, such as Y or $\mathbf{X}$, and lowercase letters to represent specific realized values of random variables such as y or $\mathbf{x}$. The probability density function of X is denoted f; the cumulative distribution function is F. We use the notation $X \sim f(x)$ to mean that X is distributed with density $f(x)$. Frequently, the dependence of $f(x)$ on one or more parameters also will be denoted with a conditioning bar, as in $f(x|\alpha, \boldsymbol{\beta})$. Because of the diversity of topics covered in this book, we want to be careful to distinguish when $f(x|\alpha)$ refers to a density function as opposed to the evaluation of that density at a point x. When the meaning is unclear from the context, we will be explicit, for example by using $f(\cdot|\alpha)$ to denote the function. When it is important to distinguish among several densities, we may adopt subscripts referring to specific random variables, so that the density functions for X and Y are f_X and f_Y respectively. We use the same notation for distributions of discrete random variables and in the Bayesian context.

The conditional distribution of X given that Y equals y (i.e., $X|Y = y$) is described by the density denoted $f(x|y)$, or $f_{X|Y}(x|y)$. In this case, we write that $X|Y$ has density $f(x|Y)$. For notational simplicity we allow density functions to be implicitly specified by their arguments, so we may use the same symbol, say f, to refer to many distinct functions, as in the equation $f(x, y|\mu) = f(x|y, \mu)f(y|\mu)$. Finally, $f(X)$ and $F(X)$ are random variables: the evaluations of the density and cumulative distribution functions, respectively, at the random argument X.

The expectation of a random variable is denoted $\mathrm{E}\{X\}$. Unless specifically mentioned, the distribution with respect to which an expectation is taken is the distribution of X or should be implicit from the context. To denote the probability of an event A, we use $P[A] = \mathrm{E}\{1_{\{A\}}\}$. The conditional expectation of $X|Y = y$ is $\mathrm{E}\{X|y\}$. When Y is unknown, $\mathrm{E}\{X|Y\}$ is a random variable that depends on Y. Other attributes of the distribution of X and Y include $\mathrm{var}\{X\}$, $\mathrm{cov}\{X, Y\}$, $\mathrm{cor}\{X, Y\}$, and $\mathrm{cv}\{X\} = \mathrm{var}\{X\}^{1/2}/\mathrm{E}\{X\}$. These quantities are the variance of X, the covariance and correlation of X and Y, and the coefficient of variation of X, respectively.

A useful result regarding expectations is *Jensen's inequality*. Let g be a *convex function* on a possibly infinite open interval I, so

$$g(\lambda x + (1 - \lambda)y) \leq \lambda g(x) + (1 - \lambda)g(y) \tag{1.12}$$

for all $x, y \in I$ and all $0 < \lambda < 1$. Then Jensen's inequality states that $\mathrm{E}\{g(X)\} \geq g(\mathrm{E}\{X\})$ for any random variable X having $P[X \in I] = 1$.

Tables 1.1, 1.2, and 1.3 provide information about many discrete and continuous distributions used throughout this book. We refer to following well-known combinatorial constants:

$$n! = n(n - 1)(n - 2) \cdots (3)(2)(1) \quad \text{with } 0! = 1, \tag{1.13}$$

$$\binom{n}{k} = \frac{n!}{k!(n - k)!}, \tag{1.14}$$

$$\begin{pmatrix} n \\ k_1 \quad \cdots \quad k_m \end{pmatrix} = \frac{n!}{\prod_{i=1}^{m} k_i!} \quad \text{where } n = \sum_{i=1}^{m} k_i, \tag{1.15}$$

$$\Gamma(r) = \begin{cases} (r-1)! & \text{for } r = 1, 2, \ldots \\ \int_0^{\infty} t^{r-1} \exp\{-t\}\, dt & \text{for general } r > 0. \end{cases} \tag{1.16}$$

It is worth knowing that $\Gamma\left(\frac{1}{2}\right) = \sqrt{\pi}$ and $\Gamma\left(n + \frac{1}{2}\right) = \frac{1 \times 3 \times 5 \times \cdots \times (2n-1)\sqrt{\pi}}{2^n}$ for positive integer n.

Many of the distributions commonly used in statistics are members of an *exponential family*. A k-parameter exponential family density can be expressed as

$$f(x|\boldsymbol{\gamma}) = c_1(x)c_2(\boldsymbol{\gamma}) \exp \left\{ \sum_{i=1}^{k} y_i(x)\theta_i(\boldsymbol{\gamma}) \right\} \tag{1.17}$$

for nonnegative functions c_1 and c_2. The vector $\boldsymbol{\gamma}$ denotes the familiar parameters, such as λ for the Poisson density and p for the binomial density. The real-valued $\theta_i(\boldsymbol{\gamma})$ are the *natural*, or *canonical, parameters*, which are usually transformations of $\boldsymbol{\gamma}$. The $y_i(x)$ are the sufficient statistics for the canonical parameters. It is straightforward to show

$$E\{\mathbf{y}(X)\} = \boldsymbol{\kappa}'(\boldsymbol{\theta}) \tag{1.18}$$

and

$$\text{var}\{\mathbf{y}(X)\} = \boldsymbol{\kappa}''(\boldsymbol{\theta}), \tag{1.19}$$

where $\kappa(\boldsymbol{\theta}) = -\log c_3(\boldsymbol{\theta})$, letting $c_3(\boldsymbol{\theta})$ denote the reexpression of $c_2(\boldsymbol{\gamma})$ in terms of the canonical parameters $\boldsymbol{\theta} = (\theta_1, \ldots, \theta_k)$, and $\mathbf{y}(X) = (y_1(X), \ldots, y_k(X))$. These results can be rewritten in terms of the original parameters $\boldsymbol{\gamma}$ as

$$E\left\{ \sum_{i=1}^{k} \frac{d\theta_i(\boldsymbol{\gamma})}{d\gamma_j} y_i(X) \right\} = -\frac{d}{d\gamma_j} \log c_2(\boldsymbol{\gamma}) \tag{1.20}$$

and

$$\text{var}\left\{ \sum_{i=1}^{k} \frac{d\theta_i(\boldsymbol{\gamma})}{d\gamma_j} y_i(X) \right\} = -\frac{d^2}{d\gamma_j^2} \log c_2(\boldsymbol{\gamma}) - E\left\{ \sum_{i=1}^{k} \frac{d^2\theta_i(\boldsymbol{\gamma})}{d\gamma_j^2} y_i(X) \right\}. \tag{1.21}$$

Example 1.1 (Poisson) The Poisson distribution belongs to the exponential family with $c_1(x) = 1/x!$, $c_2(\lambda) = \exp\{-\lambda\}$, $y(x) = x$, and $\theta(\lambda) = \log \lambda$. Deriving moments in terms of θ, we have $\kappa(\theta) = \exp\{\theta\}$, so $E\{X\} = \kappa'(\theta) = \exp\{\theta\} = \lambda$ and $\text{var}\{X\} = \kappa''(\theta) = \exp\{\theta\} = \lambda$. The same results may be obtained with (1.20) and (1.21), noting that $\frac{d\theta}{d\lambda} = \frac{1}{\lambda}$. For example, (1.20) gives $E\{\frac{X}{\lambda}\} = 1$. $\qquad\square$

Table 1.1 Notation and description for common probability distributions of discrete random variables.

Name	Notation and Parameter Space	Density and Sample Space	Mean and Variance
Bernoulli	$X \sim \text{Bernoulli}(p)$ $0 \le p \le 1$	$f(x) = p^x(1-p)^{1-x}$ $x = 0 \text{ or } 1$	$E\{X\} = p$ $\text{var}\{X\} = p(1-p)$
Binomial	$X \sim \text{Bin}(n, p)$ $0 \le p \le 1$ $n \in \{1, 2, \ldots\}$	$f(x) = \binom{n}{x} p^x(1-p)^{n-x}$ $x = 0, 1, \ldots, n$	$E\{X\} = np$ $\text{var}\{X\} = np(1-p)$
Multinomial	$\mathbf{X} \sim \text{Multinomial}(n, \mathbf{p})$ $\mathbf{p} = (p_1, \ldots, p_k)$ $0 \le p_i \le 1 \text{ and } n \in \{1, 2, \ldots\}$ $\sum_{i=1}^k p_i = 1$	$f(\mathbf{x}) = \binom{n}{x_1 \ \ldots \ x_k} \prod_{i=1}^k p_i^{x_i}$ $\mathbf{x} = (x_1, \ldots, x_k) \text{ and } x_i \in \{0, 1, \ldots, n\}$ $\sum_{i=1}^k x_i = n$	$E\{\mathbf{X}\} = n\mathbf{p}$ $\text{var}\{X_i\} = np_i(1-p_i)$ $\text{cov}\{X_i, X_j\} = -np_ip_j$
Negative Binomial	$X \sim \text{NegBin}(r, p)$ $0 \le p \le 1$ $r \in \{1, 2, \ldots\}$	$f(x) = \binom{x+r-1}{r-1} p^r(1-p)^x$ $x \in \{0, 1, \ldots\}$	$E\{X\} = r(1-p)/p$ $\text{var}\{X\} = r(1-p)/p^2$
Poisson	$X \sim \text{Poisson}(\lambda)$ $\lambda > 0$	$f(x) = \frac{\lambda^x}{x!} \exp\{-\lambda\}$ $x \in \{0, 1, \ldots\}$	$E\{X\} = \lambda$ $\text{var}\{X\} = \lambda$

Table 1.2 Notation and description for some common probability distributions of continuous random variables.

Name	Notation and Parameter Space	Density and Sample Space	Mean and Variance
Beta	$X \sim \text{Beta}(\alpha, \beta)$ $\alpha > 0$ and $\beta > 0$	$f(x) = \frac{\Gamma(\alpha+\beta)}{\Gamma(\alpha)\Gamma(\beta)} x^{\alpha-1}(1-x)^{\beta-1}$ $0 \le x \le 1$	$\text{E}\{X\} = \frac{\alpha}{\alpha+\beta}$ $\text{var}\{X\} = \frac{\alpha\beta}{(\alpha+\beta)^2(\alpha+\beta+1)}$
Cauchy	$X \sim \text{Cauchy}(\alpha, \beta)$ $\alpha \in \Re$ and $\beta > 0$	$f(x) = \frac{1}{\pi\beta\left[1+\left(\frac{x-\alpha}{\beta}\right)^2\right]}$ $x \in \Re$	$\text{E}\{X\}$ is non-existent $\text{var}\{X\}$ is non-existent
Chi-square	$X \sim \chi^2_\nu$ $\nu > 0$	$f(x) = \text{Gamma}(\nu/2, 1/2)$ $x > 0$	$\text{E}\{X\} = \nu$ $\text{var}\{X\} = 2\nu$
Dirichlet	$\mathbf{X} \sim \text{Dirichlet}(\boldsymbol{\alpha})$ $\boldsymbol{\alpha} = (\alpha_1, \ldots, \alpha_k)$ $\alpha_i > 0$ $\alpha_0 = \sum_{i=1}^k \alpha_i$	$f(\mathbf{x}) = \frac{\Gamma(\alpha_0)\prod_{i=1}^k x_i^{\alpha_i-1}}{\prod_{i=1}^k \Gamma(\alpha_i)}$ $\mathbf{x} = (x_1, \ldots, x_k)$ and $0 \le x_i \le 1$ $\sum_{i=1}^k x_i = 1$	$\text{E}\{\mathbf{X}\} = \boldsymbol{\alpha}/\alpha_0$ $\text{var}\{X_i\} = \frac{\alpha_i(\alpha_0-\alpha_i)}{\alpha_0^2(\alpha_0+1)}$ $\text{cov}\{X_i, X_j\} = \frac{-\alpha_i\alpha_j}{\alpha_0^2(\alpha_0+1)}$
Exponential	$X \sim \text{Exp}(\lambda)$ $\lambda > 0$	$f(x) = \lambda \exp\{-\lambda x\}$ $x > 0$	$\text{E}\{X\} = 1/\lambda$ $\text{var}\{X\} = 1/\lambda^2$
Gamma	$X \sim \text{Gamma}(r, \lambda)$ $\lambda > 0$ and $r > 0$	$f(x) = \frac{\lambda^r x^{r-1}}{\Gamma(r)} \exp\{-\lambda x\}$ $x > 0$	$\text{E}\{X\} = r/\lambda$ $\text{var}\{X\} = r/\lambda^2$

Table 1.3 Notation and description for more common probability distributions of continuous random variables.

Name	Notation and Parameter Space	Density and Sample Space	Mean and Variance
Lognormal	$X \sim \text{Lognormal}(\mu, \sigma^2)$ $\mu \in \Re$ and $\sigma > 0$	$f(x) = \frac{1}{x\sqrt{2\pi\sigma^2}} \exp\left\{ -\frac{1}{2}\left(\frac{\log\{x\}-\mu}{\sigma} \right)^2 \right\}$ $x \in \Re$	$\text{E}\{X\} = \exp\{\mu + \sigma^2/2\}$ $\text{var}\{X\} = \exp\{2\mu + 2\sigma^2\} - \exp\{2\mu + \sigma^2\}$
Multivariate Normal	$\mathbf{X} \sim N_k(\boldsymbol{\mu}, \boldsymbol{\Sigma})$ $\boldsymbol{\mu} = (\mu_1, \ldots, \mu_k) \in \Re^k$ $\boldsymbol{\Sigma}$ positive definite	$f(\mathbf{x}) = \frac{\exp\{-(\mathbf{x}-\boldsymbol{\mu})^T \boldsymbol{\Sigma}^{-1}(\mathbf{x}-\boldsymbol{\mu})/2\}}{(2\pi)^{k/2}|\boldsymbol{\Sigma}|^{1/2}}$ $\mathbf{x} = (x_1, \ldots, x_k) \in \Re^k$	$\text{E}\{\mathbf{X}\} = \boldsymbol{\mu}$ $\text{var}\{\mathbf{X}\} = \boldsymbol{\Sigma}$
Normal	$X \sim N(\mu, \sigma^2)$ $\mu \in \Re$ and $\sigma > 0$	$f(x) = \frac{1}{\sqrt{2\pi\sigma^2}} \exp\left\{ -\frac{1}{2}\left(\frac{x-\mu}{\sigma} \right)^2 \right\}$ $x \in \Re$	$\text{E}\{X\} = \mu$ $\text{var}\{X\} = \sigma^2$
Student's t	$X \sim t_\nu$ $\nu > 0$	$f(x) = \frac{\Gamma((\nu+1)/2)}{\Gamma(\nu/2)\sqrt{\pi\nu}} \left(1 + x^2/\nu\right)^{-(\nu+1)/2}$ $x \in \Re$	$\text{E}\{X\} = 0$ if $\nu > 1$ $\text{var}\{X\} = \frac{\nu}{\nu+2}$ if $\nu > 2$
Uniform	$X \sim \text{Unif}(a, b)$ $a, b \in \Re$ and $a < b$	$f(x) = \frac{1}{b-a}$ $x \in [a, b]$	$\text{E}\{X\} = (a+b)/2$ $\text{var}\{X\} = (b-a)^2/12$
Weibull	$X \sim \text{Weibull}(a, b)$ $a > 0$ and $b > 0$	$f(x) = abx^{b-1}\exp\{-ax^b\}$ $x > 0$	$\text{E}\{X\} = \frac{\Gamma(1+1/b)}{a^{1/b}}$ $\text{var}\{X\} = \frac{\Gamma(1+2/b) - \Gamma(1+1/b)^2}{a^{2/b}}$

It is also important to know how the distribution of a random variable changes when it is transformed. Let $\mathbf{X} = (X_1, \ldots, X_p)$ denote a p-dimensional random variable with continuous density function f. Suppose that

$$\mathbf{U} = \mathbf{g}(\mathbf{X}) = (g_1(\mathbf{X}), \ldots, g_p(\mathbf{X})) = (U_1, \ldots, U_p), \tag{1.22}$$

where $\mathbf{g}$ is a one-to-one function mapping the support region of f onto the space of all $\mathbf{u} = g(\mathbf{x})$ for which $\mathbf{x}$ satisfies $f(\mathbf{x}) > 0$. To derive the probability distribution of $\mathbf{U}$ from that of $\mathbf{X}$, we need to use the Jacobian matrix. The density of the transformed variables is

$$f(\mathbf{u}) = f(\mathbf{g}^{-1}(\mathbf{u})) \, |\mathbf{J}(\mathbf{u})| \,, \tag{1.23}$$

where $|\mathbf{J}(\mathbf{u})|$ is the absolute value of the determinant of the Jacobian matrix of $\mathbf{g}^{-1}$ evaluated at $\mathbf{u}$, having (i, j)th element $\frac{dx_i}{du_j}$, where these derivatives are assumed to be continuous over the support region of $\mathbf{U}$.

1.4   LIKELIHOOD INFERENCE

If $\mathbf{X}_1, \ldots, \mathbf{X}_n$ are independent and identically distributed (i.i.d.) each having density $f(\mathbf{x} \mid \boldsymbol{\theta})$ that depends on a vector of p unknown parameters $\boldsymbol{\theta} = (\theta_1, \ldots, \theta_p)$, then the joint likelihood function is

$$L(\boldsymbol{\theta}) = \prod_{i=1}^{n} f(\mathbf{x}_i | \boldsymbol{\theta}). \tag{1.24}$$

When the data are not i.i.d., the joint likelihood is still expressed as the joint density $f(\mathbf{x}_1, \ldots, \mathbf{x}_n | \boldsymbol{\theta})$ viewed as a function of $\boldsymbol{\theta}$.

The observed data, $\mathbf{x}_1, \ldots, \mathbf{x}_n$, might have been realized under many different values for $\boldsymbol{\theta}$. The parameters for which observing $\mathbf{x}_1, \ldots, \mathbf{x}_n$ would be most likely constitute the *maximum likelihood estimate* of $\boldsymbol{\theta}$. In other words, if $\hat{\vartheta}$ is the function of $\mathbf{x}_1, \ldots, \mathbf{x}_n$ that maximizes $L(\boldsymbol{\theta})$, then $\hat{\boldsymbol{\theta}} = \hat{\vartheta}(\mathbf{X}_1, \ldots, \mathbf{X}_n)$ is the *maximum likelihood estimator (MLE)* for $\boldsymbol{\theta}$. MLEs are invariant to transformation, so the MLE of a transformation of $\boldsymbol{\theta}$ equals the transformation of $\hat{\boldsymbol{\theta}}$.

It is typically easier to work with the *log likelihood function*,

$$l(\boldsymbol{\theta}) = \log L(\boldsymbol{\theta}), \tag{1.25}$$

which has the same maximum as the original likelihood, since log is a convex function. Furthermore, any additive constants (involving possibly $\mathbf{x}_1, \ldots, \mathbf{x}_n$ but not $\boldsymbol{\theta}$) may be omitted from the log likelihood without changing the location of its maximum or differences between log likelihoods at different $\boldsymbol{\theta}$. Note that maximizing $L(\boldsymbol{\theta})$ with respect to $\boldsymbol{\theta}$ is equivalent to solving the system of equations

$$l'(\boldsymbol{\theta}) = 0, \tag{1.26}$$

where $l'(\boldsymbol{\theta}) = \left(\frac{dl(\boldsymbol{\theta})}{d\theta_1}, \ldots, \frac{dl(\boldsymbol{\theta})}{d\theta_n} \right)$ is called the *score function*. The score function satisfies

$$E\{l'(\boldsymbol{\theta})\} = 0, \tag{1.27}$$

where the expectation is taken with respect to the distribution of $X_1, \ldots, X_n$. Sometimes an analytical solution to (1.26) provides the MLE; this book describes a variety of methods that can be used when the MLE cannot be solved for in closed form. It is worth noting that there are pathological circumstances where the MLE is not a solution of the score equation, or the MLE is not unique; see [109] for examples.

The MLE has a sampling distribution because it depends on the realization of the random variables $X_1, \ldots, X_n$. The MLE may be biased or unbiased for θ, yet under quite general conditions it is asymptotically unbiased as $n \to \infty$. The sampling variance of the MLE depends on the average curvature of the log likelihood: When the log likelihood is very pointy, the location of the maximum is more precisely known.

To make this precise, let $l''(\theta)$ denote the $p \times p$ matrix having (i,j)th element given by $\frac{d^2 l(\theta)}{d\theta_i d\theta_j}$. The *Fisher information matrix* is defined as

$$\mathbf{I}(\theta) = \mathrm{E}\{l'(\theta)l'(\theta)^T\} = -\mathrm{E}\{l''(\theta)\}, \qquad (1.28)$$

where the expectations are taken with respect to the distribution of $X_1, \ldots, X_n$. The final equality in (1.28) requires mild assumptions, which are satisfied, for example, in exponential families. $\mathbf{I}(\theta)$ may sometimes be called the *expected Fisher information* to distinguish it from $-l''(\theta)$, which is the *observed Fisher information*. There are two reasons why the observed Fisher information is quite useful. First, it can be calculated even if the expectations in (1.28) cannot easily be computed. Second, it is a good approximation to $\mathbf{I}(\theta)$ that improves as n increases.

Under regularity conditions, the asymptotic variance–covariance matrix of the MLE $\hat{\theta}$ is $\mathbf{I}(\theta^*)^{-1}$, where θ^* denotes the true value of θ. Indeed, as $n \to \infty$, the limiting distribution of $\hat{\theta}$ is $N_p(\theta^*, \mathbf{I}(\theta^*)^{-1})$. Since the true parameter values are unknown, $\mathbf{I}(\theta^*)^{-1}$ must be estimated in order to estimate the variance–covariance matrix of the MLE. An obvious approach is to use $\mathbf{I}(\hat{\theta})^{-1}$. Alternatively, it is also reasonable to use $-l''(\hat{\theta})^{-1}$. Standard errors for individual parameter MLEs can be estimated by taking the square root of the diagonal elements of the chosen estimate of $\mathbf{I}(\theta^*)^{-1}$. A thorough introduction to maximum likelihood theory and the relative merits of these estimates of $\mathbf{I}(\theta^*)^{-1}$ can be found in [109, 158, 325, 401].

Profile likelihoods provide an informative way to graph a higher-dimensional likelihood surface, to make inference about some parameters while treating others as nuisance parameters, and to facilitate various optimization strategies. The profile likelihood is obtained by constrained maximization of the full likelihood with respect to parameters to be ignored. If $\theta = (\mu, \phi)$, then the profile likelihood for ϕ is

$$L(\phi | \hat{\mu}(\phi)) = \max_{\mu} L(\mu, \phi). \qquad (1.29)$$

Thus, for each possible ϕ, a value of μ is chosen to maximize $L(\mu, \phi)$. This optimal μ is a function of ϕ. The profile likelihood is the function that maps ϕ to the value of the full likelihood evaluated at ϕ and its corresponding optimal μ. Note that the $\hat{\phi}$ that maximizes the profile likelihood $L(\phi | \hat{\mu}(\phi))$ is also the MLE for ϕ obtained from the full likelihood $L(\mu, \phi)$. Profile likelihood methods are examined in [21].

1.5 BAYESIAN INFERENCE

In the Bayesian inferential paradigm, probability distributions are associated with the parameters of the likelihood, as if the parameters were random variables. The probability distributions are used to assign subjective relative probabilities to regions of parameter space to reflect knowledge (and uncertainty) about the parameters.

Suppose that $\mathbf{X}$ has a distribution parameterized by $\boldsymbol{\theta}$. Let $f(\boldsymbol{\theta})$ represent the density assigned to $\boldsymbol{\theta}$ before observing the data. This is called a *prior distribution*. It may be based on previous data and analyses (e.g., pilot studies), it may represent a purely subjective personal belief, or it may be chosen in a way intended to have limited influence on final inference.

Bayesian inference is driven by the likelihood, often denoted $L(\boldsymbol{\theta}|\mathbf{x})$ in this context. Having established a prior distribution for $\boldsymbol{\theta}$ and subsequently observed data yielding a likelihood that is informative about $\boldsymbol{\theta}$, one's prior beliefs must be updated to reflect the information contained in the likelihood. The updating mechanism is Bayes' theorem:

$$f(\boldsymbol{\theta}|\mathbf{x}) = cf(\boldsymbol{\theta})f(\mathbf{x}|\boldsymbol{\theta}) = cf(\boldsymbol{\theta})L(\boldsymbol{\theta}|\mathbf{x}), \tag{1.30}$$

where $f(\boldsymbol{\theta}|\mathbf{x})$ is the *posterior density* of $\boldsymbol{\theta}$. The posterior distribution for $\boldsymbol{\theta}$ is used for statistical inference about $\boldsymbol{\theta}$. The constant c equals $1/\int f(\boldsymbol{\theta})L(\boldsymbol{\theta}|\mathbf{x})\,d\boldsymbol{\theta}$ and is often difficult to compute directly, although some inferences do not require c. This book describes a large variety of methods for enabling Bayesian inference, including the estimation of c.

Let $\tilde{\boldsymbol{\theta}}$ be the posterior mode, and let $\boldsymbol{\theta}^*$ be the true value of $\boldsymbol{\theta}$. The posterior distribution of $\tilde{\boldsymbol{\theta}}$ converges to $N(\boldsymbol{\theta}^*, \mathbf{I}(\boldsymbol{\theta}^*)^{-1})$ as $n \to \infty$, under regularity conditions. Note that this is the same limiting distribution as for the MLE. Thus, the posterior mode is of particular interest as a consistent estimator of $\boldsymbol{\theta}$. This convergence reflects the fundamental notion that the observed data should overwhelm any prior as $n \to \infty$.

Bayesian evaluation of hypotheses relies upon the *Bayes factor*. The ratio of posterior probabilities of two competing hypotheses or models, H_1 and H_2, is

$$\frac{P[H_2|\mathbf{x}]}{P[H_1|\mathbf{x}]} = \frac{P[H_2]}{P[H_1]}B_{2,1} \tag{1.31}$$

where $P[H_i|\mathbf{x}]$ denotes posterior probability, $P[H_i]$ denotes prior probability, and

$$B_{2,1} = \frac{f(\mathbf{x}|H_2)}{f(\mathbf{x}|H_1)} = \frac{\int f(\boldsymbol{\theta}_2|H_2)f(\mathbf{x}|\boldsymbol{\theta}_2, H_2)\,d\boldsymbol{\theta}_2}{\int f(\boldsymbol{\theta}_1|H_1)f(\mathbf{x}|\boldsymbol{\theta}_1, H_1)\,d\boldsymbol{\theta}_1} \tag{1.32}$$

with $\boldsymbol{\theta}_i$ denoting the parameters corresponding to the ith hypothesis. The quantity $B_{2,1}$ is the Bayes factor; it represents the factor by which the prior odds are multiplied to produce the posterior odds, given the data. The hypotheses H_1 and H_2 need not be nested as for likelihood ratio methods. The computation and interpretation of Bayes factors is reviewed in [321].

Bayesian interval estimation often relies on a 95% *highest posterior density* (HPD) region. The HPD region for a parameter is the region of shortest total length containing

95% of the posterior probability for that parameter for which the posterior density for every point contained in the interval is never lower than the density for every point outside the interval. For unimodal posteriors, the HPD is the narrowest possible interval containing 95% of the posterior probability. A more general interval for Bayesian inference is a *credible interval*. The $100(1 - \alpha)\%$ credible interval is the region between the $\frac{\alpha}{2}$ and $1 - \frac{\alpha}{2}$ quantiles of the posterior distribution. When the posterior density is symmetric and unimodal, the HPD and the credible interval are identical.

A primary benefit of the Bayesian approach to inference is the natural manner in which resulting credibility intervals and other inferences are interpreted. One may speak of the posterior probability that the parameter is in some range. There is also a sound philosophical basis for the Bayesian paradigm; see [25] for an introduction. Gelman et al. provide a broad survey of Bayesian theory and methods [194].

The best prior distributions are those based on prior data. A strategy that is algebraically convenient is to seek conjugacy. A *conjugate prior* distribution is one that yields a posterior distribution in the same parametric family as the prior distribution. Exponential families are the only classes of distributions that have natural conjugate prior distributions.

When prior information is poor, it is important to ensure that the chosen prior distribution does not strongly influece posterior inferences. A posterior that is strongly influenced by the prior is said to be highly *sensitive* to the prior. Several strategies are available to reduce sensitivity. The simplest approach is to use a prior whose support is dispersed over a much broader region than the parameter region supported by the data, and fairly flat over it. A more formal approach is to use a Jeffreys prior [307]. In the univariate case, the Jeffreys prior is $f(\theta) \propto I(\theta)^{-1/2}$, where $I(\theta)$ is the Fisher information; multivariate extensions are possible. In some cases, the improper prior $f(\boldsymbol{\theta}) \propto 1$ may be considered, but this can lead to improper posteriors (i.e., not integrable), and it can be unintentionally informative depending on the parameterization of the problem.

Example 1.2 (Normal–normal conjugate Bayes model) Consider Bayesian inference based on observations of i.i.d. random variables $X_1, \ldots, X_n$ with density $X_i | \theta \sim N(\theta, \sigma^2)$ where σ^2 is known. For such a likelihood, a normal prior for θ is conjugate. Suppose the prior is $\theta \sim N(\mu, \tau^2)$. The posterior density is

$$f(\theta|\mathbf{x}) \propto f(\theta) \prod_{i=1}^{n} f(x_i|\theta) \tag{1.33}$$

$$\propto \exp\left\{ -\frac{1}{2}\left(\frac{(\theta - \mu)^2}{\tau^2} + \frac{\sum_{i=1}^{n}(x_i - \theta)^2}{\sigma^2} \right) \right\} \tag{1.34}$$

$$\propto \exp\left\{ -\frac{1}{2}\left(\theta - \frac{\frac{\mu}{\tau^2} + \frac{n\bar{x}}{\sigma^2}}{\frac{1}{\tau^2} + \frac{n}{\sigma^2}} \right)^2 \bigg/ \left(\frac{1}{\frac{1}{\tau^2} + \frac{n}{\sigma^2}} \right) \right\}, \tag{1.35}$$

where $\bar{x}$ is the sample mean. Recognizing (1.35) as being in the form of a normal distribution, we conclude that $f(\theta|\mathbf{x}) = N(\mu_n, \tau_n^2)$, where

$$\tau_n^2 = \frac{1}{\frac{1}{\tau^2} + \frac{n}{\sigma^2}} \tag{1.36}$$

and

$$\mu_n = \left(\frac{\mu}{\tau^2} + \frac{n\bar{x}}{\sigma^2}\right)\tau_n^2. \tag{1.37}$$

Hence, a posterior 95% credibility interval for θ is $(\mu_n - 1.96\tau_n, \ \mu_n + 1.96\tau_n)$. Since the normal distribution is symmetric, this is also the posterior 95% HPD for θ.

For fixed σ, consider increasingly large choices for the value of τ. The posterior variance for θ converges to σ^2/n as $\tau^2 \to \infty$. In other words, the influence of the prior on the posterior vanishes as the prior variance increases. Next, note that $\lim_{n\to\infty} \frac{\tau_n^2}{\sigma^2/n} = 1$. This shows that the posterior variance for θ and the sampling variance for the MLE, $\hat{\theta} = \bar{X}$, are asymptotically equal, and the effect of any choice for τ is washed out with increasing sample size.

As an alternative to the conjugate prior, consider using the improper prior $f(\theta) \propto 1$. In this case, $f(\theta|\mathbf{x}) = N(\bar{x}, \sigma^2/n)$, and a 95% posterior credibility interval corresponds to the standard 95% confidence interval found using frequentist methods. $\qquad\square$

1.6 STATISTICAL LIMIT THEORY

Although this book is mostly concerned with a pragmatic examination of how and why various methods work, it is useful from time to time to speak more precisely about the limiting behavior of the estimators produced by some procedures. We review below some basic convergence concepts used in probability and statistics.

A sequence of random variables, $X_1, X_2, \ldots$, is said to *converge in probability* to the random variable X if $\lim_{n\to\infty} P[|X_n - X| < \epsilon] = 1$ for every $\epsilon > 0$. The sequence *converges almost surely* to X if $P[\lim_{n\to\infty} |X_n - X| < \epsilon] = 1$ for every $\epsilon > 0$. The variables converge in distribution to the distribution of X if $\lim_{n\to\infty} F_{X_n}(x) = F_X(x)$ for all points x at which $F_X(x)$ is continuous. The variable X has property A *almost everywhere* if $P[A] = \int 1_{\{A\}} f_X(x)\, dx = 1$.

Some of the best-known convergence theorems in statistics are the laws of large numbers and the central limit theorem. For i.i.d. sequences of one-dimensional random variables $X_1, X_2, \ldots$, let $\bar{X}_n = \sum_{i=1}^{n} X_i/n$. The *weak law of large numbers* states that $\bar{X}_n$ converges in probability to $\mu = E\{X_i\}$ if $E\{|X_i|\} < \infty$. The *strong law of large numbers* states that $\bar{X}_n$ converges almost surely to μ if $E\{|X_i|\} < \infty$. Both results hold under the more stringent but easily checked condition that $\text{var}\{X_i\} = \sigma^2 < \infty$.

If θ is a parameter and T_n is a statistic based on $X_1, \ldots, X_n$, then T_n is said to be *weakly* or *strongly consistent* for θ if T_n converges in probability or almost surely (respectively) to θ. T_n is *unbiased* for θ if $E\{T_n\} = \theta$; otherwise the *bias* is $E\{T_n\} - \theta$. If the bias vanishes as $n \to \infty$, then T_n is *asymptotically unbiased*.

A simple form of the *central limit theorem* is as follows. Suppose that i.i.d. random variables $X_1, \ldots, X_n$ have mean μ and finite variance σ^2, and that $E\{\exp\{tX_i\}\}$ exists in a neighborhood of $t = 0$. Then the random variable $T_n = \sqrt{n}(\bar{X}_n - \mu)/\sigma$ converges in distribution to a normal random variable with mean zero and variance one, as $n \to \infty$. There are many versions of the central limit theorem for various situations. Generally speaking, the assumption of finite variance is critical, but the assumptions of independence and identical distributions can be relaxed in certain situations.

1.7 MARKOV CHAINS

We offer here a brief introduction to univariate, discrete-time, discrete-state-space Markov chains. We will use Markov chains in Chapters 7 and 8. A thorough introduction to Markov chains is given in [467], and higher-level study is provided in [393, 460].

Consider a sequence of random variables $\{X^{(t)}\}$, $t = 0, 1, \ldots$, where each $X^{(t)}$ may equal one of a finite or countably infinite number of possible values, called *states*. The notation $X^{(t)} = j$ indicates that the process is in state j at time t. The *state space*, S, is the set of possible values of the random variable $X^{(t)}$.

A complete probabilistic specification of $X^{(0)}, \ldots, X^{(n)}$ would be to write their joint distribution as the product of conditional distributions of each random variable given its history, or

$$
\begin{aligned}
P\left[X^{(0)}, \ldots, X^{(n)}\right] = {} & P\left[X^{(n)} \mid x^{(0)}, \ldots, x^{(n-1)}\right] \\
& \times P\left[X^{(n-1)} \mid x^{(0)}, \ldots, x^{(n-2)}\right] \times \cdots \\
& \times P\left[X^{(1)} \mid x^{(0)}\right] P\left[X^{(0)}\right]. \quad (1.38)
\end{aligned}
$$

A simplification of (1.38) is possible under the conditional independence assumption that

$$
P\left[X^{(t)} \mid x^{(0)}, \ldots, x^{(t-1)}\right] = P\left[X^{(t)} \mid x^{(t-1)}\right]. \quad (1.39)
$$

Here the next state observed is only dependent upon the present state. This is the *Markov property*, sometimes called "one step memory." In this case,

$$
\begin{aligned}
P\left[X^{(0)}, \ldots, X^{(n)}\right] = {} & P\left[X^{(n)} \mid x^{(n-1)}\right] \\
& \times P\left[X^{(n-1)} \mid x^{(n-2)}\right] \cdots P\left[X^{(1)} \mid x^{(0)}\right] P\left[X^{(0)}\right]. \quad (1.40)
\end{aligned}
$$

Table 1.4 San Francisco rain data considered in Example 1.3.

	Wet today	Dry today
Wet yesterday	418	256
Dry yesterday	256	884

Let $p_{ij}^{(t)}$ be the probability that the observed state changes from state i at time t to state j at time $t+1$. The sequence $\{X^{(t)}\}, t = 0, 1, \ldots$ is a *Markov chain* if

$$p_{ij}^{(t)} = P\left[X^{(t+1)} = j \mid X^{(0)} = x^{(0)}, X^{(1)} = x^{(1)}, \ldots, X^{(t)} = i\right]$$
$$= P\left[X^{(t+1)} = j \mid X^{(t)} = i\right] \tag{1.41}$$

for all $t = 0, 1, \ldots$ and $x^{(0)}, x^{(1)}, \ldots, x^{(t-1)}$, $i, j \in S$. The quantity $p_{ij}^{(t)}$ is called the *one-step transition probability*. If none of the one-step transition probabilities change with t, then the chain is called *time-homogeneous*, and $p_{ij}^{(t)} = p_{ij}$. If any of the one-step transition probabilities change with t, then the chain is called *time-inhomogeneous*.

A Markov chain is governed by a *transition probability matrix*. Suppose that the s states in S are, without loss of generality, all integer-valued. Then $\mathbf{P}$ denotes $s \times s$ transition probability matrix of a time-homogeneous chain, and the (i, j)th element of $\mathbf{P}$ is p_{ij}. Each element in $\mathbf{P}$ must be between zero and one, and each row of the matrix must sum to one.

Example 1.3 (San Francisco weather) Let us consider daily precipitation outcomes in San Francisco. Table 1.4 gives the rainfall status for 1814 pairs of consecutive days [417]. The data are taken from the months of November through March, starting in November of 1990 and ending in March of 2002. These months are when San Francisco receives over 80% of its precipitation each year, virtually all in the form of rain. We consider a binary classification of each day. A day is considered to be wet if more than 0.01 inches of precipitation are recorded and dry otherwise. Thus, S has two elements: "wet" and "dry." The random variable corresponding to the state for the tth day is $X^{(t)}$.

Assuming time-homogeneity, an estimated transition probability matrix for $X^{(t)}$ would be

$$\hat{\mathbf{P}} = \begin{bmatrix} 0.620 & 0.380 \\ 0.224 & 0.775 \end{bmatrix}. \tag{1.42}$$

Clearly, wet and dry weather states are not independent in San Francisco, as a wet day is more likely to be followed by a wet day and pairs of dry days are highly likely. $\square$

The limiting theory of Markov chains is important for many of the methods discussed in this book. We now review some basic results.

A state to which the chain returns with probability 1 is called a *recurrent* state. A state for which the expected time until recurrence is finite is called *nonnull*. For finite state spaces, recurrent states are nonnull.

A Markov chain is *irreducible* if any state j can be reached from any state i in a finite number of steps for all i and j. In other words, for each i and j there must exist $m > 0$ such that $P\left[X^{(m+n)} = i \mid X^{(n)} = j\right] > 0$. A Markov chain is *periodic* if it can visit certain portions of the state space only at certain regularly spaced intervals. State j has period d if the probability of going from state j to state j in n steps is 0 for all n not divisible by d. If every state in a Markov chain has period 1, then the chain is called *aperiodic*. A Markov chain is *ergodic* if it is irreducible, aperiodic, and all its states are nonnull and recurrent.

Let π denote a vector of probabilities that sum to one, with ith element π_i denoting the marginal probability that $X^{(t)} = i$. Then the marginal distribution of $X^{(t+1)}$ must be $\pi^T \mathbf{P}$. Any discrete probability distribution π such that $\pi^T \mathbf{P} = \pi^T$ is called a *stationary* distribution for $\mathbf{P}$, or for the Markov chain having transition probability matrix $\mathbf{P}$. If $X^{(t)}$ follows a stationary distribution, then the marginal distributions of $X^{(t)}$ and $X^{(t+1)}$ are identical.

If a time-homogeneous Markov chain satisfies

$$\pi_i p_{ij} = \pi_j p_{ji} \tag{1.43}$$

for all $i, j \in \mathcal{S}$, then π is a stationary distribution for the chain, and the chain is called *reversible* because the joint distribution of a sequence of observations is the same whether the chain is run forwards or backwards. Equation (1.43) is called the *detailed balance* condition.

If a Markov chain with transition probability matrix $\mathbf{P}$ and stationary distribution π is irreducible and aperiodic, then π is unique and

$$\lim_{n \to \infty} P\left[X^{(t+n)} = j \mid X^{(t)} = i\right] = \pi_j, \tag{1.44}$$

where π_j is the jth element of π. The π_j are the solutions of the following set of equations:

$$\pi_j \geq 0, \quad \sum_{i \in \mathcal{S}} \pi_i = 1, \quad \text{and} \quad \pi_j = \sum_{i \in \mathcal{S}} \pi_i p_{ij} \quad \text{for each } j \in \mathcal{S}. \tag{1.45}$$

We can restate and extend (1.44) as follows. If $X^{(1)}, X^{(2)}, \ldots$ are realizations from an irreducible and aperiodic Markov chain with stationary distribution π, then $X^{(n)}$ converges in distribution to the distribution given by π, and for any function h,

$$\frac{1}{n} \sum_{t=1}^{n} h(X^{(t)}) \to \mathrm{E}_\pi\{h(X)\} \tag{1.46}$$

almost surely as $n \to \infty$, provided $\mathrm{E}_\pi\{|h(X)|\}$ exists [510]. This is one form of the *ergodic theorem*, which is a generalization of the strong law of large numbers.

We have considered here only Markov chains for discrete state spaces. In Chapters 7 and 8 we will apply these ideas to continuous state spaces. The principles and results for continuous state spaces and multivariate random variables are similar to the simple results given here.

1.8 COMPUTING

If you are new to computer programming, or wishing to learn a new language, there is no better time to start than now. Our preferred language for teaching and learning about statistical computing is S-Plus, but we avoid any language-specific limitations in this text. The language R is a free partial reimplementation of S-Plus with some additional features. Most of the methods described in this book can also be easily implemented in other high-level computer languages for mathematics and statistics, such as SAS and MATLAB. Programming in Java and low-level languages such as C++ and Fortran is also possible. The tradeoff between implementation ease for high-level languages and computation speed for low-level languages often guides this selection. Links to these and other useful software packages, including libraries of code for some of the methods described in this book, are available on the book website.

Ideally, your computer programming background includes a basic understanding of computer arithmetic: how real numbers and mathematical operations are implemented in the binary world of a computer. We focus on higher-level issues in this book, but the most meticulous implementation of the algorithms we describe can require consideration of the vagaries of computer arithmetic, or use of available routines that competently deal with such issues. Interested readers may refer to [334].

2

Optimization and Solving Nonlinear Equations

Maximum likelihood estimation is central to statistical inference. Long hours can be invested in learning about the theoretical performance of MLEs and their analytic derivation. Faced with a complex likelihood lacking analytic solution, however, many people are unsure how to proceed.

Most functions cannot be optimized analytically. For example, maximizing $g(x) = (\log x)/(1 + x)$ with respect to x by setting the derivative equal to zero and solving for x leads to an algebraic impasse because $1 + 1/x - \log x = 0$ has no analytic solution. Many realistic statistical models induce likelihoods that cannot be optimized analytically—indeed, we would argue that greater realism is strongly associated with reduced ability to find optima analytically.

Statisticians face other optimization tasks, too, aside from maximum likelihood. Minimizing risk in a Bayesian decision problem, solving nonlinear least squares problems, finding highest posterior density intervals for many distributions, and a wide variety of other tasks all involve optimization. Such diverse tasks are all versions of the following generic problem: optimize a real-valued function g with respect to its argument, a p-dimensional vector $\mathbf{x}$. In this chapter, we will limit consideration to g that are smooth and differentiable with respect to $\mathbf{x}$; in Chapter 3 we discuss optimization when g is defined over a discrete domain. There is no meaningful distinction between maximization and minimization, since maximizing a function is equivalent to minimizing its negative. As a convention, we will generally use language suggestive of seeking a maximum.

For maximum likelihood estimation, g is the log likelihood function l, and $\mathbf{x}$ is the corresponding parameter vector $\boldsymbol{\theta}$. If $\hat{\boldsymbol{\theta}}$ is a MLE, it maximizes the log likelihood.

Therefore $\hat{\boldsymbol{\theta}}$ is a solution to the score equation

$$\mathbf{l}'(\boldsymbol{\theta}) = \mathbf{0}, \tag{2.1}$$

where $\mathbf{l}'(\boldsymbol{\theta}) = \left(\frac{dl(\boldsymbol{\theta})}{d\theta_1}, \ldots, \frac{dl(\boldsymbol{\theta})}{d\theta_n} \right)^T$ and $\mathbf{0}$ is a column vector of zeros.

Immediately, we see that optimization is intimately linked with solving nonlinear equations. Indeed, one could reinterpret this chapter as an introduction to root finding rather than optimization. Finding a MLE amounts to finding a root of the score equation. The maximum of g is a solution to $\mathbf{g}'(\mathbf{x}) = \mathbf{0}$. (Conversely, one may also turn a univariate root-finding exercise into an optimization problem by minimizing $|g'(x)|$ with respect to x, where g' is the function whose root is sought.)

The solution of $\mathbf{g}'(\mathbf{x}) = \mathbf{0}$ is most difficult when this set of equations has a solution that cannot be determined analytically. In this case, the equations will be nonlinear. Solving linear equations is easy, however there is another class of difficult optimization problems where the objective function itself is linear and there are linear inequality constraints. Such problems can be solved using linear programming techniques such as the simplex method [114, 173, 217, 425] and interior point methods [304, 318, 465]. Such methods are not covered in this book.

For smooth, nonlinear functions, optima are routinely found using a variety of off-the-shelf numerical optimization software. Many of these programs are very effective, raising the question of whether optimization is a solved problem whose study here might be a low priority. For example, we have omitted the topic of uniform random number generation from this book—despite its importance in the statistical computing literature—because of the widespread use of high-quality prepackaged routines that accomplish the task. Why, then, should optimization methods be treated differently? The difference is that optimization software is confronted with a new problem every time the user presents a new function to be optimized. Even the best optimization software often initially fails to find the maximum for tricky likelihoods and requires tinkering to succeed. Therefore, the user must understand enough about how optimization works to tune the procedure successfully.

We begin by studying univariate optimization. Extensions to multivariate problems are described in Section 2.2. Optimization over discrete spaces is covered in Chapter 3, and an important special case related to missing data is covered in Chapter 4.

Useful references on optimization methods include [173, 217, 133, 405, 415, 422].

2.1 UNIVARIATE PROBLEMS

A simple univariate numerical optimization problem which we will discuss throughout this section is to maximize

$$g(x) = \frac{\log x}{1 + x} \tag{2.2}$$

with respect to x. Since no analytic solution can be found, we resort to iterative methods that rely on successive approximation of the solution. Graphing $g(x)$ in Figure 2.1, we see that the maximum is around 3. Therefore it might be reasonable

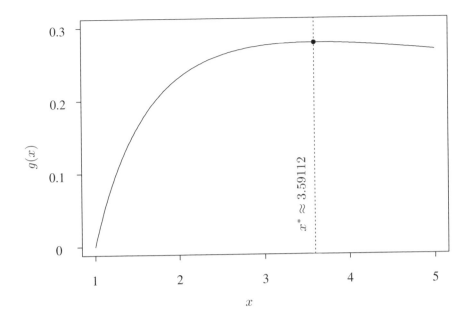

Fig. 2.1 The maximum of $g(x) = \frac{\log x}{1+x}$ occurs at $x^* \approx 3.59112$, indicated by the vertical line.

to use $x^{(0)} = 3.0$ as an initial guess, or *starting value*, for an iterative procedure. An *updating equation* will be used to produce an improved guess, $x^{(t+1)}$, from the most recent value $x^{(t)}$, for $t = 0, 1, 2, \ldots$ until iterations are stopped. The update may be based on an attempt to find the root of $g'(x) = \frac{1+1/x-\log x}{(1+x)^2}$, or on some other rationale.

The *bisection method* illustrates the main components of iterative root-finding procedures. If g' is continuous on $[a_0, b_0]$ and $g'(a_0)g'(b_0) \leq 0$, then the intermediate value theorem [473] implies that there exists at least one $x^* \in [a_0, b_0]$ for which $g'(x^*) = 0$ and hence x^* is a local optimum of g. To find it, the bisection method systematically shrinks the interval from $[a_0, b_0]$ to $[a_1, b_1]$ to $[a_2, b_2]$ and so on, where $[a_0, b_0] \supset [a_1, b_1] \supset [a_2, b_2] \supset \cdots$ and so forth.

Let $x^{(0)} = (a_0 + b_0)/2$ be the starting value. The updating equations are

$$[a_{t+1}, b_{t+1}] = \begin{cases} [a_t, x^{(t)}] & \text{if } g'(a_t)g'(x^{(t)}) \leq 0, \\ [x^{(t)}, b_t] & \text{if } g'(a_t)g'(x^{(t)}) > 0 \end{cases} \tag{2.3}$$

and

$$x^{(t+1)} = (a_{t+1} + b_{t+1})/2. \tag{2.4}$$

If g has more than one root in the starting interval, it is easy to see that bisection will find one of them, but will not find the rest.

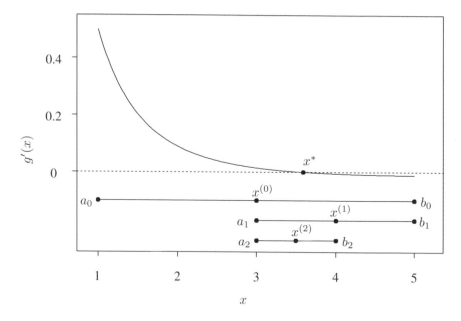

Fig. 2.2   Illustration of the bisection method from Example 2.1. The top portion of this graph shows $g'(x)$ and its root at x^*. The bottom portion shows the first three intervals obtained using the bisection method with $(a_0, b_0) = (1, 5)$. The tth estimate of the root is at the center of the tth interval.

Example 2.1 (A simple univariate optimization) To find the value of x maximizing (2.2), we might take $a_0 = 1$, $b_0 = 5$, and $x^{(0)} = 3$. Figure 2.2 illustrates the first few steps of the bisection algorithm for this simple function.   □

Suppose the true maximum of $g(x)$ with respect to x is achieved at x^*. The updating equation of any iterative procedure should be designed to encourage $x^{(t)} \rightarrow x^*$ as t increases. Of course there is no guarantee that $x^{(t)}$ will converge to anything, let alone to x^*.

In practice, we cannot allow the procedure to run indefinitely, so we require a *stopping rule*, based on some *convergence criteria*, to trigger an end to the successive approximation. At each iteration, the stopping rule should be checked. When the convergence criteria are met, the new $x^{(t+1)}$ is taken as the solution. There are two reasons to stop: if the procedure appears to have achieved satisfactory convergence or if it appears unlikely to do so soon.

It is tempting to monitor convergence by tracking the proximity of $g'(x^{(t+1)})$ to zero. However, large changes from $x^{(t)}$ to $x^{(t+1)}$ can occur even when $g'(x^{(t+1)})$ is very small; therefore a stopping rule based directly on $g'(x^{(t+1)})$ is not very reliable. On the other hand, a small change from $x^{(t)}$ to $x^{(t+1)}$ is most frequently associated with $g'(x^{(t+1)})$ near zero. Therefore, we typically assess convergence by monitoring $\left| x^{(t+1)} - x^{(t)} \right|$ and use $g'(x^{(t+1)})$ as a backup check.

The *absolute convergence criterion* mandates stopping when

$$\left| x^{(t+1)} - x^{(t)} \right| < \epsilon, \tag{2.5}$$

where ϵ is a constant chosen to indicate tolerable imprecision. For bisection, it is easy to confirm that

$$b_t - a_t = 2^{-t}(b_0 - a_0). \tag{2.6}$$

A true error tolerance of $\left| x^{(t)} - x^* \right| < \delta$ is achieved when $2^{-(t+1)}(b_0 - a_0) < \delta$, which occurs once $t > \log_2\{(b_0 - a_0)/\delta\} - 1$. Reducing δ tenfold therefore requires an increase in t of $\log_2 10 \approx 3.3$. Hence, three or four iterations are needed to achieve each extra decimal place of precision.

The *relative convergence criterion* mandates stopping when iterations have reached a point for which

$$\frac{\left| x^{(t+1)} - x^{(t)} \right|}{\left| x^{(t)} \right|} < \epsilon. \tag{2.7}$$

This criterion enables the specification of a target precision (e.g., within 1%) without worrying about the units of x.

Preference between the absolute and relative convergence criteria depends on the problem at hand. If the scale of x is huge (or tiny) relative to ϵ, an absolute convergence criterion may stop iterations too reluctantly (or too soon). The relative convergence criterion corrects for the scale of x, but can become unstable if $x^{(t)}$ values (or the true solution) lie too close to zero. In this latter case, another option is to monitor relative convergence by stopping when $\frac{\left| x^{(t+1)} - x^{(t)} \right|}{\left| x^{(t)} \right| + \epsilon} < \epsilon$.

Bisection works when g' is continuous. Taking limits on both sides of (2.6) implies $\lim_{t \to \infty} a_t = \lim_{t \to \infty} b_t$; therefore the bisection method converges to some point $x^{(\infty)}$. The method always ensures that $g'(a_t)g'(b_t) \leq 0$; continuity therefore implies that $g'(x^{(\infty)})^2 \leq 0$. Thus $g'(x^{(\infty)})$ must equal zero, which proves that $x^{(\infty)}$ is a root of g. In other words, the bisection method is—in theory—guaranteed to converge to a root in $[a_0, b_0]$.

In practice, numerical imprecision in a computer may thwart convergence. For most iterative approximation methods, it is safer to add a small correction to a previous approximation than to initiate a new approximation from scratch. The bisection method is more stable numerically when the updated endpoint is calculated as, say, $a_{t+1} = a_t + (b_t - a_t)/2$ instead of $a_{t+1} = (a_t + b_t)/2$. Yet, even carefully coded algorithms can fail, and optimization procedures more sophisticated than bisection can fail for all sorts of reasons. It is also worth noting that there are pathological circumstances where the MLE is not a solution of the score equation or the MLE is not unique; see [109] for examples.

Given such anomalies, it is important to include stopping rules that flag a failure to converge. The simplest such stopping rule is to stop after N iterations, regardless of convergence. It may also be wise to stop if one or more convergence measures like $\left| x^{(t+1)} - x^{(t)} \right|$ or $\left| x^{(t+1)} - x^{(t)} \right| / \left| x^{(t)} \right|$, or $\left| g'(x^{(t+1)}) \right|$ either fail to decrease or cycle over several iterations. The solution itself may also cycle unsatisfactorily. It is also sensible to stop if the procedure appears to be converging to a point at which $g(x)$ is inferior to another value you have already found. This prevents wasted effort when a search is converging to a known false peak or local maximum. Regardless of which such stopping rules you employ, any indication of poor convergence behavior means that $x^{(t+1)}$ must be discarded and the procedure somehow restarted in a manner more likely to yield successful convergence.

Starting is as important as stopping. In general, a bad starting value can lead to divergence, cycling, discovery of a misleading local maximum or a local minimum, or other problems. The outcome depends on g, the starting value, and the optimization algorithm tried. In general, it helps to start quite near the global optimum, as long as g is not virtually flat in the neighborhood containing $x^{(0)}$ and x^*. Methods for generating a reasonable starting value include graphing, preliminary estimates (e.g., method-of-moments estimates), educated guesses, and trial and error. If computing speed limits the total number of iterations that you can afford, it is wise not to invest them all in one long run of the optimization procedure. Using a collection of runs from multiple starting values (see the random starts local search method in Section 3.2) can be an effective way to gain confidence in your result and to avoid being fooled by local optima or stymied by convergence failure.

The bisection method is an example of a *bracketing method*, that is to say, a method that bounds a root within an sequence of nested intervals of decreasing length. Bisection is quite a slow approach: It requires a rather large number of iterations to achieve a desired precision, relative to other methods discussed below. Other bracketing methods include the secant bracket [534], which is equally slow after an initial period of greater efficiency, and the Illinois method [305], Ridders's method [454], and Brent's method [62], which are faster.

Despite the relative slowness of bracketing methods, they have one significant advantage over the methods described in the remainder of this chapter. If g' is continuous on $[a_0, b_0]$, a root can be found, regardless of the existence, behavior, or ease of deriving g''. Because they avoid worries about g'' while performing relatively robustly on most problems, bracketing methods continue to be reasonable alternatives to the methods below that rely on greater smoothness of g.

2.1.1   Newton's method

An extremely fast root-finding approach is Newton's method. This approach is also referred to as *Newton–Raphson iteration*, especially in univariate applications. Suppose that g' is continuously differentiable and that $g''(x^*) \neq 0$. At iteration t, the approach approximates $g'(x^*)$ by the linear Taylor series expansion:

$$0 = g'(x^*) \approx g'(x^{(t)}) + (x^* - x^{(t)})g''(x^{(t)}). \tag{2.8}$$

Since g' is approximated by its tangent line at $x^{(t)}$, it seems sensible to approximate the root of g' by the root of the tangent line. Thus, solving for x^* above, we obtain

$$x^* = x^{(t)} - \frac{g'(x^{(t)})}{g''(x^{(t)})} = x^{(t)} + h^{(t)}. \tag{2.9}$$

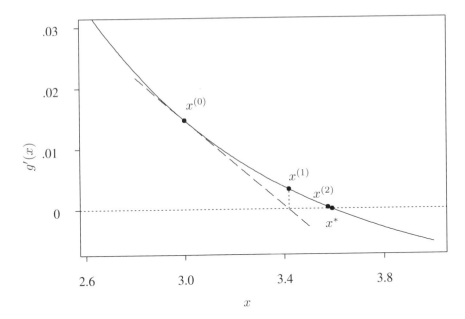

Fig. 2.3 Illustration of Newton's method applied in Example 2.2. At the first step, Newton's method approximates g' by its tangent line at $x^{(0)}$, whose root $x^{(1)}$ serves as the next approximation of the true root x^*. The next step similarly yields $x^{(2)}$, which is already quite close to x^*.

This equation describes an approximation to x^* that depends on the current guess $x^{(t)}$ and a refinement $h^{(t)}$. Iterating this strategy yields the updating equation for Newton's method:

$$x^{(t+1)} = x^{(t)} + h^{(t)}, \tag{2.10}$$

where $h^{(t)} = -g'(x^{(t)})/g''(x^{(t)})$. The same update can be motivated by analytically solving for the maximum of the quadratic Taylor series approximation to $g(x^*)$, namely $g(x^{(t)}) + (x^* - x^{(t)})g'(x^{(t)}) + (x^* - x^{(t)})^2 g''(x^{(t)})/2$. When the optimization of g corresponds to a MLE problem where $\hat{\theta}$ is a solution to $l'(\theta) = 0$, the updating equation for Newton's method is

$$\theta^{(t+1)} = \theta^{(t)} - \frac{l'(\theta^{(t)})}{l''(\theta^{(t)})}. \tag{2.11}$$

Example 2.2 (A simple univariate optimization, continued) Figure 2.3 illustrates the first several iterations of Newton's method applied to simple function in (2.2).

The Newton increment for this problem is given by

$$h^{(t)} = \frac{(x^{(t)} + 1)(1 + 1/x^{(t)} - \log x^{(t)})}{3 + 4/x^{(t)} + 1/(x^{(t)})^2 - 2\log x^{(t)}}. \tag{2.12}$$

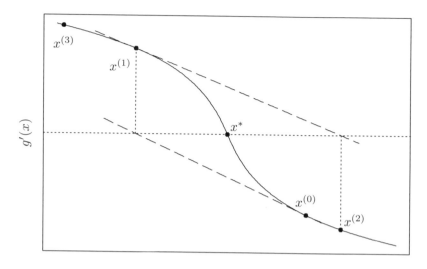

Fig. 2.4 Starting from $x^{(0)}$, Newton's method diverges by taking steps that are increasingly distant from the true root, x^*.

Starting from $x^{(0)} = 3.0$, Newton's method quickly finds $x^{(4)} \approx 3.59112$. For comparison, the first five decimal places of x^* are not correctly determined by the bisection method in Example 2.1 until iteration 19.   □

Whether Newton's method converges depends on the shape of g and the starting value. Figure 2.4 illustrates an example where the method diverges from its starting value. To better understand what ensures convergence, we must carefully analyze the errors at successive steps.

Suppose g' has two continuous derivatives and $g''(x^*) \neq 0$. Since $g''(x^*) \neq 0$ and g'' is continuous at x^*, there exists a neighborhood of x^* within which $g''(x) \neq 0$ for all x. Let us confine interest to this neighborhood, and define $\epsilon^{(t)} = x^{(t)} - x^*$.

A Taylor expansion yields

$$0 = g'(x^*) = g'(x^{(t)}) + (x^* - x^{(t)})g''(x^{(t)}) + (x^* - x^{(t)})^2 g'''(q)/2 \quad (2.13)$$

for some q between $x^{(t)}$ and x^*. Rearranging terms, we find

$$x^{(t)} + h^{(t)} - x^* = (x^* - x^{(t)})^2 \frac{g'''(q)}{2g''(x^{(t)})}, \quad (2.14)$$

where $h^{(t)}$ is the Newton update increment. Since the left hand side equals $x^{(t+1)} - x^*$, we conclude

$$\epsilon^{(t+1)} = (\epsilon^{(t)})^2 \frac{g'''(q)}{2g''(x^{(t)})}. \quad (2.15)$$

Now, consider a neighborhood of x^*, $\mathcal{N}_\delta(x^*) = [x^* - \delta, x^* + \delta]$, for $\delta > 0$. Let

$$c(\delta) = \max_{x_1, x_2 \in \mathcal{N}_\delta(x^*)} \left| \frac{g'''(x_1)}{2g''(x_2)} \right|. \tag{2.16}$$

Since $c(\delta) \to \left| \frac{g'''(x^*)}{2g''(x^*)} \right|$ as $\delta \to 0$, it follows that $\delta c(\delta) \to 0$ as $\delta \to 0$. Let us choose δ such that $\delta c(\delta) < 1$. If $x^{(t)} \in \mathcal{N}_\delta(x^*)$, then (2.15) implies that

$$\left| c(\delta) \epsilon^{(t+1)} \right| \leq \left(c(\delta) \epsilon^{(t)} \right)^2. \tag{2.17}$$

Suppose that the starting value is not too bad, in the sense that $\left| \epsilon^{(0)} \right| = \left| x^{(0)} - x^* \right| \leq \delta$. Then (2.17) implies that

$$\left| \epsilon^{(t)} \right| \leq \frac{(c(\delta)\delta)^{2^t}}{c(\delta)}, \tag{2.18}$$

which converges to zero as $t \to \infty$. Hence $x^{(t)} \to x^*$.

We have just proven the following theorem: If g''' is continuous and x^* is a simple root of g', then there exists a neighborhood of x^* for which Newton's method converges to x^* when started from any $x^{(0)}$ in that neighborhood.

In fact, when g' is twice continuously differentiable, is convex, and has a root, then Newton's method converges to the root from *any* starting point. When starting from somewhere in an interval $[a, b]$, another set of conditions one may check is as follows. If

1. $g''(x) \neq 0$ on $[a, b]$,

2. $g'''(x)$ does not change sign on $[a, b]$,

3. $g'(a)g'(b) < 0$, and

4. $|g'(a)/g''(a)| < b - a$ and $|g'(b)/g''(b)| < b - a$,

then Newton's method will converge from any $x^{(0)}$ in the interval. Results like these can be found in many introductory numerical analysis books such as [112, 173, 217, 328]. A convergence theorem with less stringent conditions is provided by [423].

2.1.1.1 *Convergence order*
The speed of a root-finding approach like Newton's method is typically measured by its order of convergence. A method has *convergence of order* β if $\lim_{t \to \infty} \epsilon^{(t)} = 0$ and

$$\lim_{t \to \infty} \frac{\left| \epsilon^{(t+1)} \right|}{\left| \epsilon^{(t)} \right|^\beta} = c \tag{2.19}$$

for some constants $c \neq 0$ and $\beta > 0$. Higher orders of convergence are better in the sense that precise approximation of the true solution is more quickly achieved. Unfortunately, high orders are sometimes achieved at the expense of robustness: Some slow algorithms are more foolproof than their faster counterparts.

For Newton's method, (2.15) shows us that

$$\frac{\epsilon^{(t+1)}}{(\epsilon^{(t)})^2} = \frac{g'''(q)}{2g''(x^{(t)})}.$$

(2.20)

If Newton's method converges, then continuity arguments allow us to note that the right hand side of this equation converges to $\frac{g'''(x^*)}{2g''(x^*)}$. Thus, Newton's method has quadratic convergence (i.e., $\beta = 2$), and $c = \left| \frac{g'''(x^*)}{2g''(x^*)} \right|$. Quadratic convergence is indeed fast: Usually the precision of the solution will double with each iteration.

For bisection, the length of the bracketing interval exhibits a property analogous to linear convergence ($\beta = 1$), in that it is halved at each iteration and $\lim_{t\to\infty} |\epsilon^{(t)}| = 0$ if there is a root in the starting interval. However, the distances $x^{(t)} - x^*$ need not shrink at every iteration, and indeed their ratio is potentially unbounded. Thus $\lim_{t\to\infty} \frac{|\epsilon^{(t+1)}|}{|\epsilon^{(t)}|^\beta}$ may not exist for any $\beta > 0$, and bisection does not formally meet the definition for determining order of convergence.

It is possible to use a foolproof bracketing method such as bisection to safeguard a faster but less reliable root-finding approach such as Newton's method. Instead of viewing the bracketing approach as a method to generate steps, view it only as a method providing an interval within which a root must lie. If Newton's method seeks a step outside these current bounds, the step must be replaced, curtailed, or (in the multivariate case) redirected. Some strategies are mentioned in Section 2.2 and in [217]. Safeguarding can reduce the convergence order of a method.

2.1.2   Fisher scoring

Recall from Section 1.4 that $I(\theta)$ can be approximated by $-l''(\theta)$. Therefore when the optimization of g corresponds to a MLE problem, it is reasonable to replace $-l''(\theta)$ in the Newton update with $I(\theta)$. This yields an updating increment of $h^{(t)} = l'(\theta^{(t)})/I(\theta^{(t)})$ where $I(\theta^{(t)})$ is the expected Fisher information evaluated at $\theta^{(t)}$. The updating equation is therefore

$$\theta^{(t+1)} = \theta^{(t)} + l'(\theta^{(t)})I(\theta^{(t)})^{-1}.$$

(2.21)

This approach is called *Fisher scoring*.

Fisher scoring and Newton's method both have the same asymptotic properties, but for individual problems one may be computationally or analytically easier than the other. Generally, Fisher scoring works better in the beginning to make rapid improvements, while Newton's method works better for refinement near the end.

2.1.3   Secant method

The updating increment for Newton's method in (2.10) relies on the second derivative, $g''(x^{(t)})$. If calculating this derivative is difficult, it might be replaced by the discrete-difference approximation, $\frac{g'(x^{(t)})-g'(x^{(t-1)})}{x^{(t)}-x^{(t-1)}}$. The result is the *secant method*, which

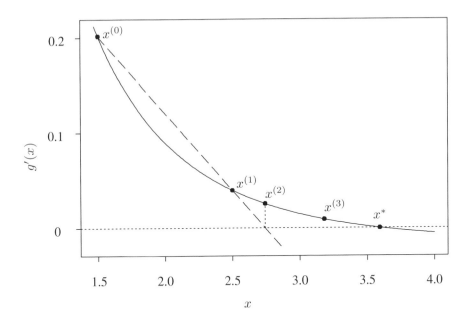

Fig. 2.5 The secant method locally approximates g' using the secant line between $x^{(0)}$ and $x^{(1)}$. The corresponding estimated root, $x^{(2)}$, is used with $x^{(1)}$ to generate the next approximation.

has updating equation

$$x^{(t+1)} = x^{(t)} - g'(x^{(t)}) \frac{x^{(t)} - x^{(t-1)}}{g'(x^{(t)}) - g'(x^{(t-1)})} \qquad (2.22)$$

for $t \geq 1$. This approach requires two starting points, $x^{(0)}$ and $x^{(1)}$. Figure 2.5 illustrates the first steps of the method for maximizing the simple function introduced in Example 2.1.

Under conditions akin to those for Newton's method, the secant method will converge to the root x^*. To find the order of convergence in this case, restrict attention to a suitably small interval $[a, b]$, containing $x^{(0)}$, $x^{(1)}$, and x^*, on which $g''(x) \neq 0$ and $g'''(x) \neq 0$. Letting $\epsilon^{(t+1)} = x^{(t+1)} - x^*$, it is straightforward to show that

$$\epsilon^{(t+1)} = \left[\frac{x^{(t)} - x^{(t-1)}}{g'(x^{(t)}) - g'(x^{(t-1)})} \right] \left[\frac{g'(x^{(t)})/\epsilon^{(t)} - g'(x^{(t-1)})/\epsilon^{(t-1)}}{x^{(t)} - x^{(t-1)}} \right] \left[\epsilon^{(t)} \epsilon^{(t-1)} \right]$$

$$= A^{(t)} B^{(t)} \epsilon^{(t)} \epsilon^{(t-1)}, \qquad (2.23)$$

where $A^{(t)} \to 1/g''(x^*)$ as $x^{(t)} \to x^*$ for continuous g''.

To deduce a limit for $B^{(t)}$, expand g' in a Taylor series about x^*:

$$g'(x^{(t)}) \approx g'(x^*) + (x^{(t)} - x^*)g''(x^*) + (x^{(t)} - x^*)^2 g'''(x^*)/2, \qquad (2.24)$$

so

$$g'(x^{(t)})/\epsilon^{(t)} \approx g''(x^*) + \epsilon^{(t)}g'''(x^*)/2. \tag{2.25}$$

Similarly, $g'(x^{(t-1)})/\epsilon^{(t-1)} \approx g''(x^*) + \epsilon^{(t-1)}g'''(x^*)/2$. Thus,

$$B^{(t)} \approx g'''(x^*)\frac{\epsilon^{(t)} - \epsilon^{(t-1)}}{2(x^{(t)} - x^{(t-1)})} = g'''(x^*)/2, \tag{2.26}$$

and a careful examination of the errors shows this approximation to be exact as $x^{(t)} \to x^*$. Thus,

$$\epsilon^{(t+1)} \approx d^{(t)}\epsilon^{(t)}\epsilon^{(t-1)}, \tag{2.27}$$

where $d^{(t)} \to \frac{g'''(x^*)}{2g''(x^*)} = d$ as $t \to \infty$.

To find the order of convergence for the secant method, we must find the β for which $\lim_{t\to\infty}\frac{|\epsilon^{(t+1)}|}{|\epsilon^{(t)}|^\beta} = c$ for some constant c. Suppose that this relationship does indeed hold, and use this proportionality expression to replace $\epsilon^{(t-1)}$ and $\epsilon^{(t+1)}$ in (2.27), leaving only terms in $\epsilon^{(t)}$. Then, after some rearrangement of terms, it suffices to find the β for which

$$\lim_{t\to\infty} |\epsilon^{(t)}|^{1-\beta+1/\beta} = \frac{c^{1+1/\beta}}{d}. \tag{2.28}$$

The right hand side of (2.28) is a positive constant. Therefore $1 - \beta + 1/\beta = 0$. The solution is $\beta = (1 + \sqrt{5})/2 \approx 1.62$. Thus, the secant method has a slower order of convergence than Newton's method.

2.1.4 Fixed-point iteration

A fixed-point of a function is a point whose evaluation by that function equals itself. The fixed-point strategy for finding roots is to determine a function G for which $g'(x) = 0$ if and only if if $G(x) = x$. This transforms the problem of finding a root of g' into a problem of finding a fixed point of G. Then the simplest way to hunt for a fixed point is to use the updating equation $x^{(t+1)} = G(x^{(t)})$.

Any suitable function G may be tried, but the most obvious choice is $G(x) = g'(x) + x$. This yields the updating equation

$$x^{(t+1)} = x^{(t)} + g'(x^{(t)}). \tag{2.29}$$

The convergence of this algorithm depends on whether G is *contractive*. To be contractive on $[a, b]$, G must satisfy:

1. $G(x) \in [a, b]$ whenever $x \in [a, b]$, and

2. $|G(x_1) - G(x_2)| \le \lambda|x_1 - x_2|$ for all $x_1, x_2 \in [a, b]$ for some $\lambda \in [0, 1)$.

The interval $[a, b]$ may be unbounded. The second requirement is a *Lipschitz condition*, and λ is called the *Lipschitz constant*. If G is contractive on $[a, b]$ then there exists a unique fixed point, x^*, in this interval, and the fixed-point algorithm will converge to it from any starting point in the interval. Furthermore, under the conditions above,

$$|x^{(t)} - x^*| \leq \frac{\lambda^t}{1 - \lambda} |x^{(1)} - x^{(0)}|. \tag{2.30}$$

Proof of a *contractive mapping theorem* like this can be found in [6, 439].

Fixed-point iteration is sometimes called *functional iteration*. Note that both Newton's method and the secant method are special cases of fixed-point iteration.

2.1.4.1 *Scaling*   If fixed-point iteration converges, the order of convergence depends on λ. Convergence is not universally assured. In particular, the Lipschitz condition holds if $|G'(x)| \leq \lambda < 1$ for all x in $[a, b]$. If $G(x) = g'(x) + x$, this amounts to requiring $|g''(x) + 1| < 1$ on $[a, b]$. When g'' is bounded and does not change sign on $[a, b]$, we can rescale nonconvergent problems by choosing $G(x) = \alpha g'(x) + x$ for $\alpha \neq 0$, since $\alpha g'(x) = 0$ if and only if $g'(x) = 0$. To permit convergence, α must be chosen to satisfy $|\alpha g''(x) + 1| < 1$ on an interval including the starting value. Although one could carefully calculate a suitable α, it may be easier just to try a few values. If the method converges quickly, then the chosen α was suitable.

Rescaling is only one of several strategies for adjusting G. In general, the effectiveness of fixed-point iteration is highly dependent on the chosen form of G. For example, consider finding the root of $g'(x) = x + \log x$. Then $G(x) = (x + e^{-x})/2$ converges quickly, whereas $G(x) = e^{-x}$ converges more slowly and $G(x) = -\log x$ fails to converge at all.

Example 2.3 (A simple univariate optimization, continued)   Figure 2.6 illustrates the first several steps of the scaled fixed-point algorithm for maximizing the function $g(x) = \frac{\log x}{1+x}$ in (2.2) using $G(x) = g'(x) + x$ and $\alpha = 4$. Note that line segments whose roots determine the next $x^{(t)}$ are parallel, with slopes equal to $-1/\alpha$. For this reason, the method is sometimes called the *method of parallel chords*. $\square$

Suppose a MLE is sought for the parameter of a quadratic log likelihood l, or one that is nearly quadratic near $\hat{\theta}$. Then the score function is locally linear, and l'' is roughly a constant, say γ. For quadratic log likelihoods, Newton's method would use the updating equation $\theta^{(t+1)} = \theta^{(t)} - l'(\theta)/\gamma$. If we use scaled fixed-point iteration with $\alpha = -1/\gamma$, we get the same updating equation. Since many log likelihoods are approximately locally quadratic, scaled fixed-point iteration can be a very effective tool. The method is also generally quite stable and easy to code.

2.2 MULTIVARIATE PROBLEMS

In a multivariate optimization problem we seek the optimum of a real-valued function g of a p-dimensional vector $\mathbf{x} = (x_1, \ldots, x_p)^T$. At iteration t, denote the estimated optimum as $\mathbf{x}^{(t)} = (x_1^{(t)}, \ldots, x_p^{(t)})^T$.

Many of the general principles discussed above for the univariate case also apply for multivariate optimization. Algorithms are still iterative. Many algorithms

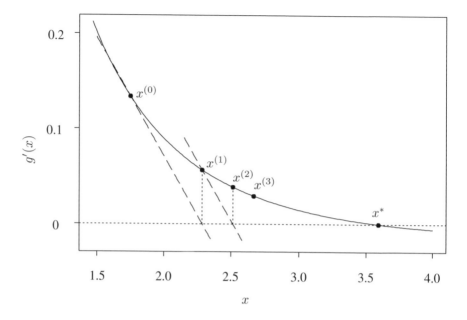

Fig. 2.6   The first three steps of scaled fixed-point iteration to maximize $g(x) = \frac{\log x}{1+x}$ using $G(x) = g'(x) + x$ and scaling with $\alpha = 4$, as in Example 2.3.

take steps based on a local linearization of g' derived from a Taylor series or secant approximation. Convergence criteria are similar in spirit despite slight changes in form. To construct convergence criteria, let $D(\mathbf{u}, \mathbf{v})$ be a distance measure for p-dimensional vectors. Two obvious choices are $D(\mathbf{u}, \mathbf{v}) = \sum_{i=1}^{p} |u_i - v_i|$ and $D(\mathbf{u}, \mathbf{v}) = \sqrt{\sum_{i=1}^{p}(u_i - v_i)^2}$. Then absolute and relative convergence criteria can be formed from the inequalities

$$D(\mathbf{x}^{(t+1)}, \mathbf{x}^{(t)}) < \epsilon, \qquad \frac{D(\mathbf{x}^{(t+1)}, \mathbf{x}^{(t)})}{D(\mathbf{x}^{(t)}, \mathbf{0})} < \epsilon, \qquad \text{or} \qquad \frac{D(\mathbf{x}^{(t+1)}, \mathbf{x}^{(t)})}{D(\mathbf{x}^{(t)}, \mathbf{0}) + \epsilon} < \epsilon.$$

2.2.1   Newton's method and Fisher scoring

To fashion the Newton's method update, we again approximate $g(\mathbf{x}^*)$ by the quadratic Taylor series expansion

$$g(\mathbf{x}^*) = g(\mathbf{x}^{(t)}) + (\mathbf{x}^* - \mathbf{x}^{(t)})^T \mathbf{g}'(\mathbf{x}^{(t)}) + (\mathbf{x}^* - \mathbf{x}^{(t)})^T \mathbf{g}''(\mathbf{x}^{(t)})(\mathbf{x}^* - \mathbf{x}^{(t)})/2 \quad (2.31)$$

and maximize this quadratic function with respect to $\mathbf{x}^*$ to find the next iterate. Setting the gradient of the right hand side of (2.31) equal to zero yields

$$\mathbf{g}'(\mathbf{x}^{(t)}) + \mathbf{g}''(\mathbf{x}^{(t)})(\mathbf{x}^* - \mathbf{x}^{(t)}) = 0. \qquad\qquad (2.32)$$

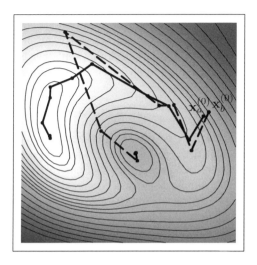

Fig. 2.7 An application of Newton's method for maximizing a complicated bivariate function, as discussed in Example 2.4. The surface of the function is indicated by shading and contours, with light shading corresponding to high values. Two runs starting from $\mathbf{x}_a^{(0)}$ and $\mathbf{x}_b^{(0)}$ are shown. These converge to the true maximum and to a local minimum, respectively.

This provides the update

$$\mathbf{x}^{(t+1)} = \mathbf{x}^{(t)} - \mathbf{g}''(\mathbf{x}^{(t)})^{-1}\mathbf{g}'(\mathbf{x}^{(t)}). \tag{2.33}$$

Alternatively, note that the left hand side of (2.32) is in fact a linear Taylor series approximation to $\mathbf{g}'(\mathbf{x}^*)$, and solving (2.32) amounts to finding the root of this linear approximation. From either viewpoint, the multivariate Newton increment is $\mathbf{h}^{(t)} = -\mathbf{g}''(\mathbf{x}^{(t)})^{-1}\mathbf{g}'(\mathbf{x}^{(t)})$.

As in the univariate case, in MLE problems we may replace the observed information at $\boldsymbol{\theta}^{(t)}$ with $\mathbf{I}(\boldsymbol{\theta}^{(t)})$, the expected Fisher information at $\boldsymbol{\theta}^{(t)}$. This yields the multivariate Fisher scoring approach with update given by

$$\boldsymbol{\theta}^{(t+1)} = \boldsymbol{\theta}^{(t)} + \mathbf{I}(\boldsymbol{\theta}^{(t)})^{-1}\mathbf{l}'(\boldsymbol{\theta}^{(t)}). \tag{2.34}$$

This method is asymptotically equivalent to Newton's method.

Example 2.4 (A bivariate optimization) Figure 2.7 illustrates the application of Newton's method to a complicated bivariate function. The surface of the function is indicated by shading and contour lines, with high values corresponding to light shading. The algorithm is started from two different starting points, $\mathbf{x}_a^{(0)}$ and $\mathbf{x}_b^{(0)}$. From $\mathbf{x}_a^{(0)}$, the algorithm converges quickly to the true maximum. Note that although steps were taken in an uphill direction, some step lengths were not ideal. From $\mathbf{x}_b^{(0)}$, which lies very close to $\mathbf{x}_a^{(0)}$, the algorithm fails to maximize the function—in fact it converges to a local minimum. One step length in this attempt was so large as to completely overshoot the uphill portion of the ridge, resulting in a step that was downhill. Near the end, the algorithm steps downhill because it has honed in on the wrong root of $\mathbf{g}'$. In Section 2.2.2, approaches for preventing such problems are discussed. ☐

2.2.1.1 *Iteratively reweighted least squares*   Consider finding the MLEs for the parameters of a logistic regression model, which is a well-known type of generalized linear model [379]. In a generalized linear model, response variables Y_i for $i = 1, \ldots, n$ are independently distributed according to a distribution parameterized by θ_i. Different types of response variables are modeled with different distributions, but the distribution is always a member of the scaled exponential family. This family has the form $f(y|\theta) = \exp\{[y\theta - b(\theta)]/a(\phi) + c(y, \phi)\}$, where θ is called the natural or canonical parameter and ϕ is the dispersion parameter. Two of the most useful properties of this family are that $E\{Y\} = b'(\theta)$ and $\mathrm{var}\{Y\} = b''(\theta)a(\phi)$ (see Section 1.3).

The distribution of each Y_i is modeled to depend on a corresponding set of observed covariates, $\mathbf{z}_i$. Specifically, we assume that some function of $E\{Y_i|\mathbf{z}_i\}$ can be related to $\mathbf{z}_i$ according to the equation $g(E\{Y_i|\mathbf{z}_i\}) = \mathbf{z}_i^T\boldsymbol{\beta}$, where $\boldsymbol{\beta}$ is a vector of parameters and g is called the *link function*.

The generalized linear model used for logistic regression is based on the Bernoulli distribution, which is a member of the exponential family. Model the response variables as $Y_i|\mathbf{z}_i \sim \mathrm{Bernoulli}(\pi_i)$ independently for $i = 1, \ldots, n$. Suppose the observed data consist of a single covariate value z_i and a response value y_i, for $i = 1, \ldots, n$. Define the column vectors $\mathbf{z}_i = (1, z_i)^T$ and $\boldsymbol{\beta} = (\beta_0, \beta_1)^T$. Then for the ith observation, the natural parameter is $\theta_i = \log\{\pi_i/(1 - \pi_i)\}$, $a(\phi) = 1$, and $b(\theta_i) = \log\{1 + \exp\{\theta_i\}\} = \log\{1 + \exp\{\mathbf{z}_i^T\boldsymbol{\beta}\}\} = -\log\{1 - \pi_i\}$. The log likelihood is

$$l(\boldsymbol{\beta}) = \mathbf{y}^T\mathbf{Z}\boldsymbol{\beta} - \mathbf{b}^T\mathbf{1}, \tag{2.35}$$

where $\mathbf{1}$ is a column vector of ones, $\mathbf{y} = (y_1 \ldots y_n)^T$, $\mathbf{b} = (b(\theta_1) \ldots b(\theta_n))^T$, and $\mathbf{Z}$ is the $n \times 2$ matrix whose ith row is $\mathbf{z}_i^T$.

Consider using Newton's method to find $\boldsymbol{\beta}$ that maximizes this likelihood. The score function is

$$l'(\boldsymbol{\beta}) = \mathbf{Z}^T(\mathbf{y} - \boldsymbol{\pi}), \tag{2.36}$$

where $\boldsymbol{\pi}$ is a column vector of the Bernoulli probabilities $\pi_1, \ldots, \pi_n$. The Hessian is given by

$$l''(\boldsymbol{\beta}) = \frac{d}{d\boldsymbol{\beta}}(\mathbf{Z}^T(\mathbf{y} - \boldsymbol{\pi})) = -\left(\frac{d\boldsymbol{\pi}}{d\boldsymbol{\beta}}\right)^T \mathbf{Z} = -\mathbf{Z}^T\mathbf{W}\mathbf{Z}, \tag{2.37}$$

where $\mathbf{W}$ is a diagonal matrix with ith diagonal entry equal to $\pi_i(1 - \pi_i)$.

Newton's update is therefore

$$\beta^{(t+1)} = \beta^{(t)} - \mathbf{l}''(\beta^{(t)})^{-1}\mathbf{l}'(\beta^{(t)}) \tag{2.38}$$

$$= \beta^{(t)} + \left(\mathbf{Z}^T\mathbf{W}^{(t)}\mathbf{Z}\right)^{-1}\left(\mathbf{Z}^T(\mathbf{y} - \pi^{(t)})\right), \tag{2.39}$$

where $\pi^{(t)}$ is the value of π corresponding to $\beta^{(t)}$, and $\mathbf{W}^{(t)}$ is the diagonal weight matrix evaluated at $\pi^{(t)}$.

Note that the Hessian does not depend on $\mathbf{y}$. Therefore, Fisher's information matrix is equal to the observed information: $\mathbf{I}(\beta) = \mathrm{E}\{-\mathbf{l}''(\beta)\} = \mathrm{E}\{\mathbf{Z}^T\mathbf{W}\mathbf{Z}\} = -\mathbf{l}''(\beta)$. Therefore, for this example the Fisher scoring approach is the same as Newton's method. For generalized linear models, this will always be true when the link function is chosen to make the natural parameter a linear function of the covariates.

Example 2.5 (Human face recognition) We will fit a logistic regression model to some data related to testing of a human face recognition algorithm. Pairs of images of faces of 1072 humans were used to train and test an automatic face recognition algorithm [580]. The experiment used the recognition software to match the first image of each person (called a probe) to one of the remaining 2143 images. Ideally, a match is made to the other image of the same person (called the target). A successful match yielded a response of $y_i = 1$ and a match to any other person yielded a response of $y_i = 0$. The predictor variable used here is the absolute difference in mean standardized eye region pixel intensity between the probe image and its corresponding target; this is a measure of whether the two images exhibit similar quality in the important distinguishing region around the eyes. Large differences in eye region pixel intensity would be expected to impede recognition (i.e., successful matches). For the data described here, there were 775 correct matches and 297 mismatches. The median and 90th percentile values of the predictor were 0.033 and 0.097, respectively, for image pairs successfully matched, and 0.060 and 0.161 for unmatched pairs. Therefore, the data appear to support the hypothesis that eye region pixel intensity discrepancies impede recognition. These data are available from the website for this book; analyses of related datasets are given in [220, 221].

To quantify the relationship between these variables, we will fit a logistic regression model. Thus, z_i is the absolute difference in eye region intensity for an image pair and y_i indicates whether the ith probe was successfully matched, for $i = 1, \ldots, 1072$. The likelihood function is composed as in (2.35), and we will apply Newton's method.

To start, we may take $\beta^{(0)} = \left(\beta_0^{(0)}, \beta_1^{(0)}\right)^T = (0.95913, 0)^T$, which means $\pi_i = 775/1072$ for all i at iteration 0. Table 2.1 shows that the approximation converges quickly, with $\beta^{(4)} = (1.73874, -13.58840)^T$. Quick convergence is also achieved from the starting value corresponding to $\pi_i = 0.5$ for all i (namely $\beta^{(0)} = \mathbf{0}$), which is a suggested rule of thumb for fitting logistic regression models with Bernoulli data [278]. Since $\hat{\beta}_1 = -13.59$ is nearly 9 marginal standard deviations below zero, these data strongly support the hypothesis that eye region intensity discrepancies impede recognition. □

Table 2.1   Parameter estimates and corresponding variance–covariance matrix estimates are shown for each Newton's method iteration for fitting a logistic regression model to the face recognition data described in Example 2.5.

Iteration, t	$\beta^{(t)}$	$-l''(\beta^{(t)})^{-1}$	
0	$\begin{pmatrix} 0.95913 \\ 0.00000 \end{pmatrix}$	$\begin{pmatrix} 0.01067 & -0.11412 \\ -0.11412 & 2.16701 \end{pmatrix}$	
1	$\begin{pmatrix} 1.70694 \\ -14.20059 \end{pmatrix}$	$\begin{pmatrix} 0.13312 & -0.14010 \\ -0.14010 & 2.36367 \end{pmatrix}$	
2	$\begin{pmatrix} 1.73725 \\ -13.56988 \end{pmatrix}$	$\begin{pmatrix} 0.01347 & -0.13941 \\ -0.13941 & 2.32090 \end{pmatrix}$	
3	$\begin{pmatrix} 1.73874 \\ -13.58839 \end{pmatrix}$	$\begin{pmatrix} 0.01349 & -0.13952 \\ -0.13952 & 2.32241 \end{pmatrix}$	
4	$\begin{pmatrix} 1.73874 \\ -13.58840 \end{pmatrix}$	$\begin{pmatrix} 0.01349 & -0.13952 \\ -0.13952 & 2.32241 \end{pmatrix}$	

The Fisher scoring approach to maximum likelihood estimation for generalized linear models is important for several reasons. First, it is an application of the method of *iteratively reweighted least squares* (IRLS). Let

$$e^{(t)} = y - \pi^{(t)} \tag{2.40}$$

and

$$x^{(t)} = Z\beta^{(t)} + (W^{(t)})^{-1}e^{(t)}. \tag{2.41}$$

Now the Fisher scoring update can be written as

$$\begin{aligned} \beta^{(t+1)} &= \beta^{(t)} + \left(Z^T W^{(t)} Z\right)^{-1} Z^T e^{(t)} \\ &= \left(Z^T W^{(t)} Z\right)^{-1} \left[Z^T W^{(t)} Z\beta^{(t)} + Z^T W^{(t)} (W^{(t)})^{-1} e^{(t)}\right] \\ &= \left(Z^T W^{(t)} Z\right)^{-1} Z^T W^{(t)} x^{(t)}. \end{aligned} \tag{2.42}$$

We call $x^{(t)}$ the *working response* because it is apparent from (2.42) that $\beta^{(t+1)}$ are the regression coefficients resulting from the weighted least squares regression of $x^{(t)}$ on Z with weights corresponding to the diagonal elements of $W^{(t)}$. At each iteration, a new working response and weight vector are calculated, and the update can be fitted via weighted least squares.

Second, IRLS for generalized linear models is a special case of the Gauss–Newton method for nonlinear least squares problems, which is introduced briefly below. IRLS therefore shows the same behavior as Gauss–Newton; in particular, it can be a slow and unreliable approach to fitting generalized linear models unless the model fits the data rather well [534].

2.2.2 Newton-like methods

Some very effective methods rely on updating equations of the form

$$\mathbf{x}^{(t+1)} = \mathbf{x}^{(t)} - (\mathbf{M}^{(t)})^{-1}\mathbf{g}'(\mathbf{x}^{(t)}) \tag{2.43}$$

where $\mathbf{M}^{(t)}$ is a $p \times p$ matrix approximating the Hessian, $\mathbf{g}''(\mathbf{x}^{(t)})$. In general optimization problems, there are several good reasons to consider replacing the Hessian by some simpler approximation. First, it may be computationally expensive to evaluate the Hessian. Second, the steps taken by Newton's method are not necessarily always uphill: At each iteration, there is no guarantee that $g(\mathbf{x}^{(t+1)}) > g(\mathbf{x}^{(t)})$. A suitable $\mathbf{M}^{(t)}$ can guarantee ascent. We already know that one possible Hessian replacement, $\mathbf{M}^{(t)} = -\mathbf{I}(\boldsymbol{\theta}^{(t)})$, yields the Fisher scoring approach. Certain other (possibly scaled) choices for $\mathbf{M}^{(t)}$ can also yield good performance while limiting computing effort.

2.2.2.1 *Ascent algorithms* To force uphill steps, one could resort to an *ascent algorithm*. (Another type of ascent algorithm is discussed in Chapter 3.) In the present context, the method of *steepest ascent* is obtained with the Hessian replacement $\mathbf{M}^{(t)} = -\mathbf{I}$, where $\mathbf{I}$ is the identity matrix. Since the gradient of g indicates the steepest direction uphill on the surface of g at the point $\mathbf{x}^{(t)}$, setting $\mathbf{x}^{(t+1)} = \mathbf{x}^{(t)} + \mathbf{g}'(\mathbf{x}^{(t)})$ amounts to taking a step in the direction of steepest ascent. Scaled steps of the form $\mathbf{x}^{(t+1)} = \mathbf{x}^{(t)} + \alpha^{(t)}\mathbf{g}'(\mathbf{x}^{(t)})$ for some $\alpha^{(t)} > 0$ can be helpful for controlling convergence, as will be discussed below.

Many forms of $\mathbf{M}^{(t)}$ will yield ascent algorithms with increments

$$\mathbf{h}^{(t)} = -\alpha^{(t)}\left[\mathbf{M}^{(t)}\right]^{-1}\mathbf{g}'(\mathbf{x}^{(t)}). \tag{2.44}$$

For any fixed $\mathbf{x}^{(t)}$ and negative definite $\mathbf{M}^{(t)}$, note that as $\alpha^{(t)} \to 0$ we have

$$\begin{aligned}
g(\mathbf{x}^{(t+1)}) - g(\mathbf{x}^{(t)}) &= g(\mathbf{x}^{(t)} + \mathbf{h}^{(t)}) - g(\mathbf{x}^{(t)}) \\
&= -\alpha^{(t)}\mathbf{g}'(\mathbf{x}^{(t)})^T(\mathbf{M}^{(t)})^{-1}\mathbf{g}'(\mathbf{x}^{(t)}) + \mathcal{O}(\alpha^{(t)}),
\end{aligned} \tag{2.45}$$

where the second equality follows from the linear Taylor expansion $g(\mathbf{x}^{(t)} + \mathbf{h}^{(t)}) = g(\mathbf{x}^{(t)}) + \mathbf{g}'(\mathbf{x}^{(t)})^T\mathbf{h}^{(t)} + \mathcal{O}(\alpha^{(t)})$. Therefore, if $-\mathbf{M}^{(t)}$ is positive definite, ascent can be assured by choosing $\alpha^{(t)}$ sufficiently small, yielding $g(\mathbf{x}^{(t+1)}) - g(\mathbf{x}^{(t)}) > 0$ from (2.45) since $\mathcal{O}(\alpha^{(t)})/\alpha^{(t)} \to 0$ as $\alpha^{(t)} \to 0$.

Typically, therefore, an ascent algorithm involves a positive definite matrix $-\mathbf{M}^{(t)}$ to approximate the negative Hessian, and a *contraction* or *step length* parameter $\alpha^{(t)} > 0$ whose value can shrink to ensure ascent at each step. For example, start each step with $\alpha^{(t)} = 1$. If the original step turns out to be downhill, $\alpha^{(t)}$ can be halved. This is called *backtracking*. If the step is still downhill, $\alpha^{(t)}$ is halved again until a sufficiently small step is found to be uphill. For Fisher scoring, $-\mathbf{M}^{(t)} = \mathbf{I}(\boldsymbol{\theta}^{(t)})$, which is positive semidefinite. Therefore backtracking with Fisher scoring would avoid stepping downhill.

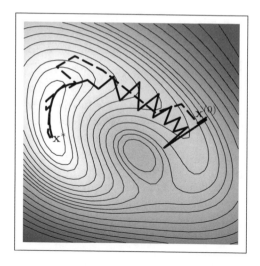

Fig. 2.8 Applications of two optimization methods for maximizing a complex bivariate function. The surface of the function is indicated by shading and contours, with light shading corresponding to high values. The two methods start at a point $\mathbf{x}^{(0)}$ and find the true maximum, $\mathbf{x}^*$. The solid line corresponds to the method of steepest ascent (Example 2.6). The dashed line corresponds to a quasi-Newton method with the BFGS update (Example 2.7). Both algorithms employed backtracking, with the initial value of $\alpha^{(t)}$ at each step set to 0.25 and 0.05, respectively.

Example 2.6 (A bivariate optimization, continued) Figure 2.8 illustrates an application of the steepest ascent algorithm to maximize the bivariate function discussed in Example 2.4, starting from $x^{(0)}$ and initialized with $\alpha^{(t)} = 1/4$ at each step. The steps taken by steepest ascent are shown by the solid line. Although the optimization was successful, it was not fast or efficient. The dashed line illustrates another method, discussed in Section 2.2.2.3. □

Step halving is only one approach to backtracking. In general methods that rely on finding an advantageous step length in the chosen direction are called *line search methods*. Backtracking with a positive definite replacement for the negative Hessian is not sufficient to ensure convergence of the algorithm, however, even when g is bounded above with a unique maximum. It is also necessary to ensure that steps make a sufficient ascent (i.e., require that $g(\mathbf{x}^{(t)}) - g(\mathbf{x}^{(t-1)})$ does not decrease too quickly as t increases) and that step directions are not nearly orthogonal to the gradient (i.e., avoid following a level contour of g). Formal versions of such requirements include the Goldstein–Armijo and Wolfe–Powell conditions, under which convergence of ascent algorithms is guaranteed [13, 239, 435, 570].

When the step direction is not uphill, approaches known as *modified Newton methods* alter the direction sufficiently to find an uphill direction [217]. A quite effective variant is the modified Cholesky decomposition approach [216]. In essence, when the negative Hessian is not positive definite, this strategy replaces it with $-\tilde{\mathbf{g}}''(\mathbf{x}^{(t)}) = -\mathbf{g}''(\mathbf{x}^{(t)}) + \mathbf{E}$, where $\mathbf{E}$ is a diagonal matrix with nonnegative elements. By crafting $\mathbf{E}$ carefully to ensure that $-\tilde{\mathbf{g}}''(\mathbf{x}^{(t)})$ is positive definite without deviating unnecessarily from the original direction $-\mathbf{g}''(\mathbf{x}^{(t)})$, a suitable uphill direction can be derived.

2.2.2.2 *Discrete Newton and fixed-point methods*

To avoid calculating the Hessian, one could resort to a secantlike method, yielding a *discrete Newton method*, or rely solely on an initial approximation, yielding a *multivariate fixed-point method*.

Multivariate fixed-point methods use an initial approximation of $\mathbf{g}''$ throughout the iterative updating. If this approximation is a matrix of constants, so $\mathbf{M}^{(t)} = \mathbf{M}$ for all t, then the updating equation is

$$\mathbf{x}^{(t+1)} = \mathbf{x}^{(t)} - \mathbf{M}^{-1}\mathbf{g}'(\mathbf{x}^{(t)}). \tag{2.46}$$

A reasonable choice for $\mathbf{M}$ is $\mathbf{g}''(\mathbf{x}^{(0)})$. Notice that if $\mathbf{M}$ is diagonal, then this amounts to applying the univariate scaled fixed-point algorithm separately to each component of $\mathbf{g}$. See Section 2.1.4 for more on the relationship between fixed-point iteration and Newton's method when maximizing log likelihoods that are locally quadratic.

Multivariate discrete Newton methods approximate the matrix $\mathbf{g}''(\mathbf{x}^{(t)})$ with a matrix $\mathbf{M}^{(t)}$ of finite-difference quotients. Let $g_i'(\mathbf{x}) = dg(\mathbf{x})/dx_i$ be the ith element of $\mathbf{g}'(\mathbf{x})$. Let $\mathbf{e}_j$ denote the p-vector with a 1 in the jth position and zeros elsewhere. Among the ways one might approximate the (i, j)th element of the Hessian using discrete differences, perhaps the most straightforward is to set the (i, j)th element of $\mathbf{M}^{(t)}$ to equal

$$\mathbf{M}_{ij}^{(t)} = \frac{g_i'\left(\mathbf{x}^{(t)} + h_{ij}^{(t)}\mathbf{e}_j\right) - g_i'\left(\mathbf{x}^{(t)}\right)}{h_{ij}^{(t)}} \tag{2.47}$$

for some constants $h_{ij}^{(t)}$. It is easiest to use $h_{ij}^{(t)} = h$ for all (i, j) and t, but this leads to a convergence order of $\beta = 1$. Alternatively, we can generally obtain an order of convergence similar to that of the univariate secant method if we set $h_{ij}^{(t)} = x_j^{(t)} - x_j^{(t-1)}$ for all i, where $x_j^{(t)}$ denotes the jth element of $\mathbf{x}^{(t)}$. It is important to average $\mathbf{M}^{(t)}$ with its transpose to ensure symmetry before proceeding with the update of $\mathbf{x}^{(t)}$ given in (2.43).

2.2.2.3 *Quasi-Newton methods*

The discrete Newton method strategy for numerically approximating the Hessian by $\mathbf{M}^{(t)}$ is a computationally burdensome one. At each step, $\mathbf{M}^{(t)}$ is wholly updated by calculating a new discrete difference for each element. A more efficient approach can be designed, based on the direction of the most recent step. When $\mathbf{x}^{(t)}$ is updated to $\mathbf{x}^{(t+1)} = \mathbf{x}^{(t)} + \mathbf{h}^{(t)}$, the opportunity is presented to learn about the curvature of $\mathbf{g}'$ in the direction of $\mathbf{h}^{(t)}$ near $\mathbf{x}^{(t)}$. Then $\mathbf{M}^{(t)}$ can be efficiently updated to incorporate this information.

To do this, we must abandon the componentwise discrete-difference approximation to $\mathbf{g}''$ used in the discrete Newton method. However, it is possible to retain a type of secant condition based on differences. Specifically, a secant condition holds for $\mathbf{M}^{(t+1)}$ if

$$g'(\mathbf{x}^{(t+1)}) - g'(\mathbf{x}^{(t)}) = \mathbf{M}^{(t+1)}(\mathbf{x}^{(t+1)} - \mathbf{x}^{(t)}). \qquad (2.48)$$

This condition suggests that we need a method to generate $\mathbf{M}^{(t+1)}$ from $\mathbf{M}^{(t)}$ in a manner that requires few calculations and satisfies (2.48). This will enable us to gain information about the curvature of g' in the direction of the most recent step. The result is a *quasi-Newton method*, sometimes called a *variable metric* approach [133, 217, 415].

There is a unique symmetric rank-one method that meets these requirements [115]. Let $\mathbf{z}^{(t)} = \mathbf{x}^{(t+1)} - \mathbf{x}^{(t)}$ and $\mathbf{y}^{(t)} = g'(\mathbf{x}^{(t+1)}) - g'(\mathbf{x}^{(t)})$. Then we can write the update to $\mathbf{M}^{(t)}$ as

$$\mathbf{M}^{(t+1)} = \mathbf{M}^{(t)} + c^{(t)}\mathbf{v}^{(t)}(\mathbf{v}^{(t)})^T, \qquad (2.49)$$

where $\mathbf{v}^{(t)} = \mathbf{y}^{(t)} - \mathbf{M}^{(t)}\mathbf{z}^{(t)}$ and $c^{(t)} = \frac{1}{(\mathbf{v}^{(t)})^T\mathbf{z}^{(t)}}$.

It is important to monitor the behavior of this update to $\mathbf{M}^{(t)}$. If $c^{(t)}$ cannot reliably be calculated because the denominator is zero or close to zero, a temporary solution is to take $\mathbf{M}^{(t+1)} = \mathbf{M}^{(t)}$ for that iteration. We may also wish to backtrack to ensure ascent. If $-\mathbf{M}^{(t)}$ is positive definite and $c^{(t)} \le 0$, then $-\mathbf{M}^{(t+1)}$ will be positive definite. We use the term *hereditary positive definiteness* to refer to the desirable situation when positive definiteness is guaranteed to be transferred from one iteration to the next. If $c^{(t)} > 0$, then it may be necessary to backtrack by shrinking $c^{(t)}$ towards zero until positive definiteness is achieved. Thus, positive definiteness is not hereditary with this update. Monitoring and backtracking techniques and method performance are further explored in [327, 349].

There are several symmetric rank-two methods for updating a Hessian approximation while retaining the secant condition. The Broyden class [71, 73] of rank-two updates to the Hessian approximation has the form

$$\mathbf{M}^{(t+1)} = \mathbf{M}^{(t)} - \frac{\mathbf{M}^{(t)}\mathbf{z}^{(t)}(\mathbf{M}^{(t)}\mathbf{z}^{(t)})^T}{(\mathbf{z}^{(t)})^T\mathbf{M}^{(t)}\mathbf{z}^{(t)}} + \frac{\mathbf{y}^{(t)}(\mathbf{y}^{(t)})^T}{(\mathbf{z}^{(t)})^T\mathbf{y}^{(t)}}$$
$$+ \delta^{(t)}\left((\mathbf{z}^{(t)})^T\mathbf{M}^{(t)}\mathbf{z}^{(t)}\right)\mathbf{d}^{(t)}(\mathbf{d}^{(t)})^T, \quad (2.50)$$

where

$$\mathbf{d}^{(t)} = \frac{\mathbf{y}^{(t)}}{(\mathbf{z}^{(t)})^T\mathbf{y}^{(t)}} - \frac{\mathbf{M}^{(t)}\mathbf{z}^{(t)}}{(\mathbf{z}^{(t)})^T\mathbf{M}^{(t)}\mathbf{z}^{(t)}}.$$

The most popular member of this class is the BFGS update [72, 172, 238, 500], which simply sets $\delta^{(t)} = 0$. An alternative, for which $\delta^{(t)} = 1$, has also been extensively studied [115, 174]. However, the BFGS update is generally accepted as superior to this, based on extensive empirical and theoretical studies. The rank-one update in (2.49) has also been shown to perform well and to be an attractive alternative to BFGS [102, 327].

The BFGS update—indeed, all members of the Broyden class—confer hereditary positive definiteness on $-\mathbf{M}^{(t)}$. Therefore, backtracking can ensure ascent. However, recall that guaranteed ascent is not equivalent to guaranteed convergence. The order of convergence of quasi-Newton methods is usually faster than linear but slower than quadratic. The loss of quadratic convergence (compared to a Newton's method) is attributable to the replacement of the Hessian by an approximation. Nevertheless, quasi-Newton methods are fast and powerful, and are among the most frequently used methods in popular software packages. Several authors suggest that the performance of (2.49) is superior to that of BFGS [102, 349].

Example 2.7 (A bivariate optimization problem, continued) Figure 2.8 illustrates an application of quasi-Newton optimization with the BFGS update and backtracking for maximizing the bivariate function introduced in Example 2.4, starting from $x^{(0)}$ and initialized with $\alpha^{(t)} = 0.05$ at each step. The steps taken in this example are shown by the dashed line. The optimization successfully (and quickly) found x^*. Recall that the solid line in this figure illustrates the steepest ascent method discussed in Section 2.2.2.1. Both quasi-Newton methods and steepest ascent require only first derivatives, and backtracking was used for both. The additional computation required by quasi-Newton approaches is almost always outweighed by its superior convergence performance, as was seen in this example. □

There has been a wide variety of research on methods to enhance the performance and stability of quasi-Newton methods. Perhaps the most important of these improvements involves the calculation of the update for $M^{(t)}$. Although (2.50) provides a relatively straightforward update equation, its direct application is frequently less numerically stable than alternatives. It is far better to update a Cholesky decomposition of $M^{(t)}$ as described in [215].

The performance of quasi-Newton methods can be extremely sensitive to the choice of the starting matrix $M^{(0)}$. The easiest choice is the negative identity matrix, but this is often inadequate when the scales of the components of $x^{(t)}$ differ greatly. In MLE problems, setting $M^{(0)} = -I(\theta^{(0)})$ is a much better choice, if calculation of the expected Fisher information is possible. In any case, it is important to rescale any quasi-Newton optimization problem so that the elements of x are on comparable scales. This should improve performance and prevent the stopping criterion from effectively depending only on those variables whose units are largest. Frequently, in poorly scaled problems, one may find that a quasi-Newton algorithm will appear to converge to a point for which some $x_i^{(t)}$ differ from the corresponding elements of the starting point but others remain unchanged.

In the context of MLE and statistical inference, the Hessian is critical because it provides estimates of standard error and covariance. Yet, quasi-Newton methods rely on the notion that the root-finding problem can be solved efficiently even using poor approximations to the Hessian. Further, if stopped at iteration t, the most recent Hessian approximation $M^{(t-1)}$ is out of date and mislocated at $\theta^{(t-1)}$ instead of at $\theta^{(t)}$. For all these reasons, the approximation may be quite bad. It is worth the extra effort, therefore, to compute a more precise approximation after iterations have stopped. Details are given in [133]. One approach is to rely on the central difference approximation, whose (i, j)th element is

$$\widehat{l''(\boldsymbol{\theta}^{(t)})} = \frac{l'_i(\boldsymbol{\theta}^{(t)} + h_{ij}\mathbf{e}_j) - l'_i(\boldsymbol{\theta}^{(t)} - h_{ij}\mathbf{e}_j)}{2h_{ij}}, \tag{2.51}$$

where $l'_i(\boldsymbol{\theta}^{(t)})$ is the ith component of the score function evaluated at $\boldsymbol{\theta}^{(t)}$. In this case, decreasing h_{ij} is associated with reduced discretization error but potentially increased computer roundoff error. One rule of thumb in this case is to take $h_{ij} = h = \varepsilon^{1/3}$ for all i and j, where ε represents the computer's floating-point precision [452].

2.2.3 Gauss–Newton method

For MLE problems, we have seen how Newton's method approximates the log likelihood function at $\boldsymbol{\theta}^{(t)}$ by a quadratic, and then maximizes this quadratic to obtain the update $\boldsymbol{\theta}^{(t+1)}$. An alternative approach can be taken in *nonlinear least squares* problems with observed data $(y_i, \mathbf{z}_i)$ for $i = 1, \ldots, n$, where one seeks to estimate $\boldsymbol{\theta}$ by maximizing an objective function $g(\boldsymbol{\theta}) = -\sum_{i=1}^n (y_i - f(\mathbf{z}_i, \boldsymbol{\theta}))^2$. Such objective functions might be sensibly used, for example, when estimating $\boldsymbol{\theta}$ to fit the model

$$Y_i = f(\mathbf{z}_i, \boldsymbol{\theta}) + \epsilon_i \tag{2.52}$$

for some nonlinear function f and random error ϵ_i.

Rather than approximate g, the Gauss–Newton approach approximates f itself by its linear Taylor series expansion about $\boldsymbol{\theta}^{(t)}$. Replacing f by its linear approximation yields a linear least squares problem, which can be solved to derive an update $\boldsymbol{\theta}^{(t+1)}$.

Specifically, the nonlinear model in (2.52) can be approximated by

$$Y_i \approx f(\mathbf{z}_i, \boldsymbol{\theta}^{(t)}) + (\boldsymbol{\theta} - \boldsymbol{\theta}^{(t)})^T \mathbf{f}'(\mathbf{z}_i, \boldsymbol{\theta}^{(t)}) + \epsilon_i = \tilde{f}(\mathbf{z}_i, \boldsymbol{\theta}^{(t)}, \boldsymbol{\theta}) + \epsilon_i, \tag{2.53}$$

where for each i, $\mathbf{f}'(\mathbf{z}_i, \boldsymbol{\theta}^{(t)})$ is the column vector of partial derivatives of $f(\mathbf{z}_i, \boldsymbol{\theta}^{(t)})$ with respect to $\theta_j^{(t)}$, for $j = 1, \ldots, p$, evaluated at $(\mathbf{z}_i, \boldsymbol{\theta}^{(t)})$. A Gauss–Newton step is derived from the maximization of $\tilde{g}(\boldsymbol{\theta}) = -\sum_{i=1}^n \left[y_i - \tilde{f}(\mathbf{z}_i, \boldsymbol{\theta}^{(t)}, \boldsymbol{\theta})\right]^2$ with respect to $\boldsymbol{\theta}$, whereas a Newton step is derived from the maximization of a quadratic approximation to g itself, namely $g(\boldsymbol{\theta}^{(t)}) + (\boldsymbol{\theta} - \boldsymbol{\theta}^{(t)})^T \mathbf{g}'(\boldsymbol{\theta}^{(t)}) + (\boldsymbol{\theta} - \boldsymbol{\theta}^{(t)})^T \mathbf{g}''(\boldsymbol{\theta}^{(t)})(\boldsymbol{\theta} - \boldsymbol{\theta}^{(t)})$.

Let $X_i^{(t)}$ denote a working response whose observed value is $x_i^{(t)} = y_i - f(\mathbf{z}_i, \boldsymbol{\theta}^{(t)})$, and define $\mathbf{a}_i^{(t)} = \mathbf{f}'(\mathbf{z}_i, \boldsymbol{\theta}^{(t)})$. Then the approximated problem can be reexpressed as minimizing the squared residuals of the linear regression model

$$\mathbf{X}^{(t)} = \mathbf{A}^{(t)}(\boldsymbol{\theta} - \boldsymbol{\theta}^{(t)}) + \boldsymbol{\epsilon}, \tag{2.54}$$

where $\mathbf{X}^{(t)}$ and $\boldsymbol{\epsilon}$ are column vectors whose ith elements consist of $X_i^{(t)}$ and ϵ_i, respectively. Similarly, $\mathbf{A}^{(t)}$ is a matrix whose ith row is $(\mathbf{a}_i^{(t)})^T$.

The minimal squared error for fitting (2.54) is achieved when

$$(\boldsymbol{\theta} - \boldsymbol{\theta}^{(t)}) = \left((\mathbf{A}^{(t)})^T \mathbf{A}^{(t)}\right)^{-1} (\mathbf{A}^{(t)})^T \mathbf{x}^{(t)}. \tag{2.55}$$

Thus, the Gauss–Newton update for $\boldsymbol{\theta}^{(t)}$ is

$$\boldsymbol{\theta}^{(t+1)} = \boldsymbol{\theta}^{(t)} + \left((\mathbf{A}^{(t)})^T \mathbf{A}^{(t)} \right)^{-1} (\mathbf{A}^{(t)})^T \mathbf{x}^{(t)}. \qquad (2.56)$$

Compared to Newton's method, the potential advantage of the Gauss–Newton method is that it does not require computation of the Hessian. It is fast when f is nearly linear or when the model fits well. In other situations, particularly when the residuals at the true solution are large because the model fits poorly, the method may converge very slowly or not at all—even from good starting values. A variant of the Gauss–Newton method has better convergence behavior in such situations [132].

2.2.4   Nonlinear Gauss–Seidel iteration and other methods

An important technique that is used frequently for fitting nonlinear statistical models, including those in Chapter 12, is *nonlinear Gauss–Seidel iteration*. This technique is alternatively referred to as *backfitting* or *cyclic coordinate ascent*.

The equation $\mathbf{g}'(\mathbf{x}) = \mathbf{0}$ is a system of p nonlinear equations in p unknowns. For $j = 1, \ldots, p$, Gauss–Seidel iteration proceeds by viewing the jth component of $\mathbf{g}'$ as a univariate real function of x_j only. Any convenient univariate optimization method can be used to solve for the one-dimensional root of $g_j'(x_j^{(t+1)}) = 0$. All p components are cycled through in succession, and at each stage of the cycle the most recent values obtained for each coordinate are used. At the end of the cycle, the complete set of most recent values constitutes $\mathbf{x}^{(t+1)}$.

The beauty of this approach lies in the way it simplifies a potentially difficult problem. The solution of univariate root-finding problems created by applying Gauss–Seidel iteration is generally easy to automate, since univariate algorithms tend to be more stable and successful than multivariate ones. Further, the univariate tasks are likely to be completed so quickly that the total number of computations may be less than would have been required for the multivariate approach. The elegance of this strategy means that it is quite easy to program.

Example 2.8 (A bivariate optimization problem, continued) Figure 2.9 illustrates an application of Gauss–Seidel iteration for finding the maximum of the bivariate function discussed in Example 2.4. Unlike other graphs in this chapter, each line segment represents a change of a single coordinate in the current solution. Thus, for example, the $\mathbf{x}^{(1)}$ is at the vertex following one horizontal step and one vertical step from $\mathbf{x}^{(0)}$. Each complete step comprises two univariate steps. A quasi-Newton method was employed for each univariate optimization. Note that the very first univariate optimization (one horizontal step left from $\mathbf{x}^{(0)}$) actually failed, finding a local univariate minimum instead of the global univariate maximum. Although this is not advised, subsequent Gauss–Seidel iterations were able to overcome this mistake, eventually finding the global multivariate maximum.   □

The optimization of continuous multivariate functions is an area of extensive research, and the references given elsewhere in this chapter include a variety of approaches not mentioned here. The *trust region* approach constrains directions and

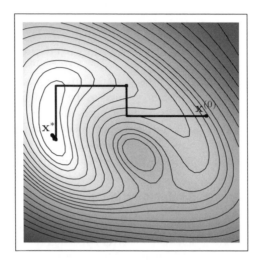

Fig. 2.9 Application of Gauss–Seidel iteration for maximizing a complex bivariate function, as discussed in Example 2.8. The surface of the function is indicated by shading and contours. From the starting point of $x^{(0)}$, several steps are required to approach the true maximum, x^*. Each line segment represents a change of a single coordinate in the current solution, so complete steps from $x^{(t)}$ to $x^{(t+1)}$ correspond to pairs of adjacent segments.

lengths of steps. The *nonlinear conjugate gradient* approach chooses search directions that deviate from the direction of the gradient with a bias toward directions not previously explored. The *polytope* or *Nelder–Mead simplex method* [411, 552] is quite popular because it requires no derivatives of the objective function g. This algorithm maintains a fixed-length list of points with corresponding objective function values, replacing the worst point at each iteration with one chosen in a direction that appears promising.

Problems

2.1   The following data are an i.i.d. sample from a Cauchy$(\theta, 1)$ distribution: 1.77, −0.23, 2.76, 3.80, 3.47, 56.75, −1.34, 4.24, −2.44, 3.29, 3.71, −2.40, 4.53, −0.07, −1.05, −13.87, −2.53, −1.75, 0.27, 43.21.

    (a) Graph the log likelihood function. Find the MLE for θ using the Newton–Raphson method. Try all of the following starting points: −11, −1, 0, 1.5, 4, 4.7, 7, 8, and 38. Discuss your results. Is the mean of the data a good starting point?

    (b) Apply the bisection method with starting points −1 and 1. Use additional runs to illustrate manners in which the bisection method may fail to find the global maximum.

(c) Apply fixed-point iterations as in (2.29), starting from -1, with scaling choices of $\alpha = 1, 0.64$, and 0.25. Investigate other choices of starting values and scaling factors.

(d) From starting values of $(\theta^{(0)}, \theta^{(1)}) = (-2, -1)$, apply the secant method to estimate θ. What happens when $(\theta^{(0)}, \theta^{(1)}) = (-3, 3)$, and for other starting choices?

(e) Use this example to compare the speed and stability of the Newton–Raphson method, bisection, fixed-point iteration, and the secant method. Do your conclusions change when you apply the methods to a random sample of size 20 from a $N(\theta, 1)$ distribution?

2.2   Consider the density $f(x) = \frac{1 - \cos\{x - \theta\}}{2\pi}$ on $0 \le x \le 2\pi$, where θ is a parameter between $-\pi$ and π. The following i.i.d. data arise from this density: 3.91, 4.85, 2.28, 4.06, 3.70, 4.04, 5.46, 3.53, 2.28, 1.96, 2.53, 3.88, 2.22, 3.47, 4.82, 2.46, 2.99, 2.54, 0.52, 2.50. We wish to estimate θ.

(a) Graph the log likelihood function between $-\pi$ and π.

(b) Find the method-of-moments estimator of θ.

(c) Find the MLE for θ using the Newton–Raphson method, using the result from (b) as the starting value. What solutions do you find when you start at -2.7 and 2.7?

(d) Repeat part (c) using 200 equally spaced starting values between $-\pi$ and π. Partition the interval between $-\pi$ and π into sets of attraction. In other words, divide the set of starting values into separate groups, with each group corresponding to a separate unique outcome of the optimization (a local mode). Discuss your results.

(e) Find two starting values, as nearly equal as you can, for which the Newton–Raphson method converges to two different solutions.

2.3   Let the survival time t for individuals in a population have density function f and cumulative distribution function F. The *survivor function* is then $S(t) = 1 - F(t)$. The *hazard function* is $h(t) = f(t)/(1 - F(t))$, which measures the instantaneous risk of dying at time t given survival to time t. A proportional hazards model posits that the hazard function depends on both time and a vector of covariates, $\mathbf{x}$, through the model

$$h(t|x) = \lambda(t) \exp\left\{\mathbf{x}^T \boldsymbol{\beta}\right\},$$

where $\boldsymbol{\beta}$ is a parameter vector.

If $\Lambda(t) = \int_{-\infty}^{t} \lambda(u)\, du$, it is easy to show that $S(t) = \exp\left\{-\Lambda(t)\exp\{\mathbf{x}^T\boldsymbol{\beta}\}\right\}$ and $f(t) = \lambda(t) \exp\left\{\mathbf{x}^T\boldsymbol{\beta} - \Lambda(t)\exp\{\mathbf{x}^T\boldsymbol{\beta}\}\right\}$.

Table 2.2 Length of remission (in weeks) for acute leukemia patients in the treatment and control groups of a clinical trial, with parentheses indicating censored values. For censored cases, patients are known to be in remission at least as long as the indicated value.

Treatment	(6)	6	6	6	7	(9)	(10)
	10	(11)	13	16	(17)	(19)	(20)
	22	23	(25)	(32)	(32)	(34)	(35)
Control	1	1	2	2	3	4	4
	5	5	8	8	8	8	11
	11	12	12	15	17	22	23

(a) Suppose that our data are censored survival times t_i for $i = 1, \ldots, n$. At the end of the study a patient is either dead (known survival time) or still alive (censored time; known to survive at least to the end of the study). Define w_i to be 1 if t_i is an uncensored time and 0 if t_i is a censored time. Prove that the log likelihood takes the form

$$\sum_{i=1}^{n} (w_i \log\{\mu_i\} - \mu_i) + \sum_{i=1}^{n} w_i \log \left\{ \frac{\lambda(t_i)}{\Lambda(t_i)} \right\},$$

where $\mu_i = \Lambda(t_i) \exp\{\mathbf{x}_i^T \boldsymbol{\beta}\}$.

(b) Consider a model for the length of remission for acute leukemia patients in a clinical trial. Patients were either treated with 6-mercaptopurine (6-MP) or a placebo [177]. One year after the start of the study, the length (weeks) of the remission period for each patient was recorded (see Table 2.2). Some outcomes were censored because remission extended beyond the study period. The goal is to determine whether the treatment lengthened time spent in remission. Suppose we set $\Lambda(t) = t^\alpha$ for $\alpha > 0$, yielding a hazard function proportional to $\alpha t^{\alpha-1}$ and a Weibull density: $f(t) = \alpha t^{\alpha-1} \exp\left\{\mathbf{x}^T\boldsymbol{\beta} - t^\alpha \exp\{\mathbf{x}^T\boldsymbol{\beta}\}\right\}$. Adopt the covariate parameterization given by $\mathbf{x}_i^T\boldsymbol{\beta} = \beta_0 + \delta_i\beta_1$ where δ_i is 1 if the ith patient was in the treatment group and 0 otherwise. Code a Newton-Raphson algorithm and find the MLEs of α, β_0, and β_1.

(c) Use any prepackaged Newton–Raphson or quasi-Newton routine to solve for the same MLEs.

(d) Estimate standard errors for your MLEs. Are any of your MLEs highly correlated? Report the pairwise correlations.

(e) Use nonlinear Gauss–Seidel iteration to find the MLEs. Comment on the implementation ease of this method compared to the multivariate Newton–Raphson method.

(f) Use the discrete Newton method to find the MLEs. Comment on the stability of this method.

2.4   A parameter θ has a Gamma$(2, 1)$ posterior distribution. Find the 95% highest posterior density interval for θ, that is, the interval containing 95% of the posterior probability for which the posterior density for every point contained in the interval is never lower than the density for every point outside the interval. Since the gamma density is unimodal, the interval is also the narrowest possible interval containing 95% of the posterior probability.

2.5   There were 46 crude oil spills of at least 1000 barrels from tankers in US waters during 1974–1999. The website for this book contains the following data: the number of spills in the ith year, N_i; the estimated amount of oil shipped through US waters as part of US import/export operations in the ith year, adjusted for spillage in international or foreign waters, b_{i1}; and the amount of oil shipped through US waters during domestic shipments in the ith year, b_{i2}. The data are adapted from [11]. Oil shipment amounts are measured in billions of barrels (Bbbl).

The volume of oil shipped is a measure of exposure to spill risk. Suppose we use the Poisson process assumption given by $N_i|b_{i1}, b_{i2} \sim$ Poisson(λ_i) where $\lambda_i = \alpha_1 b_{i1} + \alpha_2 b_{i2}$. The parameters of this model are α_1 and α_2, which represent the rate of spill occurrence per Bbbl oil shipped during import/export and domestic shipments, respectively.

(a) Derive the Newton–Raphson update for finding the MLEs of α_1 and α_2.

(b) Derive the Fisher scoring update for finding the MLEs of α_1 and α_2.

(c) Implement the Newton–Raphson and Fisher scoring methods for this problem, provide the MLEs, and compare the implementation ease and performance of the two methods.

(d) Estimate standard errors for the MLEs of α_1 and α_2.

(e) Apply the method of steepest ascent. Use step-halving backtracking as necessary.

(f) Apply quasi-Newton optimization with the Hessian approximation update given in (2.49). Compare performance with and without step-halving.

(g) Construct a graph resembling Figure 2.8 that compares the paths taken by methods used in (a)–(f). Choose the plotting region and starting point to best illustrate the features of the algorithms' performance.

2.6   Table 2.3 provides counts of a flour beetle (*Tribolium confusum*) population at various points in time [88]. Beetles in all stages of development were counted, and the food supply was carefully controlled.

Table 2.3   Counts of flour beetles in all stages of development over 154 days.

Days	0	8	28	41	63	79	97	117	135	154
Beetles	2	47	192	256	768	896	1120	896	1184	1024

An elementary model for population growth is the logistic model given by

$$\frac{dN}{dt} = rN\left(1 - \frac{N}{K}\right),$$
(2.57)

where N is population size, t is time, r is a growth rate parameter, and K is a parameter that represents the population carrying capacity of the environment. The solution to this differential equation is given by

$$N_t = f(t) = \frac{KN_0}{N_0 + (K - N_0)\exp\{-rt\}}$$
(2.58)

where N_t denotes the population size at time t.

(a) Fit the logistic growth model to the flour beetle data using the Gauss–Newton approach to minimize the sum of squared errors between model predictions and observed counts.

(b) Fit the logistic growth model to the flour beetle data using the Newton–Raphson approach to minimize the sum of squared errors between model predictions and observed counts.

(c) In many population modeling applications, an assumption of lognormality is adopted. The simplest assumption would be that the $\log N_t$ are independent and normally distributed with mean $\log f(t)$ and variance σ^2. Find the MLEs under this assumption, using both the Gauss–Newton and the Newton–Raphson methods. Provide standard errors for your parameter estimates, and an estimate of the correlation between them. Comment.

3

Combinatorial Optimization

It is humbling to learn that there are entire classes of optimization problems for which most methods—including those described previously—are utterly useless.

We will pose these problems as maximizations except in Section 3.4, although in nonstatistical contexts minimization is often customary. For statistical applications, recall that maximizing the log likelihood is equivalent to minimizing the negative log likelihood.

Let us assume that we are seeking the maximum of $f(\boldsymbol{\theta})$ with respect to $\boldsymbol{\theta} = (\theta_1, \ldots, \theta_p)$, where $\boldsymbol{\theta} \in \boldsymbol{\Theta}$ and $\boldsymbol{\Theta}$ consists of N elements for a finite positive integer N. In statistical applications, it is not uncommon for a likelihood function to depend on configuration parameters which describe the form of a statistical model and for which there are many discrete choices, as well as a small number of other parameters which could be easily optimized if the best configuration were known. In such cases, we may view $f(\boldsymbol{\theta})$ as the log profile likelihood of a configuration, $\boldsymbol{\theta}$, that is, the highest likelihood attainable using that configuration. Section 3.1.1 provides several examples.

Each $\boldsymbol{\theta} \in \boldsymbol{\Theta}$ is termed a *candidate solution*. Let $f_{\max}$ denote the globally maximum value of $f(\boldsymbol{\theta})$ achievable for $\boldsymbol{\theta} \in \boldsymbol{\Theta}$, and let the set of global maxima be $\mathcal{M} = \{\boldsymbol{\theta} \in \boldsymbol{\Theta} : f(\boldsymbol{\theta}) = f_{\max}\}$. If there are ties, $\mathcal{M}$ will contain more than one element. Despite the finiteness of $\boldsymbol{\Theta}$, finding an element of $\mathcal{M}$ may be very hard if there are distracting local maxima, plateaus, and long paths toward optima in $\boldsymbol{\Theta}$, and if N is extremely large.

3.1 HARD PROBLEMS AND NP-COMPLETENESS

Hard optimization problems are generally combinatorial in nature. In such problems, p items may be combined or sequenced in a very large number of ways, and each choice corresponds to one element in the space of possible solutions. Maximization requires a search of this very large space.

For example, consider the *traveling salesman problem*. In this problem, the salesman must visit each of p cities exactly once and return to his point of origin, using the shortest total travel distance. We seek to minimize the total travel distance over all possible routes (i.e., maximize the negative distance). If the distance between two cities does not depend on the direction traveled between them, then there are $(p-1)!/2$ possible routes (since the point of origin and direction of travel are arbitrary). Note that any tour corresponds to a permutation of the integers $1, \ldots, p$, which specifies the sequence in which the cities are visited.

To consider the difficulty of such problems, it is useful to discuss the number of steps required for an algorithm to solve it, where steps are simple operations like arithmetic, comparisons, and branching. The number of operations depends, of course, on the size of the problem posed. In general the size of a problem may be specified as the number of inputs needed to pose it. The traveling salesman problem is posed by specifying p city locations to be sequenced. The difficulty of a particular size-p problem is characterized by the number of operations required to solve it in the worst case using the best known algorithm.

The number of operations is only a rough notion, because it varies with implementation language and strategy. It is conventional, however, to bound the number of operations using the notation $\mathcal{O}(h(p))$. If $h(p)$ is polynomial in p, an algorithm is said to be polynomial.

Although the actual running time on a computer depends on the speed of the computer, we generally equate the number of operations and the execution time by relying on the simplifying assumption that all basic operations take the same amount of time (one unit). Then we may make meaningful comparisons of algorithm speeds even though the absolute scale is meaningless.

Consider two problems of size $p = 20$. Suppose that the first problem can be solved in polynomial time (say $\mathcal{O}(p^2)$ operations), and the solution requires 1 minute on your office computer. Then the size-21 problem could be solved in just a few seconds more. The size-25 problem can be solved in 1.57 minutes, size 30 in 2.25 minutes, and size 50 in 6.25 minutes. Suppose the second problem is $\mathcal{O}(p!)$ and requires 1 minute for size 20. Then it would take 21 minutes for size 21, 12.1 years (6,375,600 minutes) for size 25, 207 million years for size 30, and 2.4×10^{40} years for size 50. Similarly, if an $\mathcal{O}(p!)$ traveling salesman problem of size 20 could be solved in 1 minute, it would require far longer than the lifetime of the universe to determine the optimal path for the traveling salesman to make a tour of the 50 US state capitals. Furthermore, obtaining a computer that is 1000 times faster would barely reduce the difficulty. The conclusion is stark: Some optimization problems are simply too hard. The complexity of a polynomial problem—even for large p and high polynomial order—is dwarfed by the complexity of a quite small nonpolynomial problem.

The theory of problem complexity is discussed in [189, 425]. For us to discuss this issue further, we must make a formal distinction between *optimization* (i.e., search) problems and *decision* (i.e., recognition) problems. Thus far, we have considered optimization problems of the form: "Find the value of $\theta \in \Theta$ that maximizes $f(\theta)$." The decision counterpart to this is: "Is there a $\theta \in \Theta$ for which $f(\theta) > c$, for a fixed number c?" Clearly there is a close relationship between these two versions of the problem. In principle, we could solve the optimization problem by repeatedly solving the decision problem for strategically chosen values of c.

Decision problems that can be solved in polynomial time (e.g., $\mathcal{O}(p^k)$ operations for p inputs and constant k) are generally considered to be efficiently solvable [189]. These problems belong to the class denoted P. Once any polynomial-time algorithm has been identified for a problem, the order of the polynomial is often quickly reduced to practical levels [425]. Decision problems for which a given solution can be checked in polynomial time are called NP problems. Clearly a problem in P is in NP. However, there seem to be many decision problems, like the traveling salesman problem, that are much easier to check than they are to solve. In fact, there are many NP problems for which no polynomial-time solution has ever been developed. Many NP problems have been proven to belong to a special class for which a polynomial algorithm found to solve one such problem could be used to solve all such problems. This is the class of NP-complete problems. There are other problems at least as difficult, for which a polynomial algorithm—if found—would be known to provide a solution to all NP-complete problems, even though the problem itself is not proven to be NP-complete. These are NP-hard problems. There are also many combinatorial decision problems that are difficult and probably NP-complete or NP-hard although they haven't been proven to be in these classes. Finally, optimization problems are no easier than their decision counterparts, and we may classify optimization problems using the same categories listed above.

It has been shown that if there is a polynomial algorithm for any NP-complete problem, then there are polynomial algorithms for all NP-complete problems. The utter failure of scientists to develop a polynomial algorithm for any NP-complete problem motivates the popular conjecture that there cannot be any polynomial algorithm for any NP-complete problem. Proof (or counterexample) of this conjecture is one of the great unsolved problems in mathematics.

This leads us to the realization that there are optimization problems that are inherently too difficult to solve exactly by traditional means. Many problems in bioinformatics, experimental design, and nonparametric statistical modeling, for example, require combinatorial optimization.

3.1.1 Examples

Statisticians have been slow to realize how frequently combinatorial optimization problems are encountered in mainstream statistical model-fitting efforts. Below we give two examples. In general, when fitting a model requires optimal decisions about

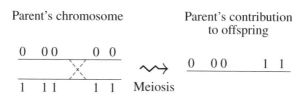

Fig. 3.1 During meiosis, a crossover occurs between the third and fourth loci. The zeros and ones indicate the origin of each allele in the contributed chromosome. Only one parental contribution is shown, for simplicity.

the inclusion, exclusion, or arrangement of a number of parameters in a set of possible parameters, combinatorial optimization problems arise frequently.

Example 3.1 (Genetic mapping) Genetic data for individuals and groups of related individuals are often analyzed in ways that present highly complex combinatorial optimization problems. For example, consider the problem of locating genes on a chromosome, known as the genetic mapping problem.

The genes, or more generally genetic markers, of interest in a chromosome can be represented as a sequence of symbols. The position of each symbol along the chromosome is called its *locus*. The symbols indicate genes or genetic markers, and the particular content stored at a locus is an *allele*.

Diploid species like humans have pairs of chromosomes, and hence two alleles at any locus. An individual is *homozygous* at a locus if the two alleles are identical at this locus; otherwise the individual is *heterozygous*. In either case, each parent contributes one allele at each locus of an offspring's chromosome pair. There are two possible contributions from any parent, because the parent has two alleles at the corresponding locus in his/her chromosome pair. Although each parent allele has a 50% chance of being contributed to the offspring, the contributions from a particular parent are not made independently at random. Instead, the contribution by a parent consists of a chromosome built during *meiosis* from segments of each chromosome in the parent's pair of chromosomes. These segments will contain several loci. When the source of the alleles on the contributed chromosome changes from one chromosome of the parent's pair to the other one, a *crossover* is said to have occurred. Figure 3.1 illustrates a crossover occurring during meiosis, forming the chromosome contributed to the offspring by one parent. This method of contribution means that alleles whose loci are closer together on one of the parent's chromosomes are more likely to appear together on the chromosome contributed by that parent.

When the alleles at two loci of a parent's chromosome appear jointly on the contributed chromosome more frequently than would be expected by chance alone, they are said to be *linked*. When the alleles at two different loci of a parent's chromosome do not both appear in the contributed chromosome, a *recombination* has occurred between the loci. The frequency of recombinations determines the degree of linkage between two loci: Infrequent recombination corresponds to strong linkage. The

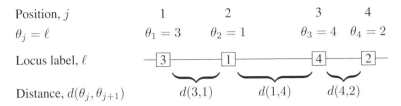

Fig. 3.2 Notation for gene mapping example with $p = 4$ loci. The loci are labeled in boxes at their positions along the chromosome. The correct sequential ordering of loci is defined by the θ_j values. Distances between loci are given by $d(\theta_j, \theta_{j+1})$ for $j = 1, \ldots, 3$.

degree of linkage, or *map distance*, between two loci corresponds to the expected number of crossovers between the two loci.

A genetic map of p markers consists of an ordering of their loci and a list of distances or probabilities of recombination between adjacent loci. Assign to each locus a label, ℓ, for $\ell = 1, \ldots, p$. The ordering component of the map, denoted $\boldsymbol{\theta} = (\theta_1, \ldots, \theta_p)$, describes the arrangement of the p locus labels in order of their positions along the chromosome, with $\theta_j = \ell$ if the locus labeled ℓ lies at the jth position along the chromosome. Thus, $\boldsymbol{\theta}$ is a permutation of the integers $1, \ldots, p$. The other component of a genetic map is a list of distances between adjacent loci. Denote the probability of recombination between adjacent loci θ_j and θ_{j+1} as $d(\theta_j, \theta_{j+1})$. This amounts to the map distance between these loci. Figure 3.2 illustrates this notation.

Such a map can be estimated by observing the alleles at the p loci for a sample of n chromosomes generated during the meiosis from a parent that is heterozygous at all p loci. Each such chromosome can be represented by a sequence of zeros and ones, indicating the origin of each allele in the contributed parent. For example, the chromosome depicted on the right side of Figure 3.1 can be denoted '00011', because the first three alleles originate from the first chromosome of the parent and the final two alleles originate from the second chromosome of the parent.

Let the random variable X_{i,θ_j} denote the origin of the allele in the locus labeled θ_j for the ith chromosome generated during meiosis. The dataset consists of observations, x_{i,θ_j}, of these random variables. Thus, a recombination for two adjacent markers has been observed in the ith case if $\left| x_{i,\theta_j} - x_{i,\theta_{j+1}} \right| = 1$, and no recombination has been observed if $\left| x_{i,\theta_j} - x_{i,\theta_{j+1}} \right| = 0$. If recombination events are assumed to occur independently in each interval, the probability of a given map is

$$
\prod_{j=1}^{p-1} \prod_{i=1}^{n} \Big\{ \big(1 - d(\theta_j, \theta_{j+1})\big) \big(1 - \left| x_{i,\theta_j} - x_{i,\theta_{j+1}} \right| \big)
$$

$$
+ d(\theta_j, \theta_{j+1}) \left| x_{i,\theta_j} - x_{i,\theta_{j+1}} \right| \Big\}. \quad (3.1)
$$

Given an ordering $\boldsymbol{\theta}$, the MLEs for the recombination probabilities are easily found to be

$$\hat{d}(\theta_j, \theta_{j+1}) = \frac{1}{n} \sum_{i=1}^{n} |x_{i,\theta_j} - x_{i,\theta_{j+1}}|. \tag{3.2}$$

Given $d(\theta_j, \theta_{j+1})$, the number of recombinations between the loci in positions j and $j+1$ is $\sum_{i=1}^{n} |X_{i,\theta_j} - X_{i,\theta_{j+1}}|$, which has a $\text{Bin}(n, d(\theta_j, \theta_{j+1}))$ distribution. We can compute the profile likelihood for $\boldsymbol{\theta}$ by adding the log likelihoods of the $p-1$ sets of adjacent loci and replacing each $d(\theta_j, \theta_{j+1})$ by its conditional maximum likelihood estimate $\hat{d}(\theta_j, \theta_{j+1})$. Let $\hat{\mathbf{d}}(\boldsymbol{\theta})$ compute these maximum likelihood estimates for any $\boldsymbol{\theta}$. Then the profile likelihood for $\boldsymbol{\theta}$ is

$$\begin{aligned} l(\boldsymbol{\theta}|\hat{\mathbf{d}}(\boldsymbol{\theta})) &= \sum_{j=1}^{p-1} n \Big\{ \hat{d}(\theta_j, \theta_{j+1}) \log\{\hat{d}(\theta_j, \theta_{j+1})\} \\ &\qquad + (1 - \hat{d}(\theta_j, \theta_{j+1})) \log\{1 - \hat{d}(\theta_j, \theta_{j+1})\} \Big\} \\ &= \sum_{j=1}^{p-1} T(\theta_j, \theta_{j+1}), \end{aligned} \tag{3.3}$$

where $T(\theta_j, \theta_{j+1})$ is defined to be zero if $\hat{d}(\theta_j, \theta_{j+1})$ is zero or one. Then the maximum likelihood genetic map is obtained by maximizing (3.3) over all permutations $\boldsymbol{\theta}$. Note that (3.3) constitutes a sum of terms $T(\theta_j, \theta_{j+1})$ whose values depend on only two loci. Suppose that all possible pairs of loci are enumerated, and the value $T(i, j)$ is computed for every i and j where $1 \le i < j \le p$. There are $p(p-1)/2$ such values of $T(i, j)$. The profile log likelihood can then be computed rapidly for any permutation $\boldsymbol{\theta}$ by summing the necessary values of $T(i, j)$.

However, finding the maximum likelihood genetic map requires maximizing the profile likelihood by searching over all $p!/2$ possible permutations. This is a variant of the traveling salesman problem, where each genetic marker corresponds to a city and the distance between cities i and j is $T(i, j)$. The salesman's tour may start at any city and terminates in the last city visited. A tour and its reverse are equivalent. There are no known algorithms for solving general traveling salesman problems in polynomial time.

Further details and extensions of this example are considered in [190, 483]. □

Example 3.2 (Variable selection in regression) Consider a multiple linear regression problem with p potential predictor variables. A fundamental step in regression is selection of a suitable model. Given a dependent variable Y and a set of candidate predictors $x_1, x_2, \ldots, x_p$, we must find the best model of the form $Y = \beta_0 + \sum_{j=1}^{s} \beta_{i_j} x_{i_j} + \epsilon$, where $\{i_1, \ldots, i_s\}$ is a subset of $\{1, \ldots, p\}$ and ϵ denotes a random error. The notion of what model is best may have any of several meanings.

Suppose that the goal is to use the Akaike information criterion (AIC) to select the best model [7, 75]. We seek to find the subset of predictors that minimizes the fitted model AIC,

$$\text{AIC} = N \log\{\text{RSS}/N\} + 2(s + 2), \tag{3.4}$$

where N is the sample size, s is the number of predictors in the model, and RSS is the sum of squared residuals. Alternatively, suppose that Bayesian regression is performed, say with the normal–gamma conjugate class of priors $\beta \sim N(\mu, \sigma^2 \mathbf{V})$ and $\nu\lambda/\sigma^2 \sim \chi_\nu^2$. In this case, one might seek to find the subset of predictors corresponding to the model that maximizes the posterior model probability [445].

In either case, the variable selection problem requires an optimization over a space of 2^{p+1} possible models, since each variable and the intercept may be included or omitted. It also requires estimating the best β_{i_j} for each of the 2^{p+1} possible models, but this step is easy for any given model. Although a search algorithm that is more efficient than exhaustive search has been developed to optimize some classical regression model selection criteria, it is practical only for fairly small p [188, 396]. We know of no efficient general algorithm to find the global optimum (i.e., the single best model) for either the AIC or the Bayesian goals.   □

3.1.2 The need for heuristics

The existence of such challenging problems requires a new perspective on optimization. It is necessary to abandon algorithms that are guaranteed to find the global maximum (under suitable conditions) but will never succeed within a practical time limit. Instead we turn to algorithms that can find a good local maximum within tolerable time.

Such algorithms are sometimes called heuristics. They are intended to find a *globally competitive* candidate solution (i.e., a nearly optimal one), with an explicit trade of global optimality for speed. The two primary features of such heuristics are

1. iterative improvement of a current candidate solution, and

2. limitation of the search to a local neighborhood at any particular iteration.

These two characteristics embody the heuristic strategy of *local search*, which we address first.

No single heuristic will work well in all problems. In fact, there is no search algorithm whose performance is better than another when performance is averaged over the set of all possible discrete functions [487, 573]. There is clearly a motivation to adopt different heuristics for different problems. Thus we continue beyond local search to examine *tabu algorithms*, *simulated annealing*, and *genetic algorithms*.

3.2 LOCAL SEARCH

Local search is a very broad optimization paradigm that arguably encompasses all of the techniques described in this chapter. In this section, we introduce some of the its simplest, most generic variations such as *k-optimization* and *random starts local search*.

Basic local search is an iterative procedure that updates a current candidate solution $\boldsymbol{\theta}^{(t)}$ at iteration t to $\boldsymbol{\theta}^{(t+1)}$. The update is termed a *move* or a *step*. One or more possible moves are identified from a neighborhood of $\boldsymbol{\theta}^{(t)}$, say $\mathcal{N}(\boldsymbol{\theta}^{(t)})$. The advantage of local search over global (i.e., exhaustive) search is that only a tiny portion of Θ need be searched at any iteration, and large portions of Θ may never be examined. The disadvantage is that the search is likely to terminate at an uncompetitive local maximum.

A neighborhood of the current candidate solution, $\mathcal{N}(\boldsymbol{\theta}^{(t)})$, contains candidate solutions that are near $\boldsymbol{\theta}^{(t)}$. Often, proximity is enforced by limiting the number of changes to the current candidate solution used to generate an alternative. In practice, simple changes to the current candidate solution are usually best, resulting in small neighborhoods that are easily searched or sampled. Complex alterations are often difficult to conceptualize, complicated to code, and slow to execute. Moreover, they rarely improve search performance, despite the intuition that larger neighborhoods would be less likely to lead to entrapment at a poor local maximum. If the neighborhood is defined by allowing k changes to the current candidate solution, then it is a *k-neighborhood*, and the alteration of k features of the current candidate is called a *k-change*.

The definition a neighborhood is intentionally vague to allow flexible usage of the term in a wide variety of problems. For the gene mapping problem introduced in Example 3.1, suppose $\boldsymbol{\theta}^{(t)}$ is a current ordering of genetic markers. A simple neighborhood might be the set of all orderings that can be obtained by swapping the locations of only two markers on the chromosome whose order is $\boldsymbol{\theta}^{(t)}$. In the regression model selection problem introduced in Example 3.2, a simple neighborhood is the set of models that either add or omit one predictor from $\boldsymbol{\theta}^{(t)}$.

A local neighborhood will usually contain several candidate solutions. An obvious strategy at each iteration is to choose the best among all candidates in the current neighborhood. This is the method of *steepest ascent*. To speed performance, one might instead select the first randomly chosen neighbor for which the objective function exceeds its previous value; this is *random ascent* or *next ascent*.

If k-neighborhoods are used for a steepest ascent algorithm, the solution is said to be *k-optimal*. Alternatively, any local search algorithm that chooses $\boldsymbol{\theta}^{(t+1)}$ uphill from $\boldsymbol{\theta}^{(t)}$ is an *ascent algorithm*, even if the ascent is not the steepest possible within $\mathcal{N}(\boldsymbol{\theta}^{(t)})$.

The sequential selection of steps that are optimal in small neighborhoods, disregarding the global problem, is reminiscent of a *greedy algorithm*. A chess player using a greedy algorithm might look for the best immediate move with total disregard to its future consequences: perhaps moving a knight to capture a pawn without recognizing that the knight will be captured on the opponent's next move. Wise selection of a new candidate solution from a neighborhood of the current candidate must balance the need for a narrow focus enabling quick moves against the need to find a globally competitive solution. To avoid entrapment in poor local maxima, it might be reasonable—every once in a while—to eschew some of the best neighbors of $\boldsymbol{\theta}^{(t)}$ in favor of a direction whose rewards are only later realized. For example, when $\boldsymbol{\theta}^{(t)}$ is a local maximum, the approach of *steepest ascent/mildest descent* [266] allows a move to the least unfavorable $\theta^{(t+1)} \in \mathcal{N}(\theta^{(t)})$ (see Section 3.3). There are also a variety of techniques in which a candidate neighbor is selected from $\mathcal{N}(\theta^{(t)})$ and a random decision rule is used to decide whether to adopt it or retain $\theta^{(t)}$. These algorithms generate Markov chains $\{\theta^{(t)}\}$ ($t = 0, 1, \ldots$) that are closely related to simulated annealing (Section 3.4) and the methods of Chapter 7.

Searching within the current neighborhood for a k-change steepest ascent move can be difficult when k is greater than 1 or 2 because the size of the neighborhood increases rapidly with k. For larger k, it can be useful to break the k-change up into smaller parts, sequentially selecting the best candidate solutions in smaller neighborhoods. To promote search diversity, breaking a k-change step into several smaller sequential changes can be coupled with the strategy of allowing one or more of the smaller steps to be suboptimal (e.g., random). Such *variable-depth local search* approaches permit a potentially better step away from the current candidate solution, even though it will not likely be optimal within the k-neighborhood.

Ascent algorithms frequently converge to local maxima that are not globally competitive. One approach to overcoming this problem is the technique of *random starts local search*. Here, a simple ascent algorithm is repeatedly run to termination from a large number of starting points. The starting points are chosen randomly. The simplest approach is to select starting points independently and uniformly at random over Θ. More sophisticated approaches may employ some type of stratified sampling where the strata are identified from some pilot runs in an effort to partition Θ into regions of qualitatively different convergence behavior.

It may seem unsatisfying to rely solely on random starts to avoid being fooled by a local maximum. In later sections we introduce methods that modify local search in ways that provide a reasonable chance of finding a globally competitive candidate solution—possibly the global maximum—on any single run. Of course, the strategy of using multiple random starts can be overlaid on any of these approaches to provide additional confidence in the best solution found.

Example 3.3 (Baseball salaries) Random starts local search can be very effective in practice because it is simple to code and fast to execute, allowing time for a large number of random starts. Here, we consider its application to a regression model selection problem.

Table 3.1 lists 27 baseball performance statistics, such as batting percentages and numbers of home runs, which were collected for 337 players (no pitchers) in 1991. Players' 1992 salaries, in thousands of dollars, may be related to these variables computed from the previous season. These data, derived from the data in [555], may be downloaded from the website for this book. We use the log of the salary variable as the response variable. The goal is to find the best subset of predictors to predict log salary using a linear regression model. Assuming that the intercept will be included in any model, there are $2^{27} = 134,217,728$ possible models in the search space.

Figure 3.3 illustrates the application of a random starts local search algorithm to minimize the AIC with respect to regression variable selection. The problem can be posed as maximizing the negative of the AIC, thus preserving our preference for uphill search. Neighborhoods were limited to 1-changes generated from the

Table 3.1   Potential predictors of baseball players' salaries.

1. Batting average	10. Strikeouts (SOs)	19. Walks per SO
2. On base pct. (OBP)	11. Stolen bases (SBs)	20. OBP / errors
3. Runs scored	12. Errors	21. Runs per error
4. Hits	13. Free agency[a]	22. Hits per error
5. Doubles	14. Arbitration[b]	23. HRs per error
6. Triples	15. Runs per SO	24. SOs × errors
7. Home runs (HRs)	16. Hits per SO	25. SBs × OBP
8. Runs batted in (RBIs)	17. HRs per SO	26. SBs × runs
9. Walks	18. RBIs per SO	27. SBs × hits

[a]Free agent, or eligible. [b]Arbitration, or eligible.

Table 3.2   Results of random starts local search model selection for Example 3.3. The bullets indicate inclusion of the corresponding predictor in each model selected, with model labels explained in the text.

Method	1	2	3	6	7	8	9	10	12	13	14	15	16	18	19	20	21	22	24	25	26	AIC
LS (2,4)		•	•	•		•		•		•	•	•	•						•	•	•	−416.95
S-Plus	•		•	•		•		•		•	•	•	•						•	•	•	−416.94
LS (1)		•	•	•		•		•		•	•				•	•	•					−414.60
LS (3)		•	•	•		•				•	•				•	•	•	•	•			−414.16
LS (5)	•			•	•	•				•	•				•	•	•	•				−413.52
Efroy.	•			•	•			•		•	•			•					•		•	−400.16

current model by either adding or deleting one predictor. Search was started from five randomly selected subsets of predictors (i.e., five starting points), and 14 additional steps were allocated to each start. Each move was made by steepest ascent. Since each steepest ascent step requires searching 27 neighbors, this small example requires 1890 evaluations of the objective function. A comparable limit to objective function evaluations was imposed on examples of other heuristic techniques that follow in the remainder of this chapter.

Figure 3.3 shows the value of the AIC for the best model at each step; some budgeted steps went unused because local maxima were quickly found. Table 3.2 summarizes the results of the search. The second and fourth random starts (labeled LS (2,4)) led to an optimal AIC of −416.95, derived from the model using predictors 2, 3, 6, 8, 10, 13, 14, 15, 16, 24, 25, and 26. The worst random start was the fifth, which led to an AIC of −413.52 for a model with ten predictors. For the sake of comparison, a greedy stepwise method (the step() procedure in S-Plus [544]) chose a model with twelve predictors, yielding an AIC of −416.94. The greedy stepwise method of Efroymson [396] chose a model with nine predictors, yielding an AIC of −400.16;

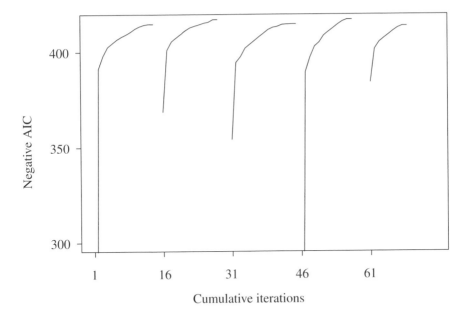

Fig. 3.3 Results of random starts local search by steepest ascent for Example 3.3, for up to 15 iterations from each of five random starts. Only AIC values between -300 and -420 are shown.

however, this method is designed to find a good parsimonious model using a criterion that differs slightly from the AIC. With default settings, neither of these off-the-shelf algorithms found a model quite as good as the one found with a simple random starts local search.   □

3.3 TABU ALGORITHMS

A tabu algorithm is a local search algorithm with a set of additional rules that guide the selection of moves in ways that are believed to promote the discovery of a global maximum. The approach employs variable neighborhoods: The rules for identifying acceptable steps change at each iteration. Detailed studies of tabu methods include [224, 225, 227, 228, 229].

In a standard ascent algorithm, entrapment in a globally uncompetitive local maximum is likely, because no downhill moves are allowed. Tabu search allows downhill moves when no uphill move can be found in the current neighborhood (and possibly in other situations too), thereby potentially escaping entrapment. An early form of tabu search, called steepest ascent/mildest descent, moved to the least unfavorable neighbor when there was no uphill move [266].

If a downhill step is chosen, care must be taken to ensure that the next step (or a future one) does not simply reverse the downhill move. Such cycling would eliminate the potential long-term benefit of the downhill move. To prevent such cycling, certain moves are temporarily forbidden, or made *tabu*, based on the recent history of the algorithm.

There are four general types of rules added to local search by tabu search methods. The first is to make certain potential moves temporarily tabu. The others involve *aspiration* to better solutions, *intensification* of search in promising areas of solution space, and *diversification* of search candidates to promote broader exploration of the solution space. These terms will be defined after we discuss tabus.

3.3.1 Basic definitions

Tabu search is an iterative algorithm initiated at time $t = 0$ with a candidate solution $\theta^{(0)}$. At the tth iteration, a new candidate solution is selected from a neighborhood of $\theta^{(t)}$. This candidate becomes $\theta^{(t+1)}$. Let $H^{(t)}$ denote the history of the algorithm through time t. It suffices for $H^{(t)}$ to be a selective history, remembering only certain matters necessary for the future operation of the algorithm.

Unlike simple local search, a tabu algorithm generates a neighborhood of the current candidate solution that depends on the search history; denote this by $\mathcal{N}(\theta^{(t)}, H^{(t)})$. Furthermore, the identification of the preferred $\theta^{(t+1)}$ in $\mathcal{N}(\theta^{(t)}, H^{(t)})$ may depend not only on f but also on the search history. Thus, we may assess neighbors using an augmented objective function, $f_{H^{(t)}}$.

A single step from $\theta^{(t)}$ to $\theta^{(t+1)}$ can be characterized by many *attributes*. Attributes will be used to describe moves or types of moves that will be forbidden, encouraged, or discouraged in future iterations of the algorithm. Examples of attributes are given in the left column of Table 3.3. Such attributes are not unique to tabu search; indeed, they can be used to characterize moves from any local search. However, tabu search explicitly adapts the current neighborhood according to the attributes of recent moves.

The attributes in Table 3.3 can be illustrated by considering a regression model selection problem. Suppose $\theta_i^{(t)} = 1$ if the ith predictor is included in the model at time t, and 0 otherwise. Suppose that 2-change neighborhoods consist of all models to which two variables separately have each been added or deleted from the current model. The right column of Table 3.3 gives one example of each generic attribute listed, in the context of these 2-change neighborhoods in the regression model selection problem from Example 3.2. These examples are labeled A_1 through A_5. Many other effective attributes can be identified from the context of specific optimization problems.

Denote the ath attribute as A_a. Note that the complement (i.e., negation) of an attribute is also an attribute, so if A_a corresponds to swapping the values of $\theta_i^{(t)}$ and $\theta_j^{(t+1)}$, then $\overline{A}_a$ corresponds to not making that swap.

As the algorithm progresses, the attributes of the tth move will vary with t, and the quality of the candidate solution will also vary. Future moves can be guided by

Table 3.3 Examples of attributes. The left column gives examples in a generic context. The right column gives corresponding attributes in the specific context of 2-change neighborhoods in a regression model selection problem.

Attribute	Model Selection Example
A change in the value of $\theta_i^{(t)}$. The attribute may be the value from which the move began, or the value at which it arrived.	A_1: Whether the ith predictor is added (or deleted) from the model.
A swap in the values of $\theta_i^{(t)}$ and $\theta_j^{(t)}$ when $\theta_i^{(t)} \neq \theta_j^{(t)}$.	A_2: Whether the absent variable is exchanged for the variable present in the model.
A change in the value of f resulting from the step, $f(\boldsymbol{\theta}^{(t+1)}) - f(\boldsymbol{\theta}^{(t)})$.	A_3: The reduction in AIC achieved by the move.
The value $g(\boldsymbol{\theta}^{(t+1)})$ of some other strategically chosen function g.	A_4: The number of predictors in the new model.
A change in the value of g resulting from the step, $g(\boldsymbol{\theta}^{(t+1)}) - g(\boldsymbol{\theta}^{(t)})$.	A_5: A change to a different variable selection criterion such as Mallows's C_p [369] or the adjusted R^2 [412].

the history of past moves, their objective function values, and their attributes. The *recency* of an attribute is the number of steps that have passed since a move most recently had that attribute. Let $R\left(A_a, H^{(t)}\right) = 0$ if the ath attribute is expressed in the move yielding $\boldsymbol{\theta}^{(t)}$, let $R\left(A_a, H^{(t)}\right) = 1$ if it is most recently expressed in the move yielding $\boldsymbol{\theta}^{(t-1)}$, and so forth.

3.3.2 The tabu list

When considering a move from $\boldsymbol{\theta}^{(t)}$, we compute the increase in the objective function achieved for each neighbor of $\boldsymbol{\theta}^{(t)}$. Ordinarily, the neighbor that provides the greatest increase would be adopted as $\boldsymbol{\theta}^{(t+1)}$. This corresponds to the steepest ascent.

Suppose, however, that no neighbor of $\boldsymbol{\theta}^{(t)}$ yields an increased objective function. Then $\boldsymbol{\theta}^{(t+1)}$ is ordinarily chosen to be the neighbor that provides the smallest decrease. This is the mildest descent.

If only these two rules were used for search, the algorithm would quickly become trapped and converge to a local maximum. After one move of mildest descent, the next move would return to the hilltop just departed. Cycling would ensue.

To avoid such cycling, a *tabu list* of temporarily forbidden moves is incorporated in the algorithm. Each time a move with attribute A_a is taken, $\overline{A}_a$ is put on a tabu list for τ iterations. When $R\left(A_a, H^{(t)}\right)$ first equals τ, the tabu expires and $\overline{A}_a$ is removed from the tabu list. Thus, moves with attributes on the tabu list are effectively excluded from the current neighborhood. The modified neighborhood is denoted

$$\mathcal{N}(\theta^{(t)}, H^{(t)}) = \left\{ \theta : \theta \in \mathcal{N}(\theta^{(t)}) \right.$$

$$\left. \text{and no attribute of } \theta \text{ is currently tabu} \right\}. \quad (3.5)$$

This prevents undoing the change for τ iterations, thereby discouraging cycling. By the time that the tabu has expired, enough other aspects of the candidate solution should have changed that reversing the move may no longer be counterproductive. Note that the tabu list is a list of attributes, not moves, so a single tabu attribute may forbid entire classes of moves.

The *tabu tenure*, τ, is the number of iterations over which an attribute is tabu. This can be a fixed number or it may vary, systematically or randomly, perhaps based on features of the attribute. For a given problem, a well-chosen tabu tenure will be long enough to prevent cycling and short enough to prevent the deterioration of candidate solution quality that occurs when too many moves are forbidden. Fixed tabu tenures between 7 and 20, or between $0.5\sqrt{p}$ and $2\sqrt{p}$, where p is the size of the problem, have been suggested for various problem types [227]. Tabu tenures that vary dynamically seem more effective in many problems [229]. Also, it will often be important to use different tenures for different attributes. If an attribute contributes tabu restrictions for a wide variety of moves, the corresponding tabu tenure should be short to ensure that future choices are not limited.

Example 3.4 (Genetic mapping, continued) We illustrate some uses of tabus, using the gene mapping problem introduced in Example 3.1.

First, consider monitoring the swap attribute. Suppose that A_a is the swap attribute corresponding to exchanging two particular loci along the chromosome. When a move A_a is taken, it is counterproductive to immediately undo the swap, so $\overline{A}_a$ is placed on the tabu list. Search progresses only among moves that do not reverse recent swaps. Such a tabu promotes search diversity by avoiding quick returns to recently searched areas.

Second, consider the attribute identifying the locus label θ_j for which $\hat{d}(\theta_j, \theta_{j+1})$ is smallest in the new move. In other words, this attribute identifies the two loci in the new chromosome that are nearest each other. If the complement of this attribute is put on the tabu list, any move to a chromosome for which other loci are closer will be forbidden moves for τ iterations. Such a tabu promotes search intensity among genetic maps for which loci θ_j and θ_j are closest.

Sometimes, it may be reasonable to place the attribute itself, rather than its complement, on the tabu list. For example, let $h(\theta)$ compute the mean $\hat{d}(\theta_j, \theta_{j+1})$ between adjacent loci in a chromosome ordered by θ. Let A_a be the attribute indicating excessive change of the mean conditional MLE map distance, so A_a equals 1 if

$|h(\boldsymbol{\theta}^{(t+1)}) - h(\boldsymbol{\theta}^{(t)})| > c$ and 0 otherwise, for some fixed threshold c. If a move with mean change greater than c is taken, we may place A_a itself on the tabu list for τ iterations. This prevents any other drastic mean changes for a period of time, allowing better exploration of the newly entered region of solution space before moving far away. $\qquad\square$

3.3.3 Aspiration criteria

Sometimes, choosing not to move to a nearby candidate solution because the move is currently tabu can be a poor decision. In these cases, we need a mechanism to override the tabu list. Such a mechanism is called an *aspiration criterion*.

The simplest and most popular aspiration criterion is to permit a tabu move if it provides a higher value of the objective function than has been found in any iteration so far. Clearly it makes no sense to overlook the best solution found so far, even if it is currently tabu. One can easily envision scenarios where this aspiration criterion is useful. For example, suppose that a swap of two components of $\boldsymbol{\theta}$ is on the tabu list and the candidate solutions at each iteration recently have drifted away from the region of solution space being explored when the tabu began. The search will now be in a new region of solution space where it is quite possible that reversing the tabu swap would lead to a drastic increase in the objective function.

Another interesting option is *aspiration by influence*. A move or attribute is influential if it is associated with a large change in the value of the objective function. There are many ways to make this idea concrete [227]. To avoid unnecessary detail about numerous specific possibilities, let us simply denote the influence of the ath attribute as $I\left(A_a, H^{(t)}\right)$ for a move yielding $\boldsymbol{\theta}^{(t)}$. In many combinatorial problems, there are a lot of neighboring moves that cause only small incremental changes to the value of the objective function, while there are a few moves that cause major shifts. Knowing the attributes of such moves can help guide search. Aspiration by influence overrides the tabu on reversing a low-influence move if a high-influence move is made prior to the reversal. The rationale for this is that the recent high-influence step may have moved the search to a new region of the solution space where further local exploration is useful. The reversal of the low-influence move will probably not induce cycling, since the intervening high-influence move likely shifted scrutiny to a portion of solution space more distant than what could be reached by the low-influence reversal.

Aspiration criteria can also be used to encourage moves that are not tabu. For example, when low-influence moves appear to provide only negligible improvement in the objective function, they can be downweighted and high-influence moves can be given preference. There are several ways to do this; one approach is to incorporate in $f_{H^{(t)}}$ either a penalty or an incentive term that depends on the relative influence of candidate moves.

3.3.4 Diversification

The *frequency* of an attribute records the number of moves that manifested that attribute since the search began. Let $C\left(A_a, H^{(t)}\right)$ represent the count of occurrences of the ath attribute thus far. Then $F\left(A_a, H^{(t)}\right)$ represents a frequency function that can be used to penalize moves that are repeated too frequently. The most direct definition is $F\left(A_a, H^{(t)}\right) = C\left(A_a, H^{(t)}\right)/t$, but the denominator may be replaced by the sum, the maximum, or the average of the counts of occurrences of various attributes.

Rules based on attribute frequencies can be used to increase the diversity of candidate solutions examined during tabu search.

Suppose the frequency of each attribute is recorded, either over the entire history or over the most recent ψ moves. Note that this frequency may be one of two types, depending on the attribute considered. If the attribute corresponds to some feature of $\theta^{(t)}$, then the frequency measures how often that feature is seen in candidate solutions considered during search. Such frequencies are termed *residence frequencies*. If, alternatively, the attribute corresponds to some change induced by moving from one candidate solution to another, then the frequency is a *transition frequency*. For example, in the regression model selection problem introduced in Example 3.2, the attribute noting the inclusion of the predictor x_i in the model would have a residence frequency. The attribute that signaled when a move reduced the AIC would have a transition frequency.

If attribute A_a has a high residence frequency and the history of the most recent ψ moves covers nearly optimal regions of solution space, this may suggest that A_a is associated with high-quality solutions. On the other hand, if the recent history reflects the search getting stuck in a low-quality region of solution space, then a high residence frequency may suggest that the attribute is associated with bad solutions. Usually, $\psi > \tau$ is a intermediate or long-term memory parameter that allows the accumulation of additional historical information to diversify future search.

If attribute A_a has a high transition frequency, this attribute may be what has been termed a crack filler. Such an attribute may be frequently visited during the search in order to fine-tune good solutions but rarely offers fundamental improvement or change [227]. In this case, the attribute has low influence.

A direct approach employing frequency to increase search diversification is to incorporate a penalty or incentive function in $f_{H^{(t)}}$. The choice

$$f_{H^{(t)}}\left(\theta^{(t+1)}\right) = \begin{cases} f(\theta^{(t+1)}) & \text{if } f(\theta^{(t+1)}) \geq f(\theta^{(t)}), \\ f(\theta^{(t+1)}) - cF\left(A_a, H^{(t)}\right) & \text{if } f(\theta^{(t+1)}) < f(\theta^{(t)}) \end{cases} \quad (3.6)$$

with $c > 0$ has been suggested [477]. If all nontabu moves are downhill, then this approach discourages moves that have the high-frequency attribute A_a. An analogous strategy can be crafted to diversify the selection of uphill moves.

Instead of incorporating a penalty or incentive in the objective function, it is possible to employ a notion of graduated tabu status, where an attribute may be only partially tabu. One way to create a tabu status that varies by degrees is to invoke probabilistic tabu decisions: an attribute can be assigned a probability of being tabu, where the probability is adjusted according to various factors, including the tabu tenure [227].

3.3.5 Intensification

In some searches it may be useful to intensify the search effort in particular areas of solution space. Frequencies can also be used to guide such intensification. Suppose that the frequencies of attributes are tabulated over the most recent v moves, and a corresponding record of objective function values is kept. By examining these data, key attributes shared by good candidate solutions can be identified. Then moves that retain such features can be rewarded and moves that remove such features can be penalized through $f_{H^{(t)}}$. The time span $v > \tau$ parameterizes the length of a long-term memory to enable search intensification in promising areas of solution space.

3.3.6 A comprehensive tabu algorithm

Below we summarize a fairly general tabu algorithm that incorporates many of the features described above. After initialization and identification of a list of problem-specific attributes, the algorithm proceeds as follows:

1. Determine an augmented objective function $f_{H^{(t)}}$ that depends on f and perhaps on

    (a) frequency-based penalties or incentives to promote diversification, and/or

    (b) frequency-based penalties or incentives to promote intensification.

2. Identify neighbors of $\theta^{(t)}$, namely the members of $\mathcal{N}(\theta^{(t)})$.

3. Rank the neighbors in decreasing order of improvement, as evaluated by $f_{H^{(t)}}$.

4. Select the highest ranking neighbor.

5. Is this neighbor currently on the tabu list? If not, go to step 8.

6. Does this neighbor pass an aspiration criterion? If so, go to step 8.

7. If all neighbors of $\theta^{(t)}$ have been considered and none have been adopted as $\theta^{(t+1)}$, then stop. Otherwise, select the next most high-ranking neighbor and go to step 5.

8. Adopt this solution as $\theta^{(t+1)}$.

9. Update the tabu list by creating new tabus based on the current move and by deleting tabus whose tenures have expired.

10. Has a stopping criterion been met? If so, stop. Otherwise, increment t and go to step 1.

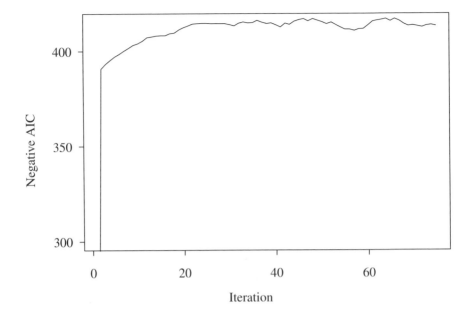

Fig. 3.4 Results of tabu search for Example 3.5.

It is sensible to stop when a maximum number of iterations has been reached, and then to take the best candidate solution yet found as the final result. Search effort can be split among a collection of random starts rather than devoting all resources to one run from a single start. By casting tabu search in a Markov chain framework, it is possible to obtain results on the limiting convergence of the approach [167].

Example 3.5 (Baseball salaries, continued) A simple tabu search was applied to the variable selection problem for regression modeling of the baseball data introduced in Example 3.3. Only attributes signaling the presence or absence of each predictor were monitored. Moves that would reverse the inclusion or removal of a predictor were made tabu for $\tau = 5$ moves, and the algorithm was run for 75 moves from a random start. The aspiration criterion permitted an otherwise tabu move if it yielded an objective function value above the best previously seen.

Figure 3.4 shows the values of the AIC for the sequence of candidate solutions generated by this tabu search. The AIC was quickly improved, and an optimum value of -416.95, derived from the model using predictors 2, 3, 6, 8, 10, 13, 14, 15, 16, 24, 25, and 26, was found on two occasions: iterations 44 and 66. This solution matches the best model found using random starts local search (Table 3.2). □

3.4 SIMULATED ANNEALING

Simulated annealing is a popular technique for combinatorial optimization because it is generic and easily implemented in its simplest form. Also, its limiting behavior is well studied. On the other hand, this limiting behavior is not easily realized in practice, the speed of convergence can be maddeningly slow, and complex esoteric tinkering may be needed to substantially improve the performance. Useful reviews of simulated annealing include [69, 543].

Annealing is the process of heating up a solid and then cooling it slowly. When a stressed solid is heated, its internal energy increases and its molecules move randomly. If the solid is then cooled slowly, the thermal energy generally decreases slowly, but there are also random increases governed by Boltzmann's probability. Namely, at temperature τ, the probability density of an increase in energy of magnitude ΔE is $\exp\{-\Delta E/k\tau\}$ where k is Boltzmann's constant. If the cooling is slow enough and deep enough, the final state is unstressed, where all the molecules are arranged to have minimal potential energy.

For consistency with the motivating physical process, we pose optimization as minimization in this section, so the minimum of $f(\boldsymbol{\theta})$ is sought over $\boldsymbol{\theta} \in \boldsymbol{\Theta}$. Then it is possible to draw an analogy between the physical cooling process and the process of solving a combinatorial minimization problem [111, 330]. For simulated annealing algorithms, $\boldsymbol{\theta}$ corresponds to the state of the material, $f(\boldsymbol{\theta})$ corresponds to its energy level, and the optimal solution corresponds to the $\boldsymbol{\theta}$ that has minimum energy. Random changes to the current state (i.e., moves from $\boldsymbol{\theta}^{(t)}$ to $\boldsymbol{\theta}^{(t+1)}$) are governed by the Boltzmann distribution given above, which depends on a parameter called temperature. When the temperature is high, acceptance of uphill moves (i.e., moves to a higher energy state) are more likely to be tolerated. This discourages convergence to the first local minimum that happens to be found, which might be premature if the space of candidate solutions has not yet been adequately explored. As search continues, the temperature is lowered. This forces increasingly concentrated search effort near the current local minimum, because few uphill moves will be allowed. If the cooling schedule is determined appropriately, the algorithm will hopefully converge to the global minimum.

The simulated annealing algorithm is an iterative procedure started at time $t = 0$ with an initial point $\boldsymbol{\theta}^{(0)}$ and a temperature τ_0. Iterations are indexed by t. The algorithm is run in stages, which we index by $j = 0, 1, 2, \ldots$, and each stage consists of several iterations. The length of the jth stage is m_j. Each iteration proceeds as follows:

1. Select a candidate solution $\boldsymbol{\theta}^*$ within the neighborhood of $\boldsymbol{\theta}^{(t)}$, say $\mathcal{N}(\boldsymbol{\theta}^{(t)})$, according to a proposal density $g^{(t)}(\cdot \mid \boldsymbol{\theta}^{(t)})$.

2. Randomly decide whether to adopt $\boldsymbol{\theta}^*$ as the next candidate solution or to keep another copy of the current solution. Specifically, let $\boldsymbol{\theta}^{(t+1)} = \boldsymbol{\theta}^*$ with probability equal to $\min\left(1, \exp\left\{[f(\boldsymbol{\theta}^{(t)}) - f(\boldsymbol{\theta}^*)]/\tau_j\right\}\right)$. Otherwise, let $\boldsymbol{\theta}^{(t+1)} = \boldsymbol{\theta}^{(t)}$.

   3. Repeat steps 1 and 2 a total of m_j times.

   4. Increment j. Update $\tau_j = \alpha(\tau_{j-1})$ and $m_j = \beta(m_{j-1})$. Go to step 1.

If the algorithm is not stopped according to a limit on the total number of iterations or a predetermined schedule of τ_j and m_j, one can monitor an absolute or relative convergence criterion (see Chapter 2). Often, however, the stopping rule is expressed as a minimum temperature. After stopping, the best candidate solution found is the estimated minimum.

   The function α should slowly decrease the temperature to zero. The number of iterations at each temperature (m_j) should be large and increasing in j. Ideally, the function β should scale the m_j exponentially in p, but in practice some compromises will be required in order to obtain tolerable computing speed.

   Although the new candidate solution is always adopted when it is superior to the current solution, note that it has some probability of being adopted even when it is inferior. In this sense, simulated annealing is a stochastic descent algorithm. Its randomness allows simulated annealing sometimes to escape uncompetitive local minima.

3.4.1   Practical issues

3.4.1.1 *Neighborhoods and proposals* Strategies for choosing neighborhoods can be very problem-specific, but the best neighborhoods are usually small and easily computed.

   Consider the traveling salesman problem. Numbering the cities $1, 2, \ldots, p$, any tour $\boldsymbol{\theta}$ can be written as a permutation of these integers. The cities are linked in this order, with an additional link between the final city visited and the original city where the tour began. A neighbor of $\boldsymbol{\theta}$ can be generated by removing two nonadjacent links and reconnecting the tour. In this case, there is only one way to obtain a valid tour through reconnection: One of the tour segments must be reversed. For example, the tour '143256' is a neighbor of the tour '123456'. Since two links are altered, the process of generating such neighbors is a 2-change, and it yields a 2-neighborhood. Any tour has $p(p-3)/2$ unique 2-change neighbors distinct from $\boldsymbol{\theta}$ itself. This neighborhood is considerably smaller than the $(p-1)!/2$ tours in the complete solution space.

   It is critical that the chosen neighborhood structure allows all solutions in Θ to *communicate*. For $\boldsymbol{\theta}_i$ and $\boldsymbol{\theta}_j$ to communicate, it must be possible to find a finite sequence of solutions $\boldsymbol{\theta}_1, \ldots, \boldsymbol{\theta}_k$ such that $\boldsymbol{\theta}_1 \in \mathcal{N}(\boldsymbol{\theta}_i)$, $\boldsymbol{\theta}_2 \in \mathcal{N}(\boldsymbol{\theta}_1)$, ..., $\boldsymbol{\theta}_k \in \mathcal{N}(\boldsymbol{\theta}_{k-1})$, and $\boldsymbol{\theta}_j \in \mathcal{N}(\boldsymbol{\theta}_k)$. The 2-neighborhoods mentioned above for the traveling salesman problem allow communication between any $\boldsymbol{\theta}_i$ and $\boldsymbol{\theta}_j$.

   The most common proposal density, $g^{(t)}(\cdot \mid \boldsymbol{\theta}^{(t)})$, is discrete uniform—a candidate is sampled completely at random from $\mathcal{N}(\boldsymbol{\theta}^{(t)})$. This has the advantage of speed and simplicity. Other, more strategic methods have also been suggested [246, 247, 560].

Rapid updating of the objective function is an important strategy for speeding simulated annealing runs. In the traveling salesman problem, sampling a 2-neighbor at random amounts to selecting two integers from which is derived a permutation of the current tour. Note also for the traveling salesman problem that $f(\boldsymbol{\theta}^*)$ can be efficiently calculated for any $\boldsymbol{\theta}^*$ in the 2-neighborhood of $\boldsymbol{\theta}^{(t)}$ when $f(\boldsymbol{\theta}^{(t)})$ has already been found. In this case, the new tour length equals the old tour length minus the distance for traveling the two broken links, plus the distance for traveling the two new links. The time to compute this does not depend on problem size p.

3.4.1.2 *Cooling schedule and convergence*

The sequence of stage lengths and temperatures is called the *cooling schedule*. Ideally, the cooling schedule should be slow.

The limiting behavior of simulated annealing follows from Markov chain theory, briefly reviewed in Chapter 1. Simulated annealing can be viewed as producing a sequence of homogeneous Markov chains (one at each temperature) or a single inhomogeneous Markov chain (with temperature decreasing between transitions). Although these views lead to different approaches to defining limiting behavior, both lead to the conclusion that the limiting distribution of draws has support only on the set of global minima.

To understand why cooling should lead to the desired convergence of the algorithm at a global minimum, first consider the temperature to be fixed at τ. Suppose further that proposing $\boldsymbol{\theta}_i$ from $\mathcal{N}(\boldsymbol{\theta}_j)$ has the same probability as proposing $\boldsymbol{\theta}_j$ from $\mathcal{N}(\boldsymbol{\theta}_i)$ for any pair of solutions $\boldsymbol{\theta}_i$ and $\boldsymbol{\theta}_j$ in Θ. In this case, the sequence of $\boldsymbol{\theta}^{(t)}$ generated by simulated annealing is a Markov chain with stationary distribution $\pi_\tau(\boldsymbol{\theta}) \propto \exp\{-f(\boldsymbol{\theta})/\tau\}$. This means that $\lim_{t\to\infty} P[\boldsymbol{\theta}^{(t)} = \boldsymbol{\theta}] = \pi_\tau(\boldsymbol{\theta})$. This approach to generating a sequence of random values is called the *Metropolis algorithm* and is discussed in Section 7.1.

In principle, we would like to run the chain at this fixed temperature long enough that the Markov chain is approximately in its stationary distribution before the temperature is reduced.

Suppose there are M global minima and the set of these solutions is $\mathcal{M}$. Denote the minimal value of f on Θ as $f_{\min}$. Then the stationary distribution of the chain for a fixed τ is given by

$$\pi_\tau(\boldsymbol{\theta}_i) = \frac{\exp\{-[f(\boldsymbol{\theta}_i) - f_{\min}]/\tau\}}{M + \sum_{j \notin \mathcal{M}} \exp\{-[f(\boldsymbol{\theta}_j) - f_{\min}]/\tau\}} \tag{3.7}$$

for each $\boldsymbol{\theta}_i \in \Theta$.

Now, as $\tau \to 0$ from above, the limit of $\exp\{-[f(\boldsymbol{\theta}_i) - f_{\min}]/\tau\}$ is 0 if $i \notin \mathcal{M}$ and 1 if $i \in \mathcal{M}$. Thus

$$\lim_{\tau \downarrow 0} \pi_\tau(\boldsymbol{\theta}_i) = \begin{cases} 1/M & \text{if } i \in \mathcal{M}, \\ 0 & \text{otherwise.} \end{cases} \tag{3.8}$$

The mathematics to make these arguments precise can be found in [61, 543].

It is also possible to relate the cooling schedule to a bound on the quality of the final solution. If one wishes any iterate to have not more than probability δ in equilibrium of being worse than the global minimum by no more than ϵ, this can be achieved if one cools until $\tau_j \leq \epsilon/\log\{(N-1)/\delta\}$, where N is the number of points in Θ [364]. In other words, this τ_j ensures that the final Markov chain configuration will in equilibrium have $P\left[f(\boldsymbol{\theta}^{(t)}) > f_{\min} + \epsilon\right] < \delta$.

Hajek has shown that if neighborhoods communicate and the depth of the deepest local (and nonglobal) minimum is c, then the cooling schedule given by $\tau = c/\log\{1+i\}$ guarantees asymptotic convergence, where i indexes iterations [255]. The depth of a local minimum is defined to be the smallest increase in the objective function needed to escape from that local minimum into the valley of any other minimum. However, mathematical bounds on the number of iterations required to achieve a high probability of having discovered at least one element of $\mathcal{M}$ often exceed the size of Θ itself. In this case, one cannot establish that simulated annealing will find the global minimum more quickly than exhaustive search [28].

If one wishes the Markov chain generated by simulated annealing to be approximately in its stationary distribution at each temperature before reducing temperature, then the length of the run ideally should be at least quadratic in the size of the solution space [1], which itself is usually exponential in problem size. Clearly, much shorter stage lengths must be chosen if simulated annealing is to require fewer iterations than exhaustive search.

In practice, many cooling schedules have been tried [543]. Recall that the temperature at stage j is $\tau_j = \alpha(\tau_{j-1})$ and the number of iterations in stage j is $m_j = \beta(m_{j-1})$. One popular approach is to set $m_j = 1$ for all j and reduce the temperature very slowly according to $\alpha(\tau_{j-1}) = \frac{\tau_{j-1}}{1+a\tau_{j-1}}$ for a small value of a. A second option is to set $\alpha(\tau_{j-1}) = a\tau_{j-1}$ for $a < 1$ (usually $a \geq 0.9$). In this case, one might increase stage lengths as temperatures decrease. For example, consider $\beta(m_{j-1}) = bm_{j-1}$ for $b > 1$, or $\beta(m_{j-1}) = b + m_{j-1}$ for $b > 0$. A third schedule uses $\alpha(\tau_{j-1}) = \frac{\tau_{j-1}}{1+\tau_{j-1}\log\{1+r\}/(3s_{\tau_{j-1}})}$ where $s_{\tau_{j-1}}^2$ is the square of the mean objective function cost at the current temperature minus the mean squared cost at the current temperature, and r is some small real number [1]. Using the temperature schedule suggested by Hajek's result is rarely practical, because it is too slow and the determination of c is difficult, with excessively large guesses for c further slowing the algorithm.

Most practitioners require lengthy experimentation to find suitable initial parameter values (e.g., τ_0 and m_0) and values of the proposed schedules (e.g., a, b, and r). While selection of the initial temperature τ_0 is usually problem-dependent, some general guidelines may be given. A useful strategy is to choose a positive τ_0 value so that $\exp\{[f(\boldsymbol{\theta}_i) - f(\boldsymbol{\theta}_j)]/\tau_0\}$ is close to 1 for any pair of solutions $\boldsymbol{\theta}_i$ and $\boldsymbol{\theta}_j$ in Θ. The rationale for this choice is that it provides any point in the parameter space with a reasonable chance of being visited in early iterations of the algorithm. Similarly, choosing m_j to be large can produce a more accurate solution, but can result in long computing times. As a general rule of thumb, larger decreases in temperature require longer runs after the decrease. Finally, a good deal of evidence suggests that running

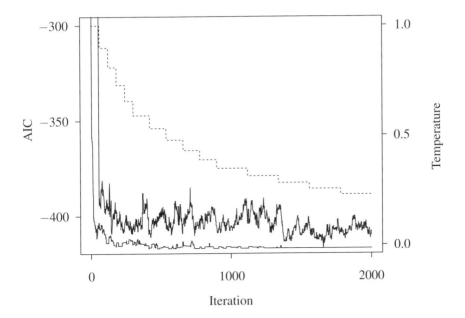

Fig. 3.5 Results of two simulated annealing minimizations of the regression model AIC for Example 3.6. The temperature for the bottom curve is shown with the dotted line and the right axis.

simulated annealing long at high temperatures is not very useful. In many problems, the barriers between local minima are sufficiently modest that jumps between them are possible even at fairly low temperatures. Good cooling schedules therefore decrease the temperature rapidly at first.

Example 3.6 (Baseball salaries, continued) To implement simulated annealing for variable selection via the AIC in the baseball salary regression problem introduced in Example 3.3, we must establish a neighborhood structure, a proposal distribution, and a temperature schedule. The simplest neighborhoods contain 1-change neighbors generated from the current model by either adding or deleting one predictor. We assigned equal probabilities to all candidates in a neighborhood. The cooling schedule had 15 stages, with stage lengths of 60 for the first five stages, 120 for the next five, and 220 for the final five. Temperatures were decreased according to $\alpha(\tau_{j-1}) = 0.9\tau_{j-1}$ after each stage.

Figure 3.5 shows the values of the AIC for the sequence of candidate solutions generated by simulated annealing, for two different choices of τ_0. The bottom curve corresponds to $\tau_0 = 1$. In this case, simulated annealing became stuck at particular candidate solutions for distinct periods because the low temperatures allowed little tolerance for uphill moves. In the particular realization shown, the algorithm quickly found good candidate solutions with low AIC values, where it became stuck frequently. However, in other cases (e.g., with a very multimodal objective function), such stickiness may result in the algorithm becoming trapped in a region far from the global minimum. A second run with $\tau_0 = 10$ (top solid line) yielded considerable mixing, with many uphill proposals accepted as moves. The temperature schedule for $\tau_0 = 1$ is shown by the dotted line and the right axis. Both runs exhibited greater mixing at higher temperatures. When $\tau_0 = 1$, the best model found was first identified in the 1419th step and dominated the simulation after that point. This model achieved an AIC of -416.95, and matched the best model found using random starts local search in Table 3.2. When $\tau_0 = 10$, the best model had an AIC of -416.94 and matched the model found by the S-Plus approach in Table 3.2. This model was found only once: on move 1718. □

3.4.2 Enhancements

There are many variations on simulated annealing that purport to improve performance. Here we list a few ideas in an order roughly corresponding to the steps in the basic algorithm.

The simplest way to start simulated annealing is to start once, anywhere. A strategy employing multiple random starts would have the dual advantages of potentially finding a better candidate solution and allowing confirmation of convergence to the particular optimum found. Purely random starts could be replaced by a stratified set of starting points chosen by strategic preprocessing to be more likely to lead to minima than simple random starts. Such strategies must have high payoffs if they are to be useful, given simulated annealing's generally slow convergence. In some cases, the extra iterations dedicated to various random starts may be better spent on a single long run with longer stage sizes and a slower cooling schedule.

The solution space, Θ, may include constraints on θ. For example, in the genetic mapping problem introduced in Example 3.1, θ must be a permutation of the integers $1, \ldots, p$ when there are p markers. When the process for generating neighbors creates solutions that violate these constraints, substantial time may be wasted fixing candidates or repeatedly sampling from $\mathcal{N}(\theta^{(t)})$ until a valid candidate is found. An alternative is to relax the constraints and introduce a penalty into f that penalizes invalid solutions. In this manner, the algorithm can be discouraged from visiting invalid solutions without dedicating much time to enforcing constraints.

In the basic algorithm, the neighborhood definition is static and the proposal distribution is the same at each iteration. Sometimes improvements can be obtained by adaptively restricting neighborhoods at each iteration. For example, it can be useful to shrink the size of the neighborhood as time increases to avoid many wasteful generations of distant candidates that are very likely to be rejected at such low temperatures. In other cases, when a penalty function is used in place of constraints, it may be useful to allow only neighborhoods composed of solutions that reduce or eliminate constraint violations embodied in the current θ.

It is handy if f can be evaluated quickly for new candidates. We noted previously that neighborhood definitions can sometimes enable this, as in the traveling salesman problem where a 2-neighborhood strategy led to a simple updating formula for f. Simple approximation of f is sometimes made, often in a problem-specific manner. At least one author suggests monitoring recent iterates and introducing a penalty term in f that discourages revisiting states like those recently visited [176].

Next consider the acceptance probability given in step 2 of the canonical simulated annealing algorithm in Section 3.4. The expression $\exp\{[f(\boldsymbol{\theta}^{(t)}) - f(\boldsymbol{\theta}^*)]/\tau_j\}$ is motivated by the Boltzmann distribution from statistical thermodynamics. Other acceptance probabilities can be used, however. The linear Taylor series expansion of the Boltzmann distribution motivates $\min\left\{1, 1 + \left([f(\boldsymbol{\theta}^{(t)}) - f(\boldsymbol{\theta}^*)] \big/ \tau_j\right)\right\}$ as a possible acceptance probability [309]. To encourage moderate moves away from local minima while preventing excessive small moves, the acceptance probability $\min\left\{1, \exp\left\{[c + f(\boldsymbol{\theta}^{(t)}) - f(\boldsymbol{\theta}^*)] \big/ \tau_j\right\}\right\}$, where $c > 0$, has been suggested for certain problems [146].

In general, there is little evidence that the shape of the cooling schedule (linear, polynomial, exponential) matters much, as long as the useful range of temperatures is covered, the range is traversed at roughly the same rate, and sufficient time is spent at each temperature (especially the low temperatures) [146]. Reheating strategies that allow sporadic, systematic, or interactive temperature increases to prevent getting stuck in a local minimum at low temperatures can be effective [146, 226, 330].

After simulated annealing is complete, one might take the final result of one or more runs and polish these with a descent algorithm. In fact, one could refine occasional accepted steps in the same way, instead of waiting until simulated annealing has terminated.

3.5 GENETIC ALGORITHMS

Annealing is not the only natural process successfully exploited as a metaphor to solve optimization problems. *Genetic algorithms* mimic the process of Darwinian natural selection. Candidate solutions to a maximization problem are envisioned as biological organisms represented by their genetic code. The fitness of an organism is analogous to the quality of a candidate solution. Breeding among highly fit organisms provides the best opportunity to pass along desirable attributes to future generations, while breeding among less fit organisms (and rare genetic mutations) ensures population diversity. Over time, the organisms in the population should evolve to become increasingly fit, thereby providing a set of increasingly good candidate solutions to the optimization problem. The pioneering development of genetic algorithms was done by Holland [291]. Other useful references include [15, 119, 175, 231, 395, 448, 450, 562].

We revert now to our standard description of optimization as maximization, where we seek the maximum of $f(\boldsymbol{\theta})$ with respect to $\boldsymbol{\theta} \in \Theta$. In statistical applications of genetic algorithms, f is often a joint log profile likelihood function.

3.5.1 Definitions and the canonical algorithm

3.5.1.1 Basic definitions In Example 3.1 above, some genetics terminology was introduced. Here we discuss additional terminology needed to study genetic algorithms.

In a genetic algorithm, every candidate solution corresponds to a *individual*, or *organism*, and every organism is completely described by its genetic code. Individuals are assumed to have one *chromosome*. A chromosome is a sequence of C symbols, each of which consists of a single choice from a predetermined alphabet. The most basic alphabet is the binary alphabet, $\{0, 1\}$, in which case a chromosome of length $C = 9$ might look like '100110001'. The C elements of the chromosome are the *genes*. The values that might be stored in a gene (i.e., the elements of the alphabet) are *alleles*. The position of a gene in the chromosome is its *locus*.

The information encoded in an individual's chromosome is its *genotype*. We will represent a chromosome or its genotype as ϑ. The expression of the genotype in the organism itself is its *phenotype*. For optimization problems, phenotypes are candidate solutions and genotypes are encodings: Each genotype, ϑ, encodes a phenotype, θ, using the chosen allele alphabet.

Genetic algorithms are iterative, with iterations indexed by t. Unlike the methods previously discussed in this chapter, genetic algorithms track more than one candidate solution simultaneously. Let the tth *generation* consist of a collection of P organisms, $\vartheta_1^{(t)}, \ldots, \vartheta_P^{(t)}$. This population of size P at generation t corresponds to a collection of candidate solutions, $\theta_1^{(t)}, \ldots, \theta_P^{(t)}$.

Darwinian natural selection favors organisms with high *fitness*. The fitness of an organism $\vartheta_i^{(t)}$ depends on the corresponding $f(\theta_i^{(t)})$. A high-quality candidate solution has a high value of the objective function and a high fitness. As generations progress, organisms inherit from their parents bits of genetic code that are associated with high fitness if fit parents are predominantly selected for breeding. An *offspring* is a new organism inserted in the $(t + 1)$th generation to replace a member of the tth generation; the offspring's chromosome is determined from those of two *parent* chromosomes belonging to the tth generation.

To illustrate some of these ideas, consider a regression model selection problem with 9 predictors. Assume that an intercept will be included in any model. The genotype of any model can then be written as a chromosome of length 9. For example, the chromosome $\vartheta_i^{(t)} = $'100110001' is a genotype corresponding the phenotype of a model containing only the fitted parameters for the intercept and predictors 1, 4, 5, and 9.

Another genotype is $\vartheta_j^{(t)} = $'110100110'. Notice that $\vartheta_i^{(t)}$ and $\vartheta_j^{(t)}$ share some common genes. A *schema* is any subcollection of genes. In this example, the two chromosomes share the schema '1*01*****', where '*' represents a wildcard: the allele in that locus is ignored. (These two chromosomes also share the schemata '**01*****', '1*01*0***', and others.) The significance of schemata is that they encode modest bits of genetic information that may be transferred as a unit from parent to offspring. If a schema is associated with a phenotypic feature that induces high values of the objective function, then the inheritance of this schema by individuals in future generations promotes optimization.

3.5.1.2 Selection mechanisms and genetic operators

Breeding drives most genetic change. The process by which parents are chosen to produce offspring is called the *selection mechanism*. One simple approach is to select one parent with probability proportional to fitness and to select the other parent completely at random. Another approach is to select each parent independently with probability proportional to fitness. Section 3.5.2.2 describes some of the most frequently used selection mechanisms.

After two parents from the tth generation have been selected for breeding, their chromosomes are combined in some way that allows schemata from each parent to be inherited by their offspring, who become members of generation $t + 1$. The methods for producing offspring chromosomes from chosen parent chromosomes are *genetic operators*.

A fundamental genetic operator is *crossover*. One of the simplest crossover methods is to select a random position between two adjacent loci and split both parent chromosomes at this position. Glue the left chromosome segment from one parent to the right segment from the other parent to form an offspring chromosome. The remaining segments can be combined to form a second offspring, or discarded. For example, suppose the two parents are '100110001' and '110100110'. If the random split point is between the third and fourth loci, then the potential offspring are '100100110' and '110110001'. Note that in this example, both offspring inherit the schema '1*01*****'. Crossover is the key to a genetic algorithm—it allows good features of two candidate solutions to be combined. Some more complicated crossover operators are discussed in Section 3.5.2.3.

Mutation is another important genetic operator. Mutation changes an offspring chromosome by randomly introducing one or more alleles in loci where those alleles are not seen in the corresponding loci of either parent chromosome. For example, if crossover produced '100100110' from the parents mentioned above, subsequent mutation might yield '101100110'. Note that the third gene was 0 in both parents and therefore crossover alone was guaranteed to retain the schema '**0******'. Mutation, however, provides a way to escape this constraint, thereby promoting search diversification and providing a way to escape from local maxima.

Mutation is usually applied after breeding. In the simplest implementation, each gene has an independent probability, μ, of mutating, and the new allele is chosen completely at random from the genetic alphabet. If μ is too low, many potentially good innovations will be missed; if μ is too high, the algorithm's ability to learn over time will be degraded, because excessive random variation will disturb the fitness selectivity of parents and the inheritance of desirable schemata.

To summarize, genetic algorithms proceed by producing generations of individuals. The $(t+1)$th generation is produced as follows. First the individuals in generation t are ranked and selected according to fitness. Then crossover and mutation are applied to these selected individuals to produce generation $t + 1$. Figure 3.6 is a small example of the production of a generation of four individuals with three chromo-

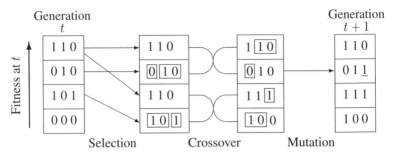

Fig. 3.6 An example of generation production in a genetic algorithm for a population of size $P = 4$ with chromosomes of length $C = 3$. Crossovers are illustrated by boxing portions of some chromosomes. Mutation is indicated by an underlined gene in the final column.

somes per individual and binary chromosome encoding. In generation t, individual '110' has the highest fitness among its generation and is chosen twice in the selection stage. In the crossover stage, the selected individuals are paired off so that each pair recombines to generate two new individuals. In the mutation stage, a low mutation rate is applied. In this example, only one mutation occurs. The completion of these steps yields the new generation.

Example 3.7 (Baseball salaries, continued) The results of applying a simple genetic algorithm to the variable selection problem for the baseball data introduced in Example 3.3 are shown in Figure 3.7. One hundred generations of size $P = 20$ were used. Binary inclusion–exclusion alleles were used for each possible predictor, yielding chromosomes of length $C = 27$. The starting generation consisted of purely random individuals. A rank-based fitness function was used; see equation (3.11) below. One parent was selected with probability proportional to this fitness; the other parent was selected independently, purely at random. Breeding employed simple crossover. A 1% mutation rate was randomly applied independently to each locus.

The horizontal axis in Figure 3.7 corresponds to generation. The AIC values for all twenty individuals in each generation are plotted. The best model found included predictors 2, 3, 6, 8, 10, 13, 14, 15, 16, 24, 25, and 26, yielding an AIC of -416.95. This matches the best model found using random starts local search (Table 3.2). Darwinian survival of the fittest is clearly illustrated in this figure: The twenty random starting individuals quickly coalesce into three effective subspecies, with the best of these slowly overwhelming the rest. The best model was first found in generation 87. □

3.5.1.3 Allele alphabets and genotypic representation The binary alphabet for alleles was introduced in the pioneering work of Holland [291] and continues to be very prevalent in recent research. The theoretical behavior of the algorithm and the relative performance of various genetic operators and other algorithmic variations are better understood for binary chromosomes than for other choices.

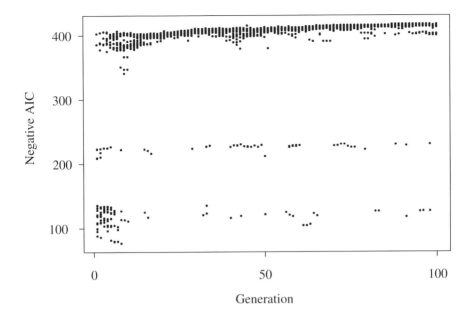

Fig. 3.7 Results of a genetic algorithm for Example 3.7.

For many optimization problems, it is possible to construct a binary encoding of solutions. For example, consider the univariate optimization of $f(\theta) = 100 - (\theta - 4)^2$ on the range $\theta \in [1, 12.999] = [a_1, a_2]$. Suppose that we represent a number in $[a_1, a_2]$ as

$$a_1 + \left(\frac{a_2 - a_1}{2^d - 1} \right) \text{decimal}(b), \tag{3.9}$$

where b is a binary number of d digits and the decimal() function converts from base 2 to base 10. If c decimal places of accuracy are required, then d must be chosen to satisfy

$$(a_2 - a_1)10^c \leq 2^d - 1. \tag{3.10}$$

In our example, 14 binary digits are required for accuracy to 3 decimal places, and $b = \text{'}01000000000000\text{'}$ maps to $\theta = 4.000$ using equation (3.9).

In some cases, such as the regression model selection problem, a binary-encoded chromosome may be very natural. In others, however, the encoding seems forced, as it does above. For $f(\theta) = 100 - (\theta - 4)^2$, the chromosome $\vartheta = \text{'}01000000000000\text{'}$ ($\theta = 4.000$) is optimal. However, chromosomes that are genetically close to this, such as '10000000000000' ($\theta = 7.000$) and '00000000000000' ($\theta = 1.000$), have phenotypes that are not close to $\theta = 4.000$. On the other hand, the genotype '00111111111111' has phenotype very close to 4.000 even though the genotype is very different than '01000000000000'. Chromosomes that are similar in genotype may have very different phenotypes. Thus, a small mutation may move to a drastically different region of solution space, and a crossover may produce offspring whose phenotypes bear little resemblance to either parent. To resolve such difficulties, a different encoding scheme or modified genetic operators may be required (see Section 3.5.2.3).

An important alternative to binary representation arises in permutation problems of size p, like the traveling salesman problem. In such cases, a natural chromosome is a permutation of the integers $1, \ldots, p$, for example, $\vartheta = $ '752631948' when $p = 9$. Since such chromosomes must obey the requirement that each integer appear in exactly one locus, some changes to standard genetic operators will be required. Strategies for dealing with permutation chromosomes are discussed in Section 3.5.2.3

3.5.1.4 Initialization, termination, and parameter values   Genetic algorithms are usually initialized with a first generation of purely random individuals.

The size of the generation, P, affects the speed, convergence behavior, and solution quality of the algorithm. Large values of P are to be preferred, if feasible, because they provide a more diverse genetic pool from which to generate offspring, thereby diversifying the search and discouraging premature convergence. For binary encoding of chromosomes, one suggestion is to choose P to satisfy $C \leq P \leq 2C$, where C is the chromosome length [8]. For permutation chromosomes, the range $2C \leq P \leq 20C$ has been suggested [293]. In most real applications, population sizes have ranged between 10 and 200 [477], although a review of empirical studies suggests that P can often be as small as 30 [448].

Mutation rates are typically very low, in the neighborhood of 1%. Theoretical work and empirical studies have supported a rate of $1/C$ [395], and another investigation suggested that the rate should be nearly proportional to $1/(P\sqrt{C})$ [482]. Nevertheless, a fixed rate independent of P and C is a common choice.

The termination criterion for a genetic algorithm is frequently just a maximum number of iterations chosen to limit computing time. One might instead consider stopping when the genetic diversity within chromosomes in the current generation is sufficiently low [15].

3.5.2 Variations

In this section we survey a number of methodological variations that may offer improved performance. These include alterations to the fitness function, selection mechanism, genetic operators, and other aspects of the basic algorithm.

3.5.2.1 Fitness   In a canonical genetic algorithm, the fitness of an organism is often taken to be the objective function value of its phenotype, perhaps scaled by the mean objective function value in its generation. It is tempting to simply equate the objective function value $f(\theta)$ to the fitness because the fittest individual then corresponds to the maximum likelihood solution. However, directly equating an organism's fitness to the objective function value for its corresponding phenotype is usually naive in that other choices yield superior optimization performance. Instead, let $\phi(\vartheta)$ denote the value of a *fitness function* that describes the fitness of a chromo-

some. The fitness function will depend on the objective function f, but will not equal it. This increased flexibility can be exploited to enhance search effectiveness.

A problem seen in some applications of genetic algorithms is excessively fast convergence to a poor local optimum. This can occur when a few of the very best individuals dominate the breeding and their offspring saturate subsequent generations. In this case, each subsequent generation consists of genetically similar individuals that lack the genetic diversity needed to produce offspring that might typify other, more profitable regions of solution space. This problem is especially troublesome if it occurs directly after initialization, when nearly all individuals have very low fitness. A few chromosomes that are more fit than the rest can then pull the algorithm to an unfavorable local maximum. This problem is analogous to entrapment near an uncompetitive local maximum, which is also a concern for the other search methods discussed earlier in this chapter.

Selective pressure must be balanced carefully, however, because genetic algorithms can be slow to find a very good optimum. It is therefore important to maintain firm selective pressure without allowing a few individuals to cause premature convergence. To do this, the fitness function can be designed to reduce the impact of large variations in f.

A common approach is to ignore the values of $f(\boldsymbol{\theta}_i^{(t)})$ and use only their ranks [16, 449, 561]. For example, one could set

$$\phi(\boldsymbol{\vartheta}_i^{(t)}) = \frac{2r_i}{P(P+1)}, \tag{3.11}$$

where r_i is the rank of $f(\boldsymbol{\theta}_i^{(t)})$ among generation t. This strategy gives the chromosome corresponding to the median quality candidate a selection probability of $1/P$, and the best chromosome has probability $2/(P+1)$, roughly double that for the median. Rank-based methods are attractive in that they retain a key feature of any successful genetic algorithm—selectivity based on relative fitness—while discouraging premature convergence and other difficulties caused by the actual form of f, which can be somewhat arbitrary [561]. Some less common fitness function formulations involving scaling and transforming f are mentioned in [231].

3.5.2.2 Selection mechanisms and updating generations
Previously, in Section 3.5.1.2, we mentioned only simple approaches to selecting parents on the basis of fitness. Selecting parents on the basis of fitness ranks (Section 3.5.2.1) is far more common than using selection probabilities proportional to fitness.

Another common approach is *tournament selection* [179, 232, 233]. In this approach, the set of chromosomes in generation t is randomly partitioned into k disjoint subsets of equal size (perhaps with a few remaining chromosomes temporarily ignored). The best individual in each group is chosen as a parent. Additional random partitionings are carried out until sufficient parents have been generated. Parents are then paired randomly for breeding. This approach ensures that the best individual will breed P times, the median individual will breed once on average, and the worst individual will not breed at all. The approaches of proportional selection, ranking, and tournament selection apply increasing selective pressure, in that order. Higher selective pressure is generally associated with superior performance, as long as premature entrapment in local optima can be avoided [15].

Populations can be partially updated. The *generation gap*, G, is a proportion of the generation to be replaced by generated offspring [126]. Thus, $G = 1$ corresponds to a canonical genetic algorithm with distinct, nonoverlapping generations. At the other extreme, $G = 1/P$ corresponds to incremental updating of the population one offspring at a time. In this case, a *steady-state* genetic algorithm produces one offspring at a time to replace the least fit (or some random relatively unfit) individual [562]. Such an process typically exhibits more variance and higher selective pressure than a standard approach.

When $G < 1$, performance can sometimes be enhanced with a selection mechanism that departs somewhat from the Darwinian analogy. For example, an *elitist* strategy would place an exact copy of the current fittest individual in the next generation, thereby ensuring the survival of the best current solution [126]. When $G = 1/P$, each offspring could replace a chromosome randomly selected from those with below-average fitness [5].

Deterministic selection strategies have been proposed to eliminate sampling variability [17, 395]. We see no compelling need to eliminate the randomness inherent in the selection mechanism.

One important consideration when generating or updating a population is whether to allow duplicate individuals in the population. Dealing with duplicate individuals wastes computing resources, and it potentially distorts the parent selection criterion by giving duplicated chromosomes more chances to produce offspring [119].

3.5.2.3 Genetic operators and permutation chromosomes
To increase genetic mixing, it is possible to choose more than one crossover point. If two crossover points are chosen, the gene sequence between them can be swapped between parents to create offspring. Such multipoint crossover can improve performance [48, 163].

Many other approaches for transferring genes from parents to offspring have been suggested. For example, each offspring gene could be filled with an allele randomly selected from the alleles expressed in that position in the parents. In this case, the parental origins of adjacent genes could be independent [4, 527] or correlated [509], with strength of correlation controlling the degree to which offspring resemble a single parent.

In some problems, a different allele alphabet may be more reasonable. Allele alphabets with many more than two elements have been investigated [12, 119, 442, 451]. For some problems, genetic algorithms using a floating-point alphabet have outperformed algorithms using the binary alphabet [119, 303, 394]. Methods known as messy genetic algorithms employ variable-length encoding with genetic operators that adapt to changing length [234, 235, 236]. Gray coding is another alternative encoding that is particularly useful for real-valued objective functions that have a bounded number of optima [563].

When a nonbinary allele alphabet is adopted, modifications to other aspects of the genetic algorithm, particularly to the genetic operators, is often necessary and even fruitful. Nowhere is this more evident than when permutation chromosomes are used. Recall that Section 3.5.1.3 introduced a special chromosome encoding for permutation optimization problems. For such problems (like the traveling salesman problem), it is natural to write a chromosome as a permutation of the integers $1, \ldots, n$. New genetic operators are needed then to ensure that each generation contains only valid permutation chromosomes.

For example, let $p = 9$, and consider the crossover operator. From two parent chromosomes '752631948' and '912386754' and a crossover point between the second and third loci, standard crossover would produce offspring '752386754' and '912631948'. Both of these are invalid permutation chromosomes, because both contain some duplicate alleles.

A remedy is *order crossover* [528]. A random collection of loci is chosen, and the order in which the alleles in these loci appear in one parent is imposed on the same alleles in the other parent to produce one offspring. The roles of the parents can be switched to produce a second offspring. This operator attempts to respect relative positions of alleles. For example, consider the parents '752631948' and '912386754', and suppose that the fourth, sixth, and seventh loci are randomly chosen. In the first parent, the alleles in these loci are 6, 1, and 9. We must rearrange the 6, 1, and 9 alleles in the second parent to impose this order. The remaining alleles in the second parent are '**238*754'. Inserting 6, 1, and 9 in this order yields '612389754' as the offspring. Reversing the roles of the parents yields a second offspring '352671948'.

Many other crossover operators for permutation chromosomes have been proposed [116, 117, 119, 237, 395, 421, 499]. Most are focused on the positions of individual genes. However, for problems like the traveling salesman problem, such operators have the undesirable tendency to destroy links between cities in the parent tours. The desirability of a candidate solution is a direct function of these links. Breaking links is effectively an unintentional source of mutation. *Edge-recombination crossover* has been proposed to produce offspring that contain only links present in at least one parent [564, 565].

We use the traveling salesman problem to explain edge-recombination crossover. The operator proceeds through the following steps.

1. We first construct an edge table that stores all the links that lead into and out of each city in either parent. For our two parents, '752631948' and '912386754', the result is shown in the leftmost portion of Table 3.4. Note that the number of links into and out of each city in either parent will always be at least two and no more than four. Also, recall that a tour returns to its starting city, so, for example, the first parent justifies listing 7 as a link from 8.

2. To begin creating an offspring, we choose between the initial cities of the two parents. In our example, the choices are cities 7 and 9. If the parents' initial cities have the same number of links, then the choice is made randomly. Otherwise, choose the initial city from the parent whose initial city has fewer links. In our example, this yields '9********'.

Table 3.4 Edge tables showing the cities linked to or from each allele in either parent for each of the first three steps of edge recombination crossover. Beneath each column is the offspring chromosome resulting from each step.

Step 1		Step 2		Step 3	
City	Links	City	Links	City	Links
1	3, 9, 2	1	3, 2	1	3, 2
2	5, 6, 1, 3	2	5, 6, 1, 3	2	5, 6, 1, 3
3	6, 1, 2, 8	3	6, 1, 2, 8	3	6, 1, 2, 8
4	9, 8, 5	4	8, 5	4	Used
5	7, 2, 4	5	7, 2, 4	5	7, 2
6	2, 3, 8, 7	6	2, 3, 8, 7	6	2, 3, 8, 7
7	8, 5, 6	7	8, 5, 6	7	8, 5, 6
8	4, 7, 3, 6	8	4, 7, 3, 6	8	7, 3, 6
9	1, 4	9	Used	9	Used
'9********'		'94*******'		'945******'	

3. We must now link onwards from allele 9. From the leftmost column of the edge table, we find that allele 9 has two links: 1 and 4. We want to chose between these by selecting the city with the fewest links. To do this, we first update the edge table by deleting all references to allele 9, yielding the center portion of Table 3.4. Since cities 1 and 4 both have two remaining links, we choose randomly between 1 and 4. If 4 is the choice, then the offspring is updated to '94*******'.

4. There are two possible links onward from city 4: cities 5 and 8. Updating the edge table to produce the rightmost portion of Table 3.4, we find that city 5 has the fewest remaining links. Therefore, we choose city 5. The partial offspring is now '945******'.

Continuing this process might yield the offspring '945786312' by the following steps: select 7; select 8; select 6; randomly select 3 from the choices of 2 and 3; randomly select 1 from the choices of 1 and 2; select 2.

Note that in each step a city is chosen among those with the fewest links. If, instead, links were chosen uniformly at random, cities would be more likely to be left without a continuing edge. Since tours are circuital, the preference for a city with few links does not introduce any sort of bias in offspring generation.

An alternative *edge assembly* strategy has been found to be extremely effective in some problems [407].

Mutation of permutation chromosomes is not as difficult as crossover. A simple mutation operator is to randomly exchange two genes in the chromosome [448]. Alternatively, the elements in a short random segment of the chromosome can be randomly permuted [119].

3.5.3   Initialization and parameter values

Although traditionally a genetic algorithm is initiated with a generation of purely random individuals, heuristic approaches to constructing individuals with good or diverse fitness have been suggested as an improvement on random starts [119, 448].

Equal sizes for subsequent generations are not required. Population fitness usually improves very rapidly during the early generations of a genetic algorithm. In order to discourage premature convergence and promote search diversity, it may be desirable to use a somewhat large generation size P for early generations. If P is fixed at too large a value, however, the entire algorithm may be too slow for practical use. Once the algorithm has made significant progress toward the optimum, important improving moves most often come from high-quality individuals; low-quality individuals are increasingly marginalized. Therefore, it has been suggested that P may be decreased progressively as iterations continue [577]. However, rank-based selection mechanisms are more commonly employed as an effective way to slow convergence.

It can be also useful to allow a variable mutation rate that is inversely proportional to the population diversity [448]. This provides a stimulus to promote search diversity as generations become less diverse. Several authors suggest other methods for allowing the probabilities of mutation and crossover and other parameters of the genetic algorithm to vary adaptively over time in manners that may encourage search diversity [48, 118, 119, 395].

3.5.4   Convergence

The convergence properties of genetic algorithms are beyond the scope of this chapter, but several important ideas are worth mentioning.

Much of the early analysis about why genetic algorithms work was based on the notion of schemata [231, 291]. Such work is based on a canonical genetic algorithm with binary chromosome encoding, selection of each parent with probability proportional to fitness, simple crossover applied every time parents are paired, and mutation randomly applied to each gene independently with probability μ. For this setting, the *schema theorem* provides a lower bound on the expected number of instances of a schema in generation $t + 1$, given that it was present in generation t.

The schema theorem shows that a short, low-order schema (i.e., one specifying only a few nearby alleles) will enjoy increased expected representation in the next generation if the average fitness of chromosomes containing that schema in the generation at time t exceeds the average fitness of all chromosomes in the generation. A longer and/or more complex schema will require higher relative fitness to have the same expectation. Proponents of schema theory argue that convergence to globally competitive candidate solutions can be explained by how genetic algorithms simultaneously juxtapose many short low-order schemata of potentially high fitness, thereby promoting propagation of advantageous schemata.

More recently, the schema theorem and convergence arguments based upon it have become more controversial. Traditional emphasis on the number of instances of a schema that propagate to the next generation and on the average fitness of chromosomes containing that schema is somewhat misguided. What matters far more is which particular chromosomes containing that schema are propagated. Further, the schema theorem overemphasizes the importance of schemata: in fact it applies equally well to any arbitrary subsets of Θ. Finally, the notion that genetic algorithms succeed because they implicitly simultaneously allocate search effort to many schemata-defined regions of Θ has been substantially discredited [549]. An authoritative exposition of the mathematical theory of genetic algorithms is given by Vose [548]. Other helpful treatments include [175, 450].

Problems

The baseball data introduced in Section 3.3 are available from the website for this book. Problems 3.1–3.4 explore the implications of various algorithm configurations. Treat these problems in the spirit of experiments, trying to identify settings where interesting differences can be observed. Increase the run lengths from those used above to suit the speed of your computer, and limit the total number of objective function evaluations in every run (effectively the search effort) to a fixed number so that different algorithms and configurations can be compared fairly. Summarize your comparisons and conclusions. Supplement your comments with graphs to illustrate key points.

3.1   Implement a random starts local search algorithm for minimizing the AIC for the baseball salary regression problem. Model your algorithm after Example 3.3.

(a) Change the move strategy from steepest descent to immediate adoption of the first randomly selected downhill neighbor.

(b) Change the algorithm to employ 2-neighborhoods, and compare the results with those of previous runs.

3.2   Implement a tabu algorithm for minimizing the AIC for the baseball salary regression problem. Model your algorithm after Example 3.5.

(a) Compare the effect of using different tabu tenures.

(b) Monitor changes in AIC from one move to the next. Define a new attribute that signals when the AIC change exceeds some value. Allow this attribute to be included on the tabu list, to promote search diversity.

(c) Implement aspiration by influence, overriding the tabu of reversing a low-influence move if a high-influence move is made prior to the reversal. Measure influence with changes in R^2.

3.3   Implement simulated annealing for minimizing the AIC for the baseball salary regression problem. Model your algorithm on Example 3.6.

(a) Compare the effects of different cooling schedules (different temperatures and different durations at each temperature).

(b) Compare the effect of a proposal distribution that is discrete uniform over 2-neighborhoods versus one that is discrete uniform over 3-neighborhoods.

3.4   Implement a genetic algorithm for minimizing the AIC for the baseball salary regression problem. Model your algorithm on Example 3.7.

(a) Compare the effects of using different mutation rates.

(b) Compare the effects of using different generation sizes.

(c) Instead of the selection mechanism used in Example 3.7, try the following three mechanisms:

 i. independent selection of one parent with probability proportional to fitness and the other completely at random,

 ii. independent selection of each parent with probability proportional to fitness, and

 iii. tournament selection with $P/5$ strata, and/or another number of strata that you prefer.

To implement some of these approaches, you may need to scale the fitness function. For example, consider the scaled fitness functions π given by

$$\phi(\vartheta_i^{(t)}) = af(\theta_i^{(t)}) + b, \qquad (3.12)$$
$$\phi(\vartheta_i^{(t)}) = f(\theta_i^{(t)}) - (\bar{f} - zs), \qquad (3.13)$$

or

$$\phi(\vartheta_i^{(t)}) = f(\theta_i^{(t)})^v, \qquad (3.14)$$

where a and b are chosen so that the mean fitness equals the mean objective function value and the maximum fitness is a user-chosen c times greater than the mean fitness, $\bar{f}$ is the mean and s is the standard deviation of the unscaled objective function values in the current generation, z is a number generally chosen between 1 and 3, and v is a number slightly larger than 1. Some scalings can sometimes produce negative values for $\vartheta_i^{(t)}$. In such situations, we may apply the transformation

$$\phi_{\text{new}}(\vartheta_i^{(t)}) = \begin{cases} \phi(\vartheta_i^{(t)}) + d^{(t)} & \text{if } \phi(\vartheta_i^{(t)}) + d^{(t)} > 0, \\ 0 & \text{otherwise,} \end{cases} \qquad (3.15)$$

where $d^{(t)}$ is the absolute value of the fitness of the worst chromosome in generation t, in the last k generations for some k, or in all preceding generations. Each of these scaling approaches has the capacity to dampen the variation in f, thereby retaining within-generation diversity and increasing the potential to find the global optimum.

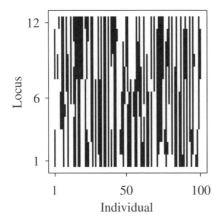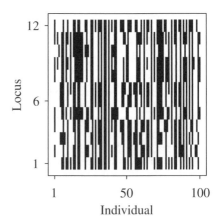

Fig. 3.8 Chromosomes for Problem 3.5. Simulated data on 12 loci are available for 100 individuals. For each locus, the source chromosome from the heterozygous parent is encoded in black or white, analogously to Figure 3.1 in Example 3.1. The left panel shows the data arranged according to the true locus ordering, whereas the right panel shows the data arranged by locus label as they would be recorded during data collection.

Compare and comment on the results for your chosen methods.

(d) Apply a steady-state genetic algorithm, with the generation gap $G = 1$. Compare with the canonical option of distinct, nonoverlapping generations.

(e) Implement the following crossover approach, termed *uniform crossover* [527]: Each locus in the offspring is filled with an allele independently selected at random from the alleles expressed in that position in the parents.

3.5 Consider the genetic mapping example introduced in Example 3.1. Figure 3.8 shows some data for 100 simulated data sequences for a chromosome of length 12. The left panel of this figure shows the data under the true genetic map ordering, and the right panel shows the actual data, with the ordering unknown to the analyst. The data are available from the website for this book.

(a) Apply a random starts local search approach to estimate the genetic map (i.e., the ordering and the genetic distances). Let neighborhoods consist of 20 orderings that differ from the current ordering by randomly swapping the placement of two alleles. Move to the best candidate in the neighborhood, thereby taking a random descent step. Begin with a small number of starts of limited length, to gauge the computational difficulty of the problem; then report the best results you obtained within reasonable limits on the computational burden. Comment on your results, the performance of the algorithm, and ideas for improved search. (Hint: Note that the orderings $(\theta_{j_1}, \theta_{j_2}, \ldots, \theta_{j_{12}})$ and $(\theta_{j_{12}}, \theta_{j_{11}}, \ldots, \theta_{j_1})$ represent identical chromosomes read from either end.)

(b) Apply an algorithm for random starts local search via steepest descent to estimate the genetic map. Comment on your results and the performance of the algorithm. This problem is computationally demanding and may require a fast computer.

3.6 Consider the genetic mapping data described in Problem 3.5.

(a) Apply a genetic algorithm to estimate the genetic map (i.e., the ordering and the genetic distances). Use the order crossover method. Begin with a small run to gauge the computational difficulty of the problem, then report your results for a run using reasonable limits on the computational burden. Comment on your results, the performance of the algorithm, and ideas for improved search.

(b) Compare the speed of fitness improvements achieved with the order crossover and the edge-recombination crossover strategies.

(c) Attempt any other heuristic search method for these data. Describe your implementation, its speed, and the results.

3.7 The website for this book also includes a second synthetic dataset for a genetic mapping problem. For these data, there are 30 chromosomes. Attempt one or more heuristic search methods for these data. Describe your implementation, the results, and the nature of any problems you encounter. The true ordering used to simulate the data is also given for this dataset. Although the true ordering may not be the MLE, how close is your best ordering to the true ordering? How much larger is this problem than the one examined in the previous problem?

3.8 Thirteen chemical measurements were carried out on each of 178 wines from three regions of Italy [47]. These data are available from the website for this book. Using one or more heuristic search methods from this chapter, partition the wines into three groups for which the total of the within-group sum of squares is minimal. Comment on your work and the results. This is a search problem of size 3^p where $p = 178$. If you have access to standard cluster analysis routines, check your results using a standard method like that of Hartigan and Wong [276].

4

EM Optimization Methods

The EM algorithm is an iterative optimization strategy motivated by a notion of missingness and by consideration of the conditional distribution of what is missing given what is observed. The strategy's statistical foundations and effectiveness in a variety of statistical problems were shown in a seminal paper by Dempster, Laird, and Rubin [130]. Other references on EM and related methods include [349, 354, 380, 387, 530]. The popularity of the EM algorithm stems from how simple it can be to implement and how reliably it can find the global optimum through stable, uphill steps.

In a frequentist setting, we may conceive of observed data generated from random variables $\mathbf{X}$ along with missing or unobserved data from random variables $\mathbf{Z}$. We envision complete data generated from $\mathbf{Y} = (\mathbf{X}, \mathbf{Z})$. Given observed data $\mathbf{x}$, we wish to maximize a likelihood $L(\boldsymbol{\theta}|\mathbf{x})$. Often it will be difficult to work with this likelihood and easier to work with the densities of $\mathbf{Y}|\boldsymbol{\theta}$ and $\mathbf{Z}|(\mathbf{x}, \boldsymbol{\theta})$. The EM algorithm sidesteps direct consideration of $L(\boldsymbol{\theta}|\mathbf{x})$ by working with these easier densities.

In a Bayesian application, interest often focuses on estimating the mode of a posterior distribution $f(\boldsymbol{\theta}|\mathbf{x})$. Again, optimization can sometimes be simplified by consideration of unobserved random variables $\boldsymbol{\psi}$ in addition to the parameters of interest, $\boldsymbol{\theta}$.

The missing data may not truly be missing: They may be only a conceptual ploy that simplifies the problem. In this case, $\mathbf{Z}$ is often referred to as *latent*. It may seem counterintuitive that optimization sometimes can be simplified by introducing this new element into the problem. However, examples in this chapter and its references illustrate the potential benefit of this approach. In some cases, the analyst must draw upon his or her creativity and cleverness to invent effective latent variables; in other cases, there is a natural choice.

4.1   MISSING DATA, MARGINALIZATION, AND NOTATION

Whether $\mathbf{Z}$ is considered latent or missing, it may viewed as having been removed from the complete $\mathbf{Y}$ through the application of some many-to-fewer mapping, $\mathbf{X} = M(\mathbf{Y})$. Let $f_{\mathbf{X}}(\mathbf{x}|\boldsymbol{\theta})$ and $f_{\mathbf{Y}}(\mathbf{y}|\boldsymbol{\theta})$ denote the densities of the observed data and the complete data, respectively. The latent- or missing-data assumption amounts to a marginalization model in which we observe $\mathbf{X}$ having density $f_{\mathbf{X}}(\mathbf{x}|\boldsymbol{\theta}) = \int_{\{\mathbf{y}:M(\mathbf{y})=\mathbf{x}\}} f_{\mathbf{Y}}(\mathbf{y}|\boldsymbol{\theta}) \, d\mathbf{y}$. Note that the conditional density of the missing data given the observed data is $f_{\mathbf{Z}|\mathbf{X}}(\mathbf{z}|\mathbf{x}, \boldsymbol{\theta}) = f_{\mathbf{Y}}(\mathbf{y}|\boldsymbol{\theta})/f_{\mathbf{X}}(\mathbf{x}|\boldsymbol{\theta})$.

In Bayesian applications focusing on the posterior density for parameters of interest, $\boldsymbol{\theta}$, there are two manners in which we may consider the posterior to represent a marginalization of a broader problem. First, it may be sensible to view the likelihood $L(\boldsymbol{\theta}|\mathbf{x})$ as a marginalization of the complete-data likelihood $L(\boldsymbol{\theta}|\mathbf{y}) = L(\boldsymbol{\theta}|\mathbf{x}, \mathbf{z})$. In this case the missing data are $\mathbf{z}$, and we use the same sort of notation as above. Secondly, we may consider there to be missing parameters $\boldsymbol{\psi}$, whose inclusion simplifies Bayesian calculations even though $\boldsymbol{\psi}$ is of no interest itself. Fortunately, under the Bayesian paradigm there is no practical distinction between these two cases. Since $\mathbf{Z}$ and $\boldsymbol{\psi}$ are both missing random quantities, it matters little whether we use notation that suggests the missing variables to be unobserved data or parameters. In cases where we adopt the frequentist notation, the reader may replace the likelihood and $\mathbf{Z}$ by the posterior and $\boldsymbol{\psi}$, respectively, to consider the Bayesian point of view.

In the literature about EM, it is traditional to adopt notation that reverses the roles of $\mathbf{X}$ and $\mathbf{Y}$ compared to our usage. We diverge from tradition, using $\mathbf{X} = \mathbf{x}$ to represent observed data as everywhere else in this book.

4.2   THE EM ALGORITHM

The EM algorithm iteratively seeks to maximize $L(\boldsymbol{\theta}|\mathbf{x})$ with respect to $\boldsymbol{\theta}$. Let $\boldsymbol{\theta}^{(t)}$ denote the estimated maximizer at iteration t, for $t = 0, 1, \ldots$. Define $Q(\boldsymbol{\theta}|\boldsymbol{\theta}^{(t)})$ to be the expectation of the joint log likelihood for the complete data, conditional on the observed data $\mathbf{X} = \mathbf{x}$. Namely,

$$Q(\boldsymbol{\theta}|\boldsymbol{\theta}^{(t)}) = \mathrm{E}\left\{ \log L(\boldsymbol{\theta}|\mathbf{Y}) \mid \mathbf{x}, \boldsymbol{\theta}^{(t)} \right\} \tag{4.1}$$

$$= \mathrm{E}\left\{ \log f_{\mathbf{Y}}(\mathbf{y}|\boldsymbol{\theta}) \mid \mathbf{x}, \boldsymbol{\theta}^{(t)} \right\} \tag{4.2}$$

$$= \int \left[\log f_{\mathbf{Y}}(\mathbf{y}|\boldsymbol{\theta}) \right] f_{\mathbf{Z}|\mathbf{X}}(\mathbf{z}|\mathbf{x}, \boldsymbol{\theta}^{(t)}) \, d\mathbf{z}, \tag{4.3}$$

where (4.3) emphasizes that $\mathbf{Z}$ is the only random part of $\mathbf{Y}$ once we are given $\mathbf{X} = \mathbf{x}$.

EM is initiated from $\boldsymbol{\theta}^{(0)}$ then alternates between two steps: E for expectation and M for maximization. The algorithm is summarized as:

1. **E step**: Compute $Q(\boldsymbol{\theta}|\boldsymbol{\theta}^{(t)})$.

2. **M step**: Maximize $Q(\boldsymbol{\theta}|\boldsymbol{\theta}^{(t)})$ with respect to $\boldsymbol{\theta}$. Set $\boldsymbol{\theta}^{(t+1)}$ equal to the maximizer of Q.

3. Return to the E step unless a stopping criterion has been met.

Stopping criteria for optimization problems are discussed in Chapter 2. In the present case, such criteria are usually built upon $(\boldsymbol{\theta}^{(t+1)} - \boldsymbol{\theta}^{(t)})^T(\boldsymbol{\theta}^{(t+1)} - \boldsymbol{\theta}^{(t)})$ or $|Q(\boldsymbol{\theta}^{(t+1)}|\boldsymbol{\theta}^{(t)}) - Q(\boldsymbol{\theta}^{(t)}|\boldsymbol{\theta}^{(t)})|$.

Example 4.1 (Simple exponential density) To understand the EM notation, consider a trivial example where $Y_1, Y_2 \sim$ i.i.d. $\mathrm{Exp}(\theta)$. Suppose $y_1 = 5$ is observed but the value y_2 is missing. The complete-data log likelihood function is $\log L(\theta|\mathbf{y}) = \log f_{\mathbf{Y}}(\mathbf{y}|\theta) = 2\log\{\theta\} - \theta y_1 - \theta y_2$. Taking the conditional expectation of $\log L(\theta|\mathbf{Y})$ yields $Q(\theta|\theta^{(t)}) = 2\log\{\theta\} - 5\theta - \theta/\theta^{(t)}$, since $\mathrm{E}\{Y_2|y_1, \theta^{(t)}\} = \mathrm{E}\{Y_2|\theta^{(t)}\} = 1/\theta^{(t)}$ follows from independence. The maximizer of $Q(\theta|\theta^{(t)})$ with respect to θ is easily found to be the root of $2/\theta - 5 - 1/\theta^{(t)} = 0$. Solving for θ provides the updating equation $\theta^{(t+1)} = \frac{2\theta^{(t)}}{5\theta^{(t)}+1}$. Note here that the E step and M step do not need to be rederived at each iteration: Iterative application of the updating formula starting from some initial value provides estimates that converge to $\hat{\theta} = 0.2$.

This example is not realistic. The maximum likelihood estimate of θ from the observed data can be determined from elementary analytic methods without reliance on any fancy numerical optimization strategy like EM. More importantly, we will learn that taking the required expectation is trickier in real applications, because one needs to know the conditional distribution of the complete data given the missing data. $\qquad\square$

Example 4.2 (Peppered moths) The peppered moth, *Biston betularia*, presents a fascinating story of evolution and industrial pollution [242]. The coloring of these moths is believed to be determined by a single gene with three possible alleles, which we denote C, I, and T. Of these, C is dominant to I, and T is recessive to I. Thus the genotypes CC, CI, and CT result in the *carbonaria* phenotype, which exhibits solid black coloring. The genotype TT results in the *typica* phenotype, which exhibits light-colored patterned wings. The genotypes II and IT produce an intermediate phenotype called *insularia*, which varies widely in appearance but is generally mottled with intermediate color. Thus, there are six possible genotypes, but only three phenotypes are measurable in field work.

In the United Kingdom and North America, the *carbonaria* phenotype nearly replaced the paler phenotypes in areas affected by coal-fired industries. This change in allele frequencies in the population is cited as an instance where we may observe microevolution occurring on a human time scale. The theory (supported by experiments) is that "differential predation by birds on moths that are variously conspicuous against backgrounds of different reflectance" induces selectivity that favors the *carbonaria* phenotype in times and regions where sooty, polluted conditions reduce the reflectance of the surface of tree bark on which the moths rest [242]. Not surprisingly, when improved environmental standards reduced pollution, the prevalence of the lighter-colored phenotypes increased and that of *carbonaria* plummeted.

Thus, it is of interest to monitor the allele frequencies of C, I, and T over time to provide insight on microevolutionary processes. Further, trends in these frequencies also provide an interesting biological marker to monitor air quality. Within a sufficiently short time period, an approximate model for allele frequencies can be built from the Hardy–Weinberg principle that each genotype frequency in a population in Hardy–Weinberg equilibrium should equal the product of the corresponding allele frequencies, or double that amount when the two alleles differ (to account for uncertainty in the parental source) [14, 275]. Thus, if the allele frequencies in the population are p_C, p_I, and p_T, then the genotype frequencies should be p_C^2, $2p_Cp_I$, $2p_Cp_T$, p_I^2, $2p_Ip_T$, and p_T^2 for genotypes CC, CI, CT, II, IT, and TT, respectively. Note that $p_C + p_I + p_T = 1$.

Suppose we capture n moths, of which there are n_C, n_I, and n_T of the *carbonaria*, *insularia*, and *typica* phenotypes, respectively. Thus $n = n_C + n_I + n_T$. Since each moth has two alleles in the gene in question, there are $2n$ total alleles in the sample. If we knew the genotype of each moth rather than merely its phenotype, we could generate genotype counts n_{CC}, n_{CI}, n_{CT}, n_{II}, n_{IT}, and n_{TT}, from which allele frequencies could easily be tabulated. For example, each moth with genotype CI contributes one C allele and one I allele, whereas a II moth contributes two I alleles. Such allele counts would immediately provide estimates of p_C, p_I, and p_T. It is far less clear how to estimate the allele frequencies from the phenotype counts alone.

In the EM notation, the observed data are $\mathbf{x} = (n_C, n_I, n_T)$ and the complete data are $\mathbf{y} = (n_{CC}, n_{CI}, n_{CT}, n_{II}, n_{IT}, n_{TT})$. The mapping from the complete data to the observed data is $\mathbf{x} = M(\mathbf{y}) = (n_{CC}+n_{CI}+n_{CT}, n_{II}+n_{IT}, n_{TT})$. We wish to estimate the allele probabilities, p_C, p_I, and p_T. Since $p_T = 1 - p_C - p_I$, the parameter vector for this problem is $\mathbf{p} = (p_C, p_I)$, but for notational brevity we often refer to p_T in what follows.

The complete data log likelihood function is multinomial:

$$\log f_{\mathbf{Y}}(\mathbf{y}|\mathbf{p}) = n_{CC} \log\{p_C^2\} + n_{CI} \log\{2p_Cp_I\} + n_{CT} \log\{2p_Cp_T\}$$
$$+ n_{II} \log\{p_I^2\} + n_{IT} \log\{2p_Ip_T\} + n_{TT} \log\{p_T^2\}$$
$$+ \log \begin{pmatrix} n \\ n_{CC} \quad n_{CI} \quad n_{CT} \quad n_{II} \quad n_{IT} \quad n_{TT} \end{pmatrix}. \qquad (4.4)$$

The complete data are not all observed. Let $\mathbf{Y} = (N_{CC}, N_{CI}, N_{CT}, N_{II}, N_{IT}, n_{TT})$, since we know $N_{TT} = n_{TT}$ but the other frequencies are not directly observed. To calculate $Q(\mathbf{p}|\mathbf{p}^{(t)})$, notice that conditional on n_C and a parameter vector $\mathbf{p}^{(t)} = (p_C^{(t)}, p_I^{(t)})$, the latent counts for the three *carbonaria* genotypes have a three-cell multinomial distribution with count parameter n_C and cell probabilities proportional to $(p_C^{(t)})^2$, $2p_C^{(t)}p_I^{(t)}$, and $2p_C^{(t)}p_T^{(t)}$. A similar result holds for the two insularia cells.

Thus the expected values of the first five random parts of (4.4) are

$$E\{N_{CC}|n_C, n_I, n_T, \mathbf{p}^{(t)}\} = n_{CC}^{(t)} = \frac{n_C(p_C^{(t)})^2}{(p_C^{(t)})^2 + 2p_C^{(t)}p_I^{(t)} + 2p_C^{(t)}p_T^{(t)}}, \tag{4.5}$$

$$E\{N_{CI}|n_C, n_I, n_T, \mathbf{p}^{(t)}\} = n_{CI}^{(t)} = \frac{2n_C p_C^{(t)}p_I^{(t)}}{(p_C^{(t)})^2 + 2p_C^{(t)}p_I^{(t)} + 2p_C^{(t)}p_T^{(t)}}, \tag{4.6}$$

$$E\{N_{CT}|n_C, n_I, n_T, \mathbf{p}^{(t)}\} = n_{CT}^{(t)} = \frac{2n_C p_C^{(t)}p_T^{(t)}}{(p_C^{(t)})^2 + 2p_C^{(t)}p_I^{(t)} + 2p_C^{(t)}p_T^{(t)}}, \tag{4.7}$$

$$E\{N_{II}|n_C, n_I, n_T, \mathbf{p}^{(t)}\} = n_{II}^{(t)} = \frac{n_I(p_I^{(t)})^2}{(p_I^{(t)})^2 + 2p_I^{(t)}p_T^{(t)}}, \tag{4.8}$$

$$E\{N_{IT}|n_C, n_I, n_T, \mathbf{p}^{(t)}\} = n_{IT}^{(t)} = \frac{2n_I p_I^{(t)}p_T^{(t)}}{(p_I^{(t)})^2 + 2p_I^{(t)}p_T^{(t)}}. \tag{4.9}$$

Finally, we know $n_{TT} = n_T$, where n_T is observed. The multinomial coefficient in the likelihood has a conditional expectation, say $k(n_I, n_I, n_T, \mathbf{p}^{(t)})$, that does not depend on $\mathbf{p}$. Thus, we have found

$$\begin{aligned} Q(\mathbf{p}|\mathbf{p}^{(t)}) = & n_{CC}^{(t)} \log\{p_C^2\} + n_{CI}^{(t)} \log\{2p_C p_I\} \\ & + n_{CT}^{(t)} \log\{2p_C p_T\} + n_{II}^{(t)} \log\{p_I^2\} \\ & + n_{IT}^{(t)} \log\{2p_I p_T\} + n_{TT} \log\{p_T^2\} + k(n_C, n_I, n_T, \mathbf{p}^{(t)}). \end{aligned} \tag{4.10}$$

Recalling $p_T = 1 - p_C - p_I$ and differentiating with respect to p_C and p_I yields

$$\frac{dQ(\mathbf{p}|\mathbf{p}^{(t)})}{dp_C} = \frac{2n_{CC}^{(t)} + n_{CI}^{(t)} + n_{CT}^{(t)}}{p_C} - \frac{2n_{TT}^{(t)} + n_{CT}^{(t)} + n_{IT}^{(t)}}{1 - p_C - p_I}, \tag{4.11}$$

$$\frac{dQ(\mathbf{p}|\mathbf{p}^{(t)})}{dp_I} = \frac{2n_{II}^{(t)} + n_{IT}^{(t)} + n_{CI}^{(t)}}{p_I} - \frac{2n_{TT}^{(t)} + n_{CT}^{(t)} + n_{IT}^{(t)}}{1 - p_C - p_I}. \tag{4.12}$$

Setting these derivatives equal to zero and solving for p_C and p_I completes the M step, yielding

$$p_C^{(t+1)} = \frac{2n_{CC}^{(t)} + n_{CI}^{(t)} + n_{CT}^{(t)}}{2n}, \tag{4.13}$$

$$p_I^{(t+1)} = \frac{2n_{II}^{(t)} + n_{IT}^{(t)} + n_{CI}^{(t)}}{2n}, \tag{4.14}$$

$$p_T^{(t+1)} = \frac{2n_{TT}^{(t)} + n_{CT}^{(t)} + n_{IT}^{(t)}}{2n}, \tag{4.15}$$

where the final expression is derived from the constraint that the probabilities sum to one. If the tth latent counts were true, the number of *carbonaria* alleles in the sample would be $2n_{CC}^{(t)} + n_{CI}^{(t)} + n_{CT}^{(t)}$. There are $2n$ total alleles in the sample. Thus, the EM

Table 4.1 EM results for peppered moth example. The diagnostic quantities $R^{(t)}$, $D_C^{(t)}$, and $D_I^{(t)}$ are defined in the text.

t	$p_C^{(t)}$	$p_I^{(t)}$	$R^{(t)}$	$D_C^{(t)}$	$D_I^{(t)}$
0	0.333333	0.333333			
1	0.081994	0.237406	5.7×10^{-1}	0.0425	0.337
2	0.071249	0.197870	1.6×10^{-1}	0.0369	0.188
3	0.070852	0.190360	3.6×10^{-2}	0.0367	0.178
4	0.070837	0.189023	6.6×10^{-3}	0.0367	0.176
5	0.070837	0.188787	1.2×10^{-3}	0.0367	0.176
6	0.070837	0.188745	2.1×10^{-4}	0.0367	0.176
7	0.070837	0.188738	3.6×10^{-5}	0.0367	0.176
8	0.070837	0.188737	6.4×10^{-6}	0.0367	0.176

update consists of setting the elements of $\mathbf{p}^{(t+1)}$ equal to the phenotypic frequencies that would result from the tth latent genotype counts.

Suppose the observed phenotype counts are $n_C = 85$, $n_I = 196$, and $n_T = 341$. Table 4.1 shows how the EM algorithm converges to the MLEs, roughly $\hat{p}_C = 0.07084$, $\hat{p}_I = 0.18874$, and $\hat{p}_T = 0.74043$. Finding a precise estimate of $\hat{p}_I$ is slower than for $\hat{p}_C$, since the likelihood is flatter along the p_I coordinate.

The last three columns of Table 4.1 show convergence diagnostics. A relative convergence criterion,

$$R^{(t)} = \frac{\|\mathbf{p}^{(t)} - \mathbf{p}^{(t-1)}\|}{\|\mathbf{p}^{(t-1)}\|}, \tag{4.16}$$

summarizes the total amount of relative change in $\mathbf{p}^{(t)}$ from one iteration to the next, where $\|\mathbf{z}\| = (\mathbf{z}^T \mathbf{z})^{1/2}$. For illustrative purposes, we also include $D_C^{(t)} = \frac{p_C^{(t)} - \hat{p}_C}{p_C^{(t-1)} - \hat{p}_C}$ and the analogous quantity $D_I^{(t)}$. These ratios quickly converge to constants, confirming that the EM rate of convergence is linear as defined by (2.19). □

Example 4.3 (Bayesian posterior mode) Consider a Bayesian problem with likelihood $L(\boldsymbol{\theta}|\mathbf{x})$, prior $f(\boldsymbol{\theta})$, and missing data or parameters $\mathbf{Z}$. To find the posterior mode, the E step requires

$$Q(\boldsymbol{\theta}|\boldsymbol{\theta}^{(t)}) = \mathrm{E}\{\log\{L(\boldsymbol{\theta}|\mathbf{Y})f(\boldsymbol{\theta})k(\mathbf{Y})\}|\mathbf{x}, \boldsymbol{\theta}^{(t)}\}$$
$$= \mathrm{E}\{\log L(\boldsymbol{\theta}|\mathbf{Y})|\mathbf{x}, \boldsymbol{\theta}^{(t)}\} + \log f(\boldsymbol{\theta}) + \mathrm{E}\{\log k(\mathbf{Y})|\mathbf{x}, \boldsymbol{\theta}^{(t)}\}, \quad (4.17)$$

where the final term in (4.17) is a normalizing constant that can be ignored because Q is to be maximized with respect to $\boldsymbol{\theta}$. This function Q is obtained by simply adding the log prior to the Q function that would be used in a maximum likelihood setting. Unfortunately, the addition of the log prior often makes it more difficult to maximize

Q during the M step. Section 4.3.2 describes a variety of methods for facilitating the M step in difficult situations.   □

4.2.1   Convergence

To investigate the convergence properties of the EM algorithm, we begin by showing that each maximization step increases the observed-data log likelihood, $l(\boldsymbol{\theta}|\mathbf{x})$. First note that the log of the observed-data density can be reexpressed as

$$\log f_{\mathbf{X}}(\mathbf{x}|\boldsymbol{\theta}) = \log f_{\mathbf{Y}}(\mathbf{y}|\boldsymbol{\theta}) - \log f_{\mathbf{Z}|\mathbf{X}}(\mathbf{z}|\mathbf{x},\boldsymbol{\theta}). \tag{4.18}$$

Therefore,

$$\mathrm{E}\{\log f_{\mathbf{X}}(\mathbf{x}|\boldsymbol{\theta})|\mathbf{x},\boldsymbol{\theta}^{(t)}\} = \mathrm{E}\{\log f_{\mathbf{Y}}(\mathbf{y}|\boldsymbol{\theta})|\mathbf{x},\boldsymbol{\theta}^{(t)}\} - \mathrm{E}\{f_{\mathbf{Z}|\mathbf{X}}(\mathbf{z}|\mathbf{x},\boldsymbol{\theta})|\mathbf{x},\boldsymbol{\theta}^{(t)}\},$$

where the expectations are taken with respect to the distribution of $\mathbf{Z}|(\mathbf{x},\boldsymbol{\theta}^{(t)})$. Thus,

$$\log f_{\mathbf{X}}(\mathbf{x}|\boldsymbol{\theta}) = Q(\boldsymbol{\theta}|\boldsymbol{\theta}^{(t)}) - H(\boldsymbol{\theta}|\boldsymbol{\theta}^{(t)}), \tag{4.19}$$

where

$$H(\boldsymbol{\theta}|\boldsymbol{\theta}^{(t)}) = \mathrm{E}\left\{\log f_{\mathbf{Z}|\mathbf{X}}(\mathbf{Z}|\mathbf{x},\boldsymbol{\theta})\,\big|\,\mathbf{x},\boldsymbol{\theta}^{(t)}\right\}. \tag{4.20}$$

The importance of (4.19) becomes apparent after we show that $H(\boldsymbol{\theta}|\boldsymbol{\theta}^{(t)})$ is maximized with respect to $\boldsymbol{\theta}$ when $\boldsymbol{\theta} = \boldsymbol{\theta}^{(t)}$. To see this, write

$$H(\boldsymbol{\theta}^{(t)}|\boldsymbol{\theta}^{(t)}) - H(\boldsymbol{\theta}|\boldsymbol{\theta}^{(t)})$$

$$= \mathrm{E}\left\{\log f_{\mathbf{Z}|\mathbf{X}}(\mathbf{Z}|\mathbf{x},\boldsymbol{\theta}^{(t)}) - \log f_{\mathbf{Z}|\mathbf{X}}(\mathbf{Z}|\mathbf{x},\boldsymbol{\theta})\,\Big|\,\mathbf{x},\boldsymbol{\theta}^{(t)}\right\}$$

$$= \int -\log\left[\frac{f_{\mathbf{Z}|\mathbf{X}}(\mathbf{z}|\mathbf{x},\boldsymbol{\theta})}{f_{\mathbf{Z}|\mathbf{X}}(\mathbf{z}|\mathbf{x},\boldsymbol{\theta}^{(t)})}\right] f_{\mathbf{Z}|\mathbf{X}}(\mathbf{z}|\mathbf{x},\boldsymbol{\theta}^{(t)})\,d\mathbf{z}$$

$$\geq -\log\int f_{\mathbf{Z}|\mathbf{X}}(\mathbf{z}|\mathbf{x},\boldsymbol{\theta})\,d\mathbf{z}$$

$$= 0. \tag{4.21}$$

Equation (4.21) follows from an application of Jensen's inequality, since $-\log u$ is strictly convex in u.

Thus, any $\boldsymbol{\theta} \neq \boldsymbol{\theta}^{(t)}$ makes $H(\boldsymbol{\theta}|\boldsymbol{\theta}^{(t)})$ smaller than $H(\boldsymbol{\theta}^{(t)}|\boldsymbol{\theta}^{(t)})$. In particular, if we choose $\boldsymbol{\theta}^{(t+1)}$ to maximize $Q(\boldsymbol{\theta}|\boldsymbol{\theta}^{(t)})$ with respect to $\boldsymbol{\theta}$, then

$$\log f_{\mathbf{X}}(\mathbf{x}|\boldsymbol{\theta}^{(t+1)}) - \log f_{\mathbf{X}}(\mathbf{x}|\boldsymbol{\theta}^{(t)}) \geq 0, \tag{4.22}$$

since Q increases and H decreases, with strict inequality when $Q(\boldsymbol{\theta}^{(t+1)}|\boldsymbol{\theta}^{(t)}) > Q(\boldsymbol{\theta}^{(t)}|\boldsymbol{\theta}^{(t)})$.

Choosing $\boldsymbol{\theta}^{(t+1)}$ at each iteration to maximize $Q(\boldsymbol{\theta}|\boldsymbol{\theta}^{(t)})$ with respect to $\boldsymbol{\theta}$ constitutes the standard EM algorithm. If instead we simply select any $\boldsymbol{\theta}^{(t+1)}$ for which

$Q(\theta^{(t+1)}|\theta^{(t)}) > Q(\theta^{(t)}|\theta^{(t)})$, then the resulting algorithm is called generalized EM, or GEM. In either case, a step that increases Q increases the log likelihood. Conditions under which this guaranteed ascent ensures convergence to a MLE are explored in [54, 576].

Having established this result, we next consider the order of convergence for the method. The EM algorithm defines a mapping $\theta^{(t+1)} = \Psi(\theta^{(t)})$ where the function $\Psi(\theta) = (\Psi_1(\theta), \ldots, \Psi_p(\theta))$ and $\theta = (\theta_1, \ldots, \theta_p)$. When EM converges, it converges to a fixed point of this mapping, so $\hat{\theta} = \Psi(\hat{\theta})$. Let $\Psi'(\theta)$ denote the Jacobian matrix whose (i, j)th element is $\frac{d\Psi_i(\theta)}{d\theta_j}$. Taylor series expansion of Ψ yields

$$\theta^{(t+1)} - \hat{\theta} \approx \Psi'(\theta^{(t)})(\theta^{(t)} - \hat{\theta}), \qquad (4.23)$$

since $\theta^{(t+1)} - \hat{\theta} = \Psi(\theta^{(t)}) - \Psi(\hat{\theta})$. Comparing this result with (2.19), we see that the EM algorithm has linear convergence when $p = 1$. For $p > 1$, convergence is still linear provided that the observed information, $-l''(\hat{\theta}|\mathbf{x})$, is positive definite. More precise details regarding convergence are given in [130, 380, 383, 386].

The global rate of EM convergence is defined as

$$\rho = \lim_{t \to \infty} \frac{\|\theta^{(t+1)} - \hat{\theta}\|}{\|\theta^{(t)} - \hat{\theta}\|}. \qquad (4.24)$$

It can be shown that ρ equals the largest eigenvalue of $\Psi'(\hat{\theta})$ when $-l''(\hat{\theta}|\mathbf{x})$ is positive definite. In Sections 4.2.3.1 and 4.2.3.2 we will examine how $\Psi'(\hat{\theta})$ is a matrix of the fractions of missing information. Therefore, ρ effectively serves as a scalar summary of the overall proportion of missing information. Conceptually, the proportion of missing information equals one minus the ratio of the observed information to the information that would be contained in the complete data. Thus, EM suffers slower convergence when the proportion of missing information is larger. The linear convergence of EM can be extremely slow compared to the quadratic convergence of, say, Newton's method, particularly when the fraction of missing information is large. However, the ease of implementation and the stable ascent of EM are often very attractive despite its slow convergence. Section 4.3.3 discusses methods for accelerating EM convergence.

To further understand how EM works, note from (4.21) that

$$l(\theta|\mathbf{x}) \geq Q(\theta|\theta^{(t)}) + l(\theta^{(t)}|\mathbf{x}) - Q(\theta^{(t)}|\theta^{(t)}) = G(\theta|\theta^{(t)}). \qquad (4.25)$$

Since the last two terms in $G(\theta|\theta^{(t)})$ are independent of θ, the functions Q and G are maximized at the same θ. Further, G is tangent to l at $\theta^{(t)}$, and lies everywhere below l. We say that G is a *minorizing function* for l. The EM strategy transfers optimization from l to the surrogate function G (effectively to Q), which is more convenient to maximize. The maximizer of G provides an increase in l. This idea is illustrated in Figure 4.1. Each E step amounts to forming the minorizing function G, and each M step amounts to maximizing it to provide an uphill step.

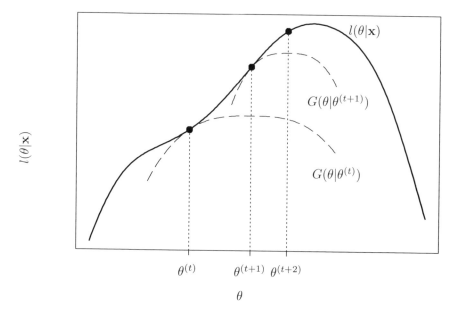

Fig. 4.1 One-dimensional illustration of EM algorithm as a minorization or optimization transfer strategy.

Temporarily replacing l by a minorizing function is an example of a more general strategy known as *optimization transfer*. Links to the EM algorithm and other statistical applications of optimization transfer are surveyed in [350]. In mathematical applications where it is standard to pose optimizations as minimizations, one typically refers to *majorization*, as one could achieve by majorizing the negative log likelihood using $-G(\theta|\theta^{(t)})$.

4.2.2   Usage in exponential families

When the complete data are modeled to have a distribution in the exponential family, the density of the data can be written as $f(\mathbf{y}|\theta) = c_1(\mathbf{y})c_2(\theta)\exp\{\theta^T\mathbf{s}(\mathbf{y})\}$, where θ is a vector of natural parameters and $\mathbf{s}(\mathbf{y})$ is a vector of sufficient statistics. In this case, the E step finds

$$Q(\theta|\theta^{(t)}) = k + \log c_2(\theta) + \int \theta^T \mathbf{s}(\mathbf{y}) f_{\mathbf{Z}|\mathbf{X}}(\mathbf{z}|\mathbf{x}, \theta^{(t)})\, d\mathbf{z}, \qquad (4.26)$$

where k is a quantity that does not depend on θ. To carry out the M step, set the gradient of $Q(\theta|\theta^{(t)})$ with respect to θ equal to zero. This yields

$$\frac{-\mathbf{c}_2'(\theta)}{c_2(\theta)} = \int \mathbf{s}(\mathbf{y}) f_{\mathbf{Z}|\mathbf{X}}(\mathbf{z}|\mathbf{x}, \theta^{(t)})\, d\mathbf{z} \qquad (4.27)$$

after rearranging terms and adopting the obvious notational shortcut to vectorize the integral of a vector. It is straightforward to show that $c'_2(\boldsymbol{\theta}) = -c_2(\boldsymbol{\theta})\mathrm{E}\{s(\mathbf{Y})|\boldsymbol{\theta}\}$. Therefore, (4.27) implies that the M step is completed by setting $\boldsymbol{\theta}^{(t+1)}$ equal to the $\boldsymbol{\theta}$ that solves

$$\mathrm{E}\{s(\mathbf{Y})|\boldsymbol{\theta}\} = \int s(\mathbf{y})f_{\mathbf{Z}|\mathbf{X}}(\mathbf{z}|\mathbf{x},\boldsymbol{\theta}^{(t)})\,d\mathbf{z}. \tag{4.28}$$

Aside from replacing $\boldsymbol{\theta}^{(t)}$ with $\boldsymbol{\theta}^{(t+1)}$, the form of $Q(\boldsymbol{\theta}|\boldsymbol{\theta}^{(t)})$ is unchanged for the next E step, and the next M step solves the same optimization problem. Therefore, the EM algorithm for exponential families consists of:

1. **E step**: Compute the expected values of the sufficient statistics for the complete data, given the observed data and using the current parameter guesses, $\boldsymbol{\theta}^{(t)}$. Let $s^{(t)} = \mathrm{E}\{s(\mathbf{Y})|\mathbf{x},\boldsymbol{\theta}^{(t)}\} = \int s(\mathbf{y})f_{\mathbf{Z}|\mathbf{X}}(\mathbf{z}|\mathbf{x},\boldsymbol{\theta}^{(t)})\,d\mathbf{z}$.

2. **M step**: Set $\boldsymbol{\theta}^{(t+1)}$ to the value which makes the unconditional expectation of the sufficient statistics for the complete data equal to $s^{(t)}$. In other words, $\boldsymbol{\theta}^{(t+1)}$ solves $\mathrm{E}\{s(\mathbf{Y})|\boldsymbol{\theta})\} = s^{(t)}$.

3. Return to the E step unless a convergence criterion has been met.

Example 4.4 (Peppered moths, continued)   The complete data in Example 4.2 arise from a multinomial distribution, which is in the exponential family. The sufficient statistics are, say, the first five genotype counts (with the sixth derived from the constraint that the counts total n), and the natural parameters are the corresponding log probabilities seen in (4.4). The first three conditional expectations for the E step are $s_{CC}^{(t)} = n_{CC}^{(t)}$, $s_{CI}^{(t)} = n_{CI}^{(t)}$, and $s_{CT}^{(t)} = n_{CT}^{(t)}$, borrowing notation from (4.5)–(4.9) and indexing the components of $s^{(t)}$ in the obvious way. The unconditional expectations of the first three sufficient statistics are np_C^2, $2np_Cp_I$, and $2np_Cp_T$. Equating these three expressions with the conditional expectations given above and solving for p_C constitutes the M step for p_C. Summing the three equations gives $np_C^2 + 2np_Cp_I + 2np_Cp_T = n_{CC}^{(t)} + n_{CI}^{(t)} + n_{CT}^{(t)}$, which reduces to the update given in (4.13). EM updates for p_I and p_T are found analogously, on noting the constraint that the three probabilities sum to 1. □

4.2.3   Variance estimation

In a maximum likelihood setting, the EM algorithm is used to find a MLE but does not automatically produce an estimate of the covariance matrix of the MLEs. Typically, we would use the asymptotic normality of the MLEs to justify seeking an estimate of the Fisher information matrix. One way to estimate the covariance matrix, therefore, is to compute the observed information, $-l''(\hat{\boldsymbol{\theta}}|\mathbf{x})$, where l'' is the Hessian matrix of second derivatives of $\log L(\boldsymbol{\theta}|\mathbf{x})$.

In a Bayesian setting, an estimate of the posterior covariance matrix for $\boldsymbol{\theta}$ can be motivated by noting the asymptotic normality of the posterior [194]. This requires the Hessian of the log posterior density.

In some cases, the Hessian may be computed analytically. In other cases, the Hessian may be difficult to derive or code. In these instances, a variety of other methods are available to simplify the estimation of the covariance matrix.

Of the options described below, the SEM algorithm is easy to implement while generally providing fast, reliable results. Even easier is bootstrapping, although for very complex problems the computational burden of nested looping may be prohibitive. These two approaches are recommended, yet the other alternatives can also be useful in some settings.

4.2.3.1 *Louis's method*

Taking second partial derivatives of (4.19) and negating both sides yields

$$-l''(\boldsymbol{\theta}|\mathbf{x}) = -\mathbf{Q}''(\boldsymbol{\theta}|\boldsymbol{\omega})|_{\boldsymbol{\omega}=\boldsymbol{\theta}} + \mathbf{H}''(\boldsymbol{\theta}|\boldsymbol{\omega})|_{\boldsymbol{\omega}=\boldsymbol{\theta}}, \tag{4.29}$$

where the primes on $\mathbf{Q}''$ and $\mathbf{H}''$ denote derivatives with respect to the first argument, namely $\boldsymbol{\theta}$.

Equation (4.29) can be rewritten as

$$\hat{\mathbf{i}}_{\mathbf{X}}(\boldsymbol{\theta}) = \hat{\mathbf{i}}_{\mathbf{Y}}(\boldsymbol{\theta}) - \hat{\mathbf{i}}_{\mathbf{Z}|\mathbf{X}}(\boldsymbol{\theta}), \tag{4.30}$$

where $\hat{\mathbf{i}}_{\mathbf{X}}(\boldsymbol{\theta}) = -l''(\boldsymbol{\theta}|\mathbf{x})$ is the observed information, and $\hat{\mathbf{i}}_{\mathbf{Y}}(\boldsymbol{\theta})$ and $\hat{\mathbf{i}}_{\mathbf{Z}|\mathbf{X}}(\boldsymbol{\theta})$ will be called the complete information and the missing information, respectively. Interchanging integration and differentiation (when possible), we have

$$\hat{\mathbf{i}}_{\mathbf{Y}}(\boldsymbol{\theta}) = -\mathbf{Q}''(\boldsymbol{\theta}|\boldsymbol{\omega})|_{\boldsymbol{\omega}=\boldsymbol{\theta}} = -\mathrm{E}\{l''(\boldsymbol{\theta}|\mathbf{Y})|\mathbf{x}, \boldsymbol{\theta}\}, \tag{4.31}$$

which is reminiscent of the Fisher information defined in (1.28). This motivates calling $\hat{\mathbf{i}}_{\mathbf{Y}}(\boldsymbol{\theta})$ the complete information. A similar argument holds for $-\mathbf{H}''$. Equation (4.30), stating that the observed information equals the complete information minus the missing information, is a result termed the *missing information principle* [363, 574].

The missing information principle can be used to obtain an estimated covariance matrix for $\hat{\boldsymbol{\theta}}$. It can be shown that

$$\hat{\mathbf{i}}_{\mathbf{Z}|\mathbf{X}}(\boldsymbol{\theta}) = \mathrm{var}\left\{\frac{d \log f_{\mathbf{Z}|\mathbf{X}}(\mathbf{Z}|\mathbf{x}, \boldsymbol{\theta})}{d\boldsymbol{\theta}}\right\} \tag{4.32}$$

where the variance is taken with respect to $f_{\mathbf{Z}|\mathbf{X}}$. Further, since the expected score is zero at $\hat{\boldsymbol{\theta}}$,

$$\hat{\mathbf{i}}_{\mathbf{Z}|\mathbf{X}}(\hat{\boldsymbol{\theta}}) = \int \mathbf{S}_{\mathbf{Z}|\mathbf{X}}(\hat{\boldsymbol{\theta}}) \mathbf{S}_{\mathbf{Z}|\mathbf{X}}(\hat{\boldsymbol{\theta}})^T f_{\mathbf{Z}|\mathbf{X}}(\mathbf{z}|\mathbf{X}, \hat{\boldsymbol{\theta}}) \, d\mathbf{z}, \tag{4.33}$$

where $\mathbf{S}_{\mathbf{Z}|\mathbf{X}}(\boldsymbol{\theta}) = \frac{d \log f_{\mathbf{Z}|\mathbf{X}}(\mathbf{z}|\mathbf{x}, \boldsymbol{\theta})}{d\boldsymbol{\theta}}$.

The missing information principle enables us to express $\hat{\mathbf{i}}_{\mathbf{X}}(\boldsymbol{\theta})$ in terms of the complete-data likelihood and the conditional density of the missing data given the observed data, while avoiding calculations involving the presumably complicated marginal likelihood of the observed data. This approach can be easier to derive and code in some instances, but it is not always significantly simpler than direct calculation of $-l''(\hat{\theta}|\mathbf{x})$.

If $\hat{\mathbf{i}}_{\mathbf{Y}}(\theta)$ or $\hat{\mathbf{i}}_{\mathbf{Z}|\mathbf{X}}(\theta)$ is difficult to compute analytically, it may be estimated via the Monte Carlo method (see Chapter 6). For example, the simplest Monte Carlo estimate of $\hat{\mathbf{i}}_{\mathbf{Y}}(\theta)$ is

$$\frac{1}{m} \sum_{i=1}^{m} -\frac{d^2 \log f_{\mathbf{Y}}(\mathbf{y}_i|\theta)}{d\theta \cdot d\theta}, \tag{4.34}$$

where for $i = 1, \ldots, m$, the $\mathbf{y}_i = (\mathbf{x}, \mathbf{z}_i)$ are simulated complete datasets consisting of the observed data and i.i.d. imputed missing data values $\mathbf{z}_i$ drawn from $f_{\mathbf{Z}|\mathbf{X}}$. Similarly, a simple Monte Carlo estimate of $\hat{\mathbf{i}}_{\mathbf{Z}|\mathbf{X}}(\theta)$ is the sample variance of $-\frac{d \log f_{\mathbf{Z}|\mathbf{X}}(\mathbf{z}_i|\mathbf{x}, \theta)}{d\theta}$ values obtained from such a collection of $\mathbf{z}_i$.

Example 4.5 (Censored exponential data) Suppose we attempt to observed complete data under the model $Y_1, \ldots, Y_n \sim$ i.i.d. Exp(λ), but some cases are right-censored. Thus, the observed data are $\mathbf{x} = (\mathbf{x}_1, \ldots, \mathbf{x}_n)$ where $\mathbf{x}_i = (\min(y_i, c_i), \delta_i)$, the c_i are the censoring levels, and $\delta_i = 1$ if $y_i \le c_i$ and $\delta_i = 0$ otherwise.
The complete-data log likelihood is $l(\lambda|y_1, \ldots, y_n) = n \log \lambda - \lambda \sum_{i=1}^{n} y_i$. Thus,

$$Q(\lambda|\lambda^{(t)}) = \mathrm{E}\{l(\lambda|Y_1, \ldots, Y_n)|\mathbf{x}, \lambda^{(t)}\} \tag{4.35}$$

$$= n \log \lambda - \lambda \sum_{i=1}^{n} \mathrm{E}\{Y_i|x_i, \lambda^{(t)}\}$$

$$= n \log \lambda - \lambda \sum_{i=1}^{n} \left[y_i \delta_i + (c_i + 1/\lambda^{(t)})(1 - \delta_i) \right] \tag{4.36}$$

$$= n \log \lambda - \lambda \sum_{i=1}^{n} [y_i \delta_i + c_i(1 - \delta_i)] - C\lambda/\lambda^{(t)}, \tag{4.37}$$

where $C = \sum_{i=1}^{n}(1 - \delta_i)$ denotes the number of censored cases. Note that (4.36) follows from the memoryless property of the exponential distribution. Therefore, $-Q''(\lambda|\lambda^{(t)}) = n/\lambda^2$.
The unobserved outcome for a censored case, Z_i, has density $f_{Z_i|X}(z_i|x, \lambda) = \lambda \exp\{-\lambda(z_i - c_i)\}1_{\{z_i > c_i\}}$. Calculating $\hat{i}_{\mathbf{Z}|\mathbf{X}}(\lambda)$ as in (4.32), we find

$$\frac{d \log f_{\mathbf{Z}|\mathbf{X}}(\mathbf{Z}|\mathbf{x}, \lambda)}{d\lambda} = C/\lambda - \sum_{\{i: \delta_i = 0\}} (Z_i - c_i). \tag{4.38}$$

The variance of this expression with respect to $f_{Z_i|X}$ is

$$\hat{i}_{\mathbf{Z}|\mathbf{X}}(\lambda) = \sum_{\{i: \delta_i = 0\}} \mathrm{var}\{Z_i - c_i\} = C/\lambda^2, \tag{4.39}$$

since $Z_i - c_i$ has an Exp(λ) distribution.

Thus, applying Louis's method,

$$\hat{i}_{\mathbf{X}}(\lambda) = n/\lambda^2 - C/\lambda^2 = U/\lambda^2, \tag{4.40}$$

where $U = \sum_{i=1}^n \delta_i$ denotes the number of uncensored cases. For this elementary example, it is easy to confirm by direct analysis that $-l''(\lambda|\mathbf{x}) = U/\lambda^2$. $\qquad\square$

4.2.3.2 SEM algorithm
Recall that Ψ denotes the EM mapping, having fixed point $\hat{\theta}$ and Jacobian matrix $\Psi'(\theta)$ with (i,j)th element equaling $\frac{d\Psi_i(\theta)}{d\theta_j}$. Dempster et al. [130] show that

$$\Psi'(\hat{\theta})^T = \hat{i}_{\mathbf{Z}|\mathbf{X}}(\hat{\theta})\hat{i}_{\mathbf{Y}}(\hat{\theta})^{-1} \tag{4.41}$$

in the terminology of (4.30).

If we reexpress the missing information principle in (4.30) as

$$\hat{i}_{\mathbf{X}}(\hat{\theta}) = \left[\mathbf{I} - \hat{i}_{\mathbf{Z}|\mathbf{X}}(\hat{\theta})\hat{i}_{\mathbf{Y}}(\hat{\theta})^{-1}\right]\hat{i}_{\mathbf{Y}}(\hat{\theta}), \tag{4.42}$$

where $\mathbf{I}$ is an identity matrix, and substitute (4.41) into (4.42), then inverting $\hat{i}_{\mathbf{X}}(\hat{\theta})$ provides the estimate

$$\widehat{\mathrm{var}}\{\hat{\theta}\} = \hat{i}_{\mathbf{Y}}(\hat{\theta})^{-1}\left(\mathbf{I} + \Psi'(\hat{\theta})^T[\mathbf{I} - \Psi'(\hat{\theta})^T]^{-1}\right). \tag{4.43}$$

This result is appealing in that it expresses the desired covariance matrix as the complete-data covariance matrix plus an incremental matrix that takes account of the uncertainty attributable to the missing data. When coupled with the following numerical differentiation strategy to estimate the increment, Meng and Rubin have termed this approach the *supplemented EM (SEM) algorithm* [384]. Since numerical imprecisions in the differentiation approach affect only the estimated increment, estimation of the covariance matrix is typically more stable than the generic numerical differentiation approach described in Section 4.2.3.5.

Estimation of $\Psi'(\hat{\theta})$ proceeds as follows. The first step of SEM is to run the EM algorithm to convergence, finding the maximizer $\hat{\theta}$. The second step is to restart the algorithm from, say, $\theta^{(0)}$. Although one may restart from the original starting point, it is preferable to choose $\theta^{(0)}$ to be closer to $\hat{\theta}$.

Having thus initialized SEM, we begin SEM iterations for $t = 0, 1, 2, \ldots$. The $(t+1)$th SEM iteration begins by taking a standard E step and M step to produce $\theta^{(t+1)}$ from $\theta^{(t)}$. Next, for $j = 1, \ldots, p$, define $\theta^{(t)}(j) = (\hat{\theta}_1, \ldots, \hat{\theta}_{j-1}, \theta_j^{(t)}, \hat{\theta}_{j+1}, \ldots, \hat{\theta}_p)$ and

$$r_{ij}^{(t)} = \frac{\Psi_i(\theta^{(t)}(j)) - \hat{\theta}_i}{\theta_j^{(t)} - \hat{\theta}_j} \tag{4.44}$$

for $i = 1, \ldots, p$, recalling that $\Psi(\hat{\theta}) = \hat{\theta}$. This ends one SEM iteration. The $\Psi_i(\theta^{(t)}(j))$ values are the estimates produced by applying one EM cycle to $\theta^{(t)}(j)$ for $j = 1, \ldots, p$.

Notice that the (i, j)th element of $\mathbf{\Psi}'(\hat{\boldsymbol{\theta}})$ equals $\lim_{t \to \infty} r_{ij}^{(t)}$. We may consider each element of this matrix to be precisely estimated when the sequence of $r_{ij}^{(t)}$ values stabilizes for $t \geq t_{ij}^*$. Note that different numbers of iterations may be needed for precise estimation of different elements of $\mathbf{\Psi}'(\hat{\boldsymbol{\theta}})$. When all elements have stabilized, SEM iterations stop and the resulting estimate of $\mathbf{\Psi}'(\hat{\boldsymbol{\theta}})$ is used to determine $\widehat{\text{var}}\{\hat{\boldsymbol{\theta}}\}$ as given in (4.43).

Numerical imprecision can cause the resulting covariance matrix to be slightly asymmetric. Such asymmetry can be used to diagnose whether the original EM procedure was run to sufficient precision and to assess how many digits are trustworthy in entries of the estimated covariance matrix. Difficulties also arise if $\mathbf{I} - \mathbf{\Psi}'(\hat{\boldsymbol{\theta}})^T$ is not positive semidefinite or cannot be inverted numerically; see [384]. It has been suggested that transforming $\boldsymbol{\theta}$ to achieve an approximately normal likelihood can lead to faster convergence and increased accuracy of the final solution.

Example 4.6 (Peppered moths, continued) The results from Example 4.2 can be supplemented using the approach of Meng and Rubin. Stable, precise results are obtained within a few SEM iterations, starting from $p_C^{(0)} = 0.07$ and $p_I^{(t)} = 0.19$. Standard errors for $\hat{p}_C, \hat{p}_I$, and $\hat{p}_T$ are 0.0074, 0.0119, and 0.132, respectively. Pairwise correlations are $\text{cor}\{\hat{p}_C, \hat{p}_I\} = -0.14$, $\text{cor}\{\hat{p}_C, \hat{p}_T\} = -0.44$, and $\text{cor}\{\hat{p}_I, \hat{p}_T\} = -0.83$. Here, SEM was used to obtain results for $\hat{p}_C$ and $\hat{p}_I$, and elementary relationships among variances, covariances, and correlations were used to extend these results for $\hat{p}_T$ since the estimated probabilities sum to one. $\square$

It may seem inefficient not to begin SEM iterations until EM iterations have ceased. An alternative would be to attempt to estimate the components of $\mathbf{\Psi}'(\hat{\boldsymbol{\theta}})$ as EM iterations progress, using

$$\tilde{r}_{ij}^{(t)} = \frac{\Psi_i(\theta_1^{(t-1)}, \ldots, \theta_{j-1}^{(t-1)}, \theta_j^{(t)}, \theta_{j+1}^{(t-1)}, \ldots, \theta_p^{(t-1)}) - \Psi_i(\boldsymbol{\theta}^{(t-1)})}{\theta_j^{(t)} - \theta_j^{(t-1)}}. \quad (4.45)$$

However, Meng and Rubin argue that this approach will not require fewer iterations overall, that the extra steps required to find $\hat{\boldsymbol{\theta}}$ first can be offset by starting SEM closer to $\hat{\boldsymbol{\theta}}$, and that the alternative is numerically less stable. Jamshidian and Jennrich survey a variety of methods for numerically differentiating $\mathbf{\Psi}$ or l' itself, including some they consider superior to SEM [302].

4.2.3.3 *Bootstrapping*
Thorough discussion of bootstrapping is given in Chapter 9. In its simplest implementation, bootstrapping to obtain an estimated covariance matrix for EM would proceed as follows for i.i.d. observed data $x_1, \ldots, x_n$:

1. Calculate $\hat{\boldsymbol{\theta}}_{\text{EM}}$ using a suitable EM approach applied to $x_1, \ldots, x_n$. Let $j = 1$ and set $\hat{\boldsymbol{\theta}}_j = \hat{\boldsymbol{\theta}}_{\text{EM}}$.

2. Increment j. Sample *pseudo-data* $\mathbf{X}_1^*, \ldots, \mathbf{X}_n^*$ completely at random from $x_1, \ldots, x_n$ with replacement.

3. Calculate $\hat{\boldsymbol{\theta}}_j$ by applying the same EM approach to the pseudo-data $\mathbf{X}_1^*, \ldots, \mathbf{X}_n^*$.

4. Stop if j is large enough; otherwise return to step 2.

For most problems, a few thousand iterations will suffice. At the end of the process, we have generated a collection of parameter estimates, $\hat{\boldsymbol{\theta}}_1, \ldots, \hat{\boldsymbol{\theta}}_B$, where B denotes the total number of iterations used. Then the sample variance of these B estimates is the estimated variance of $\hat{\boldsymbol{\theta}}$. Conveniently, other aspects of the sampling distribution of $\hat{\boldsymbol{\theta}}$, such as correlations and quantiles, can be estimated using the corresponding sample estimates based on $\hat{\boldsymbol{\theta}}_1, \ldots, \hat{\boldsymbol{\theta}}_B$. Note that bootstrapping embeds the EM loop in a second loop of B iterations. This nested looping can be computationally burdensome when solution of each EM problem is slow because of a high proportion of missing data or high dimensionality.

4.2.3.4 *Empirical information* When the data are i.i.d., note that the score function is the sum of individual scores for each observation:

$$\frac{d \log f_{\mathbf{X}}(\mathbf{x}|\boldsymbol{\theta})}{d\boldsymbol{\theta}} = \mathbf{l}'(\boldsymbol{\theta}|\mathbf{x}) = \sum_{i=1}^{n} \mathbf{l}'(\boldsymbol{\theta}|\mathbf{x}_i), \tag{4.46}$$

where we write the observed dataset as $\mathbf{x} = (\mathbf{x}_1, \ldots, \mathbf{x}_n)$. Since the Fisher information matrix is defined to be the variance of the score function, this suggests estimating the information using the sample variance of the individual scores. The *empirical information* is defined as

$$\frac{1}{n} \sum_{i=1}^{n} \mathbf{l}'(\boldsymbol{\theta}|\mathbf{x}_i)\mathbf{l}'(\boldsymbol{\theta}|\mathbf{x}_i)^T - \frac{1}{n^2}\mathbf{l}'(\boldsymbol{\theta}|\mathbf{x})\mathbf{l}'(\boldsymbol{\theta}|\mathbf{x})^T. \tag{4.47}$$

This estimate has been discussed in the EM context in [381, 447]. The appeal of this approach is that all the terms in (4.47) are by-products of the M step: No additional analysis is required. To see this, note that $\boldsymbol{\theta}^{(t)}$ maximizes $Q(\boldsymbol{\theta}|\boldsymbol{\theta}^{(t)}) - l(\boldsymbol{\theta}|\mathbf{x})$ with respect to $\boldsymbol{\theta}$. Therefore, taking derivatives with respect to $\boldsymbol{\theta}$,

$$Q'(\boldsymbol{\theta}|\boldsymbol{\theta}^{(t)})\Big|_{\boldsymbol{\theta}=\boldsymbol{\theta}^{(t)}} = \mathbf{l}'(\boldsymbol{\theta}|\mathbf{x})\Big|_{\boldsymbol{\theta}=\boldsymbol{\theta}^{(t)}}. \tag{4.48}$$

Since $\mathbf{Q}'$ is ordinarily calculated at each M step, the individual terms in (4.47) are available.

4.2.3.5 *Numerical differentiation* To estimate the Hessian, consider computing the numerical derivative of $\mathbf{l}'$ at $\hat{\boldsymbol{\theta}}$, one coordinate at a time, using (1.10). The first row of the estimated Hessian can be obtained by adding a small perturbation to the first coordinate of $\hat{\boldsymbol{\theta}}$, then computing the ratio of the difference between $\mathbf{l}'(\boldsymbol{\theta})$ at $\boldsymbol{\theta} = \hat{\boldsymbol{\theta}}$ and at the perturbed value, relative to the magnitude of the perturbation. The remaining rows of the Hessian are approximated similarly. If a perturbation is too small, estimated partial derivatives may be inaccurate due to roundoff error; if a perturbation is too big, the estimates may also be inaccurate. Such numerical differentiation can be tricky to automate, especially when the components of $\hat{\boldsymbol{\theta}}$ have different scales. More sophisticated numerical differentiation strategies are surveyed in [302].

4.3 EM VARIANTS

4.3.1 Improving the E step

The E step requires finding the expected log likelihood of the complete data conditional on the observed data. We have denoted this expectation as $Q(\boldsymbol{\theta}|\boldsymbol{\theta}^{(t)})$. When this expectation is difficult to compute analytically, it can be approximated via Monte Carlo (see Chapter 6).

4.3.1.1 Monte Carlo EM   Wei and Tanner [557] propose that the tth E step can be replaced with the following two steps:

1. Draw missing datasets $\mathbf{Z}_1^{(t)}, \dots, \mathbf{Z}_{m^{(t)}}^{(t)}$ i.i.d. from $f_{\mathbf{Z}|\mathbf{X}}(\mathbf{z}|\mathbf{x}, \boldsymbol{\theta}^{(t)})$. Each $\mathbf{Z}_j^{(t)}$ is a vector of all the missing values needed to complete the observed dataset, so $\mathbf{Y}_j = (\mathbf{x}, \mathbf{Z}_j)$ denotes a completed dataset where the missing values have been replaced by $\mathbf{Z}_j$.

2. Calculate $\hat{Q}^{(t+1)}(\boldsymbol{\theta}|\boldsymbol{\theta}^{(t)}) = \frac{1}{m^{(t)}} \sum_{j=1}^{m^{(t)}} \log f_{\mathbf{Y}}(\mathbf{Y}_j^{(t)}|\boldsymbol{\theta})$.

Then $\hat{Q}^{(t+1)}(\boldsymbol{\theta}|\boldsymbol{\theta}^{(t)})$ is a Monte Carlo estimate of $Q(\boldsymbol{\theta}|\boldsymbol{\theta}^{(t)})$. The M step is modified to maximize $\hat{Q}^{(t+1)}(\boldsymbol{\theta}|\boldsymbol{\theta}^{(t)})$.

The recommended strategy is to let $m^{(t)}$ be small during early EM iterations and to increase $m^{(t)}$ as iterations progress to reduce the Monte Carlo variability introduced in $\hat{Q}$. Nevertheless, this *Monte Carlo EM algorithm (MCEM)* will not converge in the same sense as ordinary EM. As iterations proceed, values of $\boldsymbol{\theta}^{(t)}$ will eventually bounce around the true maximum, with a precision that depends on $m^{(t)}$. Discussion of the asymptotic convergence properties of MCEM is provided in [87]. A stochastic alternative to MCEM is discussed in [129].

Example 4.7 (Censored exponential data, continued) In Example 4.5, it was easy to compute the conditional expectation of $l(\lambda|\mathbf{Y}) = n \log \lambda - \lambda \sum_{i=1}^{n} Y_i$ given the observed data. The result, given in (4.37), can be maximized to provide the ordinary EM update,

$$\lambda^{(t+1)} = \frac{n}{\sum_{i=1}^{n} x_i \delta_i + C/\lambda^{(t)}}. \tag{4.49}$$

Application of MCEM is also easy. In this case,

$$\hat{Q}^{(t+1)}(\lambda|\lambda^{(t)}) = n \log \lambda - \frac{\lambda}{m^{(t)}} \sum_{j=1}^{m^{(t)}} \mathbf{Y}_j^T \mathbf{1}, \tag{4.50}$$

where $\mathbf{1}$ is a vector of ones and $\mathbf{Y}_j$ is the jth completed dataset comprising the uncensored data and simulated data $\mathbf{Z}_j = (Z_{j1}, \dots, Z_{jC})$ with $Z_{jk} - c_k \sim$ i.i.d. $\text{Exp}(\lambda^{(t)})$ for $i = 1, \dots, C$ to replace the censored values. Setting $\hat{Q}'(\lambda|\lambda^{(t)}) = 0$ and solving for λ yields

$$\lambda^{(t+1)} = \frac{n}{\sum_{j=1}^{m^{(t)}} \mathbf{Y}_j^T \mathbf{1}/m^{(t)}} \tag{4.51}$$

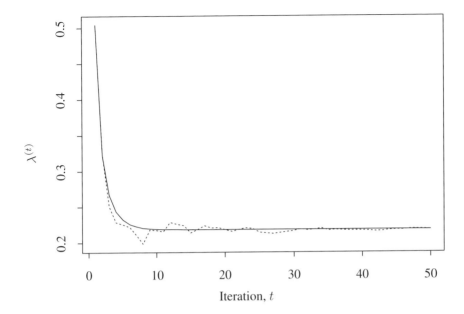

Fig. 4.2 Comparison of iterations for EM (solid) and MCEM (dotted) for the censored exponential data discussed in Example 4.7.

as the MCEM update.

The website for this book provides $n = 30$ observations, including $C = 17$ censored observations. Figure 4.2 compares the performance of MCEM and ordinary EM for estimating λ with these data. Both methods easily find the MLE $\hat{\lambda} = 0.2185$. For MCEM, we used $m^{(t)} = 5^{1+\lfloor t/10 \rfloor}$, where $\lfloor z \rfloor$ denotes the integer part of z. Fifty iterations were used altogether. Both algorithms were initiated from $\lambda^{(0)} = 0.5042$, which is the mean of all 30 data values disregarding censoring.   □

4.3.2 Improving the M step

One of the appeals of the EM algorithm is that the derivation and maximization of $Q(\boldsymbol{\theta}|\boldsymbol{\theta}^{(t)})$ is often simpler than incomplete-data maximum likelihood calculations, since $Q(\boldsymbol{\theta}|\boldsymbol{\theta}^{(t)})$ relates to the complete-data likelihood. In some cases, however, the M step cannot be carried out easily even though the E step yielding $Q(\boldsymbol{\theta}|\boldsymbol{\theta}^{(t)})$ is straightforward. Several strategies have been proposed to facilitate the M step in such cases.

4.3.2.1 *ECM algorithm* Meng and Rubin's *ECM algorithm* replaces the M step with a series of computationally simpler conditional maximization (CM) steps [385]. Each conditional maximization is designed to be a simple optimization problem which

constrains $\boldsymbol{\theta}$ to a particular subspace and permits either an analytical solution or a very elementary numerical solution.

We call the collection of simpler CM steps after the tth E step a CM *cycle*. Thus, the tth iteration of ECM is composed of the tth E step and the tth CM cycle. Let S denote the total number of CM steps in each CM cycle. For $s = 1, \ldots, S$, the sth CM step in the tth cycle requires the maximization of $Q(\boldsymbol{\theta}|\boldsymbol{\theta}^{(t)})$ subject to (or conditional on) a constraint, say

$$\mathbf{g}_s(\boldsymbol{\theta}) = \mathbf{g}_s(\boldsymbol{\theta}^{(t+(s-1)/S)}) \tag{4.52}$$

where $\boldsymbol{\theta}^{(t+(s-1)/S)}$ is the maximizer found in the $(s-1)$th CM step of the current cycle. When the entire cycle of S steps of CM has been completed, we set $\boldsymbol{\theta}^{(t+1)} = \boldsymbol{\theta}^{(t+S/S)}$ and proceed to the E step for the $(t+1)$th iteration.

Clearly any ECM is a GEM algorithm (Section 4.2.1), since each CM step increases Q. In order for ECM to be convergent, we need to ensure that each CM cycle permits search in any direction for a maximizer of $Q(\boldsymbol{\theta}|\boldsymbol{\theta}^{(t)})$, so that ECM effectively maximizes over the original parameter space for $\boldsymbol{\theta}$ and not over some subspace. Precise conditions are discussed in [383, 385]; extensions of this method include [356, 387].

The art of constructing an effective ECM algorithm lies in choosing the constraints cleverly. Usually, it is natural to partition $\boldsymbol{\theta}$ into S subvectors, $\boldsymbol{\theta} = (\boldsymbol{\theta}_1, \ldots, \boldsymbol{\theta}_S)$. Then in the sth CM step, one might seek to maximize Q with respect to $\boldsymbol{\theta}_s$ while holding all other components of $\boldsymbol{\theta}$ fixed. This amounts to the constraint induced by the function $g_s(\boldsymbol{\theta}) = (\boldsymbol{\theta}_1, \ldots, \boldsymbol{\theta}_{s-1}, \boldsymbol{\theta}_{s+1}, \ldots, \boldsymbol{\theta}_S)$. A maximization strategy of this type has previously been termed *iterated conditional modes* [30]. If the conditional maximizations are obtained by finding the roots of score functions, the CM cycle can also be viewed as a Gauss–Seidel iteration (see Section 2.2.4).

Alternatively, the sth CM step might seek to maximize Q with respect to all other components of $\boldsymbol{\theta}$ while holding $\boldsymbol{\theta}_s$ fixed. In this case, $g_s(\boldsymbol{\theta}) = \boldsymbol{\theta}_s$. Additional systems of constraints can be imagined, depending on the particular problem context. A variant of ECM inserts an E step between each pair of CM steps, thereby updating Q at every stage of the CM cycle.

Example 4.8 (Multivariate regression with missing values) A particularly illuminating example given by Meng and Rubin [385] involves multivariate regression with missing values. Let $\mathbf{U}_1, \ldots, \mathbf{U}_n$ be n independent d-dimensional vectors observed from the d-variate normal model given by

$$\mathbf{U}_i \sim N_d(\boldsymbol{\mu}_i, \boldsymbol{\Sigma}) \tag{4.53}$$

for $\mathbf{U}_i = (U_{i1}, \ldots, U_{id})$ and $\boldsymbol{\mu}_i = \mathbf{V}_i \boldsymbol{\beta}$, where the $\mathbf{V}_i$ are known $d \times p$ design matrices, $\boldsymbol{\beta}$ is a vector of p unknown parameters, and $\boldsymbol{\Sigma}$ is a $d \times d$ unknown variance–covariance matrix. There are many cases where $\boldsymbol{\Sigma}$ has some meaningful structure, but we consider $\boldsymbol{\Sigma}$ to be unstructured for simplicity. Suppose that some elements of some $\mathbf{U}_i$ are missing.

Begin by reordering the elements of $\mathbf{U}_i$, $\boldsymbol{\mu}_i$, and the rows of $\mathbf{V}_i$ so that for each i, the observed components of $\mathbf{U}_i$ are first and any missing components are last. For each $\mathbf{U}_i$, denote by $\boldsymbol{\beta}_i$ and $\boldsymbol{\Sigma}_i$ the corresponding reorganizations of the parameters. Thus $\boldsymbol{\beta}_i$ and $\boldsymbol{\Sigma}_i$ are completely determined by $\boldsymbol{\beta}$, $\boldsymbol{\Sigma}$, and the pattern of missing data: They do not represent an expansion of the parameter space.

This notational reorganization allows us to write $\mathbf{U}_i = (\mathbf{U}_{\text{obs},i}, \mathbf{U}_{\text{miss},i})$, $\boldsymbol{\mu}_i = (\boldsymbol{\mu}_{\text{obs},i}, \boldsymbol{\mu}_{\text{miss},i})$, and

$$\boldsymbol{\Sigma}_i = \begin{pmatrix} \boldsymbol{\Sigma}_{\text{obs},i} & \boldsymbol{\Sigma}_{\text{cross},i} \\ \boldsymbol{\Sigma}_{\text{cross},i}^T & \boldsymbol{\Sigma}_{\text{miss},i} \end{pmatrix}. \tag{4.54}$$

The full set of observed data can be denoted $\mathbf{U}_{\text{obs}} = (\mathbf{U}_{\text{obs},1}, \ldots, \mathbf{U}_{\text{obs},n})$.

The observed-data log likelihood function is

$$l(\boldsymbol{\beta}, \boldsymbol{\Sigma}|\mathbf{u}_{\text{obs}}) = -\frac{1}{2}\sum_{i=1}^{n}\log|\boldsymbol{\Sigma}_{\text{obs},i}| - \frac{1}{2}\sum_{i=1}^{n}(\mathbf{u}_{\text{obs},i} - \boldsymbol{\mu}_{\text{obs},i})^T\boldsymbol{\Sigma}_{\text{obs},i}^{-1}(\mathbf{u}_{\text{obs},i} - \boldsymbol{\mu}_{\text{obs},i})$$

up to an additive constant. This likelihood is quite tedious to work with or to maximize. Note, however, that the complete-data sufficient statistics are given by $\sum_{i=1}^{n} U_{ij}$ for $j = 1, \ldots, d$ and $\sum_{i=1}^{n} U_{ij}U_{ik}$ for $j, k = 1, \ldots, d$. Thus the E step amounts to finding the expected values of these sufficient statistics conditional on the observed data and current parameter values $\boldsymbol{\beta}^{(t)}$ and $\boldsymbol{\Sigma}^{(t)}$.

Now for $j = 1, \ldots, d$

$$\mathrm{E}\left\{\sum_{i=1}^{n} U_{ij} \,\middle|\, \mathbf{u}_{\text{obs}}, \boldsymbol{\beta}^{(t)}, \boldsymbol{\Sigma}^{(t)}\right\} = \sum_{i=1}^{n} a_{ij}^{(t)}, \tag{4.55}$$

where

$$a_{ij}^{(t)} = \begin{cases} \alpha_{ij}^{(t)} & \text{if } U_{ij} \text{ is missing,} \\ u_{ij} & \text{if } U_{ij} = u_{ij} \text{ is observed} \end{cases} \tag{4.56}$$

and $\alpha_{ij}^{(t)} = \mathrm{E}\{U_{ij}|\mathbf{u}_{\text{obs},i}, \boldsymbol{\beta}_i^{(t)}, \boldsymbol{\Sigma}_i^{(t)}\}$. Similarly, for $j, k = 1, \ldots, d$,

$$\mathrm{E}\left\{\sum_{i=1}^{n} U_{ij}U_{ik} \,\middle|\, \mathbf{u}_{\text{obs}}, \boldsymbol{\beta}^{(t)}, \boldsymbol{\Sigma}^{(t)}\right\} = \sum_{i=1}^{n}(a_{ij}^{(t)} a_{ik}^{(t)} + b_{ijk}^{(t)}), \tag{4.57}$$

where

$$b_{ijk}^{(t)} = \begin{cases} \gamma_{ijk}^{(t)} & \text{if } U_{ij} \text{ and } U_{ik} \text{ are both missing,} \\ 0 & \text{otherwise} \end{cases} \tag{4.58}$$

and $\gamma_{ijk}^{(t)} = \text{cov}\{U_{ij}, U_{ik}|\mathbf{u}_{\text{obs},i}, \boldsymbol{\beta}_i^{(t)}, \boldsymbol{\Sigma}_i^{(t)}\}$.

Fortunately, the derivation of the $\alpha_{ij}^{(t)}$ and $\gamma_{ijk}^{(t)}$ is fairly straightforward. The conditional distribution of $\mathbf{U}_{\text{miss},i}|(\mathbf{u}_{\text{obs},i}, \boldsymbol{\beta}_i^{(t)}, \boldsymbol{\Sigma}_i^{(t)})$ is

$$N\left(\boldsymbol{\mu}_{\text{miss},i}^{(t)} + \boldsymbol{\Sigma}_{\text{cross},i}\boldsymbol{\Sigma}_{\text{miss},i}^{-1}(\mathbf{u}_{\text{obs},i} - \boldsymbol{\mu}_{\text{obs},i}^{(t)}), \boldsymbol{\Sigma}_{\text{obs},i} - \boldsymbol{\Sigma}_{\text{cross},i}\boldsymbol{\Sigma}_{\text{miss},i}^{-1}\boldsymbol{\Sigma}_{\text{cross},i}^T\right).$$

The values for $\alpha_{ij}^{(t)}$ and $\gamma_{ijk}^{(t)}$ can be read from the mean vector and variance–covariance matrix of this distribution, respectively. Knowing these, $Q(\beta, \Sigma | \beta^{(t)}, \Sigma^{(t)})$ can be formed following (4.26).

Having thus achieved the E step, we turn now to the M step. The high dimensionality of the parameter space and the complexity of the observed-data likelihood renders difficult any direct implementation of the M step, whether by direct maximization or by reference to the exponential family setup. However, implementing an ECM strategy is straightforward using $S = 2$ conditional maximization steps in each CM cycle.

Treating β and Σ separately allows easy constrained optimizations of Q. First, if we impose the constraint that $\Sigma = \Sigma^{(t)}$, then we can maximize the constrained version of $Q(\beta, \Sigma | \beta^{(t)}, \Sigma^{(t)})$ with respect to β by using the weighted least squares estimate

$$\beta^{(t+1/2)} = \left(\sum_{i=1}^{n} \mathbf{V}_i^T (\Sigma_i^{(t)})^{-1} \mathbf{V}_i \right)^{-1} \left(\sum_{i=1}^{n} \mathbf{V}_i^T (\Sigma_i^{(t)})^{-1} \mathbf{a}_i^{(t)} \right), \qquad (4.59)$$

where $\mathbf{a}_i^{(t)} = (a_{i1}^{(t)}, \ldots, a_{id}^{(t)})^T$ and $\Sigma_i^{(t)}$ is treated as a known variance–covariance matrix. This ensures that $Q(\beta^{(t+1/2)}, \Sigma^{(t)} | \beta^{(t)}, \Sigma^{(t)}) \geq Q(\beta^{(t)}, \Sigma^{(t)} | \beta^{(t)}, \Sigma^{(t)})$. This constitutes the first of two CM steps.

The second CM step follows from the fact that setting $\Sigma^{(t+2/2)}$ equal to

$$\mathrm{E} \left\{ \frac{1}{n} \sum_{i=1}^{n} (\mathbf{U}_i - \mathbf{V}_i \beta^{(t+1/2)})(\mathbf{U}_i - \mathbf{V}_i \beta^{(t+1/2)})^T \,\middle|\, \mathbf{u}_{\mathrm{obs}}, \beta^{(t+1/2)}, \Sigma^{(t)} \right\} \quad (4.60)$$

maximizes $Q(\beta, \Sigma | \beta^{(t)}, \Sigma^{(t)})$ with respect to Σ subject to the constraint that $\beta = \beta^{(t+1/2)}$, because this amounts to plugging in $\alpha_{ij}^{(t)}$ and $\gamma_{ijk}^{(t)}$ values where necessary and computing the sample covariance matrix of the completed data. This update guarantees

$$Q(\beta^{(t+1/2)}, \Sigma^{(t+2/2)} | \beta^{(t)}, \Sigma^{(t)}) \geq Q(\beta^{(t+1/2)}, \Sigma^{(t)} | \beta^{(t)}, \Sigma^{(t)})$$
$$\geq Q(\beta^{(t)}, \Sigma^{(t)} | \beta^{(t)}, \Sigma^{(t)}). \qquad (4.61)$$

Together, the two CM steps yield $(\beta^{(t+1)}, \Sigma^{(t+1)}) = (\beta^{(t+1/2)}, \Sigma^{(t+2/2)})$ and ensure an increase in the Q function.

The E step and the CM cycle described here can each be implemented using familiar closed-form analytic results; no numerical integration or maximization is required. After updating the parameters with the CM cycle described above, we return to another E step, and so forth. In summary, ECM alternates between (i) creating updated complete data sets and (ii) sequentially estimating β and Σ in turn by fixing the other at its current value and using the current completed-data component. $\quad\square$

4.3.2.2 *EM gradient algorithm*   If maximization cannot be accomplished analytically, then one might consider carrying out each M step using an iterative numerical optimization approach like those discussed in Chapter 2. This would yield an algorithm that had nested iterative loops. The ECM algorithm inserts S conditional maximization steps within each iteration of the EM algorithm, also yielding nested iteration.

To avoid the computational burden of nested looping, Lange proposed replacing the M step with a single step of Newton's method, thereby approximating the maximum without actually solving for it exactly [347]. The M step is replaced with the update given by

$$\boldsymbol{\theta}^{(t+1)} = \boldsymbol{\theta}^{(t)} - \mathbf{Q}''(\boldsymbol{\theta}|\boldsymbol{\theta}^{(t)})^{-1}\Big|_{\boldsymbol{\theta}=\boldsymbol{\theta}^{(t)}} \ \mathbf{Q}'(\boldsymbol{\theta}|\boldsymbol{\theta}^{(t)})\Big|_{\boldsymbol{\theta}=\boldsymbol{\theta}^{(t)}} \tag{4.62}$$

$$= \boldsymbol{\theta}^{(t)} - \mathbf{Q}''(\boldsymbol{\theta}|\boldsymbol{\theta}^{(t)})^{-1}\Big|_{\boldsymbol{\theta}=\boldsymbol{\theta}^{(t)}} \ l'(\boldsymbol{\theta}^{(t)}|\mathbf{x}), \tag{4.63}$$

where $l'(\boldsymbol{\theta}^{(t)}|\mathbf{x})$ is the evaluation of the score function at the current iterate. Note that (4.63) follows from the observation in Section 4.2.3.4 that $\boldsymbol{\theta}^{(t)}$ maximizes $Q(\boldsymbol{\theta}|\boldsymbol{\theta}^{(t)}) - l(\boldsymbol{\theta}|\mathbf{x})$. This *EM gradient algorithm* has the same rate of convergence to $\hat{\boldsymbol{\theta}}$ as the full EM algorithm. Lange discusses conditions under which ascent can be ensured, and scalings of the update increment to speed convergence [347]. In particular, when $\mathbf{Y}$ has an exponential family distribution with canonical parameter $\boldsymbol{\theta}$, ascent is ensured and the method matches that of Titterington [538]. In other cases, the step can be scaled down to ensure ascent (as discussed in Section 2.2.2.1), but inflating steps speeds convergence. For problems with a high proportion of missing information, Lange suggests considering doubling the step length [347].

Example 4.9 (Peppered moths, continued) Continuing Example 4.2, we apply the EM gradient algorithm to these data. It is straightforward to show

$$\frac{d^2 Q(\mathbf{p}|\mathbf{p}^{(t)})}{dp_C^2} = -\frac{2n_{CC}^{(t)} + n_{CI}^{(t)} + n_{CT}^{(t)}}{p_C^2} - \frac{2n_{TT}^{(t)} + n_{CT}^{(t)} + n_{IT}^{(t)}}{(1 - p_C - p_I)^2}, \tag{4.64}$$

$$\frac{d^2 Q(\mathbf{p}|\mathbf{p}^{(t)})}{dp_I^2} = -\frac{2n_{II}^{(t)} + n_{IT}^{(t)} + n_{CI}^{(t)}}{p_I^2} - \frac{2n_{TT}^{(t)} + n_{CT}^{(t)} + n_{IT}^{(t)}}{(1 - p_C - p_I)^2}, \tag{4.65}$$

and

$$\frac{d^2 Q(\mathbf{p}|\mathbf{p}^{(t)})}{dp_C dp_I} = -\frac{2n_{TT}^{(t)} + n_{CT}^{(t)} + n_{IT}^{(t)}}{(1 - p_C - p_I)^2}. \tag{4.66}$$

Figure 4.3 shows the steps taken by the resulting EM gradient algorithm, starting from $p_C = p_I = p_T = 1/3$. Step halving was implemented to ensure ascent. The first step heads somewhat in the wrong direction, but in subsequent iterations the gradient steps progress quite directly uphill. The ordinary EM steps are shown for comparison in this figure.   □

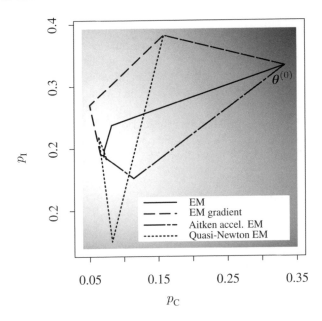

Fig. 4.3 Steps taken by the EM gradient algorithm (long dashes). Ordinary EM steps are shown with the solid line. Steps from two methods from later sections (Aitken and quasi-Newton acceleration) are also shown, as indicated in the key. The observed-data log likelihood is shown with the gray scale, with light shading corresponding to high likelihood. All algorithms were started from $p_C = p_I = 1/3$.

4.3.3   Acceleration methods

The slow convergence of the EM algorithm is a notable drawback. Several techniques have been suggested for using the relatively simple analytic setup from EM to motivate particular forms for Newton-like steps. In addition to the two approaches described below, approaches that cleverly expand the parameter space in manners that speed convergence without affecting the marginal inference about θ are topics of recent interest [360, 387].

4.3.3.1 Aitken acceleration Let $\theta_{EM}^{(t+1)}$ be the next iterate obtained by the standard EM algorithm from $\theta^{(t)}$. Recall that the Newton update to maximize the log likelihood would be

$$\theta^{(t+1)} = \theta^{(t)} - l''(\theta^{(t)}|\mathbf{x})^{-1}l'(\theta^{(t)}|\mathbf{x}). \qquad (4.67)$$

The EM framework suggests a replacement for $l'(\theta^{(t)}|\mathbf{x})$. In Section 4.2.3.4 we noted that $l'(\theta^{(t)}|\mathbf{x}) = \mathbf{Q}'(\theta|\theta^{(t)})\big|_{\theta=\theta^{(t)}}$. Expanding $\mathbf{Q}'$ around $\theta^{(t)}$, evaluated at $\theta_{EM}^{(t+1)}$, yields

$$\mathbf{Q}'(\theta|\theta^{(t)})\Big|_{\theta=\theta_{\mathrm{EM}}^{(t+1)}} \approx \mathbf{Q}'(\theta|\theta^{(t)})\Big|_{\theta=\theta^{(t)}} - \hat{\mathbf{i}}_{\mathbf{Y}}(\theta^{(t)})(\theta_{\mathrm{EM}}^{(t+1)} - \theta^{(t)}), \quad (4.68)$$

where $\hat{\mathbf{i}}_{\mathbf{Y}}(\theta^{(t)})$ is defined in (4.31). Since $\theta_{\mathrm{EM}}^{(t+1)}$ maximizes $Q(\theta|\theta^{(t)})$ with respect to θ, the left hand side of (4.68) equals zero. Therefore

$$\mathbf{Q}'(\theta|\theta^{(t)})\Big|_{\theta=\theta^{(t)}} \approx \hat{\mathbf{i}}_{\mathbf{Y}}(\theta^{(t)})(\theta_{\mathrm{EM}}^{(t+1)} - \theta^{(t)}). \quad (4.69)$$

Thus from (4.67) we arrive at

$$\theta^{(t+1)} = \theta^{(t)} - \mathbf{l}''(\theta^{(t)}|\mathbf{x})^{-1}\hat{\mathbf{i}}_{\mathbf{Y}}(\theta^{(t)})(\theta_{\mathrm{EM}}^{(t+1)} - \theta^{(t)}). \quad (4.70)$$

This update—relying on the approximation in (4.69)—is an example of a general strategy known as *Aitken acceleration* and was proposed for EM by Louis [363]. Aitken acceleration of EM is precisely the same as applying the Newton–Raphson method to find a zero of $\mathbf{\Psi}(\theta) - \theta$, where $\mathbf{\Psi}$ is the mapping defined by the ordinary EM algorithm producing $\theta^{(t+1)} = \mathbf{\Psi}(\theta^{(t)})$ [300].

Example 4.10 (Peppered moths, continued) This acceleration approach can be applied to Example 4.2. Obtaining $\mathbf{l}''$ is analytically more tedious than the simpler derivations employed for other EM approaches to this problem. Figure 4.3 shows the Aitken accelerated steps, which converge quickly to the solution. The procedure was started from $p_C = p_I = p_T = 1/3$, and step halving was used to ensure ascent. □

Aitken acceleration is sometimes criticized for potential numerical instabilities and convergence failures [133, 301]. Further, when $\mathbf{l}''(\theta|\mathbf{x})$ is difficult to compute, this approach cannot be applied without overcoming the difficulty [18, 302, 381].

Section 4.2.1 noted that the EM algorithm converges at a linear rate that depends on the fraction of missing information. The updating increment given in (4.70) is, loosely speaking, scaled by the ratio of the complete information to the observed information. Thus, when a greater proportion of the information is missing, the nominal EM steps are inflated more.

Newton's method converges quadratically, but (4.69) only becomes a precise approximation as $\theta^{(t)}$ nears $\hat{\theta}$. Therefore, we should only expect this acceleration approach to enhance convergence only as preliminary iterations hone θ sufficiently. The acceleration should not be employed without having taken some initial iterations of ordinary EM so that (4.69) holds.

4.3.3.2 *Quasi-Newton acceleration*

The quasi-Newton optimization method discussed in Section 2.2.2.3 produces updates according to

$$\theta^{(t+1)} = \theta^{(t)} - (\mathbf{M}^{(t)})^{-1}\mathbf{l}'(\theta^{(t)}|\mathbf{x}) \quad (4.71)$$

for maximizing $\mathbf{l}(\theta|\mathbf{x})$ with respect to θ, where $\mathbf{M}^{(t)}$ is an approximation to $\mathbf{l}''(\theta^{(t)}|\mathbf{x})$. Within the EM framework, one can decompose $\mathbf{l}''(\theta^{(t)}|\mathbf{x})$ into a part computed during

EM and a remainder. By taking two derivatives of (4.19), we obtain

$$l''(\theta^{(t)}|\mathbf{x}) = \mathbf{Q}''(\theta|\theta^{(t)})\Big|_{\theta=\theta^{(t)}} - \mathbf{H}''(\theta|\theta^{(t)})\Big|_{\theta=\theta^{(t)}} \qquad (4.72)$$

at iteration t. The remainder is the last term in (4.72); suppose we approximate it by $\mathbf{B}^{(t)}$. Then by using

$$\mathbf{M}^{(t)} = \mathbf{Q}''(\theta|\theta^{(t)})\Big|_{\theta=\theta^{(t)}} - \mathbf{B}^{(t)} \qquad (4.73)$$

in (4.71) we obtain a *quasi-Newton EM acceleration*.

A key feature of the approach is how $\mathbf{B}^{(t)}$ approximates $\mathbf{H}''(\theta^{(t)}|\theta^{(t)})$. The idea is to start with $\mathbf{B}^{(0)} = \mathbf{0}$ and gradually accumulate information about $\mathbf{H}''$ as iterations progress. The information is accumulated using a sequence of secant conditions, as is done in ordinary quasi-Newton approaches (Section 2.2.2.3).

Specifically, we can require that $\mathbf{B}^{(t)}$ satisfy the secant condition

$$\mathbf{B}^{(t+1)}\mathbf{a}^{(t)} = \mathbf{b}^{(t)}, \qquad (4.74)$$

where

$$\mathbf{a}^{(t)} = \theta^{(t+1)} - \theta^{(t)} \qquad (4.75)$$

and

$$\mathbf{b}^{(t)} = \mathbf{H}'(\theta|\theta^{(t+1)})\Big|_{\theta=\theta^{(t+1)}} - \mathbf{H}'(\theta|\theta^{(t+1)})\Big|_{\theta=\theta^{(t)}}. \qquad (4.76)$$

Recalling the update (2.49), we can satisfy the secant condition by setting

$$\mathbf{B}^{(t+1)} = \mathbf{B}^{(t)} + c^{(t)}\mathbf{v}^{(t)}(\mathbf{v}^{(t)})^T, \qquad (4.77)$$

where $\mathbf{v}^{(t)} = \mathbf{b}^{(t)} - \mathbf{B}^{(t)}\mathbf{a}^{(t)}$ and $c^{(t)} = \frac{1}{(\mathbf{v}^{(t)})^T\mathbf{a}^{(t)}}$.

Lange proposed this *quasi-Newton* EM algorithm, along with several suggested strategies for improving its performance [348]. First, he suggested starting with $\mathbf{B}^{(0)} = \mathbf{0}$. Note that this implies that the first increment will equal the EM gradient increment. Indeed, the EM gradient approach is exact Newton–Raphson for maximizing $Q(\theta|\theta^{(t)})$, whereas the approach described here evolves into approximate Newton–Raphson for maximizing $l(\theta|\mathbf{x})$.

Second, Davidon's update is troublesome if $(\mathbf{v}^{(t)})^T\mathbf{a}^{(t)} = 0$ or is small compared to $\|\mathbf{v}^{(t)}\| \cdot \|\mathbf{a}^{(t)}\|$. In such cases, we may simply set $\mathbf{B}^{(t+1)} = \mathbf{B}^{(t)}$.

Third, there is no guarantee that $\mathbf{M}^{(t)} = \mathbf{Q}''(\theta|\theta^{(t)})\Big|_{\theta=\theta^{(t)}} - \mathbf{B}^{(t)}$ will be negative definite, which would ensure ascent at the tth step. Therefore, we may scale $\mathbf{B}^{(t)}$ and use $\mathbf{M}^{(t)} = \mathbf{Q}''(\theta|\theta^{(t)})\Big|_{\theta=\theta^{(t)}} - \alpha^{(t)}\mathbf{B}^{(t)}$ where, for example, $\alpha^{(t)} = 2^{-m}$ for the smallest positive integer that makes $\mathbf{M}^{(t)}$ negative definite.

Finally, note that $\mathbf{y}^{(t)}$ may be expressed entirely in terms of $\mathbf{Q}'$ functions since

$$\mathbf{b}^{(t)} = \mathbf{H}'(\theta|\theta^{(t+1)})\Big|_{\theta=\theta^{(t+1)}} - \mathbf{H}'(\theta|\theta^{(t+1)})\Big|_{\theta=\theta^{(t)}} \qquad (4.78)$$

$$= 0 - \mathbf{H}'(\theta|\theta^{(t+1)})\Big|_{\theta=\theta^{(t)}} \qquad (4.79)$$

$$= \mathbf{Q}'(\theta|\theta^{(t)})\Big|_{\theta=\theta^{(t)}} - \mathbf{Q}'(\theta|\theta^{(t+1)})\Big|_{\theta=\theta^{(t)}}. \qquad (4.80)$$

Equation (4.79) follows from (4.19) and the fact that $l(\theta|x) - Q(\theta|\theta^{(t)})$ has its minimum at $\theta = \theta^{(t)}$. The derivative at this minimum must be zero, forcing $l'(\theta^{(t)}|x) = Q'(\theta|\theta^{(t)})\big|_{\theta=\theta^{(t)}}$, which allows (4.80).

Example 4.11 (Peppered moths, continued) We can apply quasi-Newton accelera-tion to Example 4.2, using the expressions for Q'' given in (4.64)–(4.66) and obtaining $b^{(t)}$ from (4.80). The procedure was started from $p_C = p_I = p_T = 1/3$ and $B^{(0)} = 0$, with step halving to ensure ascent.

The results are shown in Figure 4.3. Note that $B^{(0)} = 0$ means that the first quasi-Newton EM step will match the first EM gradient step. The second quasi-Newton EM step completely overshoots the ridge of highest likelihood, resulting in a step that is just barely uphill. In general, the quasi-Newton EM procedure behaves like other quasi-Newton methods: there can be a tendency to step beyond the solution or to converge to a local maximum rather than a local minimum. With suitable safeguards, the procedure is fast and effective in this example. □

The quasi-Newton EM requires the inversion of $M^{(t)}$ at step t. Lange et al. describe a quasi-Newton approach based on the approximation of $-l''(\theta^{(t)}|x)$ by some $M^{(t)}$ that relies on an inverse-secant update [349, 350]. In addition to avoiding computationally burdensome matrix inversions, such updates to $\theta^{(t)}$ and $B^{(t)}$ can be expressed entirely in terms of $l'(\theta^{(t)}|x)$ and ordinary EM increments when the M step is solvable.

Jamshidian and Jennrich elaborate on inverse-secant updating and discuss the more complex BFGS approach [301]. These authors also provide a useful survey of a variety of EM acceleration approaches and a comparison of effectiveness. Some of their approaches converge faster on examples than does the approach described above. In a related paper, they present a conjugate gradient acceleration of EM [300].

Problems

4.1   Recall the peppered moth analysis introduced in Example 4.2. In the field, it is quite difficult to distinguish the *insularia* and *typica* phenotypes due to variations in wing color and mottle. In addition to the 622 moths mentioned in the example, suppose the sample collected by the researchers actually included $n_U = 578$ more moths which were known to be *insularia* or *typical* but whose exact phenotypes could not be determined.

(a) Derive the EM algorithm for maximum likelihood estimation of p_C, p_I, and p_I for this modified problem having observed data n_C, n_I, n_T, and n_U as given above.

(b) Apply the algorithm to find the MLEs.

(c) Estimate the standard errors and pairwise correlations for $\hat{p}_C$, $\hat{p}_I$, and $\hat{p}_I$ using the SEM algorithm.

Table 4.2 Frequencies of respondents reporting numbers of risky sexual encounters; see Problem 4.2.

Encounters, i	0	1	2	3	4	5	6	7	8
Frequency, n_i	379	299	222	145	109	95	73	59	45

Encounters, i	9	10	11	12	13	14	15	16
Frequency, n_i	30	24	12	4	2	0	1	1

(d) Estimate the standard errors and pairwise correlations for $\hat{p}_C$, $\hat{p}_I$, and $\hat{p}_I$ by bootstrapping.

(e) Implement the EM gradient algorithm for these data. Experiment with step halving to ensure ascent, and with other step scalings that may speed convergence.

(f) Implement Aitken accelerated EM for these data. Use step halving.

(g) Implement quasi-Newton EM for these data. Compare performance with and without step halving.

(h) Compare the effectiveness and efficiency of the standard EM algorithm and the three variants in (e), (f), and (g). Use step halving to ensure ascent with the three variants. Base your comparison on a variety of starting points. Create a graph analogous to Figure 4.3.

4.2 Epidemiologists are interested in studying the sexual behavior of individuals at risk for HIV infection. Suppose 1500 gay men were surveyed and each was asked how many risky sexual encounters he had in the previous 30 days. Let n_i denote the number of respondents reporting i encounters, for $i = 1, \ldots, 16$. Table 4.2 summarizes the responses.

These data are poorly fitted by a Poisson model. It is more realistic to assume that the respondents comprise three groups. First, there is a group of people who, for whatever reason, report zero risky encounters even if this is not true. Suppose a respondent has probability α of belonging to this group.

With probability β, a respondent belongs to a second group representing typical behavior. Such people respond truthfully, and their numbers of risky encounters are assumed to follow a Poisson(μ) distribution.

Finally, with probability $1 - \alpha - \beta$, a respondent belongs to a high-risk group. Such people respond truthfully, and their numbers of risky encounters are assumed to follow a Poisson(λ) distribution.

The parameters in the model are α, β, μ, and λ. At the tth iteration of EM, we use $\theta^{(t)} = \left(\alpha^{(t)}, \beta^{(t)}, \mu^{(t)}, \lambda^{(t)}\right)$ to denote the current parameter values. The likelihood of the observed data is given by

$$L(\boldsymbol{\theta}|n_0, \ldots, n_{16}) \propto \prod_{i=0}^{16} [\pi_i(\boldsymbol{\theta})/i!]^{n_i} , \tag{4.81}$$

where

$$\pi_i(\boldsymbol{\theta}) = \alpha 1_{\{i=0\}} + \beta\mu^i \exp\{-\mu\} + (1 - \alpha - \beta)\lambda^i \exp\{-\lambda\} \tag{4.82}$$

for $i = 1, \ldots, 16$.

The observed data are $n_0, \ldots, n_{16}$. The complete data may be construed to be $n_{z,0}$, $n_{t,0}, \ldots, n_{t,16}$, and $n_{p,0}, \ldots, n_{p,16}$, where $n_{k,i}$ denotes the number of respondents in group k reporting i risky encounters and $k = z$, t, and p correspond to the zero, typical, and promiscuous groups, respectively. Thus $n_0 = n_{z,0} + n_{t,0} + n_{p,0}$ and $n_i = n_{t,i} + n_{p,i}$ for $i = 1, \ldots, 16$. Let $N = \sum_{i=0}^{16} n_i = 1500$.

Define

$$z_0(\boldsymbol{\theta}) = \frac{\alpha}{\pi_0(\boldsymbol{\theta})}, \tag{4.83}$$

$$t_i(\boldsymbol{\theta}) = \frac{\beta\mu^i \exp\{-\mu\}}{\pi_i(\boldsymbol{\theta})}, \tag{4.84}$$

$$p_i(\boldsymbol{\theta}) = \frac{(1 - \alpha - \beta)\lambda^i \exp\{-\lambda\}}{\pi_i(\boldsymbol{\theta})} \tag{4.85}$$

for $i = 0, \ldots, 16$. These correspond to probabilities that respondents with i risky encounters belong to the various groups.

(a) Show that the EM algorithm provides the following updates:

$$\alpha^{(t+1)} = n_0 z_0(\boldsymbol{\theta}^{(t)})/N, \tag{4.86}$$

$$\beta^{(t+1)} = \sum_{i=0}^{16} n_i t_i(\boldsymbol{\theta}^{(t)})/N, \tag{4.87}$$

$$\mu^{(t+1)} = \frac{\sum_{i=0}^{16} i \, n_i t_i(\boldsymbol{\theta}^{(t)})}{\sum_{i=0}^{16} n_i t_i(\boldsymbol{\theta}^{(t)})}, \tag{4.88}$$

$$\lambda^{(t+1)} = \frac{\sum_{i=0}^{16} i \, n_i p_i(\boldsymbol{\theta}^{(t)})}{\sum_{i=0}^{16} n_i p_i(\boldsymbol{\theta}^{(t)})}. \tag{4.89}$$

(b) Estimate the parameters of the model, using the observed data.

(c) Estimate the standard errors and pairwise correlations of your parameter estimates, using any available method.

4.3   The website for this book contains 50 trivariate data points drawn from the $N_3(\boldsymbol{\mu}, \boldsymbol{\Sigma})$ distribution. Some data points have missing values in one or more coordinates. Only 27 of the 50 observations are complete.

Table 4.3   Fourteen lifetimes for mining equipment gear couplings, in years. Right-censored values are in parenthesis. In these cases, we know only that the lifetime was at least as long as the given value.

(6.94)	5.50	4.54	2.14	(3.65)	(3.40)	(4.38)
10.24	4.56	9.42	(4.55)	(4.15)	5.64	(10.23)

(a) Derive the EM algorithm for joint maximum likelihood estimation of μ and Σ. It is easiest to recall that the multivariate normal density is in the exponential family.

(b) Determine the MLEs from a suitable starting point. Investigate the performance of the algorithm, and comment on your results.

(c) Consider Bayesian inference for μ when

$$\Sigma = \begin{pmatrix} 1 & 0.6 & 1.2 \\ 0.6 & 0.5 & 0.5 \\ 1.2 & 0.5 & 3.0 \end{pmatrix}$$

is known. Assume independent priors for the three elements of μ. Specifically, let the jth prior be

$$f(\mu_j) = \frac{\exp\{-(\mu_j - \alpha_j)/\beta_j\}}{\beta_j [1 + \exp\{-(\mu_j - \alpha_j)/\beta_j\}]^2},$$

where $(\alpha_1, \alpha_2, \alpha_3) = (2, 4, 6)$ and $\beta_j = 2$ for $j = 1, 2, 3$. Comment on difficulties that would be faced in implementing a standard EM algorithm for estimating the posterior mode for μ. Implement a gradient EM algorithm, and evaluate its performance.

(d) Suppose that Σ is unknown in part (c) and that an improper uniform prior is adopted, that is, $f(\Sigma) \propto 1$ for all positive definite Σ. Discuss ideas for how to estimate the posterior mode for μ and Σ.

4.4   Suppose we observe lifetimes for fourteen gear couplings in certain mining equipment, as given in Table 4.3 (in years). Some of these data are right-censored because the equipment was replaced before the gear coupling failed. The censored data are in parentheses; the actual lifetimes for these components may be viewed as missing.

Model these data with the Weibull distribution, having density function $f(x) = abx^{b-1} \exp\{-ax^b\}$ for $x > 0$ and parameters a and b. Recall that Problem 2.3 in Chapter 2 provides more details about such models. Construct an EM algorithm to estimate a and b. Since the Q function involves expectations that are analytically unavailable, adopt the MCEM strategy where necessary. Also, optimization of Q cannot be completed analytically. Therefore, incorporate the ECM strategy of conditionally maximizing with respect to each parameter separately, applying a one-dimensional Newton-like optimizer where needed. Past observations suggest $(a, b) = (0.003, 2.5)$ may be a suitable starting point. Discuss the convergence properties of the procedure you develop, and the results you obtain. What are the advantages and disadvantages of your technique compared to direct maximization of the observed-data likelihood using, say, a two-dimensional quasi-Newton approach?

4.5 A *hidden Markov model (HMM)* can be used to describe the joint probability of a sequence of unobserved (hidden) discrete state variables, $\mathbf{H} = (H_0, \ldots, H_n)$, and a sequence of corresponding observed variables $\mathbf{O} = (O_0, \ldots, O_n)$ for which O_i is dependent on H_i for each i. We say that H_i emits O_i; consideration here is limited to discrete emission variables. Let the state spaces for elements of $\mathbf{H}$ and $\mathbf{O}$ be $\mathcal{H}$ and $\mathcal{E}$, respectively.

Let $\mathbf{O}_{\leq j}$ and $\mathbf{O}_{>j}$ denote the portions of $\mathbf{O}$ with indices not exceeding j and exceeding j, respectively, and define the analogous partial sequences for $\mathbf{H}$. Under a HMM, the H_i have the Markov property

$$P[H_i | \mathbf{H}_{\leq i-1}, O_0] = P[H_i | H_{i-1}] \tag{4.90}$$

and the emissions are conditionally independent, so

$$P[O_i | \mathbf{H}, \mathbf{O}_{\leq i-1}, \mathbf{O}_{>i}] = P[O_i | H_i]. \tag{4.91}$$

Time-homogeneous transitions of the hidden states are governed by transition probabilities $p(h, h^*) = P[H_{i+1} = h^* | H_i = h]$ for $h, h^* \in \mathcal{H}$. The distribution for H_0 is parameterized by $\pi(h) = P[H_0 = h]$ for $h \in \mathcal{H}$. Finally, define emission probabilities $e(h, o) = P[O_i = o | H_i = h]$ for $h \in \mathcal{H}$ and $o \in \mathcal{E}$. Then the parameter set $\theta = (\boldsymbol{\pi}, \mathbf{P}, \mathbf{E})$ completely parameterizes the model, where $\boldsymbol{\pi}$ is a vector of initial-state probabilities, $\mathbf{P}$ is a matrix of transition probabilities, and $\mathbf{E}$ is a matrix of emission probabilities.

For an observed sequence $\mathbf{o}$, define the *forward variables* to be

$$\alpha(i, h) = P[\mathbf{O}_{\leq i} = \mathbf{o}_{\leq i}, \ H_i = h] \tag{4.92}$$

and the *backward variables* to be

$$\beta(i, h) = P[\mathbf{O}_{>i} = \mathbf{o}_{>i} | H_i = h] \tag{4.93}$$

for $i = 1, \ldots, n$ and each $h \in \mathcal{H}$. Our notation suppresses the dependence of the forward and backward variables on θ. Note that

$$P[\mathbf{O} = \mathbf{o} | \theta] = \sum_{h \in \mathcal{H}} \alpha(n, h) = \sum_{h \in \mathcal{H}} \pi(h) e(h, o_0) \beta(0, h). \tag{4.94}$$

The forward and backward variables are also useful for computing the probability that state h occurred at the ith position of the sequence given $\mathbf{O} = \mathbf{o}$ according to

$P[H_i = h | \mathbf{O} = \mathbf{o}, \boldsymbol{\theta}] = \sum_{h \in H} \alpha(i, h)\beta(i, h)/P[\mathbf{O} = \mathbf{o} | \boldsymbol{\theta}]$, and expectations of functions of the states with respect to these probabilities.

(a) Show that the following algorithms can be used to calculate $\alpha(i, h)$ and $\beta(i, h)$. The *forward algorithm* is

- Initialize $\alpha(0, h) = \pi(h)e(h, o_0)$.
- For $i = 0, \ldots, n-1$, let $\alpha(i+1, h) = \sum_{h^* \in \mathcal{H}} \alpha(i, h^*)p(h^*, h)e(h, o_{i+1})$.

The *backward algorithm* is

- Initialize $\beta(n, h) = 1$.
- For $i = n, \ldots, 1$, let $\beta(i - 1, h) = \sum_{h^* \in \mathcal{H}} p(h, h^*)e(h^*, o_i)\beta(h, i)$.

These algorithms provide very efficient methods for finding $P[\mathbf{O} = \mathbf{o} | \boldsymbol{\theta}]$ and other useful probabilities, compared to naively summing over all possible sequences of states.

(b) Let $N(h)$ denote the number of times $H_0 = h$, let $N(h, h^*)$ denote the number of transitions from h to h^*, and let $N(h, o)$ denote the number of emissions of o when the underlying state is h. Prove that these random variables have the following expectations:

$$E\{N(h)\} = \frac{\alpha(0, h)\beta(0, h)}{P[\mathbf{O} = \mathbf{o} | \boldsymbol{\theta}]}, \tag{4.95}$$

$$E\{N(h, h^*)\} = \sum_{i=0}^{n-1} \frac{\alpha(i, h)p(h, h^*)e(h^*, o_{i+1})\beta(i + 1, h^*)}{P[\mathbf{O} = \mathbf{o} | \boldsymbol{\theta}]}, \tag{4.96}$$

$$E\{N(h, o)\} = \sum_{i:\, O_i = o} \frac{\alpha(i, h)\beta(i, h)}{P[\mathbf{O} = \mathbf{o} | \boldsymbol{\theta}]}. \tag{4.97}$$

(c) The *Baum–Welch algorithm* efficiently estimates the parameters of a HMM [22]. Fitting these models has proven extremely useful in diverse applications including statistical genetics, signal processing and speech recognition, problems involving environmental time series, and Bayesian graphical networks [149, 207, 317, 342, 441]. Starting from some initial values $\boldsymbol{\theta}^{(0)}$, the Baum–Welch algorithm proceeds via iterative application of the following update formulas:

$$\pi(h)^{(t+1)} = \frac{E\{N(h) | \boldsymbol{\theta}^{(t)}\}}{\sum_{h^* \in \mathcal{H}} E\{N(h^*) | \boldsymbol{\theta}^{(t)}\}}, \tag{4.98}$$

$$p(h, h^*)^{(t+1)} = \frac{E\{N(h, h^*) | \boldsymbol{\theta}^{(t)}\}}{\sum_{h^{**} \in \mathcal{H}} E\{N(h, h^{**}) | \boldsymbol{\theta}^{(t)}\}}, \tag{4.99}$$

$$e(h, o)^{(t+1)} = \frac{E\{N(h, o) | \boldsymbol{\theta}^{(t)}\}}{\sum_{o^* \in \mathcal{E}} E\{N(h, o^*) | \boldsymbol{\theta}^{(t)}\}}. \tag{4.100}$$

Prove that the Baum–Welch algorithm is an EM algorithm. It is useful to begin by noting that the complete data likelihood is given by

$$\prod_{h \in \mathcal{H}} \pi(h)^{N(h)} \prod_{h \in \mathcal{H}} \prod_{o \in \mathcal{E}} e(h, o)^{N(h,o)} \prod_{h \in \mathcal{H}} \prod_{h^* \in \mathcal{H}} p(h, h^*)^{N(h,h^*)}. \quad (4.101)$$

(d) Consider the following scenario. In Flip's left pocket is a penny; in his right pocket is a dime. On a fair toss, the probability of showing a head is p for the penny and d for the dime. Flip randomly chooses a coin to begin, tosses it, and reports the outcome (heads or tails) without revealing which coin was tossed. Then, Flip decides whether to use the same coin for the next toss, or to switch to the other coin. He switches coins with probability s, and retains the same coin with probability $1 - s$. The outcome of the second toss is reported, again not revealing the coin used. This process is continued for a total of 200 coin tosses. The resulting sequence of heads and tails is available from the website for this book. Use the Baum–Welch algorithm to estimate p, d, and s.

(e) Only for students seeking extra challenge: Derive the Baum–Welch algorithm for the case when the dataset consists of M independent observation sequences arising from a HMM. Simulate such data, following the coin example above. (You may wish to mimic the single-sequence data, which were simulated using $p = 0.25$, $d = 0.85$, and $s = 0.1$.) Code the Baum–Welch algorithm, and test it on your simulated data.

In addition to considering multiple sequences, HMMs and the Baum–Welch algorithm can be generalized for estimation based on more general emission variables and emission and transition probabilities that have more complex parameterizations, including time inhomogeneity.

5

Numerical Integration

Consider a one-dimensional integral of the form $\int_a^b f(x)\,dx$. For only a few functions f can the value of the integral be derived analytically. For the rest, numerical approximations of the integral are often useful. Approximation methods are well known by both numerical analysts [120, 310, 328, 436] and statisticians [349, 534].

Approximation of integrals is frequently required for Bayesian inference, since a posterior distribution may not belong to a familiar distributional family. Integral approximation is also useful in some maximum likelihood inference problems when the likelihood itself is a function of one or more integrals. An example of this occurs when fitting generalized linear mixed models, as discussed in Example 5.1 below.

To initiate an approximation of $\int_a^b f(x)\,dx$, partition the interval $[a, b]$ into n subintervals, $[x_i, x_{i+1}]$ for $i = 0, \ldots, n - 1$, with $x_0 = a$ and $x_n = b$. Then $\int_a^b f(x)\,dx = \sum_{i=0}^{n-1} \int_{x_i}^{x_{i+1}} f(x)\,dx$. This *composite rule* breaks the whole integral into many smaller parts, but postpones the question of how to approximate any single part.

The approximation of a single part will be made using a *simple rule*. Within the interval $[x_i, x_{i+1}]$, insert $m + 1$ *nodes*, x_{ij}^* for $j = 0, \ldots, m$. Figure 5.1 illustrates the relationships among the interval $[a, b]$, the subintervals, and the nodes. In general, numerical integration methods require neither equal spacing of subintervals or nodes, nor equal numbers of nodes within each subinterval.

A simple rule will rely upon the approximation

$$\int_{x_i}^{x_{i+1}} f(x)\,dx \approx \sum_{j=0}^{m} A_{ij} f\left(x_{ij}^*\right) \tag{5.1}$$

121

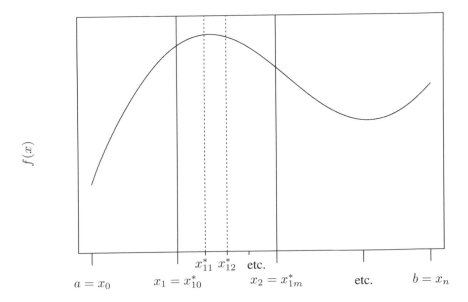

Fig. 5.1 To integrate f between a and b, the interval is partitioned into n subintervals, $[x_i, x_{i+1}]$, each of which is further partitioned using $m + 1$ nodes, $x_{i0}^*, \ldots, x_{im}^*$. Note that when $m = 0$, the subinterval $[x_i, x_{i+1}]$ contains only one interior node, $x_{i0}^* = x_i$.

for some set of constants A_{ij}. The overall integral is then approximated by the composite rule that sums (5.1) over all subintervals.

5.1   NEWTON–CÔTES QUADRATURE

A simple and flexible class of integration methods consists of the Newton–Côtes rules. In this case, the nodes are equally spaced in $[x_i, x_{i+1}]$, and the same number of nodes is used in each subinterval. The Newton–Côtes approach replaces the true integrand with a polynomial approximation on each subinterval. The constants A_{ij} are selected so that $\sum_{j=0}^{m} A_{ij} f\left(x_{ij}^*\right)$ equals the integral of an interpolating polynomial on $[x_i, x_{i+1}]$ that matches the value of f at the nodes within this subinterval. The remainder of this section reviews common Newton–Côtes rules.

5.1.1   Riemann rule

Consider the case when $m = 0$. Suppose we define $x_{i0}^* = x_i$, and $A_{i0} = x_{i+1} - x_{i0}^*$. The simple Riemann rule amounts to approximating f on each subinterval by a constant function, $f(x_i)$, whose value matches that of f at one point on the interval.

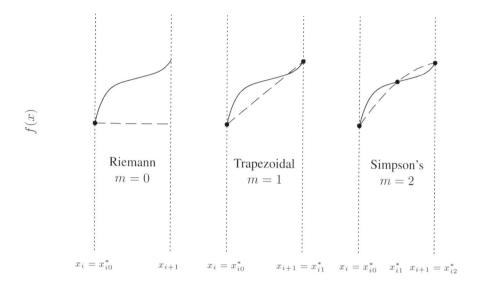

Fig. 5.2 The approximation (dashed) to f (solid) provided on the subinterval $[x_i, x_{i+1}]$, for the Riemann, trapezoidal, and Simpson's rules.

In other words,

$$\int_{x_i}^{x_{i+1}} f(x)\, dx \approx \int_{x_i}^{x_{i+1}} f(x_i)\, dx = (x_{i+1} - x_i)\, f(x_i). \tag{5.2}$$

The composite rule sums n such terms to provide an approximation to the integral over $[a, b]$.

Suppose the x_i are equally spaced so that each subinterval has the same length $h = (b - a)/n$. Then we may write $x_i = a + ih$, and the composite rule is

$$\int_{x_i}^{x_{i+1}} f(x)\, dx \approx h \sum_{i=0}^{n-1} f(a + ih) = \widehat{R}(n). \tag{5.3}$$

As Figure 5.2 shows, this corresponds to the Riemann integral studied in elementary calculus. Furthermore, there is nothing special about the left endpoints of the subintervals: We easily could have replaced $f(x_i)$ with $f(x_{i+1})$ in (5.2).

The approximation given by (5.3) converges to the true value of the integral as $n \to \infty$, by definition of the Riemann integral of an integrable function. If f is a polynomial of zero degree (i.e., a constant function), then f is constant on each subinterval, so the Riemann rule is exact.

When applying the composite Riemann rule, it makes sense to calculate a sequence of approximations, say $\widehat{R}(n_k)$, for an increasing sequence of numbers of subintervals, n_k, as $k = 1, 2, \ldots$. Then, convergence of $\widehat{R}(n_k)$ can be monitored using an absolute

Table 5.1 Words recalled on five consecutive monthly tests for 22 Alzheimer's patients receiving lecithin.

Month	Patient										
	1	2	3	4	5	6	7	8	9	10	11
1	9	6	13	9	6	11	7	8	3	4	11
2	12	7	18	10	7	11	10	18	3	10	10
3	16	10	14	12	8	12	11	19	3	11	10
4	17	15	21	14	9	14	12	19	7	17	15
5	18	16	21	15	12	16	14	22	8	18	16

Month	Patient										
	12	13	14	15	16	17	18	19	20	21	22
1	1	6	0	18	15	10	6	9	4	4	10
2	3	7	3	18	15	14	6	9	3	13	11
3	2	7	3	19	15	16	7	13	4	13	13
4	4	9	4	22	18	17	9	16	7	16	17
5	5	10	6	22	19	19	10	20	9	19	21

or relative convergence criterion as discussed in Chapter 2. It is particularly efficient to use $n_{k+1} = 2n_k$ so that half the subinterval endpoints at the next step correspond to the old endpoints from the previous step. This avoids calculations of f that are effectively redundant.

Example 5.1 (Alzheimer's disease) Data from 22 patients with Alzheimer's disease, an ailment characterized by progressive mental deterioration, are shown in Table 5.1. In each of five consecutive months, patients were asked to recall words from a standard list given previously. The number of words recalled by each patient was recorded. The patients in Table 5.1 were receiving an experimental treatment with lecithin, a dietary supplement. It is of interest to investigate whether memory improved over time. The data for these patients (and 25 control cases) are available from the website for this book and are discussed further in [134].

Consider fitting these data with a very simple generalized linear mixed model [63, 571]. Let Y_{ij} denote the number of words recalled by the ith individual in the jth month, for $i = 1, \ldots, 22$ and $j = 1, \ldots, 5$. Suppose $Y_{ij}|\lambda_{ij}$ have independent Poisson(λ_{ij}) distributions, where the mean and variance of Y_{ij} is λ_{ij}. Let $x_{ij} = (1\ j)^T$ be a covariate vector: Aside from an intercept term, only the month is used as a predictor. Let $\beta = (\beta_0\ \beta_1)^T$ be a parameter vector corresponding to x. Then we model the mean of Y_{ij} as

$$\lambda_{ij} = \exp\{x_{ij}^T\beta + \gamma_i\}, \tag{5.4}$$

where the γ_i are independent $N(0, \sigma_\gamma^2)$ random effects. This model allows separate shifts in λ_{ij} on the log scale for each patient, reflecting the assumption that there may be substantial between-patient variation in counts. This is reasonable, for example, if the baseline conditions of patients prior to the start of treatment varied.

With this model, the likelihood is

$$L\left(\boldsymbol{\beta}, \sigma_\gamma^2 | \mathbf{y}\right) = \prod_{i=1}^{22} \int \left[\phi(\gamma_i; 0, \sigma_\gamma^2) \prod_{j=1}^{5} f(y_{ij}|\lambda_{ij})\right] d\gamma_i$$

$$= \prod_{i=1}^{22} L_i\left(\boldsymbol{\beta}, \sigma_\gamma^2 | \mathbf{y}\right), \tag{5.5}$$

where $f(y_{ij}|\lambda_{ij})$ is the Poisson density, $\phi(\gamma_i; 0, \sigma_\gamma^2)$ is the normal density function with mean zero and variance σ_γ^2, and $\mathbf{Y}$ is a vector of all the observed response values. The log likelihood is therefore

$$l\left(\boldsymbol{\beta}, \sigma_\gamma^2 | \mathbf{y}\right) = \sum_{i=1}^{22} l_i\left(\boldsymbol{\beta}, \sigma_\gamma^2 | \mathbf{y}\right), \tag{5.6}$$

where l_i denotes the contribution to the log likelihood made by data from the ith patient.

To maximize the log likelihood, we must differentiate l with respect to each parameter and solve the corresponding score equations. This will require a numerical root-finding procedure, since the solution cannot be determined analytically. In this example, we look only at one small portion of this overall process: the evaluation of $\frac{dl_i}{d\beta_k}$ for particular given values of the parameters and for a single i and k. This evaluation would be repeated for the parameter values tried at each iteration of a root-finding procedure.

Let $i = 1$ and $k = 1$. The partial derivative with respect to the parameter for monthly rate of change is $\frac{dl_1}{d\beta_1} = \frac{dL_1}{d\beta_1}/L_1$, where L_1 is implicitly defined in (5.5). Further,

$$\frac{dL_1}{d\beta_1} = \frac{d}{d\beta_1} \int \left[\phi(\gamma_1; 0, \sigma_\gamma^2) \prod_{j=1}^{5} f(y_{1j}|\lambda_{1j})\right] d\gamma_1$$

$$= \int \frac{d}{d\beta_1} \left[\phi(\gamma_1; 0, \sigma_\gamma^2) \prod_{j=1}^{5} f(y_{1j}|\lambda_{1j})\right] d\gamma_1$$

$$= \int \phi(\gamma_1; 0, \sigma_\gamma^2) \left(\sum_{j=1}^{5} j(y_{1j} - \lambda_{1j})\right) \prod_{j=1}^{5} f(y_{1j}|\lambda_{1j}) \, d\gamma_1, \tag{5.7}$$

where $\lambda_{1j} = \exp\{\beta_0 + j\beta_1 + \gamma_1\}$. The last equality in (5.7) follows from standard analysis of generalized linear models [379].

Suppose, at the very first step of optimization, we start with initial values $\boldsymbol{\beta} = (1.804, \ 0.165)$ and $\sigma_\gamma^2 = 0.015^2$. These starting values were chosen using simple

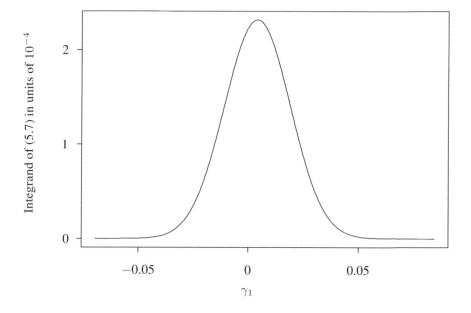

Fig. 5.3   Example 5.1 seeks to integrate this function, which arises from a generalized linear mixed model for data on patients treated for Alzheimer's disease.

exploratory analyses. Using these values for β and σ_γ^2, the integral we seek in (5.7) has the integrand shown in Figure 5.3. The desired range of integration is the entire real line, whereas we have thus far only discussed integration over a closed interval. Transformations can be used to obtain an equivalent interval over a finite range (see Section 5.4.1), but to keep things simple here we will integrate over the range $[-0.07, 0.085]$, within which nearly all of the nonnegligible values of the integrand lie.

Table 5.2 shows the results of a series of Riemann approximations, along with running relative errors. The relative errors measure the change in the new estimated value of the integral as a proportion of the previous estimate. An iterative approximation strategy could be stopped when the magnitude of these errors falls below some predetermined tolerance threshold. Since the integral is quite small, a relative convergence criterion is more intuitive than an absolute criterion.   □

5.1.2   Trapezoidal rule

Although the simple Riemann rule is exact if f is constant on $[a, b]$, it can be quite slow to converge to adequate precision in general. An obvious improvement would be to replace the piecewise constant approximation by a piecewise mth-degree polynomial approximation. We begin by introducing a class of polynomials that can be used

Table 5.2 Estimates of the integral in (5.7) using the Riemann rule with various numbers of subintervals. All estimates are multiplied by a factor of 10^5. Errors for use in a relative convergence criterion are given in the final column.

Subintervals	Estimate	Relative Error
2	3.49388458186769	
4	1.88761005959780	−0.46
8	1.72890354401971	−0.084
16	1.72889046749119	−0.0000076
32	1.72889038608621	−0.000000047
64	1.72889026784032	−0.000000068
128	1.72889018400995	−0.000000048
256	1.72889013551548	−0.000000028
512	1.72889010959701	−0.000000015
1024	1.72889009621830	−0.0000000077

for such approximations. This permits the Riemann rule to be cast as the simplest member of a family integration rules having increased precision as m increases. This family also includes the trapezoidal rule and Simpson's rule (Section 5.1.3).

Let the *fundamental polynomials* be

$$p_{ij}(x) = \prod_{k=0,k\neq j}^{m} \frac{x - x_{ik}^*}{x_{ij}^* - x_{ik}^*} \tag{5.8}$$

for $j = 0, \ldots, m$. Then the function $p_i(x) = \sum_{j=0}^{m} f\left(x_{ij}^*\right) p_{ij}(x)$ is an mth-degree polynomial that interpolates f at all the nodes $x_{i0}^*, \ldots, x_{im}^*$ in $[x_i, x_{i+1}]$. Figure 5.2 shows such interpolating polynomials for $m = 0, 1,$ and 2.

These polynomials are the basis for the simple approximation

$$\int_{x_i}^{x_{i+1}} f(x)\, dx \approx \int_{x_i}^{x_{i+1}} p_i(x)\, dx \tag{5.9}$$

$$= \sum_{j=0}^{m} f\left(x_{ij}^*\right) \int_{x_i}^{x_{i+1}} p_{ij}(x)\, dx \tag{5.10}$$

$$= \sum_{j=0}^{m} A_{ij} f\left(x_{ij}^*\right) \tag{5.11}$$

for $A_{ij} = \int_{x_i}^{x_{i+1}} p_{ij}(x)\, dx$. This approximation replaces integration of an arbitrary function f with polynomial integration. The resulting composite rule is $\int_a^b f(x)\, dx \approx \sum_{i=0}^{n-1} \sum_{j=0}^{m} A_{ij} f\left(x_{ij}^*\right)$ when there are m nodes on each subinterval.

Letting $m = 1$ with $x_{i0}^* = x_i$ and $x_{i1}^* = x_{i+1}$ yields the trapezoidal rule. In this case, $p_{i0}(x) = \frac{x - x_{i+1}}{x_i - x_{i+1}}$ and $p_{i1}(x) = \frac{x - x_i}{x_{i+1} - x_i}$. Integrating these polynomials yields

Table 5.3   Estimates of the integral in (5.7) using the trapezoidal rule with various numbers of subintervals. All estimates are multiplied by a factor of 10^5. Errors for use in a relative convergence criterion are given in the final column.

Subintervals	Estimate	Relative Error
2	3.49387751694744	
4	1.88760652713768	−0.46
8	1.72890177778965	−0.084
16	1.72888958437616	−0.0000071
32	1.72888994452869	0.00000021
64	1.72889004706156	0.000000059
128	1.72889007362057	0.000000015
256	1.72889008032079	0.0000000039
512	1.72889008199967	0.00000000097
1024	1.72889008241962	0.00000000024

$A_{i0} = A_{i1} = (x_{i+1} - x_i)/2$. Therefore, the trapezoidal rule amounts to

$$\int_a^b f(x)\,dx \approx \sum_{i=0}^{n-1} \left(\frac{x_{i+1} - x_i}{2} \right) \Big(f(x_i) + f(x_{i+1}) \Big). \tag{5.12}$$

When $[a, b]$ is partitioned into n subintervals of equal length $h = (b - a)/n$, then the trapezoidal rule estimate is

$$\int_a^b f(x)\,dx \approx \frac{h}{2} f(a) + h \sum_{i=1}^{n-1} f(a + ih) + \frac{h}{2} f(b) = \widehat{T}(n). \tag{5.13}$$

The name of this approximation arises because the area under f in each subinterval is approximated by the area of a trapezoid, as shown in Figure 5.2. Note that f is approximated in any subinterval by a first-degree polynomial (i.e., a line) whose value equals that of f at two points. Therefore, when f itself is a line on $[a, b]$, $\widehat{T}(n)$ is exact.

Example 5.2 (Alzheimer's disease, continued)   For small numbers of subintervals, applying the trapezoidal rule to the integral from Example 5.1 yields similar results to those from the Riemann rule because the integrand is nearly zero at the endpoints of the integration range. For large numbers of subintervals, the approximation is somewhat better. The results are shown in Table 5.3.   ☐

Suppose f has two continuous derivatives. Problem 5.1 asks you to show that

$$p_i(x) = f(x_i) + f'(x_i)(x - x_i) + \frac{1}{2} f''(x_i)(x_{i+1} - x_i)(x - x_i) + \mathcal{O}(n^{-3}). \tag{5.14}$$

Subtracting the Taylor expansion of f about x_i from (5.14) yields

$$p_i(x) - f(x) = \frac{1}{2} f''(x_i)(x - x_i)(x - x_{i+1}) + \mathcal{O}(n^{-3}), \qquad (5.15)$$

and integrating (5.15) over $[x_i, x_{i+1}]$ shows the approximation error of the trapezoidal rule on the ith subinterval to be $h^3 f''(x_i)/12 + \mathcal{O}(n^{-4})$. Thus

$$\widehat{T}(n) - \int_a^b f(x)\, dx = \sum_{i=1}^{n} \left(\frac{h^3 f''(x_i)}{12} + \mathcal{O}(n^{-4}) \right) \qquad (5.16)$$

$$= nh^3 f''(\xi)/12 + \mathcal{O}(n^{-3}) \qquad (5.17)$$

$$= \frac{(b-a)^3 f''(\xi)}{12n^2} + \mathcal{O}(n^{-3}) \qquad (5.18)$$

for some $\xi \in [a, b]$ by the mean value theorem for integrals. Hence the leading term of the overall error is $\mathcal{O}(n^{-2})$.

5.1.3 Simpson's rule

Letting $m = 2$, $x_{i0}^* = x_i$, $x_{i1}^* = (x_i + x_{i+1})/2$, and $x_{i2}^* = x_{i+1}$ in (5.8), we obtain Simpson's rule. Problem 5.2 asks you to show that $A_{i0} = A_{i2} = (x_{i+1} - x_i)/6$ and $A_{i1} = 2(A_{i0} + A_{i2})$. This yields the approximation

$$\int_{x_i}^{x_{i+1}} f(x)\, dx \approx \frac{x_{i+1} - x_i}{6} \left[f(x_i) + 4f\left(\frac{x_i + x_{i+1}}{2} \right) + f(x_{i+1}) \right] \qquad (5.19)$$

for the $(i+1)$th subinterval. Figure 5.2 shows how Simpson's rule provides a quadratic approximation to f on each subinterval.

Suppose the interval $[a, b]$ has been partitioned into n subintervals of equal length $h = (b - a)/n$, where n is even. To apply Simpson's rule, we need an interior node in each $[x_i, x_{i+1}]$. Since n is even, we may adjoin pairs of adjacent subintervals, with the shared endpoint serving as the interior node of the larger interval. This provides $n/2$ subintervals of length $2h$, for which

$$\int_a^b f(x)\, dx \approx \frac{h}{3} \sum_{i=1}^{n/2} \left(f(x_{2i-2}) + 4f(x_{2i-1}) + f(x_{2i}) \right) = \widehat{S}(n/2). \qquad (5.20)$$

Example 5.3 (Alzheimer's disease, continued) Table 5.4 shows the results of applying Simpson's rule to the integral from Example 5.1. One endpoint and an interior node were evaluated on each subinterval. Thus, for a fixed number of subintervals, Simpson's rule requires twice as many evaluations of f as the Riemann or the trapezoidal rule. Following this example, we show that the precision of Simpson's rule more than compensates for the increased evaluations. From another perspective, if the number of evaluations of f is fixed at n for each method, we would expect Simpson's rule to outperform the previous approaches, if n is large enough. □

Table 5.4   Estimates of the integral in (5.7) using Simpson's rule with various numbers of subintervals (and two nodes per subinterval). All estimates are multiplied by a factor of 10^5. Errors for use in a relative convergence criterion are given in the final column.

Subintervals	Estimate	Relative Error
2	1.35218286386776	
4	1.67600019467364	0.24
8	1.72888551990500	0.032
16	1.72889006457954	0.0000026
32	1.72889008123918	0.0000000096
64	1.72889008247358	0.00000000071
128	1.72889008255419	0.000000000047
256	1.72889008255929	0.0000000000029
512	1.72889008255961	0.00000000000018
1024	1.72889008255963	0.000000000000014

If f is quadratic on $[a, b]$, then it is quadratic on each subinterval. Simpson's rule approximates f on each subinterval by a second-degree polynomial that matches f at three points; therefore the polynomial is exactly f. Thus, Simpson's rule exactly integrates quadratic f.

Suppose f is smooth—but not polynomial—and we have n subintervals $[x_i, x_{i+1}]$ of equal length $2h$. To assess the degree of approximation in Simpson's rule, we begin with consideration on a single subinterval, and denote the simple Simpson's rule on that subinterval as $\widehat{S}_i(n) = \frac{h}{3}\left[f(x_i) + 4f(x_i + h) + f(x_i + 2h)\right]$. Denote the true value of the integral on that subinterval as I_i.

We use the Taylor series expansion of f about x_i, evaluated at $x = x_i + h$ and $x = x_i + 2h$, to replace terms in $\widehat{S}_i(n)$. Combining terms, this yields

$$\widehat{S}_i(n) = 2hf(x_i) + 2h^2 f'(x_i) + \frac{4}{3}h^3 f''(x_i)$$
$$+ \frac{2}{3}h^4 f'''(x_i) + \frac{100}{360}h^5 f''''(x_i) + \cdots. \quad (5.21)$$

Now let $F(x) = \int_{x_i}^{x} f(t)\, dt$. This function has the useful properties that $F(x_i) = 0$, $F(x_i + 2h) = I_i$, and $F'(x) = f(x)$. Taylor series expansion of F about x_i, evaluated at $x = x_i + 2h$, yields

$$I_i = 2hf(x_i) + 2h^2 f'(x_i) + \frac{4}{3}h^3 f''(x_i)$$
$$+ \frac{2}{3}h^4 f'''(x_i) + \frac{32}{120}h^5 f''''(x_i) + \cdots. \quad (5.22)$$

Subtracting (5.22) from (5.21) yields $\widehat{S}_i(n) - I_i = h^5 f''''(x_i)/90 + \cdots = \mathcal{O}(n^{-5})$. This is the error for Simpson's rule on a single subinterval. Over the n

subintervals that partition $[a, b]$, the error is therefore the sum of such errors, namely $\mathcal{O}\left(n^{-4}\right)$. Note that Simpson's rule therefore exactly integrates cubic functions, too.

5.1.4 General kth-degree rule

The preceding discussion raises the general question about how to determine a Newton–Côtes rule that is exact for polynomials of degree k. This would require constants $c_0, \ldots, c_k$ that satisfy

$$
\int_a^b f(x)\, dx = c_0 f(a) + c_1 f\left(a + \frac{b-a}{k}\right) + \cdots
$$
$$
+ c_i f\left(a + \frac{i(b-a)}{k}\right) + \cdots + c_k f(b) \tag{5.23}
$$

for any polynomial f. Of course, one could follow the derivations shown above for $m = k$, but there is another simple approach. If a method is to be exact for all polynomials of degree k, then it must be exact for particular—and easily integrated—choices like $1, x, x^2, \ldots, x^k$. Thus, we may set up a system of k equations in k unknowns as follows:

$$
\int_a^b 1\, dx = b - a = c_0 + \cdots + c_k,
$$
$$
\int_a^b x\, dx = \frac{b^2 - a^2}{2}
$$
$$
= c_0 a + c_1 \left(a + \frac{b-a}{k}\right) + \cdots + c_k b,
$$
$$
\vdots
$$
$$
\int_a^b x^k\, dx = \text{etc.}
$$

All that remains is to solve for the c_i to derive the algorithm. This approach is sometimes called the *method of undetermined coefficients*.

5.2 ROMBERG INTEGRATION

In general, low-degree Newton–Côtes methods are slow to converge. However, there is a very efficient mechanism to improve upon a sequence of trapezoidal rule estimates. Let $\widehat{T}(n)$ denote the trapezoidal rule estimate of $\int_a^b f(x)\, dx$ using n subintervals of equal length $h = (b-a)/n$, as given in (5.13). Without loss of generality, suppose

$a = 0$ and $b = 1$. Then

$$\widehat{T}(1) = \frac{1}{2}f(0) + \frac{1}{2}f(1),$$

$$\widehat{T}(2) = \frac{1}{4}f(0) + \frac{1}{2}f(1/2) + \frac{1}{4}f(1),$$

$$\widehat{T}(4) = \frac{1}{8}f(0) + \frac{1}{4}[f(1/4) + f(1/2) + f(3/4)] + \frac{1}{8}f(1), \qquad (5.24)$$

and so forth. Noting that

$$\widehat{T}(2) = \frac{1}{2}\widehat{T}(1) + \frac{1}{2}f(1/2),$$

$$\widehat{T}(4) = \frac{1}{2}\widehat{T}(2) + \frac{1}{4}[f(1/4) + f(3/4)], \qquad (5.25)$$

and so forth, suggests the general recursion relationship

$$\widehat{T}(2n) = \frac{1}{2}\widehat{T}(n) + \frac{h}{2}\sum_{i=1}^{n} f\big(a + (i - 1/2)h\big). \qquad (5.26)$$

The Euler–Maclaurin formula (1.8) can be used to show that

$$\widehat{T}(n) = \int_a^b f(x)\,dx + c_1 h^2 + \mathcal{O}\left(n^{-4}\right) \qquad (5.27)$$

for some constant c_1, and hence

$$\widehat{T}(2n) = \int_a^b f(x)\,dx + \frac{c_1}{4}h^2 + \mathcal{O}\left(n^{-4}\right). \qquad (5.28)$$

Therefore,

$$\frac{4\widehat{T}(2n) - \widehat{T}(n)}{3} = \int_a^b f(x)\,dx + \mathcal{O}\left(n^{-4}\right), \qquad (5.29)$$

so the h^2 error terms in (5.27) and (5.28) cancel. With this simple adjustment we have made a striking improvement in the estimate. In fact, the estimate given in (5.29) turns out to be Simpson's rule with subintervals of width $h/2$. Moreover, this strategy can be iterated for even greater gains.

Begin by defining $\widehat{T}_{i,0} = \widehat{T}\left(2^i\right)$ for $i = 0, \ldots, m$. Then define a triangular array of estimates like

$$
\begin{array}{cccccc}
\widehat{T}_{0,0} & & & & & \\
\widehat{T}_{1,0} & \widehat{T}_{1,1} & & & & \\
\widehat{T}_{2,0} & \widehat{T}_{2,1} & \widehat{T}_{2,2} & & & \\
\widehat{T}_{3,0} & \widehat{T}_{3,1} & \widehat{T}_{3,2} & \widehat{T}_{3,3} & & \\
\widehat{T}_{4,0} & \widehat{T}_{4,1} & \widehat{T}_{4,2} & \widehat{T}_{4,3} & \widehat{T}_{4,4} & \\
\vdots & \vdots & \vdots & \vdots & \vdots & \ddots
\end{array}
$$

using the relationship

$$\widehat{T}_{i,j} = \frac{4^j \widehat{T}_{i,j-1} - \widehat{T}_{i-1,j-1}}{4^j - 1} \tag{5.30}$$

for $j = 1, \ldots, i$ and $i = 1, \ldots, m$. Note that (5.30) can be reexpressed to calculate $\widehat{T}_{i,j}$ by adding an increment equal to $\frac{1}{4^j-1}$ times the difference $\widehat{T}_{i,j-1} - \widehat{T}_{i-1,j-1}$ to the estimate given by $\widehat{T}_{i,j-1}$.

If f has $2m$ continuous derivatives on $[a, b]$, then the entries in the mth row of the array have error $\widehat{T}_{m,j} - \int_a^b f(x)\, dx = \mathcal{O}\left(2^{-2mj}\right)$ for $j \leq m$ [103, 328]. This is such fast convergence that very small m will often suffice.

It is important to check that the Romberg calculations do not deteriorate as m is increased. To do this, consider the quotient

$$Q_{ij} = \frac{\widehat{T}_{i,j} - \widehat{T}_{i-1,j}}{\widehat{T}_{i+1,j} - \widehat{T}_{i,j}}. \tag{5.31}$$

The error in $\widehat{T}_{i,j}$ is attributable partially to the approximation strategy itself, and partially to numerical imprecision introduced by computer roundoff. As long as the former source dominates, the Q_{ij} values should approach 4^{j+1} as i increases. However, when computer roundoff error is substantial relative to approximation error, the Q_{ij} values will become erratic. The columns of the triangular array of $\widehat{T}_{i,j}$ can be examined to determine the largest j for which the quotients appear to approach 4^{j+1} before deteriorating. No further column should be used to calculate an update via (5.30). The following example illustrates the approach.

Example 5.4 (Alzheimer's disease, continued) Table 5.5 shows the results of applying Romberg integration to the integral from Example 5.1. The right columns of this table are used to diagnose the stability of the Romberg calculations. The top portion of the table corresponds to $j = 0$, and the $\widehat{T}_{i,j}$ are the trapezoidal rule estimates given in Table 5.3. After some initial steps, the quotients in the top portion of the table converge nicely to 4. Therefore, it is safe and advisable to apply (5.30) to generate a second column of the triangular array. It is safe because the convergence of the quotients to 4 implies that computer roundoff error has not yet become a dominant source of error. It is advisable because incrementing one of the current integral estimates by one-third of the corresponding difference would yield a noticeably different updated estimate.

The second column of the triangular array is shown in the middle portion of Table 5.5. The quotients in this portion also appear reasonable, so the third column is calculated and shown in the bottom portion of the table. The values of Q_{i2} approach 64, allowing more tolerance for larger j. At $i = 10$, computer roundoff error appears to dominate approximation error, because the quotient departs from near 64. However, note that incrementing the integral estimate by 1/63 of the difference at this point would have a negligible impact on the updated estimate itself. Had we proceeded one more step with the reasoning that the growing amount of roundoff error will cause little harm at this point, we would have found that the estimate was not improved

Table 5.5   Estimates of the integral in (5.7) using Romberg integration. All estimates and differences are multiplied by a factor of 10^5. The final two columns provide performance evaluation measures discussed in the text.

i	j	Subintervals	$\widehat{T}_{i,0}$	$\widehat{T}_{i,j} - \widehat{T}_{i-1,j}$	Q_{ij}
1	0	2	3.49387751694744		
2	0	4	1.88760652713768	−1.60627098980976	
3	0	8	1.72890177778965	−0.15870474934803	10.12
4	0	16	1.72888958437616	−0.00001219341349	13015.61
5	0	32	1.72888994452869	0.00000036015254	−33.86
6	0	64	1.72889004706156	0.00000010253287	3.51
7	0	128	1.72889007362057	0.00000002655901	3.86
8	0	256	1.72889008032079	0.00000000670022	3.96
9	0	512	1.72889008199967	0.00000000167888	3.99
10	0	1024	1.72889008241962	0.00000000041996	4.00
1	1	2			
2	1	4	1.35218286386776		
3	1	8	1.67600019467364	0.32381733080589	
4	1	16	1.72888551990500	0.05288532523136	6.12
5	1	32	1.72889006457954	0.00000454467454	11636.77
6	1	64	1.72889008123918	0.00000001665964	272.80
7	1	128	1.72889008247358	0.00000000123439	13.50
8	1	256	1.72889008255420	0.00000000008062	15.31
9	1	512	1.72889008255929	0.00000000000510	15.82
10	1	1024	1.72889008255961	0.00000000000032	16.14
1	2	2			
2	2	4			
3	2	8	1.69758801672736		
4	2	16	1.73241120825375	0.03482319152639	
5	2	32	1.72889036755784	−0.00352084069591	−9.89
6	2	64	1.72889008234983	−0.00000028520802	12344.82
7	2	128	1.72889008255587	0.00000000020604	−1384.21
8	2	256	1.72889008255957	0.00000000000370	55.66
9	2	512	1.72889008255963	0.00000000000006	59.38
10	2	1024	1.72889008255963	<0.00000000000001	−20.44

and the resulting quotients clearly indicated that no further extrapolations should be considered.

Thus, we may take $\widehat{T}_{9,2} = 1.72889008255963 \times 10^{-5}$ to be the estimated value of the integral. In this example, we calculated the triangular array one column at a time, for $m = 10$. However, in implementation it makes more sense to generate the array one row at a time. In this case, we would have stopped after $i = 9$, obtaining a precise estimate with fewer subintervals—and fewer evaluations of f—than in any of the previous examples.   □

The Romberg strategy can be applied to other Newton–Côtes integration rules. For example, if $\widehat{S}(n)$ is the Simpson's rule estimate of $\int_a^b f(x)\,dx$ using n subintervals of equal length, then the analogous result to (5.29) is

$$\frac{16\widehat{S}(2n) - \widehat{S}(n)}{15} = \int_a^b f(x)\,dx + \mathcal{O}\left(n^{-6}\right). \tag{5.32}$$

Romberg integration is a form of a more general strategy called Richardson extrapolation [283, 436].

5.3 GAUSSIAN QUADRATURE

All the Newton–Côtes rules discussed above are based on subintervals of equal length. The estimated integral is a sum of weighted evaluations of the integrand on a regular grid of points. For a fixed number of subintervals and nodes, only the weights may be flexibly chosen; we have limited attention to choices of weights that yield exact integration of polynomials. Using $m + 1$ nodes per subinterval allowed mth-degree polynomials to be integrated exactly.

An important question is the amount of improvement that can be achieved if the constraint of evenly spaced nodes and subintervals is removed. By allowing both the weights and the nodes to be freely chosen, we have twice as many parameters to use in the approximation of f. If we consider that the value of an integral is predominantly determined by regions where the magnitude of the integrand is large, then it makes sense to put more nodes in such regions. With a suitably flexible choice of $m + 1$ nodes, $x_0, \ldots, x_m$, and corresponding weights, $A_0, \ldots, A_m$, exact integration of $2(m+1)$th-degree polynomials can be obtained using $\int_a^b f(x)\,dx = \sum_{i=0}^{m} A_i f(x_i)$.

This approach, called *Gaussian quadrature*, can be extremely effective for integrals like $\int_a^b f(x)w(x)\,dx$ where w is a nonnegative function and $\int_a^b x^k w(x)\,dx < \infty$ for all $k \geq 0$. These requirements are reminiscent of density function with finite moments. Indeed, it is often useful to think of w as a density, in which case integrals like expected values and Bayesian posterior normalizing constants are natural candidates for Gaussian quadrature. The method is more generally applicable, however, by defining $f^*(x) = f(x)/w(x)$ and applying the method to $\int_a^b f^*(x)w(x)\,dx$.

The best node locations turn out to be the roots of a set of orthogonal polynomials that is determined by w.

5.3.1   Orthogonal polynomials

Some background on orthogonal polynomials is needed to develop Gaussian quadrature methods [2, 120, 343, 525]. Let $p_k(x)$ denote a generic polynomial of degree k. For convenience in what follows, assume that the leading coefficient of $p_k(x)$ is positive.

If $\int_a^b f(x)^2 w(x)\,dx < \infty$, then the function f is said to be *square-integrable* with respect to w on $[a, b]$. In this case we will write $f \in \mathcal{L}^2_{w,[a,b]}$. For any f and g in $\mathcal{L}^2_{w,[a,b]}$, their *inner product* with respect to w on $[a, b]$ is defined to be

$$\langle f, g \rangle_{w,[a,b]} = \int_a^b f(x)g(x)w(x)\,dx. \tag{5.33}$$

If $\langle f, g \rangle_{w,[a,b]} = 0$, then f and g are said to be *orthogonal* with respect to w on $[a, b]$. If also f and g are scaled so that $\langle f, f \rangle_{w,[a,b]} = \langle g, g \rangle_{w,[a,b]} = 1$, then f and g are *orthonormal* with respect to w on $[a, b]$.

Given any w that is nonnegative on $[a, b]$, there exists a sequence of polynomials $\{p_k(x)\}_{k=0}^{\infty}$ that are orthogonal with respect to w on $[a, b]$. This sequence is not unique without some form of standardization, because $\langle f, g \rangle_{w,[a,b]} = 0$ implies $\langle cf, g \rangle_{w,[a,b]} = 0$ for any constant c. The canonical standardization for a set of orthogonal polynomials depends on w and will be discussed later; a common choice is to set the leading coefficient of $p_k(x)$ equal to 1. For use in Gaussian quadrature, the range of integration is also customarily transformed from $[a, b]$ to a range $[a^*, b^*]$ whose choice depends on w.

A set of standardized, orthogonal polynomials can be summarized by a recurrence relation

$$p_k(x) = (\alpha_k + x\beta_k)\,p_{k-1}(x) - \gamma_k p_{k-2}(x) \tag{5.34}$$

for appropriate choices of α_k, β_k, and γ_k that vary with k and w.

The roots of any polynomial in such a standardized set are all in (a^*, b^*). These roots will serve as nodes for Gaussian quadrature. Table 5.6 lists several sets of orthogonal polynomials, their standardizations, and their correspondences to common density functions.

5.3.2   The Gaussian quadrature rule

Standardized orthogonal polynomials like (5.34) are important because they determine both the weights and the nodes for a Gaussian quadrature rule based on a chosen w. Let $\{p_k(x)\}_{k=0}^{\infty}$ be a sequence of orthonormal polynomials with respect to w on $[a, b]$ for a function w that meets the conditions previously discussed. Denote the roots of $p_{m+1}(x)$ by $a < x_0 < \ldots < x_m < b$. Then there exist weights $A_0, \ldots, A_m$ such that:

1. $A_i > 0$ for $i = 0, \ldots, m$.

2. $A_i = -c_{m+2} \,/\, \left[c_{m+1} p_{m+2}(x_i)\, p'_{m+1}(x_i) \right]$, where c_k is the leading coefficient of $p_k(x)$.

Table 5.6 Orthogonal polynomials, their standardizations, their correspondence to common density functions, and the terms used for their recursive generation. The leading coefficient of a polynomial is denoted c_k. In some cases, variants of standard definitions are chosen for best correspondence with familiar densities.

Name (Density)	$w(x)$	Standardization (a^*, b^*)	α_k β_k γ_k
Jacobi[a] (Beta)	$(1-x)^{p-q}x^{q-1}$	$c_k = 1$ $(0,1)$	See [2, 436]
Legendre[a] (Uniform)	1	$p_k(1) = 1$ $(0,1)$	$(1-2k)/k$ $(4k-2)/k$ $(k-1)/k$
Laguerre (Exponential)	$\exp\{-x\}$	$c_k = (-1)^k/k!$ $(0,\infty)$	$(2k-1)/k$ $-1/k$ $(k-1)/k$
Laguerre[b] (Gamma)	$x^r \exp\{-x\}$	$c_k = (-1)^k/k!$ $(0,\infty)$	$(2k-1+r)/k$ $-1/k$ $(k-1+r)/k$
Hermite[c] (Normal)	$\exp\{-x^2/2\}$	$c_k = 1$ $(-\infty,\infty)$	0 1 $k-1$

[a]Shifted. [b]Generalized. [c]Alternative form.

3. $\int_a^b f(x)w(x)\,dx = \sum_{i=0}^m A_i f(x_i)$ whenever f is a polynomial of degree not exceeding $2m+1$. In other words, the method is exact for the expectation of any such polynomial with respect to w.

4. If f is $2(m+1)$ times continuously differentiable, then

$$\int_a^b f(x)w(x)\,dx - \sum_{i=0}^m A_i f(x_i) = \frac{f^{(2m+2)}(\xi)}{(2m+2)!c_{m+1}^2} \tag{5.35}$$

for some $\xi \in (a,b)$.

The proof of this result may be found in [120].

Although this result and Table 5.6 provide the means by which the nodes and weights for an $(m+1)$-point Gaussian quadrature rule can be calculated, one should be hesitant to derive these directly, due to potential numerical imprecision. Numerically

Table 5.7   Estimates of the integral in (5.7) using Gauss–Hermite quadrature with various numbers of nodes. All estimates are multiplied by a factor of 10^5. Errors for use in a relative convergence criterion are given in the final column.

Nodes	Estimate	Relative Error
2	1.72893306163335	
3	1.72889399083898	−0.000023
4	1.72889068827101	−0.0000019
5	1.72889070910131	0.000000012
6	1.72889070914313	0.000000000024
7	1.72889070914166	−0.00000000000085
8	1.72889070914167	−0.0000000000000071

stable calculations of these quantities can be obtained from publicly available software [199, 418]. Alternatively, one can draw the nodes and weights from published tables like those in [2, 337]. Lists of other published tables are given in [120, 534].

Of the choices in Table 5.6, Gauss–Hermite quadrature is particularly useful because it enables integration over the entire real line. The prominence of normal distributions in statistical practice and limiting theory means that many integrals resemble the product of a smooth function and a normal density; the usefulness of Gauss–Hermite quadrature in Bayesian applications is demonstrated in [408].

Example 5.5 (Alzheimer's disease, continued)   Table 5.7 shows the results of applying *Gauss–Hermite quadrature* to estimate the integral from Example 5.1. Using the Hermite polynomials in this case is particularly appealing because the integrand from Example 5.1 really should be integrated over the entire real line, rather than the interval $(−0.07, 0.085)$. Convergence was extremely fast: With 8 nodes we obtained a relative error half the magnitude of that achieved by Simpson's rule with 1024 nodes. The estimate in Table 5.7 differs from previous examples because the range of integration differs. Applying *Gauss–Legendre quadrature* to estimate the integral over the interval $(−0.07, 0.085)$ yields an estimate of $1.72889008255962 \times 10^{-5}$ using 26 nodes.   □

Gaussian quadrature is quite different from the Newton–Côtes rules discussed previously. Whereas the latter rely on potentially enormous numbers of nodes to achieve sufficient precision, Gaussian quadrature is often very precise with a remarkably small number of nodes. However, for Gaussian quadrature the nodes for an m-point rule are not usually shared by an $(m + k)$-point rule for $k \geq 1$. Recall the strategy discussed for Newton–Côtes rules where the number of subintervals is sequentially doubled so that half the new nodes correspond to old nodes. This is not effective for Gaussian quadrature, because each increase in the number of nodes will require a separate effort to generate the nodes and the weights.

5.4 FREQUENTLY ENCOUNTERED PROBLEMS

This section briefly addresses strategies to try when you are faced with a problem more complex than a one-dimensional integral of a smooth function with no singularities on a finite range.

5.4.1 Range of integration

Integrals over infinite ranges can be transformed to a finite range. Some useful transformations include $1/x$, $\frac{\exp\{x\}}{1+\exp\{x\}}$, $\exp\{-x\}$, and $\frac{x}{1+x}$. Any cumulative distribution function is a potential basis for transformation, too. For example, the exponential cumulative distribution function transforms the positive half line to the unit interval. Cumulative distribution functions for real-valued random variables transform doubly infinite ranges to the unit interval. Of course, transformations to remove an infinite range may introduce other types of problems such as singularities. Thus, among the options available, it is important to choose a good transformation. Roughly speaking, a good choice is one that produces an integrand that is as nearly constant as can be managed.

Infinite ranges can be dealt with in other ways, too. Example 5.5 illustrates the use of Gauss–Hermite quadrature to integrate over the real line. Alternatively, when the integrand vanishes near the extremes of the integration range, integration can be truncated with a controllable amount of error. Truncation was used in Example 5.1.

Further discussion of transformations and strategies for selecting a suitable one are given in [120, 534]

5.4.2 Integrands with singularities or other extreme behavior

Several strategies can be employed to eliminate or control the effect of singularities that would otherwise impair the performance of an integration rule.

Transformation is one approach. For example, consider $\int_0^1 \frac{\exp\{x\}}{\sqrt{x}}\, dx$, which has a singularity at 0. The integral is easily fixed using the transformation $u = \sqrt{x}$, yielding $2\int_0^1 \exp\{u^2\}\, du$.

The integral $\int_0^1 x^{999} \exp\{x\}\, dx$ has no singularity on $[0,1]$ but is very difficult to estimate directly with a Newton–Côtes approach. Transformation is helpful in such cases, too. Letting $u = x^{1000}$ yields $\int_0^e \exp\{u^{1/1000}\}\, du$, whose integrand is nearly constant on $[0, e]$. The transformed integral is much easier to estimate reliably.

Another approach is to subtract out the singularity. For example, consider integrating $\int_{-\pi/2}^{\pi/2} \log\{\sin^2 x\}\, dx$, which has a singularity at 0. By adding and subtracting away the square of the log singularity at zero, we obtain $\int_{-\pi/2}^{\pi/2} \log\{(\sin^2 x)/x^2\}\, dx + \int_{-\pi/2}^{\pi/2} \log x^2\, dx$. The first term is then suitable for quadrature, and elementary methods can be used to derive that the second term equals $2\pi(\log\frac{\pi}{2} - 1)$.

Refer to [120, 436, 534] for more detailed discussions of how to formulate an appropriate strategy to address singularities.

5.4.3 Multiple integrals

The most obvious extensions of univariate quadrature techniques to multiple integrals are *product formulas*. This entails, for example, writing $\int_a^b \int_c^d f(x,y)\,dy\,dx$ as $\int_a^b g(x)\,dx$ where $g(x) = \int_c^d f(x,y)\,dy$. Values of $g(x)$ could be obtained via univariate quadrature approximations to $\int_c^d f(x,y)\,dy$ for a grid of x values. Univariate quadrature could then be completed for g. Using n subintervals in each univariate quadrature would require n^p evaluations of f, where p is the dimension of the integral. Thus, this approach is not feasible for large p. Even for small p, care must be taken to avoid the accumulation of a large number of small errors, since each exterior integral depends on the values obtained for each interior integral at a set of points. Also, product formulas can only be implemented directly for regions of integration that have simple geometry, such as hyperrectangles.

To cope with higher dimensions and general multivariate regions, one may develop specialized grids over the region of integration, search for one or more dimensions that can be integrated analytically to reduce the complexity of the problem, or turn to multivariate adaptive quadrature techniques. Multivariate methods are discussed in more detail in [120, 253, 436, 524].

Monte Carlo methods discussed in Chapters 6 and 7 can be employed to estimate integrals over high-dimensional regions efficiently. For estimating a one-dimensional integral based on n points, a Monte Carlo estimate will typically have a convergence rate of $\mathcal{O}(n^{-1/2})$, whereas the quadrature methods discussed in this chapter converge at $\mathcal{O}(n^{-2})$ or faster. In higher dimensions, however, the story changes. Quadrature approaches are then much more difficult to implement and slower to converge, whereas Monte Carlo approaches generally retain their implementation ease and their convergence performance. Accordingly, Monte Carlo approaches are generally preferred for high-dimensional integration.

5.4.4 Adaptive quadrature

The principle of adaptive quadrature is to choose subinterval lengths based on the local behavior of the integrand. For example, one may recursively subdivide those existing subintervals where the integral estimate has not yet stabilized. This can be a very effective approach if the bad behavior of the integrand is confined to a small portion of the region of integration. It also suggests a way to reduce the effort expended for multiple integrals, because much of the integration region may be adequately covered by a very coarse grid of subintervals. A variety of ideas is covered in [103, 328, 534].

5.4.5 Software for exact integration

This chapter has focused on integrals that do not have analytic solutions. For most of us, there is a class of integrals that have analytic solutions that are so complex as to be beyond our skills, patience, or cleverness to derive. Numerical approximation will work for such integrals, but so will symbolic integration tools. Software packages such as Mathematica [572] and Maple [335] allow the user to type integrands in a syntax resembling many computer languages. The software interprets these algebraic expressions. With deft use of commands for integrating and manipulating terms, the user can derive exact expressions for analytic integrals. The software does the algebra. Such software is particularly helpful for difficult indefinite integrals.

Problems

5.1 For the trapezoidal rule, express $p_i(x)$ as

$$f(x_i) + (x - x_i)\frac{f(x_{i+1}) - f(x_i)}{x_{i+1} - x_i}.$$

Expand f in Taylor series about x_i and evaluate this at $x = x_{i+1}$. Use the resulting expression in order to prove (5.14).

5.2 Following the approach in (5.8)–(5.11), derive A_{ij} for $j = 0, 1, 2$ for Simpson's rule.

5.3 Suppose the data $(x_1, \ldots, x_7) = (6.52, 8.32, 0.31, 2.82, 9.96, 0.14, 9.64)$ are observed. Consider Bayesian estimation of μ based on a $N(\mu, 3^2/7)$ likelihood for the minimally sufficient $\bar{x} \mid \mu$, and a Cauchy(5,2) prior.

(a) Using a numerical integration method of your choice, show that the proportionality constant is roughly 7.84654. (In other words, find k such that $\int k \times (\text{prior}) \times (\text{likelihood})\, d\mu = 1$.)

(b) Using the value 7.84654 from (a), determine the posterior probability that $2 \le \mu \le 8$ using the Riemann, trapezoidal, and Simpson's rules over the range of integration (implementing Simpson's rule as in (5.20) by pairing adjacent subintervals). Compute the estimates until relative convergence within 0.0001 is achieved for the slowest method. Table the results. How close are your estimates to the correct answer of 0.99605?

(c) Find the posterior probability that $\mu \ge 3$ in the following two ways. Since the range of integration is infinite, use the transformation $u = \frac{\exp\{\mu\}}{1+\exp\{\mu\}}$. First, ignore the singularity at 1 and find the value of the integral using one or more quadrature methods. Second, fix the singularity at 1 using one or more appropriate strategies, and find the value of the integral. Compare your results. How close are the estimates to the correct answer of 0.99086?

(d) Use the transformation $u = 1/\mu$, and obtain a good estimate for the integral in part (c).

5.4 Let $X \sim \text{Unif}[1, a]$ and $Y = (a - 1)/X$, for $a > 1$. Compute $E\{Y\} = \log a$ using Romberg's algorithm for $m = 6$. Table the resulting triangular array. Comment on your results.

5.5 The Gaussian quadrature rule having $w(x) = 1$ for integrals on $[-1, 1]$ (cf. Table 5.6) is called *Gauss–Legendre quadrature* because it relies on the Legendre

Table 5.8 Nodes and weights for 10-point Gauss–Legendre quadrature on the range $[-1, 1]$.

$\pm x_i$	A_i
0.148874338981631	0.295524224714753
0.433395394129247	0.269266719309996
0.679409568299024	0.219086362515982
0.865063366688985	0.149451394150581
0.973906528517172	0.066671344308688

polynomials. The nodes and weights for the 10-point Gauss–Legendre rule are given in Table 5.8.

(a) Plot the weights versus the nodes.

(b) Find the area under the curve $y = x^2$ between -1 and 1. Compare this with the exact answer and comment on the precision of this quadrature technique.

5.6 Suppose 10 i.i.d. observations result in $\bar{x} = 47$. Let the likelihood for μ correspond to the model $\bar{X} \mid \mu \sim N(\mu, 50/10)$, and the prior for $(\mu - 50)/8$ be Student's t with 1 degree of freedom.

(a) Show that the five-point Gauss–Hermite quadrature rule relies on the Hermite polynomial $H_5(x) = c(x^5 - 10x^3 + 15x)$.

(b) Show that the normalization of $H_5(x)$ (namely, $\langle H_5(x), H_5(x) \rangle = 1$) requires $c = 1/\sqrt{120\sqrt{2\pi}}$. You may wish to recall that a standard normal distribution has odd moments equal to zero and rth moments equal to $\frac{r!}{(r/2)!2^{r/2}}$ when r is even.

(c) Using your favorite root finder, estimate the nodes of the five-point Gauss–Hermite quadrature rule. (Recall that finding a root of f is equivalent to finding a local minimum of $|f|$.) Plot $H_5(x)$ from -3 to 3, and indicate the roots.

(d) Find the quadrature weights. Plot the weights versus the nodes. You may appreciate knowing that the normalizing constant for $H_6(x)$ is $1/\sqrt{720\sqrt{2\pi}}$.

(e) Using the nodes and weights found above for five-point Gauss–Hermite integration, estimate the posterior variance of μ. (Remember to account for the normalizing constant in the posterior before taking posterior expectations.)

6

Simulation and Monte Carlo Integration

This chapter addresses the simulation of random draws $\mathbf{X}_1, \ldots, \mathbf{X}_n$ from a target distribution f. The most frequent use of such draws is to perform *Monte Carlo integration*, which is the statistical estimation of the value of an integral using evaluations of an integrand at a set of points drawn randomly from a distribution with support over the range of integration [392].

Estimation of integrals via Monte Carlo simulation can be useful in a wide variety of settings. In Bayesian analyses, posterior moments can be written in the form of an integral but typically cannot be evaluated analytically. Posterior probabilities can also be written as the expectation of an indicator function with respect to the posterior. The calculation of risk in Bayesian decision theory relies on integration. Integration is also an important component in frequentist likelihood analyses. For example, marginalization of a joint density relies upon integration. Example 5.1 illustrates an integration problem arising from the maximum likelihood fit of a generalized linear mixed model. A variety of other integration problems are discussed here and in Chapter 7.

Aside from its application to Monte Carlo integration, simulation of random draws from a target density f is important in many other contexts. Indeed, Chapter 7 is devoted to a specific strategy for Monte Carlo integration called Markov chain Monte Carlo. Bootstrap methods, stochastic search algorithms, and a wide variety of other statistical tools also rely on generation of random deviates.

Further details about the topics discussed in this chapter can be found in [91, 137, 166, 326, 334, 357, 366, 400, 456, 466, 468].

6.1   INTRODUCTION TO THE MONTE CARLO METHOD

Many quantities of interest in inferential statistical analyses can be expressed as the expectation of a function of a random variable, say $\mathrm{E}\{h(\mathbf{X})\}$. Let f denote the density of $\mathbf{X}$, and μ denote the expectation of $h(\mathbf{X})$ with respect to f. When an i.i.d. random sample $\mathbf{X}_1, \ldots, \mathbf{X}_n$ is obtained from f, we can approximate μ by a sample average:

$$\hat{\mu}_{\mathrm{MC}} = \frac{1}{n} \sum_{i=1}^{n} h(\mathbf{X}_i) \rightarrow \int h(\mathbf{x}) f(\mathbf{x}) \, d\mathbf{x} = \mu \qquad (6.1)$$

as $n \rightarrow \infty$, by the strong law of large numbers (see Section 1.6). Further, let $v(\mathbf{x}) = [h(\mathbf{x}) - \mu]^2$, and assume that $h(\mathbf{X})^2$ has finite expectation under f. Then the sampling variance of $\hat{\mu}_{\mathrm{MC}}$ is $\sigma^2/n = \mathrm{E}\{v(\mathbf{X})/n\}$, where the expectation is taken with respect to f. A similar Monte Carlo approach can be used to estimate σ^2 by

$$\widehat{\mathrm{var}}\{\hat{\mu}_{\mathrm{MC}}\} = \frac{1}{n-1} \sum_{i=1}^{n} [h(\mathbf{X}_i) - \hat{\mu}_{\mathrm{MC}}]^2 . \qquad (6.2)$$

When σ^2 exists, the central limit theorem implies that $\hat{\mu}_{\mathrm{MC}}$ has an approximate normal distribution for large n, so approximate confidence bounds and statistical inference for μ follow. Generally, it is straightforward to extend (6.1), (6.2), and most of the methods in this chapter to cases when the quantity of interest is multivariate, so it suffices hereafter to consider μ to be scalar.

Monte Carlo integration provides slow $\mathcal{O}(n^{-1/2})$ convergence. With n nodes, the quadrature methods described in Chapter 5 offer convergence of order $\mathcal{O}(n^{-2})$ or better. There are several reasons why Monte Carlo integration is nonetheless a very powerful tool.

Most importantly, quadrature methods are difficult to extend to multidimensional problems because general p-dimensional space is so vast. Straightforward product rules creating quadrature grids of size n^p quickly succumb to the curse of dimensionality (discussed in Section 10.4.1), becoming harder to implement and slower to converge. Monte Carlo integration samples randomly from f over the p-dimensional support region of f, but does not attempt any systematic exploration of this region. Thus, implementation of Monte Carlo integration is less hampered by high dimensionality than is quadrature. However, when p is large, a very large sample size may still be required to obtain an acceptable standard error for $\hat{\mu}_{\mathrm{MC}}$. Quadrature methods also perform best when h is smooth, even when $p = 1$. In contrast, the Monte Carlo integration approach ignores smoothness. Further comparisons are offered in [166].

Monte Carlo integration replaces the systematic grid of quadrature nodes with a set of points chosen randomly from a probability distribution. The first step, therefore, is to study how to generate such draws. This topic is addressed in Section 6.2. Methods for improving upon the standard estimator given in equation (6.1) are described in Section 6.3.

6.2   SIMULATION

The primary focus of this section is simulation of random variables that do not follow a familiar parametric distribution. We refer to the desired sampling density f as the *target distribution*. When the target distribution comes from a standard parametric family, abundant software exists to easily generate random deviates. At some level, all of this code relies on the generation of standard uniform random deviates. Given the deterministic nature of the computer, such draws are not really random, but a good generator will produce a sequence of values that are statistically indistinguishable from independent standard uniform variates. Generation of standard uniform random deviates is a classic problem studied in [171, 198, 334, 455, 456, 468].

Rather than rehash the theory of uniform random number generation, we focus on the practical quandary faced by those with good software: what should be done when the target density is not one easily sampled using the software. For example, nearly all Bayesian posterior distributions are not members of standard parametric families. Posteriors obtained when using conjugate priors in exponential families are exceptions.

There can be additional difficulties beyond the absence of an obvious method to sample f. In many cases—especially in Bayesian analyses—the target density may be known only up to a multiplicative proportionality constant. In such cases, f cannot be sampled and can only be evaluated up to that constant. Fortunately, there are a variety of simulation approaches that still work in this setting.

Finally, it may be possible to evaluate f, but computationally expensive. If each computation of $f(x)$ requires an optimization, an integration, or other time-consuming computations, we may seek simulation strategies that avoid direct evaluation of f as much as possible.

6.2.1   Generating from standard parametric families

Before discussing sampling from difficult target distributions, we survey some strategies for producing random variates from familiar distributions using uniform random variates. We omit justifications for these approaches, which are given in the references cited above. Table 6.1 summarizes a variety of approaches. Although the tabled approaches are not necessarily state-of-the-art, they illustrate some of the underlying principles exploited by sophisticated generators.

6.2.2   Inverse cumulative distribution function

The methods for the Cauchy and exponential distributions in Table 6.1 are justified by the *inverse cumulative distribution function* or *probability integral transform* approach. For any continuous distribution function F, if $U \sim \text{Unif}(0, 1)$, then $X = F^{-1}(U) = \inf\{x : F(x) \geq U\}$ has cumulative distribution function equal to F.

If F^{-1} is available for the target density, then this strategy is probably the simplest option. If F^{-1} is not available but F is either available or easily approximated, then

Table 6.1 Some methods for generating a random variable X from familiar distributions.

Distribution	Method
Uniform	See [171, 198, 334, 455, 456, 468]. For $X \sim \text{Unif}(a,b)$; draw $U \sim \text{Unif}(0,1)$; then let $X = a+(b-a)U$.
Normal(μ, σ^2) and lognormal(μ, σ^2)	Draw $U_1, U_2 \sim$ i.i.d. Unif(0,1); then $X_1 = \mu + \sigma\sqrt{-2\log U_1}\sin\{2\pi U_2\}$ and $X_2 = \mu + \sigma\sqrt{-2\log U_1}\cos\{2\pi U_2\}$ are independent $N(\mu, \sigma^2)$. If $X \sim N(\mu, \sigma^2)$ then $\exp\{X\} \sim$ lognormal(μ, σ^2).
Multivariate $N(\boldsymbol{\mu}, \boldsymbol{\Sigma})$	Generate standard multivariate normal vector, $\mathbf{Y}$, coordinatewise; then $\mathbf{X} = \boldsymbol{\Sigma}^{-1/2}\mathbf{Y} + \boldsymbol{\mu}$.
Cauchy(α, β)	Draw $U \sim \text{Unif}(0,1)$; then $X = \alpha + \beta\tan\{\pi(U-1/2)\}$.
Exponential(λ)	Draw $U \sim \text{Unif}(0,1)$; then $X = -(\log U)/\lambda$.
Poisson(λ)	Draw $U_1, U_2, \ldots \sim$ i.i.d. Unif(0,1); then $X = j-1$, where j is the lowest index for which $\prod_{i=1}^{j} U_i < e^{-\lambda}$.
Gamma(r, λ)	See Example 6.1, references, or for integer r, $X = -\frac{1}{\lambda}\sum_{i=1}^{r} \log U_i$ for $U_1, \ldots, U_r \sim$ i.i.d. Unif(0,1).
Chi-square (df= k)	Draw $Y_1, \ldots, Y_k \sim$ i.i.d. $N(0,1)$, then $X = \sum_{i=1}^{k} Y_i^2$; or draw $X \sim$ Gamma($k/2, 1/2$).
Student's t (df= k) and $F_{k,m}$ distribution	Draw $Y \sim N(0,1)$, $Z \sim \chi_k^2$, $W \sim \chi_m^2$ independently; then $X = Y/\sqrt{Z/k}$ has the t distribution and $F = (Z/k)/(W/m)$ has the F distribution.
Beta(a, b)	Draw $Y \sim$ Gamma($a, 1$) and $Z \sim$ Gamma($b, 1$) independently; then $X = Y/(Y + Z)$.
Bernoulli(p) and Binomial(n, p)	Draw $U \sim \text{Unif}(0,1)$; then $X = 1_{\{U<p\}}$ is Bernoulli(p). The sum of n independent Bernoulli(p) draws has a Binomial(n, p) distribution.
Negative Binomial(r, p)	Draw $U_1, \ldots, U_r \sim$ i.i.d. Unif(0,1); then $X = \sum_{i=1}^{r}\lfloor(\log U_i)/\log\{1-p\}\rfloor$, and $\lfloor\cdot\rfloor$ means greatest integer.
Multinomial($1, (p_1, \ldots, p_k)$)	Partition $[0,1]$ into k segments so the ith segment has length p_i. Draw $U \sim$ Unif(0,1); then let X equal the index of the segment into which U falls. Tally such draws for Multinomial($n, (p_1, \ldots, p_k)$).
Dirichlet($\alpha_1, \ldots, \alpha_k$)	Draw independent $Y_i \sim$ Gamma($\alpha_i, 1$) for $i = 1, \ldots, k$; then $\mathbf{X}^T = \left(Y_1/\sum_{i=1}^{k} Y_i, \ldots, Y_k/\sum_{i=1}^{k} Y_i\right)$.

a crude approach can be built upon linear interpolation. Using a grid of $x_1, \ldots, x_m$ spanning the region of support of f, calculate or approximate $u_i = F(x_i)$ at each grid point. Then, draw $U \sim \text{Unif}(0, 1)$, and linearly interpolate between the two nearest grid points for which $u_i \le U \le u_j$ according to

$$X = \frac{u_j - U}{u_j - u_i} x_i + \frac{U - u_i}{u_j - u_i} x_j. \tag{6.3}$$

This approach is not appealing, because it requires a complete approximation to F regardless of the desired sample size, it does not generalize to multiple dimensions, and it is less efficient than other approaches.

6.2.3 Rejection sampling

If $f(x)$ can be calculated, at least up to a proportionality constant, then we can use *rejection sampling* to obtain a random draw from exactly the target distribution. This strategy relies on sampling candidates from an easier distribution and then correcting the sampling probability through random rejection of some candidates.

Let g denote another density from which we know how to sample and for which we can easily calculate $g(x)$. Let $e(\cdot)$ denote an *envelope*, having the property $e(x) = g(x)/\alpha \ge f(x)$ for all x for which $f(x) > 0$ for a given constant $\alpha \le 1$. Rejection sampling proceeds as follows:

1. Sample $Y \sim g$.

2. Sample $U \sim \text{Unif}(0, 1)$.

3. Reject Y if $U > f(Y)/e(Y)$. In this case, do not record the value of Y as an element in the target random sample. Instead, return to step 1.

4. Otherwise, keep the value of Y. Set $X = Y$, and consider X to be an element of the target random sample. Return to step 1 until you have accumulated a sample of the desired size.

The draws kept using this algorithm constitute an i.i.d. sample from the target density f; there is no approximation involved. To see this, note that the probability that a kept draw falls at or below a value y is

$$
\begin{aligned}
P[X \le y] &= P\left[Y \le y \,\middle|\, U \le \frac{f(Y)}{e(Y)} \right] \\
&= P\left[Y \le y \text{ and } U \le \frac{f(Y)}{e(Y)} \right] \Big/ P\left[U \le \frac{f(Y)}{e(Y)} \right] \\
&= \int_{-\infty}^{y} \int_{0}^{f(z)/e(z)} du \, g(z) \, dz \Big/ \int_{-\infty}^{\infty} \int_{0}^{f(z)/e(z)} du \, g(z) \, dz \tag{6.4} \\
&= \int_{-\infty}^{y} f(z) \, dz, \tag{6.5}
\end{aligned}
$$

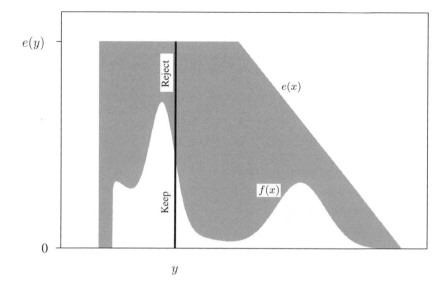

Fig. 6.1 Illustration of rejection sampling for a target distribution f using a rejection sampling envelope e.

which is the desired probability. Thus, the sampling distribution is exact, and α can be interpreted as the expected proportion of candidates that are accepted. Hence α is an measure of the efficiency of the algorithm. We may continue the rejection sampling procedure until it yields exactly the desired number of sampled points, but this requires a random total number of iterations that will depend on the proportion of rejections.

Recall the rejection rule in step 3 for determining the fate of a candidate draw, $Y = y$. Sampling $U \sim \text{Unif}(0, 1)$ and obeying this rule is equivalent to sampling $U|y \sim \text{Unif}(0, e(y))$ and keeping the value y if $U < f(y)$. Consider Figure 6.1. Suppose the value y falls at the point indicated by the vertical bar. Then imagine sampling $U|Y = y$ uniformly along the vertical bar. The rejection rule eliminates this Y draw with probability proportional to the length of the bar above $f(y)$ relative to the overall bar length. Therefore, one can view rejection sampling as sampling uniformly from the two-dimensional region under the curve e and then throwing away any draws falling above f and below e. Since sampling from f is equivalent to sampling uniformly from the two-dimensional region under the curve labeled $f(x)$ and then ignoring the vertical coordinate, rejection sampling provides draws exactly from f.

The shaded region in Figure 6.1 above f and below e indicates the waste. The draw $Y = y$ is very likely to be rejected when $e(y)$ is far larger than $f(y)$. Envelopes that exceed f everywhere by at most a slim margin produce fewer wasted (i.e., rejected) draws and correspond to α values near 1.

Suppose now that the target distribution f is only known up to a proportionality constant c. That is, suppose we are only able to compute easily $q(x) = f(x)/c$, where c is unknown. Such densities arise, for example, in Bayesian inference when f is a posterior distribution known to equal the product of the prior and the likelihood scaled by some normalizing constant. Fortunately, rejection sampling can be applied in such cases. We find an envelope e such that $e(x) \geq q(x)$ for all x for which $q(x) > 0$. A draw $Y = y$ is rejected when $U > q(y)/e(y)$. The sampling probability remains correct because the unknown constant c cancels out in the numerator and denominator of (6.4) when f is replaced by q. The proportion of kept draws is α/c.

Multivariate targets can also be sampled using rejection sampling, provided that a suitable multivariate envelope can be constructed. The rejection sampling algorithm is conceptually unchanged.

To produce an envelope we must know enough about the target to bound it. This may require optimization or a clever approximation to f or q in order to ensure that e can be constructed to exceed the target everywhere. Note that when the target is continuous and log-concave, it is unimodal. If we select x_1 and x_2 on opposite sides of that mode, then the function obtained by connecting the line segments that are tangent to $\log f$ or $\log q$ at x_1 and x_2 yields a piecewise exponential envelope with exponential tails. Deriving this envelope does not require knowing the maximum of the target density; it merely requires checking that x_1 and x_2 lie on opposite sides of it. The adaptive rejection sampling method described in Section 6.2.3.2 exploits this idea to generate good envelopes.

To summarize, good rejection sampling envelopes have three properties: They are easily constructed or confirmed to exceed the target everywhere, they are easy to sample, and they generate few rejected draws.

Example 6.1 (Gamma deviates) Consider the problem of generating a Gamma$(r, 1)$ random variable when $r \geq 1$. When Y is generated according to the density

$$f(y) = t(y)^{r-1} t'(y) \exp\{-t(y)\}/\Gamma(r) \tag{6.6}$$

for $t(y) = a(1 + by)^3$ for $-1/b < y < \infty$, $a = r - 1/3$, and $b = 1/\sqrt{9a}$, then $X = t(Y)$ will have a Gamma$(r, 1)$ distribution [376]. Marsaglia and Tsang describe how to use this fact in a rejection sampling framework [377]. Adopt (6.6) as the target distribution, because transforming draws from f gives the desired gamma draws.

Simplifying f and ignoring the normalizing constant, we wish to generate from the density that is proportional to $q(y) = \exp\{a \log\{t(y)/a\} - t(y) + a\}$. Conveniently, q fits snugly under the function $e(y) = \exp\{-y^2/2\}$, which is the unscaled standard normal density. Therefore, rejection sampling amounts to sampling a standard normal random variable, Z, and a standard uniform random variable, U, then setting $X = t(Z)$ if

$$U \leq q(Z)/e(Z) = \exp\{Z^2/2 + a \log\{t(Z)/a\} - t(Z) + a\} \tag{6.7}$$

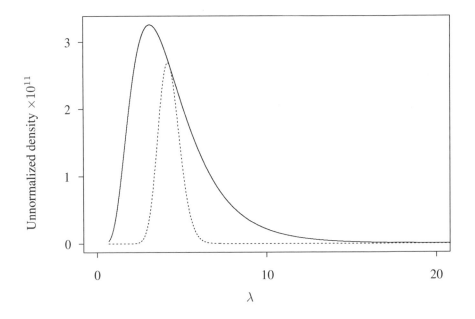

Fig. 6.2   Unnormalized target (dotted) and envelope (solid) for rejection sampling in Example 6.2.

and $t(Z) > 0$. Otherwise, the draw is rejected and the process begun anew. An accepted draw has density $\text{Gamma}(r, 1)$. Draws from $\text{Gamma}(r, 1)$ can be rescaled to obtain draws from $\text{Gamma}(r, \lambda)$.

In a simulation when $r = 4$, over 99% of candidate draws are accepted and a plot of $e(y)$ and $q(y)$ against y shows that the two curves are nearly superimposed. Even in the worst case ($r = 1$), the envelope is excellent, with less than 5% waste.   □

Example 6.2 (Sampling a Bayesian posterior)   Suppose ten independent observations $(8, 3, 4, 3, 1, 7, 2, 6, 2, 7)$ are observed from the model $X_i|\lambda \sim \text{Poisson}(\lambda)$. A lognormal prior distribution for λ is assumed: $\log \lambda \sim N(4, 0.5^2)$. Denote the likelihood as $L(\lambda|\mathbf{x})$ and the prior as $f(\lambda)$. We know that $\hat{\lambda} = \bar{x} = 4.3$ maximizes $L(\lambda|\mathbf{x})$ with respect to λ; therefore the unnormalized posterior, $q(\lambda|\mathbf{x}) = f(\lambda)L(\lambda|\mathbf{x})$ is bounded above by $e(\lambda) = f(\lambda)L(4.3|\mathbf{x})$. Figure 6.2 shows q and e. Note that the prior is proportional to e. Thus, rejection sampling begins by sampling λ_i from the lognormal prior and U_i from a standard uniform distribution. Then λ_i is kept if $U_i < q(\lambda_i|\mathbf{x})/e(\lambda_i) = L(\lambda_i|\mathbf{x})/L(4.3|\mathbf{x})$. Otherwise, λ_i is rejected and the process is begun anew. Any kept λ_i is a draw from the posterior. Although not efficient—only about 30% of candidate draws are kept—this approach is easy and exact.   □

6.2.3.1 *Squeezed rejection sampling*   Ordinary rejection sampling requires one evaluation of f for every candidate draw Y. In cases where evaluating f is computationally expensive but rejection sampling is otherwise appealing, improved simulation speed is achieved by *squeezed rejection sampling* [334, 374, 375].

This strategy preempts the evaluation of f in some instances by employing a nonnegative *squeezing function*, s. For s to be a suitable squeezing function, $s(x)$ must not exceed $f(x)$ anywhere on the support of f. An envelope, e, is also used; as with ordinary rejection sampling, $e(x) = g(x)/\alpha \geq f(x)$ on the support of f.

The algorithm proceeds as follows:

1. Sample $Y \sim g$.

2. Sample $U \sim \text{Unif}(0, 1)$.

3. If $U \leq s(Y)/e(Y)$, keep the value of Y. Set $X = Y$ and consider X to be an element in the target random sample. Then go to step 6.

4. Otherwise, determine whether $U \leq f(Y)/e(Y)$. If this inequality holds, keep the value of Y, setting $X = Y$. Consider X to be an element in the target random sample; then go to step 6.

5. If Y has not yet been kept, reject it as an element in the target random sample.

6. Return to step 1 until you have accumulated a sample of the desired size.

Note that when $Y = y$, this candidate draw is kept with overall probability $f(y)/e(y)$, and rejected with probability $[e(y) - f(y)]/e(y)$. These are the same probabilities as with simple rejection sampling. Step 3 allows a decision to keep Y to be made on the basis of an evaluation of s, rather than of f. When s nestles up just underneath f everywhere, we achieve the largest decrease in the number of evaluations of f.

Figure 6.3 illustrates the procedure. When a candidate $Y = y$ is sampled, the algorithm proceeds in a manner equivalent to sampling a $\text{Unif}(0, e(y))$ random variable. If this uniform variate falls below $s(y)$, the candidate is kept immediately. The lighter shaded region indicates where candidates are immediately kept. If the candidate is not immediately kept, then a second test must be employed to determine whether the uniform variate falls under $f(y)$ or not. Finally, the darker shaded region indicates where candidates are ultimately rejected.

As with rejection sampling, the proportion of candidate draws kept is α. The proportion of iterations in which evaluation of f is avoided is $\int s(x)\, dx / \int e(x)\, dx$.

Squeezed rejection sampling can also be carried out when the target is known only up to a proportionality constant. In this case, the envelope and squeezing function sandwich the unnormalized target. The method is still exact, and the same efficiency considerations apply.

Generalizations for sampling multivariate targets are straightforward.

6.2.3.2 *Adaptive rejection sampling*

Clearly the most challenging aspect of the rejection sampling strategy is the construction of a suitable envelope. Gilks and Wild proposed an automatic envelope generation strategy for squeezed rejection sampling for a continuous, differentiable, log-concave density on a connected region of support [214].

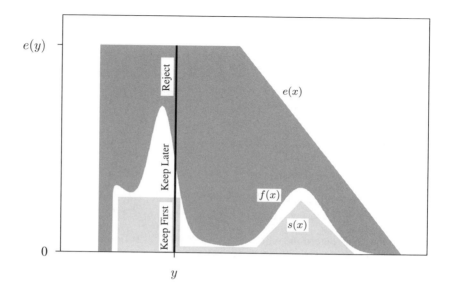

Fig. 6.3 Illustration of squeezed rejection sampling for a target distribution, f, using envelope e and squeezing function s. 'Keep first' and 'Keep later' correspond to steps 3 and 4 of the algorithm, respectively.

The approach is termed *adaptive rejection sampling* because the envelope and squeezing function are iteratively refined concurrently with the generation of sample draws. The amount of waste and the frequency with which f must be evaluated both shrink as iterations increase.

Let $\ell(x) = \log f(x)$, and assume $f(x) > 0$ on a (possibly infinite) interval of the real line. Let f be log-concave in the sense that $\ell(a) - 2\ell(b) + \ell(c) < 0$ for any three points in the support region of f for which $a < b < c$. Under the additional assumptions that f is continuous and differentiable, note that $\ell'(x)$ exists and decreases monotonically with increasing x, but may have discontinuities.

The algorithm is initiated by evaluating ℓ and ℓ' at k points, $x_1 < x_2 < \cdots < x_k$. Let $T_k = \{x_1, \ldots, x_k\}$. If the support of f extends to $-\infty$, choose x_1 such that $\ell'(x_1) > 0$. Similarly, if the support of f extends to ∞, choose x_k such that $\ell'(x_k) < 0$.

Define the rejection envelope on T_k to be the exponential of the piecewise linear upper hull of ℓ formed by the tangents to ℓ at each point in T_k. If we denote the upper hull of ℓ as e_k^*, then the rejection envelope is $e_k(x) = \exp\{e_k^*(x)\}$. To understand the concept of an upper hull, consider Figure 6.4. This figure shows ℓ with a solid line and illustrates the case when $k = 5$. The dashed line shows the piecewise upper hull, e^*. It is tangent to ℓ at each x_i, and the concavity of ℓ ensures that e_k^* lies completely

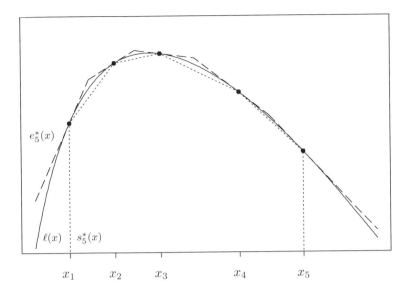

Fig. 6.4 Piecewise linear outer and inner hulls for $\ell(x) = \log f(x)$ used in adaptive rejection sampling when $k = 5$.

above ℓ everywhere else. One can show that the tangents at x_i and x_{i+1} intersect at

$$z_i = \frac{\ell(x_{i+1}) - \ell(x_i) - x_{i+1}\ell'(x_{i+1}) + x_i\ell'(x_i)}{\ell'(x_i) - \ell'(x_{i+1})} \qquad (6.8)$$

for $i = 1, \ldots, k - 1$. Therefore,

$$e_k^*(x) = \ell(x_i) + (x - x_i)\ell'(x_i) \qquad \text{for } x \in [z_{i-1}, z_i] \qquad (6.9)$$

and $i = 1, \ldots, k$, with z_0 and z_k defined respectively to equal the (possibly infinite) lower and upper bounds of the support region for f. Figure 6.5 shows the envelope e_k exponentiated to the original scale.

Define the squeezing function on T_k to be the exponential of the piecewise linear lower hull of ℓ formed by the chords between adjacent points in T_k. This lower hull is given by

$$s_k^*(x) = \frac{(x_{i+1} - x)\ell(x_i) + (x - x_i)\ell(x_{i+1})}{x_{i+1} - x_i} \qquad \text{for } x \in [x_i, x_{i+1}] \qquad (6.10)$$

and $i = 1, \ldots, k - 1$. When $x < x_1$ or $x > x_k$, let $s_k^*(x) = -\infty$. Thus the squeezing function is $s_k(x) = \exp\{s_k^*(x)\}$. Figure 6.4 shows a piecewise linear lower hull, $s_k^*(x)$, when $k = 5$. Figure 6.5 shows the squeezing function s_k on the original scale.

Figures 6.4 and 6.5 illustrate several important features of the approach. Both the rejection envelope and the squeezing function are piecewise exponential functions.

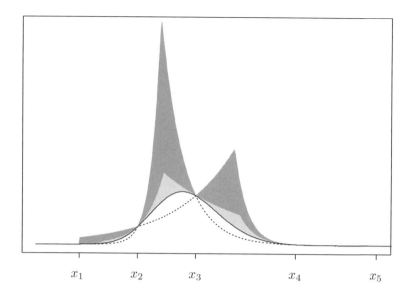

Fig. 6.5 Envelopes and squeezing function for adaptive rejection sampling. The target density is the smooth, nearly bell-shaped curve. The first method discussed in the text, using the derivative of ℓ, produces the envelope shown as the upper boundary of the lighter shaded region. This corresponds to equation (6.9) and Figure 6.4. Later in the text, a derivative-free method is presented. That envelope is the upper bound of the darker shaded region and corresponds to (6.11) and Figure 6.6. The squeezing function for both approaches is given by the dotted curve.

The envelope has exponential tails that lie above the tails of f. The squeezing function has bounded support.

Adaptive rejection sampling is initialized by choosing a modest k and a corresponding suitable grid T_k. The first iteration of the algorithm proceeds as for squeezed rejection sampling, using e_k and s_k as the envelope and squeezing function, respectively. When a candidate draw is accepted, it may be accepted without evaluating ℓ and ℓ' at the candidate if the squeezing criterion was met. However, it may also be accepted at the second stage, where evaluation of ℓ and ℓ' at the candidate is required. When a candidate is accepted at this second stage, the accepted point is added to the set T_k, creating T_{k+1}. Updated functions e_{k+1} and s_{k+1} are also calculated. Then iterations continue. When a candidate draw is rejected, no update to T_k, e_k, or s_k is made. Further, we see now that a new point that matches any existing member of T_k provides no meaningful update to T_k, e_k, or s_k.

Candidate draws are taken from the density obtained by scaling the piecewise exponential envelope e_k so that it integrates to 1. Since each accepted draw is made using a rejection sampling approach, the draws are an i.i.d. sample precisely from f. If f is known only up to a multiplicative constant, the adaptive rejection sampling approach may still be used, since the proportionality constant merely shifts ℓ, e_k^*, and s_k^*.

Gilks and coauthors have developed a similar approach that does not require evaluation of ℓ' [208, 210]. We retain the assumptions that f is log-concave with a connected support region, along with the basic notation and setup for the tangent-based approach above.

For the set of points T_k, define $L_i(\cdot)$ to be the straight line function connecting $(x_i, \ell(x_i))$ and $(x_{i+1}, \ell(x_{i+1}))$ for $i = 1, \ldots, k - 1$. Define

$$e_k^*(x) = \begin{cases} \min\{L_{i-1}(x), L_{i+1}(x)\} & \text{for } x \in [x_i, x_{i+1}], \\ L_1(x) & \text{for } x < x_1, \\ L_{k-1}(x) & \text{for } x > x_k, \end{cases} \qquad (6.11)$$

with the convention that $L_0(x) = L_k(x) = \infty$. Then e_k^* is a piecewise linear upper hull for ℓ, because the concavity of ℓ ensures that $L_i(x)$ lies below $\ell(x)$ on (x_i, x_{i+1}) and above $\ell(x)$ when $x < x_i$ or $x > x_{i+1}$. The rejection sampling envelope is then $e_k(x) = \exp\{e_k^*(x)\}$.

The squeezing function remains as in (6.10). Iterations of the derivative-free adaptive rejection sampling algorithm proceed analogously to the previous approach, with T_k, the envelope, and the squeezing function updated each time a new point is kept.

Figure 6.6 illustrates the derivative-free adaptive rejection sampling algorithm for the same target shown in Figure 6.4. The envelope is not as efficient as when ℓ' is used. Figure 6.5 shows the envelope on the original scale. The lost efficiency is seen on this scale, too.

Regardless of the method used to construct e_k, notice that one would prefer the T_k grid to be most dense in regions where $f(x)$ is largest, near the mode of f. Fortunately, this will happen automatically, since such points are most likely to be kept in subsequent iterations and included in updates to T_k. Grid points too far in the tails of f, such as x_5, are not very helpful.

Software for the tangent-based approach is available in [209]. The derivative-free approach has been popularized by its use in the WinBUGS software for carrying out Markov chain Monte Carlo algorithms to facilitate Bayesian analyses [211, 213, 515]. Adaptive rejection sampling can also be extended to densities that are not log-concave, for example by applying Markov chain Monte Carlo methods like those in Chapter 7 to further correct the sampling probabilities. One strategy is given in [210].

6.2.4 The sampling importance resampling algorithm

The *sampling importance resampling (SIR)* algorithm simulates realizations approximately from some target distribution. SIR is based upon the notion of importance sampling, discussed in detail in Section 6.3.1. Briefly, importance sampling proceeds by drawing a sample from an *importance sampling function*, g. Informally, we will call g an envelope. Each point in the sample is weighted to correct the sampling probabilities so that the weighted sample can be related to a target density f. For example, the weighted sample can be used to estimate expectations under f.

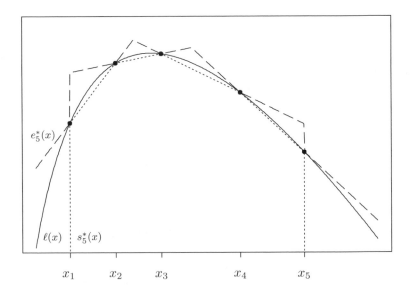

Fig. 6.6 Piecewise linear outer and inner hulls for $\ell(x) = \log f(x)$ used in derivative-free adaptive rejection sampling when $k = 5$.

Having graphed some univariate targets and envelopes in the early part of this chapter to illustrate basic concepts, we shift now to multivariate notation to emphasize the full generality of techniques. Thus, $\mathbf{X} = (X_1, \ldots, X_p)$ denotes a random vector with density $f(\mathbf{x})$, and $g(\mathbf{x})$ denotes the density corresponding to a multivariate envelope for f.

For the target density f, the weights used to correct sampling probabilities are called the *standardized importance weights*, and are defined as

$$w(\mathbf{x}_i) = \frac{f(\mathbf{x}_i)/g(\mathbf{x}_i)}{\sum_{i=1}^{m} f(\mathbf{x}_i)/g(\mathbf{x}_i)} \tag{6.12}$$

for a collection of values $\mathbf{x}_1, \ldots, \mathbf{x}_m$ drawn i.i.d. from an envelope g. Although not necessary for general importance sampling, it is useful to standardize the weights as in (6.12) so they sum to 1. When $f = cq$ for some unknown proportionality constant c, the unknown c cancels in the numerator and denominator of (6.12).

We may view importance sampling as approximating f by the discrete distribution having mass $w(\mathbf{x}_i)$ on each observed point $\mathbf{x}_i$ for $i = 1, \ldots, m$. Rubin proposed sampling from this distribution to provide an approximate sample from f [470, 471]. The SIR algorithm therefore proceeds as follows:

1. Sample candidates $\mathbf{Y}_1, \ldots, \mathbf{Y}_m$ i.i.d. from g.

2. Calculate the standardized importance weights, $w(\mathbf{Y}_1), \ldots, w(\mathbf{Y}_m)$.

3. Resample $\mathbf{X}_1, \ldots, \mathbf{X}_n$ from $\mathbf{Y}_1, \ldots, \mathbf{Y}_m$ with replacement with probabilities $w(\mathbf{Y}_1), \ldots, w(\mathbf{Y}_m)$.

A random variable $\mathbf{X}$ drawn with the SIR algorithm has distribution that converges to f as $m \to \infty$. To see this, define $w^*(\mathbf{y}) = f(\mathbf{y})/g(\mathbf{y})$, let $\mathbf{Y}_1, \ldots, \mathbf{Y}_m \sim$ i.i.d. g, and consider a set $\mathcal{A}$. Then

$$P[\mathbf{X} \in \mathcal{A}|\mathbf{Y}_1, \ldots, \mathbf{Y}_m] = \sum_{i=1}^{m} 1_{\{\mathbf{Y}_i \in \mathcal{A}\}} w^*(\mathbf{Y}_i) \bigg/ \sum_{i=1}^{m} w^*(\mathbf{Y}_i). \qquad (6.13)$$

The strong law of large numbers gives

$$\frac{1}{m} \sum_{i=1}^{m} 1_{\{\mathbf{Y}_i \in \mathcal{A}\}} w^*(\mathbf{Y}_i) \to \mathrm{E}\left\{ 1_{\{\mathbf{Y}_i \in \mathcal{A}\}} w^*(\mathbf{Y}_i) \right\} = \int_{\mathcal{A}} w^*(\mathbf{y}) g(\mathbf{y})\, d\mathbf{y} \qquad (6.14)$$

as $m \to \infty$. Further,

$$\frac{1}{m} \sum_{i=1}^{m} w^*(\mathbf{Y}_i) \to \mathrm{E}\left\{ w^*(\mathbf{Y}_i) \right\} = 1 \qquad (6.15)$$

as $m \to \infty$. Hence,

$$P[\mathbf{X} \in \mathcal{A}|\mathbf{Y}_1, \ldots, \mathbf{Y}_m] \to \int_{\mathcal{A}} w^*(\mathbf{y}) g(\mathbf{y})\, d\mathbf{y} = \int_{\mathcal{A}} f(\mathbf{y})\, d\mathbf{y} \qquad (6.16)$$

as $m \to \infty$. Finally, we note that

$$P[\mathbf{X} \in \mathcal{A}] = \mathrm{E}\left\{ P[\mathbf{X} \in \mathcal{A}|\mathbf{Y}_1, \ldots, \mathbf{Y}_m] \right\} \to \int_{\mathcal{A}} f(\mathbf{y})\, d\mathbf{y} \qquad (6.17)$$

by Lebesgue's dominated convergence theorem [43, 504]. The proof is similar when the target and envelope are known only up to a constant [466].

Although both SIR and rejection sampling rely on the ratio of target to envelope, they differ in an important way. Rejection sampling is perfect, in the sense that the distribution of a generated draw is exactly f, but it requires a random number of draws to obtain a sample of size n. In contrast, the SIR algorithm uses a predetermined number of draws to generate an n-sample but permits a random degree of approximation to f in the distribution of the sampled points.

When conducting SIR, it is important to consider the relative sizes of the initial sample and the resample. These sample sizes are m and n, respectively. In principle, we require $n/m \to 0$ for distributional convergence of the sample. In the context of asymptotic analysis of Monte Carlo estimates based on SIR, where $n \to \infty$, this condition means that $m \to \infty$ even faster than $n \to \infty$. For fixed n, distributional convergence of the sample occurs as $m \to \infty$, therefore in practice one obviously wants to initiate SIR with the largest possible m. However, one faces the competing desire to choose n as large as possible to increase the inferential precision. The maximum tolerable ratio n/m depends on the quality of the envelope. We have sometimes found $n/m \leq 1/10$ tolerable so long as the resulting resample does not contain too many replicates of any initial draw.

The SIR algorithm can be sensitive to the choice of g. First, the support of g must include the entire support of f if a reweighted sample from g is to approximate a sample from f. Further, g should have heavier tails than f, or more generally g should be chosen to ensure that $f(\mathbf{x})/g(\mathbf{x})$ never grows too large. If $g(\mathbf{x})$ is nearly zero anywhere where $f(\mathbf{x})$ is positive, then a draw from this region will happen only extremely rarely, but when it does it will receive a huge weight.

When this problem arises, the SIR algorithm exhibits the symptom that one or a few standardized importance weights are enormous compared to the other weights, and the secondary sample consists nearly entirely of replicated values of one or a few initial draws. When the problem is not too severe, taking the secondary resample without replacement has been suggested [193]. This is asymptotically equivalent to sampling with replacement, but has the practical advantage that it prevents excessive duplication. The disadvantage is that it introduces some additional distributional approximation in the final sample. When the distribution of weights is found to be highly skewed, it is probably wiser to switch to a different envelope or a different sampling strategy altogether.

Since SIR generates $\mathbf{X}_1, \ldots, \mathbf{X}_n$ approximately i.i.d. from f, one may proceed with Monte Carlo integration such as estimating the expectation of $h(\mathbf{X})$ by $\hat{\mu}_{\text{SIR}} = \sum_{i=1}^{n} h(\mathbf{X}_i)/n$ as in (6.1). However, in Section 6.3 we will introduce superior ways to use the initial weighted importance sample, along with other powerful methods to improve Monte Carlo estimation of integrals.

Example 6.3 (Slash distribution) The random variable Y has a *slash distribution* if $Y = X/U$ where $X \sim N(0,1)$ and $U \sim \text{Unif}(0,1)$ independently. Consider using the slash distribution as a SIR envelope to generate standard normal variates, and conversely using the normal distribution as a SIR envelope to generate slash variates. Since it is easy to simulate from both densities using standard methods, SIR is not needed in either case, but examining the results is instructive.

The slash density function is

$$f(y) = \begin{cases} \frac{1-\exp\{-y^2/2\}}{y^2\sqrt{2\pi}}, & y \neq 0, \\ \frac{1}{2\sqrt{2\pi}}, & y = 0. \end{cases}$$

This density has very heavy tails. Therefore, it is a fine importance sampling function for generating draws from a standard normal distribution using SIR. The left panel of Figure 6.7 illustrates the results when $m = 100,000$ and $n = 5000$. The true normal density is superimposed for comparison.

On the other hand, the normal density is not a suitable importance sampling function for SIR use when generating draws from the slash distribution, because the envelope's tails are far lighter than the target's. The right panel of Figure 6.7 (where, again, $m = 100,000$ and $n = 5000$) illustrates the problems that arise. Although the tails of the slash density assign appreciable probability as far as 10 units from the origin, no candidate draws from the normal density exceeded 5 units from the origin.

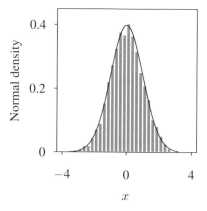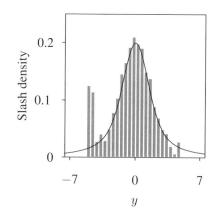

Fig. 6.7 The left panel shows a histogram of approximate draws from a standard normal density obtained via SIR with a slash distribution envelope. The right panel shows a histogram of approximate draws from a slash density obtained via SIR using a normal envelope. The solid lines show the target densities.

Therefore, beyond these limits, the simulated tails of the target have been completely truncated. Further, the most extreme candidate draws generated have far less density under the normal envelope than they do under the slash target, so their importance ratios are extremely high. This leads to abundant resampling of these points in the tails. Indeed, 528 of the 5000 values selected by SIR are replicates of the three lowest unique values in the histogram.   □

Example 6.4 (Bayesian inference) Suppose that we seek a sample from the posterior distribution from a Bayesian analysis. Such a sample could be used to provide Monte Carlo estimates of posterior moments, probabilities, or highest posterior density intervals, for example. Let $f(\boldsymbol{\theta})$ denote the prior, and $L(\boldsymbol{\theta}|\mathbf{x})$ the likelihood, so the posterior is $f(\boldsymbol{\theta}|\mathbf{x}) = cf(\boldsymbol{\theta})L(\boldsymbol{\theta}|\mathbf{x})$ for some constant c that may be difficult to determine. If the prior does not seriously restrict the parameter region favored by the data via the likelihood function, then perhaps the prior can serve as a useful importance sampling function. Sample $\boldsymbol{\theta}_1, \ldots, \boldsymbol{\theta}_m$ i.i.d. from $f(\boldsymbol{\theta})$. Since the target density is the posterior, the ith unstandardized weight equals $L(\boldsymbol{\theta}_i|\mathbf{x})$. Thus the SIR algorithm has a very simple form: Sample from the prior, weight by the likelihood, and resample.

For instance, recall Example 6.2. In this case, importance sampling could begin by drawing $\lambda_1, \ldots, \lambda_m \sim$ i.i.d. lognormal$(4, 0.5^2)$. The importance weights would be proportional to $L(\lambda_i|\mathbf{x})$. Resampling from $\lambda_1, \ldots, \lambda_m$ with replacement with these weights yields an approximate sample from the posterior.   □

6.2.4.1 *Adaptive importance, bridge, and path sampling* In some circumstances, one initially may be able to specify only a very poor importance sampling envelope. This may occur, for example, when the target density has support nearly limited to a lower-dimensional space or surface due to strong dependencies between variables not well understood by the analyst. In other situations, one may wish to conduct importance sampling for a variety of related problems, but no single envelope may be suitable for all the target densities of interest. In situations like this, it is possible to adapt the importance sampling envelope.

One collection of ideas for envelope improvement is termed *adaptive importance sampling*. An initial sample of size m_1 is taken from an initial envelope e_1. This sample is weighted (and possibly resampled) to obtain an initial estimate of quantities of interest or an initial view of f itself. Based on the information obtained, the envelope is improved, yielding e_2. Further importance sampling and envelope improvement steps are taken as needed. When such steps are terminated, it is most efficient to use the draws from all the steps, along with their weights, to formulate suitable inference. Alternatively, one can aim to conduct quick envelope refinement during several initial steps, withholding the majority of simulation effort to the final stage and limiting inference to this final sample for simplicity.

In parametric adaptive importance sampling, the envelope is typically assumed to belong to some family of densities indexed by a low-dimensional parameter. The best choice for the parameter is estimated at each iteration, and the importance sampling steps are iterated until estimates of this indexing parameter stabilize [165, 332, 419, 420, 511]. In nonparametric adaptive importance sampling, the envelope is often assumed to be a mixture distribution, such as is generated with the kernel density estimation approach in Chapter 10. Again, importance sampling steps are alternated with envelope updating steps, adding, deleting, or modifying mixture components. Examples include [222, 558, 559, 579]. Although potentially useful in some circumstances, these approaches are overshadowed by Markov chain Monte Carlo approaches like those described in Chapter 7, because the latter are usually simpler and at least as effective.

A second collection of ideas for envelope improvement is relevant when a single envelope is inadequate for the consideration of several densities. In Bayesian statistics and certain marginal likelihood and missing data problems, one is often interested in estimating a ratio of normalizing constants for a pair of densities. For example, if $f_i(\boldsymbol{\theta}|\mathbf{x}) = c_i q_i(\boldsymbol{\theta}|\mathbf{x})$ is the ith posterior density for $\boldsymbol{\theta}$ (for $i = 1, 2$) under two competing models, where q_i is known but c_i is unknown, then $r = c_2/c_1$ is the posterior odds for model 1 compared to model 2. The Bayes factor is the ratio of r to the prior odds.

Since it is often difficult to find good importance sampling envelopes for both f_1 and f_2, one standard importance sampling approach is to use only a single envelope to estimate r. For example, in the convenient case when the support of f_2 contains that of f_1 and we are able to use f_2 as the envelope, $r = \mathrm{E}\{q_1(\boldsymbol{\theta}|\mathbf{x})/q_2(\boldsymbol{\theta}|\mathbf{x})\}$. However, when f_1 and f_2 differ greatly, such a strategy will perform poorly because no single envelope can be sufficiently informative about both c_1 and c_2. The strategy of bridge sampling employs an unnormalized density, q_{bridge} that is, in some sense, between

q_1 and q_2 [388]. Then noting that

$$r = \frac{\mathrm{E}_{f_2}\{q_{\mathrm{bridge}}(\boldsymbol{\theta}|\mathbf{x})/q_2(\boldsymbol{\theta}|\mathbf{x})\}}{\mathrm{E}_{f_1}\{q_{\mathrm{bridge}}(\boldsymbol{\theta}|\mathbf{x})/q_1(\boldsymbol{\theta}|\mathbf{x})\}}, \qquad (6.18)$$

we may employ importance sampling to estimate the numerator and the denominator, thus halving the difficulty of each task, since q_{bridge} is nearer to each q_i than the q_i are to each other.

In principle, the idea of bridging can be extended by iterating the strategy employed in (6.18) with a nested sequence of intermediate densities between q_1 and q_2. Each neighboring pair of densities in the sequence between q_1 and q_2 would be close enough to enable reliable estimation of the corresponding ratio of normalizing constants, and from those ratios one could estimate r. In practice, it turns out that the limit of such a strategy amounts to a very simple algorithm termed *path sampling*. Details are given in [195].

6.2.4.2 *Sequential importance sampling* *Sequential importance sampling* is a strategy for constructing high-dimensional envelopes one dimension at a time. Let $\mathbf{X}_{\leq i} = (X_1, \ldots, X_i)$ denote the first i coordinates of a p-dimensional variable $\mathbf{X} = (X_1, \ldots, X_p)$, and consider the decomposition of the target density given by

$$f(\mathbf{x}) = f(x_1)f(x_2|\mathbf{x}_{\leq 1})f(x_3|\mathbf{x}_{\leq 2}) \cdots f(x_p|\mathbf{x}_{\leq p-1}). \qquad (6.19)$$

Decomposing the envelope g in the same way yields

$$w^*(\mathbf{x}) = \frac{f(x_1)f(x_2|\mathbf{x}_{\leq 1})f(x_3|\mathbf{x}_{\leq 2}) \cdots f(x_p|\mathbf{x}_{\leq p-1})}{g(x_1)g(x_2|\mathbf{x}_{\leq 1})g(x_3|\mathbf{x}_{\leq 2}) \cdots f(x_p|\mathbf{x}_{\leq p-1})} \qquad (6.20)$$

as an expression for the unstandardized importance weight. Note that this suggests sequentially drawing the components of $\mathbf{X}$ from $g(x_1)$, $g(x_2|\mathbf{x}_{\leq 1})$, $g(x_3|\mathbf{x}_{\leq 2})$, and so forth. In this case, consider setting $w_1(x_1) = f(x_1)/g(x_1)$ and applying the recursive expression

$$w_i^*(\mathbf{x}_{\leq i}) = w_{i-1}^*(\mathbf{x}_{\leq i-1})\frac{f(x_i|\mathbf{x}_{\leq i-1})}{g(x_i|\mathbf{x}_{\leq i-1})} \qquad (6.21)$$

for $i = 2, \ldots, p$ to find $w_p^*(\mathbf{x}_{\leq p}) = w^*(\mathbf{x})$. Equation (6.21) would seem to provide a way to accumulate the overall importance weight $w^*(\mathbf{x})$ one dimension at a time, but is impractical, since the conditional distributions $f(x_i|\mathbf{x}_{\leq i-1})$ are typically unavailable.

Suppose, however, that we can construct densities that reasonably approximate the marginal density of $\mathbf{X}_{\leq i}$, namely $f(\mathbf{x}_{\leq i})$, for $i = 1, \ldots, p$. Let $\{\tilde{f}(\mathbf{x}_{\leq 1}), \ldots, \tilde{f}(\mathbf{x}_{\leq p})\}$ be any sequence of marginal densities approximating $\{f(\mathbf{x}_{\leq 1}), \ldots, f(\mathbf{x}_{\leq p})\}$ satisfying $\tilde{f}(\mathbf{x}_{\leq p}) = f(\mathbf{x})$. Then $\tilde{f}(\mathbf{x}_{\leq i})/\tilde{f}(\mathbf{x}_{\leq i-1})$ is an approximation to $f(x_i|\mathbf{x}_{\leq i-1})$, albeit a potentially crude one. Nevertheless, we can use the $\tilde{f}$ functions to reweight sequential draws from the conditional forms of g in the spirit of (6.21) while avoiding reliance on the $f(x_i|\mathbf{x}_{\leq i-1})$.

Define $u_1(x_1) = \tilde{f}(x_1)/g(x_1)$ and

$$u_i(\mathbf{x}_{\leq i}) = \frac{\tilde{f}(\mathbf{x}_{\leq i})}{\tilde{f}(\mathbf{x}_{\leq i-1})g(x_i|\mathbf{x}_{\leq i-1})} \tag{6.22}$$

for $i = 2, \ldots, p$. Then

$$\prod_{i=1}^{p} u_i(\mathbf{x}_{\leq i}) = f(\mathbf{x})/g(\mathbf{x}) = w^*(\mathbf{x}). \tag{6.23}$$

Thus we may use the following algorithm to generate a draw from g and a corresponding importance weight:

1. Initialize by sampling X_1 from $g(x_1)$ and setting $\tilde{w}_1^*(X_1) = \tilde{f}(X_1)/g(X_1)$ and $i = 2$.

2. Given $\mathbf{X}_{\leq i-1} = \mathbf{x}_{\leq i-1}$, sample $X_i \sim g(x_i|\mathbf{x}_{\leq i-1})$.

3. Set $\mathbf{X}_{\leq i} = (\mathbf{X}_{\leq i-1}, X_i)$, and define

$$\tilde{w}_i^*(\mathbf{X}_{\leq i}) = \tilde{w}_{i-1}^*(\mathbf{X}_{\leq i-1})u_i(\mathbf{X}_{\leq i}). \tag{6.24}$$

4. Increment i and return to step 2 until all p components of $\mathbf{X}$ have been drawn.

At the end of these steps, $\mathbf{X} = \mathbf{X}_{\leq p}$ and $w^*(\mathbf{X}) = \tilde{w}_p^*(\mathbf{X}_{\leq p})$ constitute a sequentially generated draw from g and an importance weight that corrects for inference about the target f.

Notice in (6.24) that the approximating functions $\tilde{f}$ only appear in ratios. Therefore, the $\tilde{f}$ need only be specified up to a proportionality constant. Further, since they all cancel out eventually, they need only approximate the true marginals of the target to an extent necessary to guide weight calculation suitably.

The appeal of this strategy is most apparent when the $\tilde{f}$ can be used to improve the overall envelope over what could be achieved through ordinary importance sampling. For example, knowledge about the marginal or conditional distribution of X_i under f for some i can be exploited to improve the generation of draws. Further, the partial weights $\tilde{w}_i^*(\mathbf{X}_{\leq i})$ can be monitored to trigger remedies when the partially generated sample point is sufficiently poor that the full draw will likely have negligible importance weight.

Additional details about implementing sequential importance sampling, including remedies for diminishing partial weights, are given in [336, 357, 358]. A particularly attractive application to the difficult problem of sampling sparse contingency tables is given in [92].

6.3 VARIANCE REDUCTION TECHNIQUES

The simple Monte Carlo estimator of $\int h(\mathbf{x})f(\mathbf{x})\, d\mathbf{x}$ is $\hat{\mu}_{\mathrm{MC}} = \frac{1}{n}\sum_{i=1}^{n} h(\mathbf{X}_i)$ where the variables $\mathbf{X}_1, \ldots, \mathbf{X}_n$ are randomly sampled from f. This approach is intuitively appealing, and we have thus far focused on methods to generate draws from f. In some situations, however, better Monte Carlo estimators can be derived. These approaches are still based on the principle of averaging Monte Carlo draws, but they employ clever sampling strategies and different forms of estimators to yield integral estimates with lower variance than the simplest Monte Carlo approach.

6.3.1 Importance sampling

Suppose we wish to estimate the probability that a die roll will yield a one. If we roll the die n times, we would expect to see about $n/6$ ones, and our point estimate of the true probability would be the proportion of ones in the sample. The variance of this estimator is $\frac{5}{36n}$ if the die is fair. To achieve an estimate with a coefficient of variation of, say, 5%, one should expect to have to roll the die 2000 times.

To reduce the number of rolls required, consider biasing the die by replacing the faces bearing 2 and 3 with additional 1 faces. This increases the probability of rolling a one to 0.5, but we are no longer sampling from the target distribution provided by a fair die. To correct for this, we should weight each roll of a one by 1/3. In other words, let $Y_i = 1/3$ if the roll is a one and $Y_i = 0$ otherwise. Then the expectation of the sample mean of the Y_i is 1/6, and the variance of the sample mean is $\frac{1}{36n}$. To achieve a coefficient of variation of 5% for this estimator, one expects to need only 400 rolls.

This improved accuracy is achieved by causing the event of interest to occur more frequently than it would in the naive Monte Carlo sampling framework, thereby enabling more precise estimation of it. Using importance sampling terminology, the die-rolling example is successful because an *importance sampling distribution* (corresponding to rolling the die with three ones) is used to oversample a portion of the state space that receives lower probability under the target distribution (for the outcome of a fair die). An *importance weighting* corrects for this bias and can provide an improved estimator. For very rare events, extremely large reductions in Monte Carlo variance are possible.

The *importance sampling* approach is based upon the principle that the expectation of $h(\mathbf{X})$ with respect to its density f can be written in the alternative form

$$\mu = \int h(\mathbf{x}) f(\mathbf{x}) \, d\mathbf{x} = \int h(\mathbf{x}) \frac{f(\mathbf{x})}{g(\mathbf{x})} g(\mathbf{x}) \, d\mathbf{x} \qquad (6.25)$$

or even

$$\mu = \frac{\int h(\mathbf{x}) f(\mathbf{x}) \, d\mathbf{x}}{\int f(\mathbf{x}) \, d\mathbf{x}} = \frac{\int h(\mathbf{x}) \frac{f(\mathbf{x})}{g(\mathbf{x})} g(\mathbf{x}) \, d\mathbf{x}}{\int \frac{f(\mathbf{x})}{g(\mathbf{x})} g(\mathbf{x}) \, d\mathbf{x}}, \qquad (6.26)$$

where g is another density function, called the *importance sampling function* or *envelope*.

Equation (6.25) suggests that a Monte Carlo approach to estimating $\mathrm{E}\{h(\mathbf{X})\}$ is to draw $\mathbf{X}_1, \ldots, \mathbf{X}_n$ i.i.d. from g and use the estimator

$$\hat{\mu}_{\mathrm{IS}}^* = \frac{1}{n} \sum_{i=1}^{n} h(\mathbf{X}_i) w^*(\mathbf{X}_i), \qquad (6.27)$$

where $w^*(\mathbf{X}_i) = f(\mathbf{X}_i)/g(\mathbf{X}_i)$ are unstandardized weights, also called *importance ratios*. For this strategy to be convenient, it must be easy to sample from g and to evaluate f, even when it is not easy to sample from f.

Equation (6.26) suggests drawing $\mathbf{X}_1, \ldots, \mathbf{X}_n$ i.i.d. from g and using the estimator

$$\hat{\mu}_{\mathrm{IS}} = \sum_{i=1}^{n} h(\mathbf{X}_i) w(\mathbf{X}_i), \tag{6.28}$$

where $w(\mathbf{X}_i) = w^*(\mathbf{X}_i)/\sum_{i=1}^{n} w^*(\mathbf{X}_i)$ are standardized weights. This second approach is particularly important in that it can be used when f is known only up to a proportionality constant, as is frequently the case when f is a posterior density in a Bayesian analyses.

Both estimators converge by the same argument applied to the simple Monte Carlo estimator given in (6.1), as long as the support of the envelope includes all of the support of f. In order for the estimators to avoid excess variability, it is important that $f(\mathbf{x})/g(\mathbf{x})$ be bounded and that g have heavier tails than f. If this requirement is not met, then some standardized importance weights will be huge. A rare draw from g with much higher density under f than under g will receive huge weight and inflate the variance of the estimator.

Naturally, $g(\mathbf{X})$ often will be larger than $f(\mathbf{X})$ when $\mathbf{X} \sim g$, yet it is easy to show that $E\{f(\mathbf{X})/g(\mathbf{X})\} = 1$. Therefore, if $f(\mathbf{X})/g(\mathbf{X})$ is to have mean 1, this ratio must sometimes be quite large to counterbalance the predominance of values between 0 and 1. Thus, the variance of $f(\mathbf{X})/g(\mathbf{X})$ will tend to be large. Hence, we should expect the variance of $h(\mathbf{X})f(\mathbf{X})/g(\mathbf{X})$ to be large, too. For an importance sampling estimate of μ to have low variance, therefore, we should choose the function g so that $f(\mathbf{x})/g(\mathbf{x})$ is large only when $h(\mathbf{x})$ is very small. For example, when h is an indicator function that equals 1 only for a very rare event, we can choose g to sample in a way that makes that event occur much more frequently, at the expense of failing to sample adequately uninteresting outcomes for which $h(\mathbf{x}) = 0$. This strategy works very well in cases where estimation of a small probability is of interest, such as in estimation of statistical power, probabilities of failure or exceedance, and likelihoods over combinatorial spaces like those that arise frequently with genetic data.

An informal measure of the *effective sample size* can be used to measure the efficiency of an importance sampling strategy using envelope g. When f is known exactly and the unstandardized weights are used as in (6.27), the effective sample size is

$$\hat{N}(g, f) = \frac{n}{1 + \widehat{\mathrm{var}}\{w^*(\mathbf{X})\}}, \tag{6.29}$$

where $\widehat{\mathrm{var}}\{w^*(\mathbf{X})\}$ is the sample variance of the $w^*(\mathbf{X}_i)$. When f is only known up to a proportionality constant and the standardized weights are used as in (6.28), we may use

$$\hat{N}(g, f) = \frac{n}{1 + \widehat{\mathrm{cv}}^2\{w(\mathbf{X})\}}, \tag{6.30}$$

where $\widehat{\mathrm{cv}}\{w(\mathbf{X})\}$ is the sample standard deviation of the standardized importance weights divided by their sample mean. The effective sample size is a measure of how much g differs from f. It can be interpreted as indicating that the n weighted samples used in an importance sampling estimate are worth $\hat{N}(g, f)$ unweighted i.i.d. samples drawn exactly from f and used in a simple Monte Carlo estimate [336, 357].

The choice between using the unstandardized and the standardized weights depends on several considerations. First consider the estimator $\hat{\mu}_{\mathrm{IS}}^*$ defined in (6.27) using the unstandardized weights. Let $t(\mathbf{x}) = h(\mathbf{x})w^*(\mathbf{x})$. When $\mathbf{X}_1, \ldots, \mathbf{X}_n$ are drawn i.i.d. from g, let $\bar{w}^*$ and $\bar{t}$ denote averages of the $w^*(\mathbf{X}_i)$ and $t(\mathbf{X}_i)$, respectively. Note $\mathrm{E}\{\bar{w}^*\} = \mathrm{E}\{w^*(\mathbf{X})\} = 1$. Now,

$$\mathrm{E}\{\hat{\mu}_{\mathrm{IS}}^*\} = \frac{1}{n} \sum_{i=1}^{n} \mathrm{E}\{t(\mathbf{X}_i)\} = \mu \tag{6.31}$$

and

$$\mathrm{var}\{\hat{\mu}_{\mathrm{IS}}^*\} = \frac{1}{n^2} \sum_{i=1}^{n} \mathrm{var}\{t(\mathbf{X}_i)\} = \frac{1}{n}\mathrm{var}\{t(\mathbf{X})\}. \tag{6.32}$$

Thus $\hat{\mu}_{\mathrm{IS}}^*$ is unbiased, and an estimator of its Monte Carlo standard error is the sample standard deviation of $t(\mathbf{X}_1), \ldots, t(\mathbf{X}_n)$ divided by n.

Now consider the estimator $\hat{\mu}_{\mathrm{IS}}$ defined in (6.28) that employs importance weight standardization. Note that $\hat{\mu}_{\mathrm{IS}} = \bar{t}/\bar{w}^*$. Taylor series approximations yield

$$\begin{aligned} \mathrm{E}\{\hat{\mu}_{\mathrm{IS}}\} &= \mathrm{E}\left\{ \bar{t}\,[1 - (\bar{w}^* - 1) + (\bar{w}^* - 1)^2 + \cdots] \right\} \\ &= \mathrm{E}\left\{ \bar{t} - (\bar{t} - \mu)(\bar{w}^* - 1) - \mu(\bar{w}^* - 1) + \bar{t}(\bar{w}^* - 1)^2 + \cdots \right\} \\ &= \mu - \frac{1}{n}\mathrm{cov}\{t(\mathbf{X}), w^*(\mathbf{X})\} + \frac{\mu}{n}\mathrm{var}\{w^*(\mathbf{X})\} + \mathcal{O}(1/n^2). \end{aligned} \tag{6.33}$$

Thus, standardizing the importance weights introduces a slight bias in the estimator $\hat{\mu}_{\mathrm{IS}}$. The bias can be estimated by replacing the variance and covariance terms in (6.33) with sample estimates obtained from the Monte Carlo draws; see also Example 6.8.

The variance of $\hat{\mu}_{\mathrm{IS}}$ is similarly found to be

$$\begin{aligned} \mathrm{var}\{\hat{\mu}_{\mathrm{IS}}\} = &\frac{1}{n}\left[\mathrm{var}\{t(\mathbf{X})\} + \mu^2\,\mathrm{var}\{w^*(\mathbf{X})\} - 2\mu\,\mathrm{cov}\{t(\mathbf{X}), w^*(\mathbf{X})\}\right] \\ &+ \mathcal{O}(1/n^2). \end{aligned} \tag{6.34}$$

Again, a variance estimate for $\hat{\mu}_{\mathrm{IS}}$ can be computed by replacing the variances and covariances in (6.34) with sample estimates obtained from the Monte Carlo draws.

Finally, consider the mean squared errors of $\hat{\mu}_{\mathrm{IS}}^*$ and $\hat{\mu}_{\mathrm{IS}}$. Combining the bias and variance estimates derived above, we find

$$\begin{aligned} \mathrm{MSE}&\{\hat{\mu}_{\mathrm{IS}}\} - \mathrm{MSE}\{\hat{\mu}_{\mathrm{IS}}^*\} \\ &= \frac{1}{n}\left(\mu^2\,\mathrm{var}\{w^*(\mathbf{X})\} - 2\mu\,\mathrm{cov}\{t(\mathbf{X}), w^*(\mathbf{X})\}\right) + \mathcal{O}(1/n^2). \end{aligned} \tag{6.35}$$

Assuming without loss of generality that $\mu > 0$, the leading terms in (6.35) suggest that the approximate difference in mean squared errors is negative when

$$\mathrm{cor}\{t(\mathbf{X}), w^*(\mathbf{X})\} > \frac{\mathrm{cv}\{w^*(\mathbf{X})\}}{2\,\mathrm{cv}\{t(\mathbf{X})\}}, \tag{6.36}$$

where cv$\{\cdot\}$ denotes a coefficient of variation. This condition can be checked using sample-based estimators as discussed above. Thus, using the standardized weights should provide a better estimator when $w^*(\mathbf{X})$ and $h(\mathbf{X})w^*(\mathbf{X})$ are strongly correlated. In addition to these considerations, a major advantage to using the standardized weights is that it does not require knowing the proportionality constant for f. Hesterberg warns that using the standardized weights can be inferior to using the raw weights in many settings, especially when estimating small probabilities, and recommends consideration of an improved importance sampling strategy we describe below in Example 6.8 [284]. Casella and Robert also discuss a variety of uses of the importance weights [85].

Using the standardized weights is reminiscent of the SIR algorithm (Section 6.2.4), and it is sensible to compare the estimation properties of $\hat{\mu}_{\text{IS}}$ with those of the sample mean of the SIR draws. Suppose that an initial sample $\mathbf{Y}_1, \ldots, \mathbf{Y}_m$ with corresponding weights $w(\mathbf{Y}_1), \ldots, w(\mathbf{Y}_m)$ is resampled to provide n SIR draws $\mathbf{X}_1, \ldots, \mathbf{X}_n$, where $n < m$. Let $\hat{\mu}_{\text{SIR}} = \frac{1}{n}\sum_{i=1}^n h(X_i)$ denote the SIR estimate of μ.

When interest is limited to estimation of μ, the importance sampling estimator $\hat{\mu}_{\text{IS}}$ ordinarily should be preferred over $\hat{\mu}_{\text{SIR}}$. To see this, note $\text{E}\{\hat{\mu}_{\text{SIR}}\} = \text{E}\{h(\mathbf{X}_i)\} = \text{E}\left\{\text{E}\{h(\mathbf{X}_i)|\mathbf{Y}_1, \ldots, \mathbf{Y}_m\}\right\} = \text{E}\left\{\frac{\sum_{i=1}^m h(\mathbf{Y}_i)w^*(\mathbf{Y}_i)}{\sum_{i=1}^m w^*(\mathbf{Y}_i)}\right\} = \text{E}\{\hat{\mu}_{\text{IS}}\}$. Therefore the SIR estimator has the same bias as $\hat{\mu}_{\text{IS}}$. However, the variance of $\hat{\mu}_{\text{SIR}}$ is

$$
\begin{aligned}
\text{var}\{\hat{\mu}_{\text{SIR}}\} &= \text{E}\left\{\text{var}\{\hat{\mu}_{\text{SIR}}|\mathbf{Y}_1, \ldots, \mathbf{Y}_m\}\right\} + \text{var}\left\{\text{E}\{\hat{\mu}_{\text{SIR}}|\mathbf{Y}_1, \ldots, \mathbf{Y}_m\}\right\} \\
&= \text{E}\left\{\text{var}\{\hat{\mu}_{\text{SIR}}|\mathbf{Y}_1, \ldots, \mathbf{Y}_m\}\right\} + \text{var}\left\{\frac{\sum_{i=1}^m h(\mathbf{Y}_i)w^*(\mathbf{Y}_i)}{\sum_{i=1}^m w^*(\mathbf{Y}_i)}\right\} \\
&\geq \text{var}\{\hat{\mu}_{\text{IS}}\}.
\end{aligned}
\tag{6.37}
$$

Thus the SIR estimator provides convenience at the expense of precision.

An attractive feature of any importance sampling method is the possibility of reusing the simulations. The same sampled points and weights can be used to compute a variety of Monte Carlo integral estimates of different quantities. The weights can be changed to reflect an alternative importance sampling envelope, to assess or improve performance of the estimator itself. The weights can also be changed to reflect an alternative target distribution, thereby estimating the expectation of $h(\mathbf{X})$ with respect to a different density.

For example, in a Bayesian analysis, one can efficiently update estimates based on a revised posterior distribution in order to carry out Bayesian sensitivity analysis or sequentially to update previous results via Bayes' theorem in light of new information. Such updates can be carried out by multiplying each existing weight $w(\mathbf{X}_i)$ by an adjustment factor. For example, if f is a posterior distribution for $\mathbf{X}$ using prior p_1, then weights equal to $w(\mathbf{X}_i)p_2(\mathbf{X}_i)/p_1(\mathbf{X}_i)$ for $i = 1, \ldots, n$ can be used with the existing sample to provide inference from the posterior distribution using prior p_2.

Example 6.5 (Network failure probability) Many systems can be represented by connected graphs like Figure 6.8. These graphs are composed of nodes (circles) and edges (line segments). A signal sent from A to B must follow a path along any

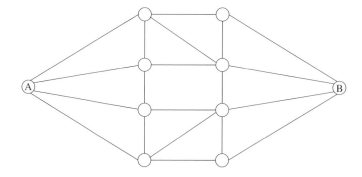

Fig. 6.8 The network connecting A and B described in Example 6.5.

available edges. Imperfect network reliability means that the signal may fail to be transmitted correctly between any pair of connected nodes—in other words, some edges may be broken. In order for the signal to successfully reach B, a connected path from A to B must exist. For example, Figure 6.9 shows a degraded network where only a few routes remain from A to B. If the lowest horizontal edge in this figure were broken, the network would fail.

Network graphs can be used to model many systems. Naturally, such a network can model transmission of diverse types of signals such as analog voice transmission, electromagnetic digital signals, and optical transmission of digital data. The model may also be more conceptual, with each edge representing different machines or people whose participation may be needed to achieve some outcome. Usually, an important quantity of interest is the probability of network failure given specific probabilities for the failure of each edge.

Consider the simplest case, where each edge is assumed to fail independently with the same probability, p. In many signal processing applications p can be quite small. Bit error rates for many types of signal transmission range from 10^{-10} to 10^{-3} [513]. Let $\mathbf{X}$ denote a network, summarizing random outcomes for each edge: intact or failed. The network considered in our example has 20 potential edges, so $\mathbf{X} = (X_1, \ldots, X_{20})$. Let $b(\mathbf{X})$ denote the number of broken edges in $\mathbf{X}$. The network in Figure 6.8 has $b(\mathbf{X}) = 0$; the network in Figure 6.9 has $b(\mathbf{X}) = 10$. Let $h(\mathbf{X})$ indicate network failure, so $h(\mathbf{X}) = 1$ if A is not connected to B, and $h(\mathbf{X}) = 0$ if A and B are connected. The probability of network failure, then, is $\mu = \mathrm{E}\{h(\mathbf{X})\}$. Computing μ for a network of any realistic size can be a very difficult combinatorial problem.

The naive Monte Carlo estimate of μ is obtained by drawing $\mathbf{X}_1, \ldots, \mathbf{X}_n$ independently and uniformly at random from the set of all possible network configurations whose edges fail independently with probability p. The estimator is computed as

$$\hat{\mu}_{\mathrm{MC}} = \frac{1}{n} \sum_{i=1}^{n} h(\mathbf{X}_i). \tag{6.38}$$

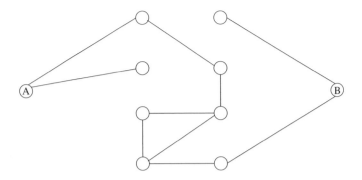

Fig. 6.9 The network connecting A and B described in Example 6.5, with some edges broken.

Notice that this estimator has variance $\mu(1-\mu)/n$. For $n = 100,000$ and $p = 0.05$, simulation yields $\hat{\mu}_{\text{MC}} = 2.00 \times 10^{-5}$ with a Monte Carlo standard error of about 1.41×10^{-5}.

The problem with $\hat{\mu}_{\text{MC}}$ is that $h(\mathbf{X})$ is very rarely 1 unless p is unrealistically large. Thus, a huge number of networks may need to be simulated in order to estimate μ with sufficient precision. Instead, we can use importance sampling to focus on simulation of $\mathbf{X}$ for which $h(\mathbf{X}) = 1$, compensating for this bias through the assignment of importance weights. The calculations that follow adopt this strategy, using the nonstandardized importance weights as in (6.27).

Suppose we simulate $\mathbf{X}_1^*, \ldots, \mathbf{X}_n^*$ by generating network configurations formed by breaking edges in Figure 6.8, assuming independent edge failure with probability $p^* > p$. The importance weight for $\mathbf{X}_i^*$ can be written as

$$w^*(\mathbf{X}_i^*) = \left(\frac{1-p}{1-p^*}\right)^{20} \left(\frac{p(1-p^*)}{p^*(1-p)}\right)^{b(\mathbf{X}_i^*)}, \tag{6.39}$$

and the importance sampling estimator of μ is

$$\hat{\mu}_{\text{IS}}^* = \frac{1}{n} \sum_{i=1}^{n} h(\mathbf{X}_i^*) w^*(\mathbf{X}_i^*). \tag{6.40}$$

Let $\mathcal{C}$ denote the set of all possible network configurations, and let $\mathcal{F}$ denote the subset of configurations for which A and B are not connected. Then

$$\text{var}\{\hat{\mu}_{\text{IS}}^*\} = \frac{1}{n} \text{var}\{h(\mathbf{X}_i^*) w^*(\mathbf{X}_i^*)\} \tag{6.41}$$

$$= \frac{1}{n} \left(\text{E}\left\{ [h(\mathbf{X}_i^*) w^*(\mathbf{X}_i^*)]^2 \right\} - \left[\text{E}\left\{ h(\mathbf{X}_i^*) w^*(\mathbf{X}_i^*) \right\} \right]^2 \right) \tag{6.42}$$

$$= \frac{1}{n} \left(\sum_{\mathbf{x} \in \mathcal{F}} \left(w^*(\mathbf{x}) p^{b(\mathbf{x})} (1-p)^{20-b(\mathbf{x})} \right) - \mu^2 \right). \tag{6.43}$$

Now, for a network derived from Figure 6.8, failure only occurs when $b(\mathbf{X}) \geq 4$. Therefore,

$$w^*(\mathbf{X}) \leq \left(\frac{1-p}{1-p^*}\right)^{20} \left(\frac{p(1-p^*)}{p^*(1-p)}\right)^4. \tag{6.44}$$

When $p^* = 0.25$ and $p = 0.05$, we find $w^*(\mathbf{X}) \leq 0.07$. In this case,

$$\mathrm{var}\{\hat{\mu}_{\mathrm{IS}}^*\} \leq \frac{1}{n}\left(0.07 \sum_{\mathbf{x}\in\mathcal{F}} p^{b(\mathbf{x})}(1-p)^{20-b(\mathbf{x})} - \mu^2\right) \tag{6.45}$$

$$= \frac{1}{n}\left(0.07 \sum_{\mathbf{x}\in\mathcal{C}} h(\mathbf{x})p^{b(\mathbf{x})}(1-p)^{20-b(\mathbf{x})} - \mu^2\right) \tag{6.46}$$

$$= \frac{0.07\mu - \mu^2}{n}. \tag{6.47}$$

Thus $\mathrm{var}\{\hat{\mu}_{\mathrm{IS}}^*\}$ is substantially smaller than $\mathrm{var}\{\hat{\mu}_{\mathrm{MC}}\}$. Under the approximation that $c\mu - \mu^2 \approx c\mu$ for small μ and relatively larger c, we see that $\mathrm{var}\{\hat{\mu}_{\mathrm{MC}}\}/\mathrm{var}\{\hat{\mu}_{\mathrm{IS}}^*\} \approx 14$.

With the naive simulation strategy using $p = 0.05$, only 2 of 100,000 simulated networks failed. However, the importance sampling strategy with $p^* = 0.25$ sampled 497 failing networks, producing an estimate of $\hat{\mu}_{\mathrm{IS}}^* = 1.01 \times 10^{-5}$ with a Monte Carlo standard error of 1.56×10^{-6}.

Related Monte Carlo variance reduction techniques for network reliability problems are discussed in [366]. □

6.3.2 Antithetic sampling

A second approach to variance reduction for Monte Carlo integration relies on finding two identically distributed unbiased estimators, say $\hat{\mu}_1$ and $\hat{\mu}_2$, that are negatively correlated. Averaging these estimators will be superior to using either estimator alone with double the sample size, since the estimator

$$\hat{\mu}_{\mathrm{AS}} = (\hat{\mu}_1 + \hat{\mu}_2)/2 \tag{6.48}$$

has variance equal to

$$\mathrm{var}\{\hat{\mu}_{\mathrm{AS}}\} = \frac{1}{4}\left(\mathrm{var}\{\hat{\mu}_1\} + \mathrm{var}\{\hat{\mu}_2\}\right) + \frac{1}{2}\mathrm{cov}\{\hat{\mu}_1, \hat{\mu}_2\} = \frac{(1+\rho)\sigma^2}{2n}, \tag{6.49}$$

where ρ is the correlation between the two estimators and σ^2/n is the variance of either estimator using a sample of size n. Such pairs of estimators can be generated using the *antithetic sampling* approach [264, 466].

Given an initial estimator, $\hat{\mu}_1$, the question is how to construct a second, identically distributed estimator $\hat{\mu}_2$ that is negatively correlated with $\hat{\mu}_1$. In many situations, there is a convenient way to create such estimators while reusing one simulation sample of size n rather than generating a second sample from scratch. To describe the strategy, we must first introduce some notation. Let $\mathbf{X}$ denote a set of i.i.d. random variables, $\{\mathbf{X}_1, \ldots, \mathbf{X}_n\}$. Suppose $\hat{\mu}_1(\mathbf{X}) = \sum_{i=1}^{n} h_1(\mathbf{X}_i)/n$, where h_1 is a real-valued function of m arguments, so $h_1(\mathbf{X}_i) = h_1(X_{i1}, \ldots, X_{im})$. Assume $\mathrm{E}\{h_1(\mathbf{X}_i)\} = \mu$. Let $\hat{\mu}_2(\mathbf{X}) = \sum_{i=1}^{n} h_2(\mathbf{X}_i)/n$ be a second estimator, with the analogous assumptions about h_2.

We will now prove that if h_1 and h_2 are both increasing in each argument (or both decreasing), then $\mathrm{cov}\{h_1(\mathbf{X}_i), h_2(\mathbf{X}_i)\}$ is positive. From this result, we will be able to determine requirements for h_1 and h_2 that ensure that $\mathrm{cor}\{\hat{\mu}_1, \hat{\mu}_2\}$ is negative.

The proof proceeds via induction. Suppose the above hypotheses hold and $m = 1$. Then

$$[h_1(X) - h_1(Y)][h_2(X) - h_2(Y)] \geq 0 \tag{6.50}$$

for any random variables X and Y. Hence, the expectation of the left hand side of (6.50) is also nonnegative. Therefore, when X and Y are independent and identically distributed, this nonnegative expectation implies

$$\mathrm{cov}\{h_1(X_i), h_2(X_i)\} \geq 0. \tag{6.51}$$

Now, suppose that the desired result holds when $\mathbf{X}_i$ is a random vector of length $m - 1$, and consider the case when $\mathbf{X}_i = (X_{i1}, \ldots, X_{im})$. Then, by hypothesis, the random variable

$$\mathrm{cov}\{h_1(\mathbf{X}_i), h_2(\mathbf{X}_i)| X_{im}\} \geq 0. \tag{6.52}$$

Taking the expectation of this inequality gives

$$\begin{aligned}
0 &\leq \mathrm{E}\{\mathrm{E}\{h_1(\mathbf{X}_i)h_2(\mathbf{X}_i)| X_{im}\}\} - \mathrm{E}\{\mathrm{E}\{h_1(\mathbf{X}_i)| X_{im}\}\mathrm{E}\{h_2(\mathbf{X}_i)| X_{im}\}\} \\
&\leq \mathrm{E}\{h_1(\mathbf{X}_i)h_2(\mathbf{X}_i)\} - \mathrm{E}\{\mathrm{E}\{h_1(\mathbf{X}_i)| X_{im}\}\}\mathrm{E}\{\mathrm{E}\{h_2(\mathbf{X}_i)| X_{im}\}\} \\
&= \mathrm{cov}\{h_1(\mathbf{X}_i), h_2(\mathbf{X}_i)\},
\end{aligned} \tag{6.53}$$

where the substitution of terms in the product on the right side of (6.53) follows from the fact that each $\mathrm{E}\{h_j(\mathbf{X}_i)| X_{im}\}$ for $j = 1, 2$ is a function of the single random argument X_{im}, for which the result (6.51) applies.

Thus, we have proven by induction that $h_1(\mathbf{X}_i)$ and $h_2(\mathbf{X}_i)$ will be positively correlated in these circumstances; it follows that $\hat{\mu}_1$ and $\hat{\mu}_2$ will also be positively correlated. We leave it to the reader to show the following key implication: If h_1 and h_2 are functions of m random variables $U_1, \ldots, U_m$, and if each function is monotone in each argument, then $\mathrm{cov}\{h_1(U_1, \ldots, U_m), h_2(1 - U_1, \ldots, 1 - U_m)\} \leq 0$. This result follows simply from our previous proof after redefining h_1 and h_2 to create two functions increasing in their arguments that satisfy the previous hypotheses. See Problem 6.5.

Now the *antithetic sampling* strategy becomes apparent. The Monte Carlo integral estimate $\hat{\mu}_1(\mathbf{X})$ can be written as

$$\hat{\mu}_1(\mathbf{X}) = \frac{1}{n} \sum_{i=1}^{n} h_1\left(F_1^{-1}(U_{i1}), \ldots, F_m^{-1}(U_{im})\right), \tag{6.54}$$

where F_j is the cumulative distribution function of each X_{ij} ($j = 1, \ldots, m$) and the U_{ij} are independent Unif(0, 1) random variables. Since F_j is a cumulative distribution function, its inverse is nondecreasing. Therefore, $h_1(F_1^{-1}(U_{i1}), \ldots, F_m^{-1}(U_{im}))$ is monotone in each U_{ij} for $j = 1, \ldots, m$ whenever h_1 is monotone in its arguments. Moreover, if $U_{ij} \sim \text{Unif}(0, 1)$, then $1 - U_{ij} \sim \text{Unif}(0, 1)$. Hence, $h_2(1 - \mathbf{U}_{i1}, \ldots, 1 - \mathbf{U}_{im}) = h_1(F_1^{-1}(1 - U_{i1}), \ldots, F_m^{-1}(1 - U_{im}))$ is monotone in each argument and has the same distribution as $h_1(F_1^{-1}(U_{i1}), \ldots, F_m^{-1}(U_{im}))$. Therefore,

$$\hat{\mu}_2(\mathbf{X}) = \frac{1}{n} \sum_{i=1}^{n} h_1\left(F_1^{-1}(1 - U_{i1}), \ldots, F_m^{-1}(1 - U_{im})\right) \qquad (6.55)$$

is a second estimator of μ having the same distribution as $\hat{\mu}_1(\mathbf{X})$. Our analysis above allows us to conclude that

$$\text{cov}\{\hat{\mu}_1(\mathbf{X}), \hat{\mu}_2(\mathbf{X})\} \leq 0. \qquad (6.56)$$

Therefore, the estimator $\hat{\mu}_{\text{AS}} = (\hat{\mu}_1 + \hat{\mu}_2)/2$ will have smaller variance than $\hat{\mu}_1$ would have with a sample of size $2n$. Equation (6.49) quantifies the amount of improvement. We accomplish this improvement while generating only a single set of n random numbers, with the other n derived from the antithetic principle.

Example 6.6 (Normal expectation) Suppose X has a standard normal distribution and we wish to estimate $\mu = \text{E}\{h(X)\}$ where $h(x) = x/(2^x - 1)$. A standard Monte Carlo estimator can be computed as the sample mean of $n = 100,000$ values of $h(X_i)$ where $X_1, \ldots, X_n \sim$ i.i.d. $N(0, 1)$. An antithetic estimator can be constructed using the first $n = 50,000$ draws. The antithetic variate for X_i is simply $-X_i$, so the antithetic estimator is $\hat{\mu}_{\text{AS}} = \sum_{i=1}^{50,000} [h(X_i) + h(-X_i)]/100,000$. In the simulation, $\widehat{\text{cor}}\{h(X_i), h(-X_i)\} = -0.95$, so the antithetic approach is profitable. The standard approach yielded $\hat{\mu}_{\text{MC}} = 1.4993$ with a Monte Carlo standard error of 0.0016, whereas the antithetic approach gave $\hat{\mu}_{\text{AS}} = 1.4992$ with a standard error of 0.0003 (estimated via (6.49) using the sample variance and correlation). Further simulation confirms a more than fourfold reduction in standard error for the antithetic approach. $\qquad \square$

Example 6.7 (Network failure probability, continued) Recalling Example 6.5, let the ith simulated network, $\mathbf{X}_i$, be determined by standard uniform random variables $U_{i1}, \ldots, U_{im}$, where $m = 20$. The jth edge in the ith simulated network is broken if $U_{ij} < p$. Now $h(\mathbf{X}_i) = h(U_{i1}, \ldots, U_{im})$ equals 1 if A and B are not connected, and 0 if they are connected. Note that h is nondecreasing in each U_{ij}; therefore the antithetic approach will be profitable. Since $\mathbf{X}_i$ is obtained by breaking the jth edge when $U_{ij} < p$ for $j = 1, \ldots, m$, the antithetic network draw, say $\mathbf{X}_i^*$, is obtained by breaking the jth edge when $U_{ij} > 1 - p$, for the same set of U_{ij} used to generate $\mathbf{X}_i$. The negative correlation induced by this strategy will ensure that $\frac{1}{2n}\left(\sum_{i=1}^{n} h(\mathbf{X}_i) + h(\mathbf{X}_i^*)\right)$ is a superior estimator to $\frac{1}{2n}\sum_{i=1}^{2n} h(\mathbf{X}_i)$. $\qquad \square$

6.3.3   Control variates

The control variate strategy improves estimation of an unknown integral by relating the estimate to some correlated estimator of an integral whose value is known. Suppose we wish to estimate the unknown quantity $\mu = E\{h(\mathbf{X})\}$ and we know of a related quantity, $\theta = E\{c(\mathbf{Y})\}$, whose value can be determined analytically. Let $(\mathbf{X}_1, \mathbf{Y}_1), \ldots, (\mathbf{X}_n, \mathbf{Y}_n)$ denote pairs of random variables observed independently as simulation outcomes, so $\text{cov}\{\mathbf{X}_i, \mathbf{X}_j\} = \text{cov}\{\mathbf{Y}_i, \mathbf{Y}_j\} = \text{cov}\{\mathbf{X}_i, \mathbf{Y}_j\} = 0$ when $i \neq j$. The simple Monte Carlo estimators are $\hat{\mu}_{\text{MC}} = \frac{1}{n} \sum_{i=1}^{n} h(\mathbf{X}_i)$ and $\hat{\theta}_{\text{MC}} = \frac{1}{n} \sum_{i=1}^{n} c(\mathbf{Y}_i)$. Of course, $\hat{\theta}_{\text{MC}}$ is unnecessary, since θ can be found analytically. However, note that $\hat{\mu}_{\text{MC}}$ will be correlated with $\hat{\theta}_{\text{MC}}$ when $\text{cor}\{h(\mathbf{X}_i), c(\mathbf{Y}_i)\} \neq 0$. For example, if the correlation is positive, an unusually high outcome for $\hat{\theta}_{\text{MC}}$ should tend to be associated with an unusually high outcome for $\hat{\mu}_{\text{MC}}$. If comparison of $\hat{\theta}_{\text{MC}}$ with θ suggests such an outcome, then we should adjust $\hat{\mu}_{\text{MC}}$ downward accordingly. The opposite adjustment should be made when the correlation is negative.

This reasoning suggests the *control variate* estimator

$$\hat{\mu}_{\text{CV}} = \hat{\mu}_{\text{MC}} + \lambda(\hat{\theta}_{\text{MC}} - \theta), \tag{6.57}$$

where λ is a parameter to be chosen by the user. It is straightforward to show that

$$\text{var}\{\hat{\mu}_{\text{CV}}\} = \text{var}\{\hat{\mu}_{\text{MC}}\} + \lambda^2 \, \text{var}\{\hat{\theta}_{\text{MC}}\} + 2\lambda \, \text{cov}\{\hat{\mu}_{\text{MC}}, \hat{\theta}_{\text{MC}}\}. \tag{6.58}$$

Minimizing this quantity with respect to λ shows that the minimal variance,

$$\min_{\lambda} \left(\text{var}\{\hat{\mu}_{\text{CV}}\}\right) = \text{var}\{\hat{\mu}_{\text{MC}}\} - \frac{\left(\text{cov}\{\hat{\mu}_{\text{MC}}, \hat{\theta}_{\text{MC}}\}\right)^2}{\text{var}\{\hat{\theta}_{\text{MC}}\}}, \tag{6.59}$$

is obtained when

$$\lambda = \frac{-\text{cov}\{\hat{\mu}_{\text{MC}}, \hat{\theta}_{\text{MC}}\}}{\text{var}\{\hat{\theta}_{\text{MC}}\}}. \tag{6.60}$$

This optimal λ depends on unknown moments of $h(\mathbf{X}_i)$ and $c(\mathbf{Y}_i)$, but these can be estimated using the sample $(\mathbf{X}_1, \mathbf{Y}_1), \ldots, (\mathbf{X}_n, \mathbf{Y}_n)$. Specifically, using

$$\widehat{\text{var}}\{\hat{\theta}_{\text{MC}}\} = \sum_{i=1}^{n} \frac{[c(\mathbf{Y}_i) - \bar{c}]^2}{n(n-1)} \tag{6.61}$$

and

$$\widehat{\text{cov}}\{\hat{\mu}_{\text{MC}}, \hat{\theta}_{\text{MC}}\} = \sum_{i=1}^{n} \frac{\left[h(\mathbf{X}_i) - \bar{h}\right]\left[c(\mathbf{Y}_i) - \bar{c}\right]}{n(n-1)} \tag{6.62}$$

in (6.60) provides an estimator $\hat{\lambda}$, where $\bar{c} = \frac{1}{n} \sum_{i=1}^{n} c(\mathbf{Y}_i)$ and $\bar{h} = \frac{1}{n} \sum_{i=1}^{n} h(\mathbf{Y}_i)$. Further, plugging such sample variance and covariance estimates into the right hand side of (6.59) provides a variance estimate for $\hat{\mu}_{\text{CV}}$.

In practice, $\hat{\mu}_{MC}$ and $\hat{\theta}_{MC}$ often depend on the same random variables, so $\mathbf{X}_i = \mathbf{Y}_i$. Also, it is possible to use more than one control variate. In this case, we may write the estimator as $\hat{\mu}_{CV} = \hat{\mu}_{MC} + \sum_{j=1}^{m} \lambda_j (\hat{\theta}_{MC,j} - \theta_j)$ when using m control variates.

Equation (6.59) shows that the proportional reduction in variance obtained by using $\hat{\mu}_{CV}$ instead of $\hat{\mu}_{MC}$ is equal to the square of the correlation between $\hat{\mu}_{MC}$ and $\hat{\theta}_{MC}$. If this result sounds familiar, you have astutely noted a parallel with simple linear regression. Consider the regression model $E\{h(\mathbf{X}_i)|\mathbf{Y}_i = \mathbf{y}_i\} = \beta_0 + \beta_1 c(\mathbf{y}_i)$ with the usual regression assumptions and estimators. Then $\hat{\lambda} = -\hat{\beta}_1$ and $\hat{\mu}_{MC} + \hat{\lambda}(\hat{\theta}_{MC} - \theta) = \hat{\beta}_0 + \hat{\beta}_1\theta$. In other words, the control variate estimator is the fitted value on the regression line at the mean value of the predictor (i.e., at θ), and the standard error of this control variate estimator is the standard error for the fitted value from the regression. Thus, linear regression software may be used to obtain the control variate estimator and a corresponding confidence interval. When more than one control variate is used, multiple linear regression can be used to obtain $\hat{\lambda}_i$ ($i = 1, \ldots, m$) and $\hat{\mu}_{CV}$ [466].

Problem 6.5 asks you to show that the antithetic approach to variance reduction can be viewed as a special case of the control variate method.

Example 6.8 (A control variate for importance sampling) Hesterberg suggests using a control variate estimator to improve importance sampling [284]. Recall that importance sampling is built upon the idea of sampling from an envelope that induces a correlation between $h(\mathbf{X})w^*(\mathbf{X})$ and $w^*(\mathbf{X})$. Further, we know $E\{w^*(\mathbf{X})\} = 1$. Thus, the situation is well suited for using the control variate $\bar{w}^* = \sum_{i=1}^{n} w^*(\mathbf{X}_i)/n$. If the average weight exceeds 1, then the average value of $h(\mathbf{X})w^*(\mathbf{X})$ is also probably unusually high, in which case $\hat{\mu}_{IS}$ probably differs from its expectation, μ. Thus, the importance sampling control variate estimator is

$$\hat{\mu}_{ISCV} = \hat{\mu}_{IS}^* + \lambda(\bar{w}^* - 1). \tag{6.63}$$

The value for λ and the standard error of $\hat{\mu}_{ISCV}$ can be estimated from a regression of $h(\mathbf{X})w^*(\mathbf{X})$ on $w^*(\mathbf{X})$ as previously described. Like $\hat{\mu}_{IS}$, which uses the standardized weights, the estimator $\hat{\mu}_{ISCV}$ has bias of order $\mathcal{O}(1/n)$, but will often have lower mean squared error than the importance sampling estimator with unstandardized weights $\hat{\mu}_{IS}^*$ given in (6.27). □

Example 6.9 (Option pricing) A *call option* is a financial instrument that gives its holder the right—but not the obligation—to buy a specified amount of a financial asset, on or by a specified maturity date, for a specified price. For a European call option, the option can be exercised only at the maturity date. The *strike price* is the price at which the transaction is completed if the option is exercised. Let $S^{(t)}$ denote the price of the underlying financial asset (say, a stock) at time t. Denote the strike price by K, and let T denote the maturity date. When time T arrives, the holder of the call option will not wish to exercise his option if $K > S^{(T)}$, because he can obtain the stock more cheaply on the open market. However, the option will be valuable if $K < S^{(T)}$, because he can buy the stock at the low price K and immediately sell it at the higher market price $S^{(T)}$. It is of interest to determine how much the buyer of this call option should pay at time $t = 0$ for this option with strike price K at maturity date T.

The Nobel Prize-winning model introduced by Black, Scholes, and Merton in 1973 provides a popular approach to determining the fair price of an option using a stochastic differential equation [46, 390]. Further background on option pricing and the stochastic calculus of finance includes [160, 346, 498, 566].

The fair price of an option is the amount to be paid at time $t = 0$ that would exactly counterbalance the expected payoff at maturity. We'll consider the simplest case: a European call option on a stock that pays no dividends. The fair price of such an option can be determined analytically under the Black–Scholes model, but estimation of the fair price via Monte Carlo is an instructive starting point. According to the Black–Scholes model, the value of the stock at day T can be simulated as

$$S^{(T)} = S^{(0)} \exp\left\{ \left(r - \frac{\sigma^2}{2} \right) \frac{T}{365} + \sigma Z \sqrt{\frac{T}{365}} \right\}, \tag{6.64}$$

where r is the risk-free rate of return (typically the return rate of the US Treasury bill that matures on day $T - 1$), σ is the stock's volatility (an annualized estimate of the standard deviation of $\log(S^{(t+1)}/S^{(t)})$ under a lognormal price model), and Z is a standard normal deviate. If we knew that the price of the stock at day T would equal $S^{(T)}$, then the fair price of the call option would be

$$C = \exp\{-rT/365\} \max\{0, S^{(T)} - K\}, \tag{6.65}$$

discounting the payoff to present value. Since $S^{(T)}$ is unknown to the buyer of the call, the fair price to pay at $t = 0$ is the expected value of the discounted payoff, namely $E\{C\}$. Thus, a Monte Carlo estimate of the fair price to pay at $t = 0$ is

$$\bar{C} = \frac{1}{n} \sum_{i=1}^{n} C_i, \tag{6.66}$$

where the C_i are simulated from (6.64) and (6.65) for $i = 1, \ldots, n$ using an i.i.d. sample of standard normal deviates, $Z_1, \ldots, Z_n$.

Since the true fair price, $E\{C\}$, can be computed analytically in this instance, there is no need to apply the Monte Carlo approach. However, a special type of European call option, called an Asian, path-dependent, or average-price option, has a payoff based on the average price of the underlying stock throughout the holding period. Such options are attractive for consumers of energies and commodities, because they tend to be exposed to average prices over time. Since the averaging process reduces volatility, Asian options also tend to be cheaper than standard options. Control variate and many other variance reduction approaches for the Monte Carlo pricing of options like these are examined in [53].

To simulate the fair price of an Asian call option, simulation of stock value at maturity is carried out by applying (6.64) sequentially T times, each time advancing the stock price one day and recording the simulated closing price for that day, so

$$S^{(t+1)} = S^{(t)} \exp\left\{ \frac{r - \sigma^2/2}{365} + \frac{\sigma Z^{(t)}}{\sqrt{365}} \right\} \tag{6.67}$$

for a sequence of standard normal deviates, $\{Z^{(t)}\}$, where $t = 0, \ldots, T - 1$. The discounted payoff at day T of the Asian call option on a stock with current price $S^{(0)}$ is defined as

$$A = \exp\{-rT/365\} \max\{0, \bar{S} - K\}, \tag{6.68}$$

where $\bar{S} = \sum_{t=1}^{T} S^{(t)}/T$ and the $S^{(t)}$ for $t = 1, \ldots, T$ are the random variables representing future stock prices at the averaging times. The fair price to pay at $t = 0$ is $E\{A\}$, but in this case there is no known analytic solution for it. Denote the standard Monte Carlo estimator for the fair price of an Asian call option as

$$\hat{\mu}_{\text{MC}} = \bar{A} = \frac{1}{n} \sum_{i=1}^{n} A_i, \tag{6.69}$$

where the A_i are simulated independently as described above.

If $\bar{S}$ is replaced in (6.68) by the geometric average of the price of the underlying stock throughout the holding period, an analytic solution for $E\{A\}$ can be found [324]. The fair price is then

$$\theta = S^{(0)} \Phi(c_1) \exp\left\{ -T \left(r + \frac{c_3 \sigma^2}{6} \right) \frac{1 - 1/N}{730} \right\}$$
$$- K \Phi(c_1 - c_2) \exp\{-rT/365\}, \tag{6.70}$$

where

$$c_1 = \frac{1}{c_2} \left[\log\left\{ \frac{S^{(0)}}{K} \right\} + \left(\frac{c_3 T}{730} \right) \left(r - \frac{\sigma^2}{2} \right) + \frac{c_3 \sigma^2 T}{1095} \left(1 + \frac{1}{2N} \right) \right],$$
$$c_2 = \sigma \left[\frac{c_3 T}{1095} \left(1 + \frac{1}{2N} \right) \right]^{1/2},$$
$$c_3 = 1 + 1/N,$$

Φ is the standard normal cumulative distribution function, and N is the number of prices in the average. Alternatively, one could estimate the fair price of an Asian call option with geometric averaging using the same sort of Monte Carlo strategy described above. Denote this Monte Carlo estimator as $\hat{\theta}_{\text{MC}}$.

The estimator $\hat{\theta}_{\text{MC}}$ makes an excellent control variate for estimation of μ. Let $\hat{\mu}_{\text{CV}} = \hat{\mu}_{\text{MC}} + \lambda(\hat{\theta}_{\text{MC}} - \theta)$. Since we expect the fair price of the two Asian options (arithmetic and geometric mean pricing) to be very highly correlated, a reasonable initial guess is to use $\lambda = -1$.

Consider a European call option with payoff based on the arithmetic mean price during the holding period. Suppose that the underlying stock has current price $S^{(0)} =$

100, strike price $K = 102$, and volatility $\sigma = 0.3$. Suppose there are $N = 50$ days to maturity, so simulation of the maturity price requires 50 iterations of (6.67). Assume the risk-free rate of return is $r = 0.05$. Then the analogous geometric mean price option has a fair price of 1.82. Simulations show that the true fair value of the arithmetic mean price option is roughly $\mu = 1.876$. Using $n = 100,000$ simulations, we can estimate μ using either $\hat{\mu}_{MC}$ or $\hat{\mu}_{CV}$, and both estimators tend to give answers in the vicinity of μ. But what is of interest is the standard error of estimates of μ. We replicated the entire Monte Carlo estimation process 100 times, obtaining 100 values for $\hat{\mu}_{MC}$ and for $\hat{\mu}_{CV}$. The sample standard deviation of the values obtained for $\hat{\mu}_{MC}$ was 0.0107, whereas that of the $\hat{\mu}_{CV}$ values was 0.000295. Thus, the control variate approach provided an estimator with 36 times smaller standard error.

Finally, consider estimating λ from the simulations using (6.60). Repeating the same experiment as above, the typical correlation between $\hat{\mu}_{MC}$ and $\hat{\theta}_{MC}$ was 0.9999. The mean of $\hat{\lambda}$ was -1.0217 with sample standard deviation 0.0001. Using the $\hat{\lambda}$ found in each simulation to produce each $\hat{\mu}_{CV}$ yielded a set of 100 $\hat{\mu}_{CV}$ values whose standard deviation was 0.000168. This represents a 63-fold improvement in the standard error over $\hat{\mu}_{MC}$.   □

6.3.4   Rao–Blackwellization

We have been considering the estimation of $\mu = E\{h(\mathbf{X})\}$ using a random sample $\mathbf{X}_1, \ldots, \mathbf{X}_n$ drawn from f. Suppose that each $\mathbf{X}_i = (\mathbf{X}_{i1}, \mathbf{X}_{i2})$ and that the conditional expectation $E\{h(\mathbf{X}_i)|\mathbf{x}_{i2}\}$ can be solved for analytically. To motivate an alternate estimator to $\hat{\mu}_{MC}$, we may use the fact that $E\{h(\mathbf{X}_i)\} = E\{E\{h(\mathbf{X}_i)|\mathbf{X}_{i2}\}\}$, where the outer expectation is taken with respect to the distribution of $\mathbf{X}_{i2}$. The *Rao–Blackwellized estimator* can be defined as

$$\hat{\mu}_{RB} = \frac{1}{n}\sum_{i=1}^{n} E\{h(\mathbf{X}_i)|\mathbf{X}_{i2}\} \tag{6.71}$$

and has the same mean as the ordinary Monte Carlo estimator $\hat{\mu}_{MC}$. Notice that

$$\text{var}\{\hat{\mu}_{MC}\} = \frac{1}{n}\text{var}\{E\{h(\mathbf{X}_i)|\mathbf{X}_{i2}\}\} + \frac{1}{n}E\{\text{var}\{h(\mathbf{X}_i)|\mathbf{X}_{i2}\}\} \geq \text{var}\{\hat{\mu}_{RB}\} \tag{6.72}$$

follows from the conditional variance formula. Thus, $\hat{\mu}_{RB}$ is superior to $\hat{\mu}_{MC}$ in terms of mean squared error. This conditioning process is often called Rao–Blackwellization due to its use of the Rao–Blackwell theorem, which states that one can reduce the variance of an unbiased estimator by conditioning it on the sufficient statistics [81]. Further study of Rao–Blackwellization for Monte Carlo methods is given in [84, 191, 431, 459, 460].

Example 6.10 (Rao–Blackwellization of rejection sampling)   A generic approach that Rao–Blackwellizes rejection sampling is described by Casella and Robert [84]. In ordinary rejection sampling, candidates $Y_1, \ldots, Y_M$ are generated sequentially, and some are rejected. The uniform random variables $U_1, \ldots, U_M$ provide the rejection decisions, with Y_i being rejected if $U_i > w^*(Y_i)$, where $w^*(Y_i) = f(Y_i)/e(Y_i)$. Rejection sampling stops at a random time M with the acceptance of the nth draw, yielding $X_1, \ldots, X_n$. The ordinary Monte Carlo estimator of $\mu = \mathrm{E}\{h(X)\}$ can then be reexpressed as

$$\hat{\mu}_{\mathrm{MC}} = \frac{1}{n} \sum_{i=1}^{M} h(Y_i) 1_{\{U_i \leq w^*(Y_i)\}}, \tag{6.73}$$

which presents the intriguing possibility that $\hat{\mu}_{\mathrm{MC}}$ somehow can be improved by using all the candidate Y_i draws (suitably weighted), rather than merely the accepted draws.

Rao–Blackwellization of (6.73) yields the estimator

$$\hat{\mu}_{\mathrm{RB}} = \frac{1}{n} \sum_{i=1}^{M} h(Y_i) t_i(\mathbf{Y}), \tag{6.74}$$

where the $t_i(\mathbf{Y})$ are random quantities that depend on $\mathbf{Y} = (Y_1, \ldots, Y_M)$ and M according to

$$\begin{aligned} t_i(\mathbf{Y}) &= \mathrm{E}\left\{ 1_{\{U_i \leq w^*(Y_i)\}} | M, Y_1, \ldots, Y_M \right\} \\ &= P[U_i < w^*(Y_i) | M, Y_1, \ldots, Y_M]. \end{aligned} \tag{6.75}$$

Now $t_M(\mathbf{Y}) = 1$, since the final candidate was accepted. For previous candidates, the probability in (6.75) can be found by averaging over permutations of subsets of the realized sample [84]. We obtain

$$t_i(\mathbf{Y}) = \frac{w^*(Y_i) \sum_{A \in \mathcal{A}_i} \prod_{j \in A} w^*(Y_j) \prod_{j \notin A} [1 - w^*(Y_j)]}{\sum_{B \in \mathcal{B}} \prod_{j \in B} w^*(Y_j) \prod_{j \notin B} [1 - w^*(Y_j)]}, \tag{6.76}$$

where $\mathcal{A}_i$ is the set of all subsets of $\{1, \ldots, i-1, i+1, \ldots, M-1\}$ containing $n-2$ elements, and $\mathcal{B}$ is the set of all subsets of $\{1, \ldots, M-1\}$ containing $n-1$ elements. Casella and Robert offer a recursion formula for computing the $t_i(\mathbf{Y})$, but it is difficult to implement unless n is fairly small.

Notice that the conditioning variables used here are statistically sufficient, since the conditional distribution of $U_1, \ldots, U_M$ does not depend on f. Both $\hat{\mu}_{\mathrm{RB}}$ and $\hat{\mu}_{\mathrm{MC}}$ are unbiased; thus, the Rao–Blackwell theorem implies that $\hat{\mu}_{\mathrm{RB}}$ will have smaller variance than $\hat{\mu}_{\mathrm{MC}}$. □

Problems

6.1   Consider the integral sought in Example 5.1, equation (5.7), for the parameter values given there. Find a simple rejection sampling envelope that will produce extremely few rejections when used to generate draws from the density proportional to that integrand.

6.2   Consider the piecewise exponential envelopes for adaptive rejection sampling of the standard normal density, which is log-concave. For the tangent-based envelope, suppose you are limited to an even number of nodes at $\pm c_1, \ldots, \pm c_n$. For the envelope that does not require tangent information, suppose you are limited to an odd number of nodes at $0, \pm d_1, \ldots, \pm d_n$. The problems below will require optimization using strategies like those in Chapter 2.

(a) For $n = 1, 2, 3, 4, 5$, find the optimal placement of nodes for the tangent-based envelope.

(b) For $n = 1, 2, 3, 4, 5$, find the optimal placement of nodes for the tangent-free envelope.

(c) Plot these collections of envelopes; also plot rejection sampling waste against number of nodes for both envelopes. Comment on your results.

6.3   Consider finding $\sigma^2 = E\{X^2\}$ when X has the density that is proportional to $q(x) = \exp\{-|x|^3/3\}$.

(a) Estimate σ^2 using importance sampling with standardized weights.

(b) Repeat the estimation using rejection sampling.

(c) Philippe and Robert describe an alternative to importance-weighted averaging that employs a Riemann sum strategy with random nodes [430, 431]. When draws $X_1, \ldots, X_n$ originate from f, an estimator of $E\{h(X)\}$ is

$$\sum_{i=1}^{n-1} (X_{[i+1]} - X_{[i]}) h(X_{[i]}) f(X_{[i]}), \qquad (6.77)$$

where $X_{[1]} \leq \cdots \leq X_{[n]}$ is the ordered sample associated with $X_1, \ldots, X_n$. This estimator has faster convergence than the simple Monte Carlo estimator. When $f = cq$ and the normalization constant c is not known, then

$$\frac{\sum_{i=1}^{n-1} (X_{[i+1]} - X_{[i]}) h(X_{[i]}) q(X_{[i]})}{\sum_{i=1}^{n-1} (X_{[i+1]} - X_{[i]}) q(X_{[i]})} \qquad (6.78)$$

estimates $E\{h(X)\}$, noting that the denominator estimates $1/c$. Use this strategy to estimate σ^2, applying it post hoc to the output obtained in part (b).

(d) Carry out a replicated simulation experiment to compare the performance of the two estimators in parts (b) and (c). Discuss your results.

6.4   Figure 6.10 shows some data on the number of coal-mining disasters per year between 1851 and 1962, available from the website for this book. These data originally appeared in [368] and were corrected in [306]. The form of the data we consider is given in [79]. Other analyses of these data include [378, 443].

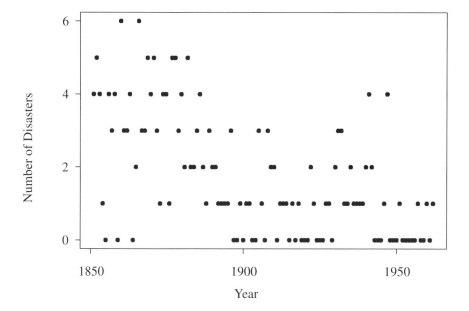

Fig. 6.10 Number of coal-mining disasters per year between 1851 and 1962.

The rate of accidents per year appears to decrease around 1900, so we consider a change-point model for these data. Let $j = 1$ in 1851, and index each year thereafter, so $j = 112$ in 1962. Let X_j be the number of accidents in year j, with $X_1, \ldots, X_\theta \sim$ i.i.d. Poisson(λ_1) and $X_{\theta+1}, \ldots, X_{112} \sim$ i.i.d. Poisson(λ_2). Thus the change-point occurs after the θth year in the series, where $\theta \in \{1, \ldots, 111\}$. This model has parameters θ, λ_1, and λ_2. Below are three sets of priors for a Bayesian analysis of this model. In each case, consider sampling from the priors as the first step of applying the SIR algorithm for simulating from the posterior for the model parameters. Of primary interest is inference about θ.

(a) Assume a discrete uniform prior for θ on $\{1, 2, \ldots, 111\}$, and priors $\lambda_i | a_i \sim$ Gamma($3, a_i$) and $a_i \sim$ Gamma($10, 10$) independently for $i = 1, 2$. Using the SIR approach, estimate the posterior mean for θ, and provide a histogram and a credible interval for θ. Provide similar information for estimating λ_1 and λ_2. Make a scatterplot of λ_1 against λ_2 for the initial SIR sample, highlighting the points resampled at the second stage of SIR. Also report your initial and resampling sample sizes, the number of unique points and highest observed frequency in your resample, and a measure of the effective sample size for importance sampling in this case. Discuss your results.

(b) Assume that $\lambda_2 = \alpha \lambda_1$. Use the same discrete uniform prior for θ and $\lambda_1 | a \sim$ Gamma($3, a$), $a \sim$ Gamma($10, 10$), and $\log \alpha \sim$ Unif($\log 1/8, \log 2$). Provide the same results listed in part (a), and discuss your results.

(c) Markov chain Monte Carlo approaches (see Chapter 7) are often applied in the analysis of these data. A set of priors that resembles the improper diffuse priors used in some such analyses is: θ having the discrete uniform prior, $\lambda_i | a_i \sim$ Gamma$(3, a_i)$, and $a_i \sim$ Unif$(0, 100)$ independently for $i = 1, 2$. Provide the same result listed in part (a), and discuss your results, including reasons why this analysis is more difficult than the previous two.

6.5   Prove the following results.

(a) If h_1 and h_2 are functions of m random variables $U_1, \ldots, U_m$, and if each function is monotone in each argument, then

$$\text{cov}\{h_1(U_1, \ldots, U_m), h_2(1 - U_1, \ldots, 1 - U_m)\} \leq 0.$$

(b) Let $\hat{\mu}_1(\mathbf{X})$ estimate a quantity of interest, μ, and let $\hat{\mu}_2(\mathbf{Y})$ be constructed from realizations $\mathbf{Y}_1, \ldots, \mathbf{Y}_n$ chosen to be antithetic to $\mathbf{X}_1, \ldots, \mathbf{X}_n$. Assume that both estimators are unbiased for μ and are negatively correlated. Find a control variate for $\hat{\mu}_1$, say Z, with mean zero, for which the control variate estimator $\hat{\mu}_{\text{CV}} = \hat{\mu}_1(\mathbf{X}) + \lambda Z$ corresponds to the antithetic estimator based on $\hat{\mu}_1$ and $\hat{\mu}_2$ when the optimal λ is used. Include your derivation of the optimal λ.

6.6   Consider testing the hypotheses $H_0 : \lambda = 2$ versus $H_a : \lambda > 2$ using 25 observations from a Poisson(λ) model. Rote application of the central limit theorem would suggest rejecting H_0 at $\alpha = 0.05$ when $Z \geq 1.645$, where $Z = \frac{\bar{X} - 2}{\sqrt{2/25}}$.

(a) Estimate the size of this test (i.e., the type I error rate) using five Monte Carlo approaches: standard, antithetic, importance sampling with unstandardized and standardized weights, and importance sampling with a control variate as in Example 6.8. Provide a confidence interval for each estimate. Discuss the relative merits of each variance reduction technique, and compare the importance sampling strategies with each other. For the importance sampling approaches, use a Poisson envelope with mean equal to the H_0 rejection threshold, namely $\lambda = 2.4653$.

(b) Draw the power curve for this test for $\lambda \in [2.2, 4]$, using the same five techniques. Provide pointwise confidence bands in each case. Discuss the relative merits of each technique in this setting. Compare the performances of the importance sampling strategies with their performance in part (a).

6.7   Consider pricing a European call option on an underlying stock with current price $S^{(0)} = 50$, strike price $K = 52$, and volatility $\sigma = 0.5$. Suppose that there are $N = 30$ days to maturity and that the risk-free rate of return is $r = 0.05$.

(a) Confirm that the fair price for this option is 2.10 when the payoff is based on $S^{(30)}$ (i.e., a standard option with payoff as in (6.65)).

(b) Consider the analogous Asian option (same $S^{(0)}$, K, σ, N, and r) with payoff based on the arithmetic mean stock price during the holding period, as in (6.68). Using simple Monte Carlo, estimate the fair price for this option.

(c) Improve upon the estimate in (b) using the control variate strategy described in Example 6.9.

(d) Try an antithetic approach to estimate the fair price for the option described in part (b).

(e) Using simulation and/or analysis, compare the sampling distributions of the estimators in (b), (c), and (d).

6.8   Consider the model given by $X \sim$ lognormal$(0, 1)$ and $\log Y = 9 + 3 \log X + \epsilon$, where $\epsilon \sim N(0, 1)$. We wish to estimate $E\{Y/X\}$. Compare the performance of the standard Monte Carlo estimator and the Rao–Blackwellized estimator.

7

Markov Chain Monte Carlo

When a target density f can be evaluated but not easily sampled, the methods from Chapter 6 can be applied to obtain an approximate or exact sample. The primary use of such a sample is to estimate the expectation of a function of $\mathbf{X} \sim f(\mathbf{x})$. The Markov chain Monte Carlo (MCMC) methods introduced in this chapter can also be used to generate a draw from a distribution that approximates f, but they are more properly viewed as methods for generating a sample from which expectations of functions of $\mathbf{X}$ can reliably be estimated. MCMC methods are distinguished from the simulation techniques in Chapter 6 by their iterative nature and the ease with which they can be customized to very diverse and difficult problems. Viewed as an integration method, MCMC has several advantages over the approaches in Chapter 5: Increasing problem dimensionality usually does not slow convergence or make implementation more complex.

A quick review of discrete-state-space Markov chain theory is provided in Section 1.7. Let the sequence $\{\mathbf{X}^{(t)}\}$ denote a Markov chain for $t = 0, 1, 2, \ldots$, where $\mathbf{X}^{(t)} = \left(X_1^{(t)}, \ldots, X_p^{(t)} \right)$ and the state space is either continuous or discrete. For the types of Markov chains introduced in this chapter, the distribution of $\mathbf{X}^{(t)}$ converges to the limiting stationary distribution of the chain when the chain is irreducible and aperiodic. The MCMC sampling strategy is to construct an irreducible, aperiodic Markov chain for which the stationary distribution equals the target distribution f. For sufficiently large t, a realization $\mathbf{X}^{(t)}$ from this chain will have approximate marginal distribution f. A very popular application of MCMC methods is to facilitate Bayesian inference where f is a Bayesian posterior distribution for parameters $\mathbf{X}$; a short review of Bayesian inference is given in Section 1.5.

The art of MCMC lies in the construction of a suitable chain. A wide variety of algorithms has been proposed. The dilemma lies in how to determine the degree of distributional approximation that is inherent in realizations from the chain as well as estimators derived from these realizations. This question arises because the distribution of $X^{(t)}$ may differ substantially from f when t is too small (note that t is always limited in computer simulations), and because the $X^{(t)}$ are serially dependent.

MCMC theory and applications are areas of active research interest. Our emphasis here is on introducing some basic MCMC algorithms that are easy to implement and broadly applicable. In Chapter 8, we address several more sophisticated MCMC techniques. Some comprehensive expositions of MCMC and helpful tutorials include [64, 82, 91, 93, 460, 537]

7.1 METROPOLIS–HASTINGS ALGORITHM

A very general method for constructing a Markov chain is the Metropolis–Hastings algorithm [282, 391]. The method begins at $t = 0$ with the selection of $X^{(0)} = x^{(0)}$ drawn at random from some starting distribution g, with the requirement that $f\left(x^{(0)}\right) > 0$. Given $X^{(t)} = x^{(t)}$, the algorithm generates $X^{(t+1)}$ as follows:

1. Sample a candidate value X^* from a *proposal distribution* $g\left(\cdot \mid x^{(t)}\right)$.

2. Compute the *Metropolis–Hastings ratio* $R\left(x^{(t)}, X^*\right)$, where

$$R(u, v) = \frac{f(v)\, g(u \mid v)}{f(u)\, g(v \mid u)}. \tag{7.1}$$

Note that $R\left(x^{(t)}, X^*\right)$ is always defined, because the proposal $X^* = x^*$ can only occur if $f\left(x^{(t)}\right) > 0$ and $g\left(x^* \mid x^{(t)}\right) > 0$.

3. Sample a value for $X^{(t+1)}$ according to the following:

$$X^{(t+1)} = \begin{cases} X^* & \text{with probability } \min\left\{R\left(x^{(t)}, X^*\right), 1\right\}, \\ x^{(t)} & \text{otherwise.} \end{cases} \tag{7.2}$$

4. Increment t and return to step 1.

We will call the tth iteration the process that generates $X^{(t)} = x^{(t)}$.

As with optimization algorithms, it is a good idea to run MCMC procedures like the Metropolis–Hastings algorithm from multiple starting points to check for consistent output. When the proposal distribution is symmetric so that $g\left(x^{(t)} \mid x^*\right) = g\left(x^* \mid x^{(t)}\right)$, the method is known as the Metropolis algorithm [391].

Clearly, a chain constructed via the Metropolis–Hastings algorithm is Markov, since $X^{(t+1)}$ is only dependent on $X^{(t)}$. Whether the chain is irreducible and aperiodic depends on the choice of proposal distribution; the user *must* check these conditions diligently for any implementation. If this check confirms irreducibility and aperiodicity, then the chain generated by the Metropolis–Hastings algorithm has a unique limiting stationary distribution. This result would seem to follow from equation (1.44). However, we are now considering both continuous and discrete state space Markov chains. Nevertheless, irreducibility and aperiodicity remain sufficient conditions for convergence of Metropolis–Hastings chains. Additional theory is provided in [393, 460].

To find the unique stationary distribution of an irreducible aperiodic Metropolis–Hastings chain, suppose $\mathbf{X}^{(t)} \sim f(\mathbf{x})$, and consider two points in the state space of the chain, say $\mathbf{x}_1$ and $\mathbf{x}_2$, for which $f(\mathbf{x}_1) > 0$ and $f(\mathbf{x}_2) > 0$. Without loss of generality, label these points in the manner such that $f(\mathbf{x}_2)g(\mathbf{x}_1|\mathbf{x}_2) \geq f(\mathbf{x}_1)g(\mathbf{x}_2|\mathbf{x}_1)$.

It follows that the unconditional joint density of $\mathbf{X}^{(t)} = \mathbf{x}_1$ and $\mathbf{X}^{(t+1)} = \mathbf{x}_2$ is $f(\mathbf{x}_1)g(\mathbf{x}_2|\mathbf{x}_1)$, because if $\mathbf{X}^{(t)} = \mathbf{x}_1$ and $\mathbf{X}^* = \mathbf{x}_2$ then $R(\mathbf{x}_1, \mathbf{x}_2) \geq 1$ so $\mathbf{X}^{(t)} = \mathbf{x}_2$. The unconditional joint density of $\mathbf{X}^{(t)} = \mathbf{x}_2$ and $\mathbf{X}^{(t+1)} = \mathbf{x}_1$ is

$$f(\mathbf{x}_2)g(\mathbf{x}_1|\mathbf{x}_2)\frac{f(\mathbf{x}_1)g(\mathbf{x}_2|\mathbf{x}_1)}{f(\mathbf{x}_2)g(\mathbf{x}_1|\mathbf{x}_2)}, \tag{7.3}$$

because we need to start with $\mathbf{X}^{(t)} = \mathbf{x}_2$, to propose $\mathbf{X}^* = \mathbf{x}_1$, and then to set $\mathbf{X}^{(t+1)}$ equal to $\mathbf{X}^*$ with probability $R(\mathbf{x}_1, \mathbf{x}_2)$. Note that (7.3) reduces to $f(\mathbf{x}_1)g(\mathbf{x}_2|\mathbf{x}_1)$, which matches the joint density of $\mathbf{X}^{(t)} = \mathbf{x}_1$ and $\mathbf{X}^{(t+1)} = \mathbf{x}_2$. Therefore the joint distribution of $\mathbf{X}^{(t)}$ and $\mathbf{X}^{(t+1)}$ is symmetric. Hence $\mathbf{X}^{(t)}$ and $\mathbf{X}^{(t+1)}$ have the same marginal distributions. Thus the marginal distribution of $\mathbf{X}^{(t+1)}$ is f, and f must be the stationary distribution of the chain.

Recall from equation (1.46) that we can approximate the expectation of a function of a random variable by averaging realizations from the stationary distribution of a Metropolis–Hastings chain. The distribution of realizations from the Metropolis–Hastings chain approximates the stationary distribution of the chain as t progresses; therefore $E\{h(\mathbf{X})\} \approx \frac{1}{n}\sum_{i=1}^{n} h\left(\mathbf{x}^{(i)}\right)$. Some of the useful quantities that can be estimated this way include means $E\{h(\mathbf{X})\}$, variances $E\{[h(\mathbf{X}) - E\{h(\mathbf{X})\}]^2\}$, and tail probabilities $E\{1_{\{h(\mathbf{X}) \leq q\}}\}$ for constant q, where $1_{\{A\}} = 1$ if A is true and 0 otherwise. Using the density estimation methods of Chapter 10, estimates of f itself can also be obtained. Due to the limiting properties of the Markov chain, estimates of all these quantities based on sample averages are strongly consistent. Note that the sequence $\mathbf{x}^{(0)}, \mathbf{x}^{(1)}, \ldots$ will likely include multiple copies of some points in the state space. This occurs when $\mathbf{X}^{(t+1)}$ retains the previous value $\mathbf{x}^{(t)}$ rather than jumping to the proposed value $\mathbf{x}^*$. It is important to include these copies in the chain and in any sample averages, since the frequencies of sampled points are used to correct for the fact that the proposal density differs from the target density. In most applications we can never be certain that the chain has converged to its stationary distribution, so it may be sensible to omit some of the initial realizations of the chain when computing a sample average.

Specific features of good proposal distributions can greatly enhance the performance of the Metropolis–Hastings algorithm. A well-chosen proposal distribution produces candidate values that cover the support of the stationary distribution in a reasonable number of iterations and, similarly, produces candidate values that are not accepted or rejected too frequently [93]. Both of these factors are related to the spread of the proposal distribution. If the proposal distribution is too diffuse relative to the target distribution, the candidate values will be rejected frequently and thus the chain will require many iterations to adequately explore the space of the target distribution. If the proposal distribution is too focused (e.g., has too small a variance), then the chain will remain in one small region of the target distribution for many iterations while other regions of the target distribution will not be adequately explored. Thus a proposal distribution whose spread is either too small or too large can produce a chain that requires many iterations to adequately sample the regions supported by the target distribution. Section 7.3.1 further discusses this and related issues.

Below we introduce several Metropolis–Hastings variants obtained by using different classes of proposal distributions.

7.1.1   Independence chains

Suppose that the proposal distribution for the Metropolis–Hastings algorithm is chosen such that $g\left(\mathbf{x}^* \mid \mathbf{x}^{(t)}\right) = g(\mathbf{x}^*)$ for some fixed density g. This yields an independence chain, where each candidate value is drawn independently of the past. In this case, the Metropolis–Hastings ratio is

$$R\left(\mathbf{x}^{(t)}, \mathbf{X}^*\right) = \frac{f\left(\mathbf{X}^*\right) g\left(\mathbf{x}^{(t)}\right)}{f\left(\mathbf{x}^{(t)}\right) g\left(\mathbf{X}^*\right)}. \tag{7.4}$$

The resulting Markov chain is irreducible and aperiodic if $g\left(\mathbf{x}\right) > 0$ whenever $f\left(\mathbf{x}\right) > 0$.

Notice that the Metropolis–Hastings ratio in (7.4) can be reexpressed as the ratio of importance ratios (see Section 6.3.1) where f is the target and g is the envelope: If $w^* = f\left(\mathbf{X}^*\right)/g\left(\mathbf{X}^*\right)$ and $w^{(t)} = f\left(\mathbf{x}^{(t)}\right)/g\left(\mathbf{x}^{(t)}\right)$, then $R\left(\mathbf{x}^{(t)}, \mathbf{X}^*\right) = w^*/w^{(t)}$. This reexpression indicates that when $w^{(t)}$ is much larger than typical w^* values, then the chain will tend to get stuck for long periods at the current value. Therefore, the criteria discussed in Section 6.2.4 for choosing importance sampling envelopes are also relevant here for choosing proposal distributions: The proposal distribution g should resemble the target distribution f, but should cover f in the tails.

Example 7.1 (Bayesian inference) MCMC methods like the Metropolis–Hastings algorithm are particularly popular tools for Bayesian inference, where some data $\mathbf{y}$ are observed with likelihood function $L(\boldsymbol{\theta}|\mathbf{y})$ for parameters $\boldsymbol{\theta}$ which have prior distribution $p(\boldsymbol{\theta})$. Bayesian inference is based on the posterior distribution $p(\boldsymbol{\theta}|\mathbf{y}) = c\, p(\boldsymbol{\theta})L(\boldsymbol{\theta}|\mathbf{y})$, where c is an unknown constant. The difficulty of computing c and other features of the posterior prevents most direct inferential strategies. However, if we can obtain a sample from a Markov chain whose stationary distribution is the target posterior, this sample may be used to estimate posterior moments, tail probabilities, and many other useful quantities, including the posterior density itself. MCMC methods typically allow easy generation of such a sample in the Bayesian context.

A very simple strategy is to use the prior as a proposal distribution in an independence chain. In our Metropolis–Hastings notation, $f(\boldsymbol{\theta}) = p(\boldsymbol{\theta}|\mathbf{y})$ and $g(\boldsymbol{\theta}^*) =$

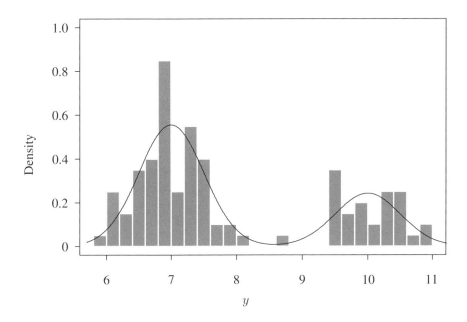

Fig. 7.1 Histogram of 100 observations simulated from the mixture distribution (7.6) in Example 7.2.

$p(\boldsymbol{\theta}^*)$. Conveniently, this means

$$R(\boldsymbol{\theta}^{(t)}, \boldsymbol{\theta}^*) = \frac{L(\boldsymbol{\theta}^* \mid \mathbf{y})}{L(\boldsymbol{\theta}^{(t)} \mid \mathbf{y})}. \tag{7.5}$$

In other words, we propose from the prior, and the Metropolis–Hastings ratio equals the likelihood ratio. By definition, the support of the prior covers the support of the target posterior, so the stationary distribution of this chain is the desired posterior. There are often more specialized MCMC algorithms to sample various types of posteriors in more efficient manners, but this is perhaps the simplest generic approach. □

Example 7.2 (Estimating a mixture parameter) Suppose we have observed data $y_1, y_2, \ldots, y_{100}$ sampled independently and identically distributed from the mixture distribution

$$\delta N(7, 0.5^2) + (1 - \delta)N(10, 0.5^2). \tag{7.6}$$

Figure 7.1 shows a histogram of the data, which are available from the website for this book. Mixture densities are common in real-life applications where, for example, the data may come from more than one population. We will use MCMC techniques to construct a chain whose stationary distribution equals the posterior density of δ

assuming a Unif(0,1) prior distribution for δ. The data were generated with $\delta = 0.7$, so we should find that the posterior density is concentrated in this area.

In this example, we try two different independence chains. In the first case we use a Beta(1,1) density as the proposal density, and in the second case we use a Beta(2,10) density. The first proposal distribution is equivalent to a Unif(0,1) distribution, while the second is skewed right with mean approximately equal to 0.167. In this second case values of δ near 0.7 are unlikely to be generated from the proposal distribution.

Figure 7.2 shows the sample paths for 10,000 iterations of both chains. A *sample path* is a plot of the chain realizations $\delta^{(t)}$ against the iteration number t. This plot is useful for investigating the behavior of the Markov chain and is discussed further in Section 7.3.1. The top panel of Figure 7.2 corresponds to the chain generated using the Beta(1,1) proposal density. This panel shows a Markov chain that moves quickly away from its starting value and seems easily able to sample values from all portions of the parameter space supported by the posterior for δ. Such behavior is called good *mixing*. The lower panel corresponds to the chain using a Beta(2,10) proposal density. The resulting chain moves slowly from its starting value and does a poor job of exploring the region of posterior support (i.e., poor mixing). This chain has clearly not converged to its stationary distribution, since drift is still visible. Of course, the long-run behavior of the chain will in principle allow estimation of aspects of the posterior distribution for δ since the posterior is still the limiting distribution of the chain. Yet, chain behavior like that shown in the bottom panel of Figure 7.2 does not inspire confidence: The chain seems nonstationary, only a few unique values of $\delta^{(t)}$ were accepted, and the starting value does not appear to have washed out. A plot like the lower plot in Figure 7.2 should make the MCMC user reconsider the proposal density and other aspects of the MCMC implementation.

Figure 7.3 shows histograms of the realizations from the chains, after the first 200 iterations were omitted to reduce the effect of the starting value (see the discussion of burn-in periods in Section 7.3.1.5). The top and bottom panels again correspond to the Beta(1,1) and Beta(2,10) proposal distributions, respectively. This plot shows that the chain with the Beta(1,1) proposal density produced a sample for δ whose mean well approximates the true value (and posterior mean) of $\delta = 0.7$. On the other hand, the chain with the Beta(2,10) proposal density would not yield reliable estimates for the posterior or the true value of δ based on the first 10,000 iterations.   □

7.1.2   Random walk chains

A random walk chain is another type of Markov chain produced via a simple variant of the Metropolis–Hastings algorithm. Let $\mathbf{X}^*$ be generated by drawing $\epsilon \sim h(\epsilon)$ for some density h and then setting $\mathbf{X}^* = \mathbf{x}^{(t)} + \epsilon$. This yields a random walk chain. In this case, $g\left(\mathbf{x}^*|\mathbf{x}^{(t)}\right) = h\left(\mathbf{x}^* - \mathbf{x}^{(t)}\right)$. Common choices for h include a uniform distribution over a ball centered at the origin, a scaled standard normal distribution, and a scaled Student's t distribution. If the support region of f is connected and h is positive in a neighborhood of 0, the resulting chain is irreducible and aperiodic [460].

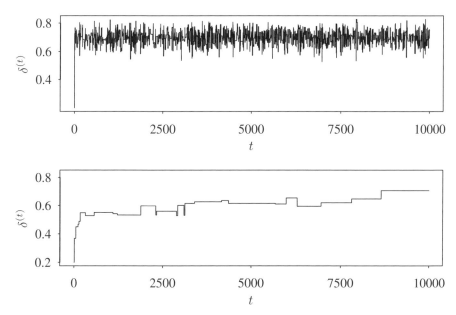

Fig. 7.2 Sample paths for δ from independence chains with proposal densities Beta(1,1) (top) and Beta(2,10) (bottom) considered in Example 7.2.

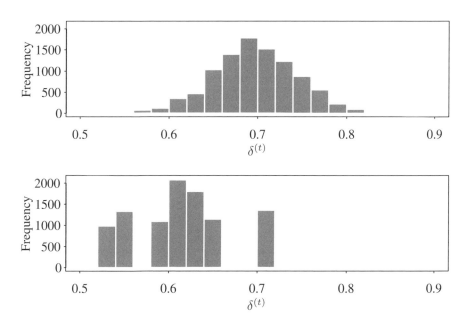

Fig. 7.3 Histograms of $\delta^{(t)}$ for iterations 201–10,000 of independence chains with proposal densities Beta(1,1) (top) and Beta(2,10) (bottom) considered in Example 7.2.

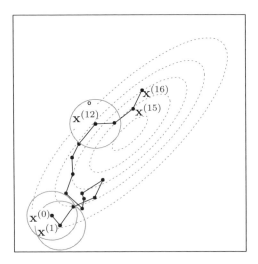

Fig. 7.4 Hypothetical random walk chain for sampling a two-dimensional target distribution (dotted contours) using proposed increments sampled uniformly from a disk centered at the current value. See text for details.

Figure 7.4 illustrates how a random walk chain might progress in a two-dimensional problem. The figure shows a contour plot of a two-dimensional target distribution (dotted lines) along with the first steps of a random walk MCMC procedure. The sample path is shown by the solid line connecting successive values in the chain (dots). The chain starts at $\mathbf{x}^{(0)}$. The second candidate value is accepted to yield $\mathbf{x}^{(1)}$. The circles around $\mathbf{x}^{(0)}$ and $\mathbf{x}^{(1)}$ show the proposal densities, where h is a uniform distribution over a disk centered at the origin. In a random walk chain, the proposal density at iteration $t + 1$ is centered around $\mathbf{x}^{(t)}$. Some candidate values are rejected. For example, the 13th candidate value, denoted by $\circ$, is not accepted, so $\mathbf{x}^{(13)} = \mathbf{x}^{(12)}$. Note how the chain frequently moves up the contours of the target distribution, while also allowing some downhill moves. The move from $\mathbf{x}^{(15)}$ to $\mathbf{x}^{(16)}$ is one instance where the chain moves downhill.

Example 7.3 (Estimating a mixture parameter, continued) As a continuation of Example 7.2, consider using a random walk chain to learn about the posterior for δ under a Unif(0,1) prior. Suppose we generate proposals by adding a Unif$(-a, a)$ random increment to the current $\delta^{(t)}$. Clearly it is likely that some proposals will be generated outside the interval $[0, 1]$ during the progression of the chain. An inelegant approach is to note that the posterior is zero for any $\delta \notin [0, 1]$, thereby forbidding steps to such points. An approach that usually is better involves reparameterizing the problem. Let $U = \text{logit}\{\delta\} = \log\left\{\frac{\delta}{1-\delta}\right\}$. We may now run a random walk chain on U, generating a proposal by adding, say, a Unif$(-b, b)$ random increment to $u^{(t)}$.

There are two ways to view the reparameterization. First, we may run the chain in δ-space. In this case, the proposal density $g(\cdot|u^{(t)})$ must be transformed into a proposal density in δ-space, taking account of the Jacobian. The Metropolis–Hastings ratio for a proposed value δ^* is then

$$\frac{f\left(\delta^*\right) g\left(\operatorname{logit}\left\{\delta^{(t)}\right\} \mid \operatorname{logit}\left\{\delta^*\right\}\right) \left|J(\delta^{(t)})\right|}{f\left(\delta^{(t)}\right) g\left(\operatorname{logit}\left\{\delta^*\right\} \mid \operatorname{logit}\left\{\delta^{(t)}\right\}\right) \left|J(\delta^*)\right|}, \tag{7.7}$$

where, for example, $|J(\delta^{(t)})|$ is the absolute value of the (determinant of the) Jacobian for the transformation from δ to u, evaluated at $\delta^{(t)}$. The second option is to run the chain in u-space. In this case, the target density for δ must be transformed into a density for u, where $\delta = \operatorname{logit}^{-1}\{U\} = \frac{\exp\{U\}}{1+\exp\{U\}}$. For $U^* = u^*$, this yields the Metropolis–Hasting ratio

$$\frac{f\left(\operatorname{logit}^{-1}\{u^*\}\right) |J(u^*)| g\left(u^{(t)} \mid u^*\right)}{f\left(\operatorname{logit}^{-1}\{u^{(t)}\}\right) |J(u^{(t)})| g\left(u^* \mid u^{(t)}\right)}. \tag{7.8}$$

Since $|J(u^*)| = 1/|J(\delta^*)|$, we can see that these two viewpoints produce equivalent chains.

The random walk chain run with uniform increments in a reparameterized space may have quite different properties than one generated from uniform increments in the original space. Reparameterization is a useful approach to improving the performance of MCMC methods, and is discussed further in Section 7.3.1.4.

Figure 7.5 shows sample paths for δ from two random walk chains run in u-space. The top panel corresponds to a chain generated by drawing $\epsilon \sim \text{Unif}(-1,1)$, setting $U^* = u^{(t)} + \epsilon$, and then using (7.8) to compute the Metropolis–Hastings ratio. The top panel in Figure 7.5 shows a Markov chain that moves quickly away from its starting value and seems easily able to sample values from all portions of the parameter space supported by the posterior for δ. The lower panel corresponds to the chain using $\epsilon \sim \text{Unif}(-0.01, 0.01)$, which yields very poor mixing. The resulting chain moves slowly from its starting value and takes very small steps in δ-space at each iteration. $\square$

7.1.3 Hit-and-run algorithm

As given above, the Metropolis–Hastings algorithm is time-homogeneous in the sense that the proposal distribution does not change as t increases. It is possible to construct MCMC approaches that rely on time-varying proposal distributions, $g^{(t)}\left(\cdot|\mathbf{x}^{(t)}\right)$. Such methods can be very effective, but their convergence properties are generally more difficult to ascertain due to the time inhomogeneity [460].

One such strategy that resembles a random walk chain is known as the *hit-and-run algorithm* [90]. In this approach, the proposed move away from $\mathbf{x}^{(t)}$ is generated in two stages: by choosing a direction to move, and then a distance to move in the chosen direction. After initialization at $\mathbf{x}^{(0)}$, the chain proceeds from $t = 0$ with the following steps.

1. Draw a random direction $\boldsymbol{\rho}^{(t)} \sim h(\boldsymbol{\rho})$, where h is a density defined over the surface of the unit p-sphere.

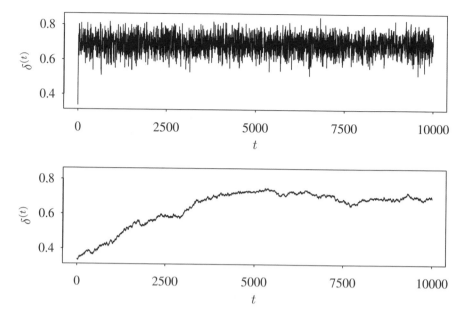

Fig. 7.5 Sample paths for δ from random walk chains in Example 7.3, run in u-space with $b = 1$ (top) and $b = 0.01$ (bottom).

2. Find the set of all real numbers λ for which $\mathbf{x}^{(t)} + \lambda\boldsymbol{\rho}^{(t)}$ is in the state space of $\mathbf{X}$. Denote this set of signed lengths as $\Lambda^{(t)}$.

3. Draw a random signed length $\lambda^{(t)} | \left(\mathbf{x}^{(t)}, \boldsymbol{\rho}^{(t)}\right) \sim g_\lambda^{(t)}\left(\lambda | \mathbf{x}^{(t)}, \boldsymbol{\rho}^{(t)}\right)$, where the density $g_\lambda^{(t)}\left(\lambda | \mathbf{x}^{(t)}, \boldsymbol{\rho}^{(t)}\right) = g^{(t)}\left(\mathbf{x}^{(t)} + \lambda\boldsymbol{\rho}^{(t)}\right)$ is defined over $\Lambda^{(t)}$. The proposal distribution may differ from one iteration to the next only through a dependence on $\Lambda^{(t)}$.

4. For the proposal $\mathbf{X}^* = \mathbf{x}^{(t)} + \lambda^{(t)}\boldsymbol{\rho}^{(t)}$, compute the Metropolis–Hastings ratio

$$R\left(\mathbf{x}^{(t)}, \mathbf{X}^*\right) = \frac{f\left(\mathbf{X}^*\right) g^{(t)}\left(\mathbf{x}^{(t)}\right)}{f\left(\mathbf{x}^{(t)}\right) g^{(t)}\left(\mathbf{X}^*\right)}.$$

5. Set

$$\mathbf{X}^{(t+1)} = \begin{cases} \mathbf{X}^* & \text{with probability } \min\left\{R\left(\mathbf{x}^{(t)}, \mathbf{X}^*\right), 1\right\}, \\ \mathbf{x}^{(t)} & \text{otherwise.} \end{cases}$$

6. Increment t and go to step 1.

The above algorithm is one variant of several general hit-and-run approaches [90].

The direction distribution h is frequently taken to be uniform over the surface of the unit sphere. In p dimensions, a random variable may be drawn from this distribution by sampling a p-dimensional standard normal variable $\mathbf{Y} \sim N(\mathbf{0}, \mathbf{I})$ and making the transformation $\rho = \mathbf{Y}/\sqrt{\mathbf{Y}^T\mathbf{Y}}$.

The performance of this approach has been compared with that of other simple MCMC methods [89]. It has been noted that the hit-and-run algorithm can offer particular advantage when the state space of $\mathbf{X}$ is sharply constrained [26], thereby making it difficult to explore all regions of the space effectively with other methods. The choice of h has a strong effect on the performance and convergence rate of the algorithm, with the best choice often depending on the shape of f and the geometry of the state space (including constraints and the chosen units for the coordinates of $\mathbf{X}$) [322].

7.1.4 Langevin Metropolis–Hastings algorithm

A random walk with drift can be generated using the proposal

$$\mathbf{X}^* = \mathbf{x}^{(t)} + \mathbf{d}^{(t)} + \sigma\boldsymbol{\epsilon}^{(t)}, \tag{7.9}$$

where

$$\mathbf{d}^{(t)} = \left(\frac{\sigma^2}{2}\right) \left.\frac{\partial \log f(\mathbf{x})}{\partial \mathbf{x}}\right|_{\mathbf{x}=\mathbf{x}^{(t)}} \tag{7.10}$$

and $\boldsymbol{\epsilon}^{(t)}$ is a p-dimensional standard normal random variable. The scalar σ is a tuning parameter whose fixed value is chosen by the user to control the magnitude of proposed steps. The standard Metropolis–Hastings ratio is used to decide whether to accept this proposal, using $g(\mathbf{x}^*|\mathbf{x}^{(t)}) \propto \exp\left\{-\frac{1}{2\sigma^2}\left(\mathbf{x}^* - \mathbf{x}^{(t)} - \mathbf{d}^{(t)}\right)^T\left(\mathbf{x}^* - \mathbf{x}^{(t)} - \mathbf{d}^{(t)}\right)\right\}$.

The proposal distribution for this method is motivated by a stochastic differential equation that produces a diffusion (i.e., a continuous-time stochastic process) with f as its stationary distribution [248, 432]. To ensure that the discretization of this process given by the discrete-time Markov chain described here shares the correct stationary distribution, Besag overlaid the Metropolis–Hastings acceptance strategy [31].

The requirement to know the gradient of the target is not as burdensome as it may seem. Any unknown multiplicative constant in f drops out when the derivative is taken. Also, when exact derivatives are difficult to obtain, they can be replaced with numerical approximations.

Unlike a random walk, this algorithm introduces a drift that favors proposals that move towards modes of the target distribution. Ordinary Metropolis–Hastings algorithms—including the random walk chain and the independence chain—generally are driven by proposals that are made independently of the shape of f, thereby being easy to implement but sometimes slow to approach stationarity or adequately explore the support region of f. When performance of a generic algorithm is poor, problem-specific Metropolis–Hastings algorithms are frequently employed with specialized proposal distributions crafted in ways that are believed to exploit features of the target. Langevin Metropolis–Hastings algorithms also provide proposal distributions motivated by the shape of f, but the *self-targeting* is done generically through the use of the gradient. These methods can provide better exploration of the target distribution and faster convergence.

In some applications, the update given by (7.10) can yield Markov chains that fail to approach convergence in runs of reasonable length, and fail to explore more than one mode of f. Stramer and Tweedie [523] generalize (7.10) somewhat with different drift and scaling terms that yield improved performance. Further study of Langevin methods is given in [464, 522, 523].

7.1.5 Multiple-try Metropolis–Hastings algorithm

If a Metropolis–Hastings algorithm is not successful in some problem, it is probably because the chain is slow to converge or trapped in a local mode of f. To overcome such difficulties, it may pay to expand the region of likely proposals characterized by $g(\cdot|\mathbf{x}^{(t)})$. However, this strategy often leads to very small Metropolis–Hastings ratios, and therefore to poor mixing. Liu, Liang, and Wong proposed an alternative strategy known as *multiple-try Metropolis–Hastings sampling* for effectively expanding the proposal region to improve performance without impeding mixing [359].

The approach is to generate a larger number of candidates, thereby improving exploration of f near $\mathbf{x}^{(t)}$. One of these proposals is then selected in a manner that ensures that the chain retains the correct limiting stationary distribution. We still use a proposal distribution g, along with optional nonnegative weights $\lambda(\mathbf{x}^{(t)}, \mathbf{x}^*)$, where the symmetric function λ is discussed further below. To ensure the correct limiting stationary distribution, it is necessary to require that $g(\mathbf{x}^*|\mathbf{x}^{(t)}) > 0$ if and only if $g(\mathbf{x}^{(t)}|\mathbf{x}^*) > 0$, and that $\lambda(\mathbf{x}^{(t)}, \mathbf{x}^*) > 0$ whenever $g(\mathbf{x}^*|\mathbf{x}^{(t)}) > 0$.

Let $\mathbf{x}^{(0)}$ denote the starting value, and define

$$w(\mathbf{u}, \mathbf{v}) = f(\mathbf{v})g(\mathbf{u}|\mathbf{v})\lambda(\mathbf{u}, \mathbf{v}). \tag{7.11}$$

Then, for $t = 0, 1, \ldots$, the algorithm proceeds as follows:

1. Sample k proposals $\mathbf{X}_1^*, \ldots, \mathbf{X}_k^*$ i.i.d. from $g(\cdot|\mathbf{x}^{(t)})$.

2. Randomly select a single proposal $\mathbf{X}_j^*$ from the set of proposals, with probability proportional to $w(\mathbf{x}^{(t)}, \mathbf{X}_j^*)$ for $j = 1, \ldots, k$.

3. Given $\mathbf{X}_j^* = \mathbf{x}_j^*$, sample $k - 1$ random variables $\mathbf{X}_1^{**}, \ldots, \mathbf{X}_{k-1}^{**}$ i.i.d. from the proposal density $g(\cdot|\mathbf{x}_j^*)$. Set $\mathbf{X}_k^{**} = \mathbf{x}^{(t)}$.

4. Calculate the *generalized Metropolis–Hastings ratio*

$$R_g = \sum_{i=1}^{k} w(\mathbf{x}^{(t)}, \mathbf{X}_i^*) \bigg/ \sum_{i=1}^{k} w(\mathbf{X}_j^*, \mathbf{X}_i^{**}). \tag{7.12}$$

5. Set

$$\mathbf{X}^{(t+1)} = \begin{cases} \mathbf{X}_j^* & \text{with probability } \min\{R_g, 1\}, \\ \mathbf{x}^{(t)} & \text{otherwise.} \end{cases} \tag{7.13}$$

6. Increment t and go to step 1.

It is straightforward to show that this algorithm yields a reversible Markov chain with limiting stationary distribution equal to f. The efficiency of this approach depends on k, the shape of f, and the spread of g relative to f. In practice, selecting from one of many proposals at each iteration can lead to chains with lower serial correlation. This leads to better mixing in the sense that larger steps can be made to find other local modes or to promote movement in certain advantageous directions when we are unable to encourage such steps through other means.

The weighting function λ can be used to further encourage certain types of proposals. The simplest choice is $\lambda(\mathbf{x}^{(t)}, \mathbf{x}^*) = 1$. An "orientational-biased" method with $\lambda(\mathbf{x}^{(t)}, \mathbf{x}^*) = \left\{ \left[g(\mathbf{x}^*|\mathbf{x}^{(t)}) + g(\mathbf{x}^{(t)}|\mathbf{x}^*) \right] /2 \right\}^{-1}$ was suggested in [178]. Another interesting choice is $\lambda(\mathbf{x}^{(t)}, \mathbf{x}^*) = \left[g(\mathbf{x}^*|\mathbf{x}^{(t)}) g(\mathbf{x}^{(t)}|\mathbf{x}^*) \right]^{-\alpha}$, defined on the region where $g(\mathbf{x}^*|\mathbf{x}^{(t)}) > 0$. When $\alpha = 1$, the weight $w(\mathbf{x}^{(t)}, \mathbf{x}^*)$ corresponds to the importance weight $f(\mathbf{x}^*)/g(\mathbf{x}^*|\mathbf{x}^{(t)})$ assigned to $\mathbf{x}^*$ when attempting to sample from f using g and the importance sampling envelope (see Section 6.3.1).

7.2 GIBBS SAMPLING

Thus far we have treated $\mathbf{X}^{(t)}$ with little regard to its dimensionality. The Gibbs sampler is specifically adapted for multidimensional target distributions. The goal is to construct a Markov chain whose stationary distribution—or some marginalization thereof—equals the target distribution f. The Gibbs sampler does this by sequentially sampling from univariate conditional distributions, which are often available in closed form.

7.2.1 Basic Gibbs sampler

Recall $\mathbf{X} = (X_1, \ldots, X_p)^T$, and denote $\mathbf{X}_{-i} = (X_1, \ldots, X_{i-1}, X_{i+1}, \ldots, X_p)^T$. Suppose that the univariate conditional density of $X_i|\mathbf{X}_{-i} = \mathbf{x}_{-i}$, denoted $f(x_i|\mathbf{x}_{-i})$, is easily sampled for $i = 1, \ldots, p$. A general Gibbs sampling procedure can be described as follows

1. Select starting values $\mathbf{x}^{(0)}$, and set $t = 0$.

2. Generate, in turn,

$$X_1^{(t+1)}|\cdot \sim f\left(x_1|x_2^{(t)}, \ldots, x_p^{(t)}\right),$$
$$X_2^{(t+1)}|\cdot \sim f\left(x_2|x_1^{(t+1)}, x_3^{(t)}, \ldots, x_p^{(t)}\right),$$
$$\vdots \tag{7.14}$$
$$X_{p-1}^{(t+1)}|\cdot \sim f\left(x_{p-1}|x_1^{(t+1)}, x_2^{(t+1)}, \ldots, x_{p-2}^{(t+1)}, x_p^{(t)}\right),$$
$$X_p^{(t+1)}|\cdot \sim f\left(x_p|x_1^{(t+1)}, x_2^{(t+1)}, \ldots, x_{p-1}^{(t+1)}\right),$$

where $|\cdot|$ denotes conditioning on the most recent updates to all other elements of $\mathbf{X}$.

3. Increment t and go to step 2.

The completion of step 2 for all components of $\mathbf{X}$ is called a *cycle*. Several methods for improving and generalizing the basic Gibbs sampler are discussed in Sections 7.2.3–7.2.6. In subsequent discussion of the Gibbs sampler, we frequently refer to the term $\mathbf{x}_{-i}^{(t)}$ which represents all the components of $\mathbf{x}$, except for x_i, at their current values, so

$$\mathbf{x}_{-i}^{(t)} = \left(x_1^{(t+1)}, \ldots, x_{i-1}^{(t+1)}, x_{i+1}^{(t)}, \ldots, x_p^{(t)} \right).$$

Example 7.4 (Stream ecology monitoring) Stream insects called benthic invertebrates are an effective indicator for monitoring stream ecology, because their relatively stationary substrate habitation provides constant exposure to contamination and easy sampling of large numbers of individuals. Imagine that at many sites along a stream, insects are collected and classified into several categories based on some ecologically significant criterion. At a particular site, let $Y_1, \ldots, Y_c$ denote the counts of insects in each of c different classes.

The probability that an insect is classified in each category varies randomly from site to site, as does the total number collected at each site. For a given site, let $P_1, \ldots, P_c$ denote the class probabilities, and let N denote the random total number of insects collected. Suppose, further, that the $P_1, \ldots, P_c$ depend on a set of site-specific features summarized by parameters $\alpha_1, \ldots, \alpha_c$, respectively. Let N depend on a site-specific parameter, λ.

Suppose two competing statistics, $T_1(Y_1, \ldots, Y_c)$ and $T_2(Y_1, \ldots, Y_c)$, are used to monitor streams for negative environmental events. An alarm is triggered if the value of T_1 or T_2 exceeds some threshold. To compare the performance of these statistics across multiple sites within the same stream or across different types of streams, a Monte Carlo simulation experiment is designed. The experiment is designed by choosing a collection of parameter sets $(\lambda, \alpha_1, \ldots, \alpha_c)$ which are believed to encompass the range of sampling effort and characteristics of sites and streams likely to be monitored. Each parameter set corresponds to a hypothetical sampling effort at a simulated site.

Let $c = 3$. For a given simulated site, we can establish the model:

$$(Y_1, Y_2, Y_3) | (N = n, P_1 = p_1, P_2 = p_2, P_3 = p_3) \sim \text{Multinomial}(n; p_1, p_2, p_3),$$
$$(P_1, P_2, P_3) \sim \text{Dirichlet}(\alpha_1, \alpha_2, \alpha_3),$$
$$N \sim \text{Poisson}(\lambda),$$

where N is viewed as random because it varies across sites. This model is overspecified, because we require $Y_1 + Y_2 + Y_3 = N$ and $P_1 + P_2 + P_3 = 1$. Therefore, we can write the state of the model as $\mathbf{X} = (Y_1, Y_2, P_1, P_2, N)$, where the remaining variables can be determined analytically for any value of $\mathbf{X}$. Cassella and George offer a related model for the hatching of insect eggs [82]. More sophisticated models of stream ecology data are given in [308].

To complete the simulation experiment, it is necessary to sample from the marginal distribution of (Y_1, Y_2, Y_3) so that the performance of the statistics T_1 and T_2 may be compared for a simulated site of the current type. Having repeated this process over the designed set of simulated sites, comparative conclusions about T_1 and T_2 can be drawn.

It is impossible to get a closed-form expression for the marginal distribution of (Y_1, Y_2, Y_3) given the parameters λ, α_1, α_2, and α_3. For purposes of illustration we use Gibbs sampling to simulate from this distribution, but a more direct simulation approach is also available. The simplest way to summarize the Gibbs sampling scheme for this problem is given by

$$
\begin{aligned}
(Y_1, Y_2, Y_3)|\cdot &\sim \text{Multinomial}\,(n; p_1, p_2, p_3)\,, \\
(P_1, P_2, P_3)|\cdot &\sim \text{Dirichlet}\,(y_1 + \alpha_1, y_2 + \alpha_2, n - y_1 - y_2 + \alpha_3)\,, \qquad (7.15) \\
N - y_1 - y_2|\cdot &\sim \text{Poisson}\,(\lambda\,(1 - p_1 - p_2))\,,
\end{aligned}
$$

where $|\cdot$ denotes that the distribution is conditional on the variables remaining from the complete set of variables $\{N, Y_1, Y_2, Y_3, P_1, P_2, P_3\}$. Problem 7.4 asks you to derive these distributions.

At first glance, equation (7.15) does not seem to resemble the univariate sampling strategy inherent in a Gibbs sampler. It is straightforward to show that (7.15) amounts to the following sampling scheme based on univariate conditional distributions of components of $\mathbf{X}$:

$$
Y_1^{(t+1)}|\cdot \sim \text{Bin}\left(n^{(t)} - y_2^{(t)}, \frac{p_1^{(t)}}{1 - p_2^{(t)}}\right),
$$

$$
Y_2^{(t+1)}|\cdot \sim \text{Bin}\left(n^{(t)} - y_1^{(t+1)}, \frac{p_2^{(t)}}{1 - p_1^{(t)}}\right),
$$

$$
\frac{P_1^{(t+1)}}{1 - p_2^{(t)}}\bigg|\cdot \sim \text{Beta}\left(y_1^{(t+1)} + \alpha_1, n^{(t)} - y_1^{(t+1)} - y_2^{(t+1)} + \alpha_3\right),
$$

$$
\frac{P_2^{(t+1)}}{1 - p_1^{(t+1)}}\bigg|\cdot \sim \text{Beta}\left(y_2^{(t+1)} + \alpha_2, n^{(t)} - y_1^{(t+1)} - y_2^{(t+1)} + \alpha_3\right),
$$

and

$$
N^{(t+1)} - y_1^{(t+1)} - y_2^{(t+1)}|\cdot \sim \text{Poisson}\left(\lambda\big(1 - p_1^{(t+1)} - p_2^{(t+1)}\big)\right).
$$

We will see below that it is not necessary to specify this more elaborate scheme which relies exclusively on univariate conditionals. □

The Gibbs sampler is most frequently used for Bayesian applications when the goal is to make inferences based on the posterior distribution of some parameters. Recall Example 7.1 where the parameter vector $\boldsymbol{\theta}$ has prior distribution $p(\boldsymbol{\theta})$ and likelihood function $L(\boldsymbol{\theta}|\mathbf{y})$ arising from observed data $\mathbf{y}$. Bayesian inference is based on the posterior distribution $p(\boldsymbol{\theta}|\mathbf{y}) = c\,p(\boldsymbol{\theta})L(\boldsymbol{\theta}|\mathbf{y})$, where c is an unknown constant. When the requisite univariate conditional densities are easily sampled, the Gibbs sampler can be applied and does not require evaluation of the constant $c = \int p(\boldsymbol{\theta})L(\boldsymbol{\theta}|\mathbf{y})\,d\boldsymbol{\theta}$. In this case the ith step in a cycle of the Gibbs sampler at iteration t is given by draws from $\theta_i^{(t+1)}|\theta_{-i}^{(t)}, \mathbf{y} \sim f(\theta_i|\theta_{-i}^{(t)}, \mathbf{y})$ where f is the univariate conditional posterior of θ_i given the remaining parameters.

7.2.2   Properties of the Gibbs sampler

Clearly the chain produced by a Gibbs sampler is Markov. Under rather mild conditions, Geman and Geman [197] showed that the stationary distribution of the Gibbs sampler chain is f. It also follows that the limiting marginal distribution of $X_i^{(t)}$ equals the univariate marginalization of the target distribution along the ith coordinate. As with the Metropolis–Hastings algorithm, we can use realizations from the chain to estimate the expectation of any function of $\mathbf{X}$.

It is possible to relate the Gibbs sampler to the Metropolis–Hastings algorithm, allowing for a proposal distribution in the Metropolis–Hastings algorithm that varies over time. Each Gibbs cycle consists of p Metropolis–Hastings steps. To see this, note that the ith Gibbs step in a cycle effectively proposes the candidate vector $\mathbf{X}^* = (x_1^{(t+1)}, \ldots, x_{i-1}^{(t+1)}, X_i^*, x_{i+1}^{(t)}, \ldots, x_p^{(t)})$ given the current state of the chain $(x_1^{(t+1)}, \ldots, x_{i-1}^{(t+1)}, x_i^{(t)}, \ldots, x_p^{(t)})$. Thus, the ith univariate Gibbs update can be viewed as a Metropolis–Hastings step drawing

$$\mathbf{X}^*|x_1^{(t+1)}, \ldots, x_{i-1}^{(t+1)}, x_i^{(t)}, \ldots, x_p^{(t)} \sim g_i\left(\cdot\,\middle|\,x_1^{(t+1)}, \ldots, x_{i-1}^{(t+1)}, x_i^{(t)}, \ldots, x_p^{(t)}\right),$$

where

$$g_i\left(\mathbf{x}^*\middle|x_1^{(t+1)}, \ldots, x_{i-1}^{(t+1)}, x_i^{(t)}, \ldots, x_p^{(t)}\right) = \begin{cases} f\left(x_i^*\middle|\mathbf{x}_{-i}^{(t)}\right) & \text{if } \mathbf{X}_{-i}^* = \mathbf{x}_{-i}^{(t)}, \\ 0 & \text{otherwise.} \end{cases}$$

$$(7.16)$$

It is easy to show that in this case the Metropolis–Hastings ratio equals 1, which means that the candidate is always accepted.

The Gibbs sampler should not be applied when the dimensionality of $\mathbf{X}$ changes. Section 8.2 gives methods for constructing a suitable Markov chain with the correct stationary distribution.

The "Gibbs sampler" is actually a generic name for a rich family of very adaptable algorithms. In the following subsections we describe various strategies that have been developed to improve the performance of the general algorithm described above.

7.2.3   Update ordering

The ordering of updates made to the components of $\mathbf{X}$ in (7.14) can change from one cycle to the next. Sometimes a random ordering for each cycle is sensible. This is called *random scan Gibbs sampling* [460]. In fact, it is not even necessary to update every component at each cycle, as long as each component is updated sufficiently frequently.

7.2.4 Blocking

Another modification to the Gibbs sampler is called *blocking* or *grouping*. In the Gibbs algorithm, it is not necessary to treat each element of $\mathbf{X}$ individually. In Example 7.4, we saw that the stream ecology parameters were naturally grouped into a conditionally multinomial group, a conditionally Dirichlet group, and a single conditionally Poisson element. In the generic case of (7.16) above, with $p = 4$, it would be allowable for each cycle to proceed, for example, with the following sequence of updates:

$$X_1^{(t+1)} \Big| \cdot \sim f\left(x_1 \Big| x_2^{(t)}, x_3^{(t)}, x_4^{(t)}\right),$$
$$X_2^{(t+1)}, X_3^{(t+1)} \Big| \cdot \sim f\left(x_2, x_3 \Big| x_1^{(t+1)}, x_4^{(t)}\right),$$
$$X_4^{(t+1)} \Big| \cdot \sim f\left(x_4 \Big| x_1^{(t+1)}, x_2^{(t+1)}, x_3^{(t+1)}\right).$$

Blocking is typically useful when elements of $\mathbf{X}$ are correlated, with the algorithm constructed so that more correlated elements are sampled together in one block. Roberts and Sahu compare convergence rates for various blocking and update ordering strategies [463]. The structured Markov chain Monte Carlo method of Sargent et al. offers a systematic approach to blocking which is directly motivated by the model structure [480]. This method has been shown to offer faster convergence for problems with a large number of parameters, such as Bayesian analyses for geostatistical models [106].

7.2.5 Hybrid Gibbs sampling

Since each step in a cycle of the Gibbs sampler is itself a Metropolis–Hastings step, it is also permissible to use different Metropolis–Hastings variants where convenient. For example, for $p = 6$, a *hybrid MCMC* algorithm might proceed with the following sequence of updates:

1. Update $X_1^{(t+1)} \Big| \left(x_2^{(t)}, x_3^{(t)}, x_4^{(t)}, x_5^{(t)}, x_6^{(t)}\right)$ with a Gibbs step.

2. Update $\left(X_2^{(t+1)}, X_3^{(t+1)}\right) \Big| \left(x_1^{(t+1)}, x_4^{(t)}, x_5^{(t)}, x_6^{(t)}\right)$ with a Metropolis step.

3. Update $X_4^{(t+1)} \Big| \left(x_1^{(t+1)}, x_2^{(t+1)}, x_3^{(t+1)}, x_5^{(t)}, x_6^{(t)}\right)$ with a step from a random walk chain.

4. Update $\left(X_5^{(t+1)}, X_6^{(t+1)}\right) \Big| \left(x_1^{(t+1)}, x_2^{(t+1)}, x_3^{(t+1)}, x_4^{(t+1)}\right)$ with a Gibbs step.

The Metropolis–Hastings steps within a Gibbs algorithm are typically useful when the univariate conditional density for one or more elements of $\mathbf{X}$ is not available in closed form.

7.2.6 Alternative univariate proposal methods

Hybrid methods such as embedding Metropolis–Hastings steps within a Gibbs algorithm are one way to construct a Gibbs-like chain when not all the univariate conditionals are easily sampled. Other strategies, evolved from techniques in Chapter 6, can be used to sample difficult univariate conditionals.

One such method is the griddy Gibbs sampler [458, 529]. Suppose that it is difficult to sample from the univariate conditional density for $X_k|\mathbf{x}_{-k}$ for a particular k. To implement a griddy Gibbs step, select some grid points $z_1, \ldots, z_n$ over the range of support of $f(\cdot|\mathbf{x}_{-k})$. Let $w_j^{(t)} = f(z_j|\mathbf{x}_{-k}^{(t)})$ for $j = 1, \ldots, n$. Using these weights and the corresponding grid, one can approximate the density function $f(\cdot|\mathbf{x}_{-k})$ or, equivalently, its inverse cumulative distribution function. Generate $X_k^{(t+1)}|\mathbf{x}_{-k}^{(t)}$ from this approximation, and proceed with remainder of the MCMC algorithm. The approximation to the kth univariate conditional can be refined as iterations proceed. The simplest approach for the approximation and sampling step is to draw $X_k^{(t+1)}|\mathbf{x}_{-k}^{(t)}$ from the discrete distribution on $z_1, \ldots, z_n$ with probabilities proportional to $w_1^{(t)}, \ldots, w_n^{(t)}$, using the inverse cumulative distribution function method (Section 6.2.2). A piecewise linear cumulative distribution function could be generated from an approximating density function that is piecewise constant between the midpoints of any two adjacent grid values with a density height set to ensure that the total probability on the segment containing z_i is proportional to $w_i^{(t)}$. Other approaches could be based on the density estimation ideas presented in Chapter 10.

If the approximation to $f(\cdot|\mathbf{x}_{-k})$ is updated from time to time by improving the grid, then the chain is not time-homogeneous. In this case, reference to convergence results for Metropolis–Hastings or Gibbs chains is not sufficient to guarantee that a griddy Gibbs chain has a limiting stationary distribution equal to f. One way to ensure time homogeneity is to resist making any improvements to the approximating univariate distribution as iterations progress. In this case, however, the limiting distribution of the chain is still not correct, because it relies on an approximation to $f(\cdot|\mathbf{x}_{-k})$ rather than the true density. This can be corrected by reverting to a hybrid Metropolis–with-Gibbs framework where the variable generated from the approximation to $f(\cdot|\mathbf{x}_{-k})$ is viewed as a proposal, which is then randomly kept or discarded based on the Metropolis–Hastings ratio. Tanner discusses a wide variety of potential enhancements to the basic griddy Gibbs strategy [529].

7.3 IMPLEMENTATION

The goal of a MCMC analysis is to estimate features of the target distribution f. The reliability of such estimates depends on the extent to which sample averages computed using realizations of the chain correspond to their expectation under the limiting stationary distribution of the chain. All of the MCMC methods described above have the correct limiting stationary distribution. In practice, however, it is necessary to determine when the chain has run sufficiently long so that it is reasonable to believe that the output adequately represents the target distribution and can be used reliably for estimation. Unfortunately, MCMC methods can sometimes be quite slow to converge, requiring extremely long runs, especially if the dimensionality of $\mathbf{X}$ is large. Further, it is often easy to be misled when using MCMC algorithm output to judge whether convergence has approximately been obtained.

In this section, we examine questions about the long-run behavior of the chain. Has the chain run long enough? Is the first portion of the chain highly influenced by the starting value? Should the chain be run from several different starting values? Has the chain traversed all portions of the region of support of f? Are the sampled values approximate draws from f? How shall the chain output be used to produce estimates and assess their precision? Useful reviews of MCMC diagnostic methods include [70, 107, 320, 389, 459]. We end with some practical advice for coding MCMC algorithms.

7.3.1   Ensuring good mixing and convergence

It is important to consider how efficiently a MCMC algorithm provides useful information about a problem of interest. Efficiency can take on several meanings in this context, but here we will focus on how quickly the chain forgets its starting value and how quickly the chain fully explores the support of the target distribution. A related concern is how far apart in a sequence observations need to be before they can be considered to be approximately independent. These qualities can be described as the *mixing* properties of the chain.

We must also be concerned whether the chain has approximately reached its stationary distribution. There is substantial overlap between the goals of diagnosing convergence to the stationary distribution and investigating the mixing properties of the chain. Many of the same diagnostics can be used to investigate both mixing and convergence. In addition, no diagnostic is fail-safe; some methods can suggest that a chain has converged when it has not. For these reasons, we combine the discussion of mixing and convergence in the following subsections, and we recommend that a variety of diagnostic techniques be used.

7.3.1.1   Choice of proposal   As mentioned in Section 7.1, mixing is strongly affected by features of the proposal distribution, especially its spread. Further, advice on desirable features of a proposal distribution depends on the type of MCMC algorithm employed.

For a Gibbs sampler, performance is enhanced when components of $\mathbf{X}$ are as independent as possible. Reparameterization is the primary strategy for reducing dependence. A variety of strategies are discussed in [212, 287]. See Section 7.3.1.4 and Problem 7.7 for examples.

For a general Metropolis–Hastings chain such as an independence chain, it seems intuitively clear that we wish the proposal distribution g to approximate the target distribution f very well, which in turn suggests that a very high rate of accepting proposals is desirable. Although we would like g to resemble f, the tail behavior of g is more important than its resemblance to f in regions of high density. In particular, if f/g is bounded, the convergence of the Markov chain to its stationary distribution is faster overall [460]. Thus, it is wiser to aim for a proposal distribution that is somewhat more diffuse than f.

In practice, the variance of the proposal distribution can be selected through an informal iterative process. Start a chain, and monitor the proportion of proposals that have been accepted; then adjust the spread of the proposal distribution accordingly. After the some predetermined acceptance rate is achieved, restart the chain using the appropriately scaled proposal distribution. For a Metropolis algorithm with normal target and proposal distributions, it has been suggested that an acceptance rate of between 25% and 45% should be preferred, with the best choice being about 45% for one- or two-dimensional problems and decreasing to about 23% for higher-dimensional problems [194, 461]. For the multiple-try Metropolis–Hastings algorithm of Section 7.1.5, an acceptance rate of 40% to 50% is suggested [359]. To apply such rules, care are must be taken to ensure that the target and proposal distributions are roughly normally distributed or at least simple, unimodal distributions. If, for example, the target distribution is multimodal, the chain may get stuck in one mode without adequate exploration of the other portions of the parameter space. In this case the acceptance rate may very high, but the probability of jumping from one mode to another may be low. This suggests one difficult issue with most MCMC methods; it is useful to have as much knowledge as possible about the target distribution, even though that distribution is typically unknown.

A completely automated approach to specifying g is proposed in [460], generalizing the adaptive rejection sampling method (see Section 6.2.3.2). This approach can be useful when there is very little available information about the shape of f.

7.3.1.2 *Number of chains*

One of the most difficult problems to diagnose is whether or not the chain has become stuck in one or more modes of the target distribution. In this case, all convergence diagnostics may indicate that the chain has converged, though the chain does not fully represent the target distribution. A partial solution to this problem is to run multiple chains from diverse starting values and then compare the within- and between-chain behavior. A formal approach for doing this is described in Section 7.3.1.5.

The general notion of running multiple chains to study between-chain performance is surprisingly contentious. One of the most vigorous debates during the early statistical development of MCMC methods centered around whether it was more important to invest limited computing time in lengthening the run of a single chain, or in running several shorter chains from diverse starting points to check performance [204, 196, 389]. The motivation for trying multiple runs is the hope that all interesting features (e.g., modes) of the target distribution will be explored by at least one chain, and that the failure of individual chains to find such features or to wash out the influence of their starting values can be detected, in which case chains must be lengthened or the problem reparameterized to encourage better mixing.

Arguments for one long chain include the following. Many short runs are more informative than one long run only when they indicate poor convergence behavior. In this case, the simulated values from the many short chains remain unusable. Second, the effectiveness of using many short runs to diagnose poor convergence is mainly limited to unrealistically simple problems or problems where the features of f are already well understood. Third, splitting computing effort into many short runs may yield an indication of poor convergence that would not have occurred if the total computing effort had been devoted to one longer run.

We do not find the single-chain arguments entirely convincing from a practical point of view. Starting a number of shorter chains from diverse starting points is an essential component of thorough debugging of computer code. Some primary features of f (e.g., multimodality, highly constrained support region) are often broadly known—even in complex realistic problems—notwithstanding uncertainty about specific details of these features. Results from diverse starts can also provide information about key features of f, which in turn helps determine whether the MCMC method and problem parameterization are suitable. Poor convergence of several short chains can help determine what aspects of chain performance will be most important to monitor when a longer run is made. Finally, CPU cycles are more abundant and less expensive than they were a decade ago. We can have diverse short runs *and* one longer run. Exploratory work can be carried out using several shorter chains started from various points covering the believed support of f. Diagnosis of chain behavior can be made using a variety of informal and formal techniques, many of which are described below. After building confidence that the implementation is a promising one, it is advisable to run one final very long run from a good starting point to calculate and publish results.

7.3.1.3 *Simple graphs to assess mixing and convergence*   After programming and running the MCMC algorithm from multiple starting points, users should perform various diagnostics to investigate the properties of the MCMC algorithm for the particular problem. Three simple diagnostics are discussed below.

A *sample path* is a plot of the iteration number versus the realizations of $\mathbf{X}^{(t)}$ for $t = 0, 1, \ldots$. Sample paths are sometimes called trace or history plots. If a chain is mixing poorly, it will remain at or near the same value for many iterations, as in the lower panel in Figure 7.2. A chain that is mixing well will quickly move away from its starting value—no matter where it started from—and the sample path will wiggle about vigorously in the region supported by f.

The cumulative sum (cusum) diagnostic assesses the convergence of an estimator of a one-dimensional parameter $\theta = \mathrm{E}\{h(\mathbf{X})\}$ [578]. For n realizations of the chain after discarding some initial iterates, the estimator is given by $\hat{\theta}_n = \frac{1}{n}\sum_{j=1}^{n} h(\mathbf{X}^{(j)})$. The cusum diagnostic is a plot of $\sum_{i=1}^{t}\left[h(\mathbf{X}^{(i)}) - \hat{\theta}_n\right]$ versus t. If the final estimator will be computed using only the iterations of the chain that remain after removing some burn-in values, discussed below, then the estimator and cusum plot should be based only on the values to be used in the final estimator. Yu and Mykland [578] suggest that cusum plots that are very wiggly and have smaller excursions from 0 indicate that the chain is mixing well. Plots that have large excursions from 0 and are smoother suggest slower mixing speeds. The cusum plot shares one drawback with many other convergence diagnostics: For a multimodal distribution where the chain

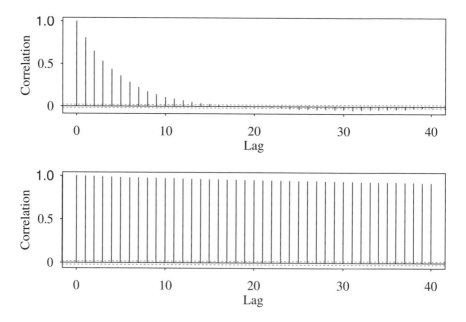

Fig. 7.6 Autocorrelation plots for independence chain of Example 7.2 with proposal densities Beta(1,1) (top) and Beta(2,10) (bottom).

is stuck in one of the modes, the cusum plot may indicate good performance when, in fact, the chain is not performing well.

An autocorrelation plot summarizes the correlation in the sequence of $\mathbf{X}^{(t)}$ at different iteration lags. The autocorrelation at lag i is the correlation between iterates that are i iterations apart [187]. A chain that has poor mixing properties will exhibit slow decay of the autocorrelation as the lag between iterations increases. For problems with more than one parameter it may also be of use to consider cross-correlations between parameters that might be related, since high cross-correlations may also indicate poor mixing of the chain.

Example 7.5 (Estimating a mixture parameter, continued) Figure 7.6 shows autocorrelation plots for the independence chain described in Example 7.2. In the top panel, the more appropriate proposal distribution yields a chain for which the autocorrelations decrease rather quickly. In the lower panel, the bad proposal distribution yields a chain for which autocorrelations are very high, with a correlation of 0.92 for observations that are 40 iterations apart. This panel clearly indicates poor mixing. □

7.3.1.4 *Reparameterization* The mixing of the Gibbs sampler and Metropolis–Hastings algorithms can be improved via reparameterization of the model. High correlation between the elements of $\mathbf{X}$ can lead to slow convergence of the Gibbs sampler, but reparameterization of the model can reduce the correlation and thus speed convergence. For example, if f is a bivariate normal distribution with very strong positive correlation, both univariate conditionals will allow only small steps away from $\mathbf{X}^{(t)} = \mathbf{x}^{(t)}$ along one axis. Therefore, the Gibbs sampler will explore f very slowly. However, suppose $\mathbf{Y} = (\mathbf{X}_1 + \mathbf{X}_2, \mathbf{X}_1 - \mathbf{X}_2)$. This transformation yields one univariate conditional on the axis of maximal variation in $\mathbf{X}$ and the second on an orthogonal axis. If we view the support of f as cigar-shaped, then the univariate conditionals for $\mathbf{Y}$ allow one step along the length of the cigar, followed by one across its width. Therefore, the parameterization inherent in $\mathbf{Y}$ makes it far easier to move from one point supported by the target distribution to any other point in a single move (or a few moves).

If there are continuous covariates in a linear model, it is useful to center and scale the covariates to reduce correlations between the parameters in the model. Another strategy is called hierarchical centering [91]. This strategy is particularly useful in models with random effects. See Problem 7.7 for an example.

Unfortunately, reparameterization approaches are typically adapted for specific models, so it is difficult to provide generic advice. Another way to improve mixing and accelerate convergence of MCMC algorithms is to augment the problem using so-called auxiliary variables; see Chapter 8. A variety of reparameterizations and acceleration techniques are described in [91, 460] and the references therein.

7.3.1.5 *Burn-in and run length*   Key considerations in the diagnosis of convergence are the *burn-in* period and run length. Recall that it is only in the limit that a MCMC algorithm yields $X^{(t)} \sim f$. For any implementation, the iterates will not have exactly the correct marginal distribution, and the dependence on the initial point (or distribution) from which the chain was started may remain strong. To reduce the severity of this problem, the first D values from the chain are typically discarded as a *burn-in period*.

The determination of an appropriate burn-in period and run length is an active area of research. A commonly used approach is that of Gelman and Rubin [194, 196]. In this approach, the MCMC algorithm is run from J separate, equal-length chains ($J \geq 2$) with starting values dispersed over the support of the target density. Let L denote the length of each chain after discarding D burn-in iterates. Suppose that the variable (e.g., parameter) of interest is X, and its value at the tth iteration of the jth chain is $x_j^{(t)}$. Thus, for the jth chain, the D values $x_j^{(0)}, \ldots, x_j^{(D-1)}$ are discarded and the L values $x_j^{(D)}, \ldots, x_j^{(D+L-1)}$ are retained.

Let

$$\bar{x}_j = \frac{1}{L} \sum_{t=D}^{D+L-1} x_j^{(t)} \quad \text{and} \quad \bar{x}_{\cdot\cdot} = \frac{1}{J} \sum_{j=1}^{J} \bar{x}_j, \tag{7.17}$$

and define the between-chain variance as

$$B = \frac{L}{J-1} \sum_{j=1}^{J} \left(\bar{x}_j - \bar{x}_{\cdot\cdot}\right)^2. \tag{7.18}$$

Now if the within-chain variance for the jth chain is $s_j^2 = \frac{1}{L-1} \sum_{t=D}^{D+L-1} \left(x_j^{(t)} - \bar{x}_j \right)^2$, then let

$$W = \frac{1}{J} \sum_{j=1}^{J} s_j^2 \tag{7.19}$$

represent the mean of the J within-chain estimated variances. Finally, let

$$R = \frac{\frac{L-1}{L}W + \frac{1}{L}B}{W}. \tag{7.20}$$

If all the chains are stationary, then both the numerator and the denominator should estimate the marginal variance of X. If, however, there are notable differences between the chains, then the numerator will exceed the denominator. As $L \to \infty$, $\sqrt{R} \to 1$. In practice, some authors suggest that $\sqrt{R} < 1.2$ is acceptable [194]. If the chosen burn-in period did not yield an acceptable result, then D should be increased, L should be increased, or preferably both. A conservative choice is to use one-half of the iterations for burn-in. The performance of this diagnostic is improved if the iterates $x_j^{(t)}$ are transformed so that their distribution is approximately normal. Alternatively, a reparameterization of the model could be undertaken and the chain rerun.

There are several potential difficulties with this approach. Selecting suitable starting values in cases of multimodal f may be difficult, and the procedure will not work if all of the chains become stuck in the same subregion or mode. Due to its unidimensionality, the method may also give a misleading impression of convergence for multidimensional target distributions. Enhancements of the Gelman–Rubin statistic are described in [196], and an extension for multidimensional target distributions is given in [65].

Raftery and Lewis [444] propose a very different quantitative strategy for estimating run length and burn-in period. Some researchers advocate no burn-in [202].

7.3.2 Practical implementation advice

The discussion above raises the question of what values should be used for the number of chains, the number of iterations for burn-in, and the length of the chain after burn-in. Most authors are reluctant to recommend generic values because appropriate choices are highly dependent on the problem at hand and the rate and efficiency with which the chain explores the region supported by f. Similarly, the choices are limited by how much computing time is available. In the last few years, published analyses have used burn-ins from zero to tens of thousands and chain lengths from the thousands to the millions. Diagnostics usually rely on at least three, and typically more, multiple chains. Five to ten years ago, burn-ins and chain lengths were shorter by a factor of 10. As computing power continues to grow, so too will the scope and intensity of MCMC efforts.

In summary, we reiterate our advice from Section 7.3.1.2 here, which in turn echoes [108]. First, create multiple trial runs of the chain from diverse starting values. Next, carry out a suite of diagnostic procedures like those discussed above to ensure that the chain appears to be well mixing and has approximately converged to the stationary distribution. Then, restart the chain for a final long run using a new seed for the random number generator.

For learning about MCMC methods and chain behavior, nothing beats programming these algorithms from scratch. For easier implementation, various software packages have been developed to automate the development of MCMC algorithms and the related diagnostics. The most comprehensive software to date is WinBUGS (Bayesian inference Using Gibbs Sampling) [515]. Software like BOA (Bayesian Output Analysis) [512] allows users to easily construct the relevant convergence diagnostics using either the S-Plus [476] or R [199] statistical packages. Most of this software is freely available via the internet.

7.3.3 Using the results

We consider here some of the common summaries of MCMC algorithm output; see the example in Section 7.4 for further illustration.

The first topic to consider is marginalization. If $\{\mathbf{X}^{(t)}\}$ represents a p-dimensional Markov chain, then $\{X_i^{(t)}\}$ is a Markov chain whose limiting distribution is the ith marginal of f. If you are focused only on a property of this marginal, discard the rest of the simulation and analyze the realizations of $X_i^{(t)}$.

Standard one-number summary statistics such as means and variances are commonly desired (see Section 7.1). The most commonly used estimator is based on an empirical average. Discard the burn-in; then calculate the desired statistic by taking

$$\frac{1}{L} \sum_{t=D}^{D+L-1} h\left(\mathbf{X}^{(t)}\right) \tag{7.21}$$

as the estimator of $\mathrm{E}\{h(\mathbf{X})\}$, where L denotes the length of each chain after discarding D burn-in iterates. This estimator is consistent even though the $\mathbf{X}^{(t)}$ are serially correlated. There are asymptotic arguments in favor of using no burn-in (so $D = 1$) [202]. However, since a finite number of iterations is used to compute the estimator in (7.21), most researcher employ a burn-in to reduce the influence of the initial iterates sampled from a distribution that may be far from the target distribution. Note that it is not necessary to run a chain for every quantity of interest. Post hoc inference about any quantity can be obtained from the realizations of $\mathbf{X}^{(t)}$ generated by the chain. In particular, the probability for any event can be estimated by the frequency of that event in the chain.

Other estimators have been developed. The Riemann sum estimator in (6.77) has been shown to have faster convergence than the standard estimator given above. Other variance reduction techniques discussed in Section 6.3, such as Rao–Blackwellization, can also be used to reduce the Monte Carlo variability of estimators based on the chain output [431].

The Monte Carlo, or simulation, standard error of an estimator is also of interest. The naive estimate of the standard error for an estimator like (7.21) is the sample standard deviation of the L realizations after burn-in divided by $\sqrt{L}$. However, MCMC realizations are typically positively correlated, so this can underestimate the standard error. An obvious correction is to compute the standard error based on a systematic subsample of, say, every kth iterate after burn-in. However, this approach is inefficient [365]. A simple estimator of the standard error is the so-called *batch method* [80, 282]. Separate the L iterates into batches with, say, 50 consecutive iterations in each batch. Compute the mean of each batch. Then the estimated standard error is the standard deviation of these means divided by the square root of the number of batches. Other strategies to estimate Monte Carlo variance are surveyed in [91, 204, 456, 514].

Quantile estimates and other interval estimates are also commonly desired. Estimates of various quantiles such as the median or the fifth percentile can be computed using the corresponding percentile of the realizations of the chain. This is simply implementing (7.21) for tail probabilities and inverting the relationship to find the quantile.

For Bayesian analyses, computation of the highest posterior density (HPD) interval is often of interest (see Section 1.5). For a symmetric posterior distribution, the $(1 - \alpha)\%$ HPD interval is given by the $(\alpha/2)$th and $(1 - \alpha/2)$th percentiles of the iterates. For nonsymmetric posterior distributions, more care must be taken to find the appropriate interval.

Chen et al. give a comprehensive review of more sophisticated alternatives for many of the simple summaries described here [91].

Simple graphical summaries of MCMC output should not be overlooked. Histograms of quantities of interest are standard practice, such as a histogram of the realizations of $h(\mathbf{X}^{(t)})$ for any h of interest. Alternatively, one can apply one of the density estimation techniques from Chapter 10 to summarize the collection of values. It is also common practice to display pairwise scatterplots and other descriptive plots to illustrate key features of f.

7.4 EXAMPLE: FUR SEAL PUP CAPTURE–RECAPTURE DATA

We conclude with an example that incorporates many of the ideas discussed in this chapter.

After centuries of severe population reductions due to commercial and subsistence hunting, the abundance of fur seals in New Zealand has been increasing in recent years. This increase has been of great interest to scientists, and these animals have been studied extensively [55, 56, 345].

Our goal is to estimate the number of pups in a fur seal colony using a capture–recapture approach [496]. In such studies, separate repeated efforts are made to count a population of unknown size. In our case, the population to be counted is the population of pups. No single census attempt is likely to provide a complete enumeration of the population, nor is it even necessary to try to capture most of the individuals. The individuals captured during each census are released with a marker indicating their capture. A capture of a marked individual during any subsequent

Table 7.1 Fur seal data for seven census efforts in one season.

		\multicolumn{7}{c}{Census attempt, i}
		1	2	3	4	5	6	7
Number captured	c_i	30	22	29	26	31	32	35
Number newly caught	m_i	30	8	17	7	9	8	5

census is termed a recapture. Population size can be estimated on the basis of the history of capture and recapture data. High recapture rates suggest that the true population size does not greatly exceed the total number of unique individuals ever captured.

Let N be the unknown population size to be estimated using I census attempts yielding total numbers of captures (including recaptures) equaling $\mathbf{c} = (c_1, \ldots, c_I)$. We assume that the population is closed during the period of the sampling, which means that deaths, births, and migrations are inconsequential during this period. The total number of distinct animals captured during the study is denoted by r.

We consider a model with separate, unknown capture probabilities for each census effort, $\boldsymbol{\alpha} = (\alpha_1, \ldots, \alpha_I)$. This model assumes that all animals are equally catchable on any one capture occasion, but capture probabilities may change over time. The likelihood for this model is

$$L(N, \boldsymbol{\alpha}|\mathbf{c}, r) \propto \frac{N!}{(N-r)!} \prod_{i=1}^{I} \alpha_i^{c_i} (1 - \alpha_i)^{N-c_i}. \tag{7.22}$$

This model is often called the $M(t)$ model [49].

In a capture–recapture study conducted on the Otago Pennisula on the South Island of New Zealand, fur seal pups were marked and released during $I = 7$ census attempts during one season. It is reasonable to assume the population of pups was closed during the study period. Table 7.1 shows the number of pups captured (c_i) and the number of these captures corresponding to pups never previously caught (m_i), for census attempts $i = 1, \ldots, 7$. A total of $r = \sum_{i=1}^{7} m_i = 84$ unique fur seals were observed during the sampling period.

For estimation, one might adopt a hierarchical Bayesian framework where N and $\boldsymbol{\alpha}$ are assumed to be a priori independent with the following priors: the noninformative Jeffreys prior $f(N) \propto 1/N$ for N; and $f(\alpha_i|\theta_1, \theta_2) = \text{Beta}(\theta_1, \theta_2)$ for $i = 1, \ldots, 7$ for the capture probabilities, assumed to be a priori exchangeable. Previous analyses with the $M(t)$ model have indicated sensitivity to the prior distribution for the capture probabilities [201]. To mitigate this sensitivity, we introduce a hyperprior for (θ_1, θ_2): $f(\theta_1, \theta_2) \propto \exp\{-(\theta_1 + \theta_2)/1000\}$, with (θ_1, θ_2) assumed to be a priori independent of the remaining parameters. A Gibbs sampler can then be constructed

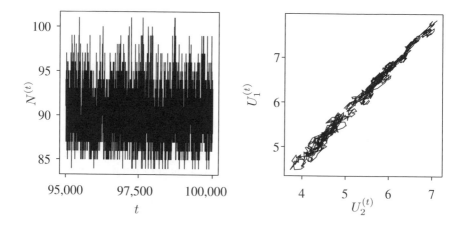

Fig. 7.7   Sample paths for N (left panel) and $\mathbf{U}$ (right panel) for final 5000 iterations in the seal pup example.

by simulating from the conditional posterior distributions

$$N - 84|\cdot \sim \text{NegBin}\left(84, 1 - \prod_{i=1}^{7}(1 - \alpha_i)\right), \tag{7.23}$$

$$\alpha_i|\cdot \sim \text{Beta}(c_i + \theta_1, N - c_i + \theta_2) \qquad \text{for } i = 1, \ldots, 7, \tag{7.24}$$

$$\theta_1, \theta_2|\cdot \sim k \left[\frac{\Gamma(\theta_1 + \theta_2)}{\Gamma(\theta_1)\Gamma(\theta_2)}\right]^7 \prod_{i=1}^{7} \alpha_i^{\theta_1}(1 - \alpha_i)^{\theta_2} \exp\left\{-\frac{\theta_1 + \theta_2}{1000}\right\}, \tag{7.25}$$

where $|\cdot$ denotes conditioning on the remaining parameters from $\{N, \boldsymbol{\alpha}, \theta_1, \theta_2\}$ as well as the data in Table 7.1, NegBin denotes the negative binomial distribution, and k is an unknown constant. Note that (7.25) is not easy to sample. This suggests using a hybrid Gibbs sampler with a Metropolis–Hastings step for (7.25).

Unfortunately, it is very difficult to produce a chain for (θ_1, θ_2) with adequate mixing and convergence behavior. To improve performance, transform (θ_1, θ_2) to $\mathbf{U} = (U_1, U_2) = (\log \theta_1, \log \theta_2)$. This permits a random walk step to update $\mathbf{U}$ effectively. Specifically, proposal values $\mathbf{U}^*$ can be generated by drawing $\epsilon \sim N(0, 0.085^2 \mathbf{I})$ (where $\mathbf{I}$ is the 2×2 identity matrix) and then setting $\mathbf{U}^* = \mathbf{u}^{(t)} + \epsilon$. The standard deviation 0.085 was chosen to get an acceptance rate of about 23% for the $\mathbf{U}$ updates. Recalling equation (7.8) in Example 7.2, it is necessary to transform (7.24) and (7.25) to reflect the change of variable. Thus,

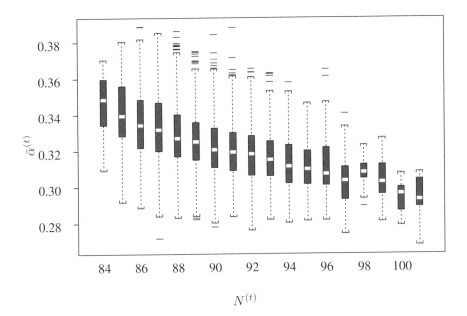

Fig. 7.8 Split boxplot of $\bar{\alpha}^{(t)}$ against $N^{(t)}$ for the seal pup example.

(7.24) becomes

$$\alpha_i|\cdot \sim \text{Beta}(c_i + \exp\{u_1\}, N - c_i + \exp\{u_2\}) \qquad \text{for } i = 1, \ldots, 7,$$

and (7.25) becomes

$$U_1, U_2|\cdot \sim k_u \exp\{u_1 + u_2\} \left[\frac{\Gamma(\exp\{u_1\} + \exp\{u_2\})}{\Gamma(\exp\{u_1\})\Gamma(\{\exp\{u_2\})} \right]^7$$

$$\times \prod_{i=1}^{7} \alpha_i^{\exp\{u_1\}} (1 - \alpha_i)^{\exp\{u_2\}} \exp\left\{ -\frac{\exp\{u_1\} + \exp\{u_2\}}{1000} \right\},$$

where k_u is an unknown constant.

The results below are based on a chain of 100,000 iterations with the first 1000 iterations discarded for burn-in. Sample paths for the last 5000 iterations are shown in Figure 7.7. The right panel of Figure 7.7 shows the bivariate sample path of $U_1^{(t)}$ and $U_2^{(t)}$. Based on five runs of 100,000 iterations each, the Gelman–Rubin statistic (7.20) for N is equal to 1.00047, which suggests the $N^{(t)}$ chain is roughly stationary.

Figure 7.8 shows split boxplots of the realizations of mean capture probability, $\bar{\alpha}^{(t)} = \frac{1}{7} \sum_{i=1}^{7} \alpha_i^{(t)}$, against $N^{(t)}$. As expected, the population size increases as the mean probability of capture decreases. Figure 7.9 show a histogram of the $N^{(t)}$ realizations, upon which posterior inference about N may be based. The posterior mean of N is 90 with a 95% HPD interval of (84, 95).

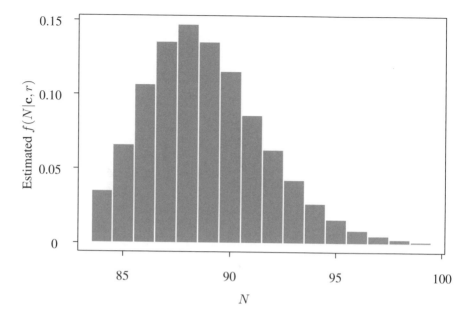

Fig. 7.9 Estimated marginal posterior probabilities for N for the seal pup example.

The likelihood given in (7.22) is just one of the many forms of capture–recapture models that could have been considered. For example, a model with a common capture probability may be more appropriate. Other parameterizations of the problem might also be investigated to improve MCMC convergence and mixing, which is strongly dependent on the parameterization and updating of (θ_1, θ_2).

Problems

7.1   The goal of this problem is to investigate the role of the proposal distribution in a Metropolis–Hastings algorithm designed to simulate from the posterior distribution of a parameter δ. In part (a), you are asked to simulate data from a distribution with δ known. For parts (b)–(d), assume δ is unknown with a Unif(0,1) prior distribution for δ. For parts (b)–(d), provide an appropriate plot and a table summarizing the output of the algorithm. To facilitate comparisons, use the same number of iterations, random seed, starting values, and burn-in period for all implementations of the algorithm.

(a) Simulate 200 realizations from the mixture distribution in equation (7.6) with $\delta = 0.7$. Draw a histogram of these data.

(b) Implement an independence chain MCMC procedure to simulate from the posterior distribution of δ, using your data from part (a).

(c) Implement a random walk chain with $\delta^* = \delta^{(t)} + \epsilon$ with $\epsilon \sim$ Unif$(-1,1)$.

(d) Reparameterize the problem letting $U = \log\{\delta/(1 - \delta)\}$ and $U^* = u^{(t)} + \epsilon$. Implement a random walk chain in U-space as in equation (7.8).

(e) Compare the estimates and convergence behavior of the three algorithms.

7.2   Simulating from the mixture distribution in equation (7.6) is straightforward (see part (a) of Problem 7.1). However, using the Metropolis–Hastings algorithm to simulate realizations from this distribution is useful for exploring the role of the proposal distribution.

(a) Implement a Metropolis–Hastings algorithm to simulate from equation (7.6) with $\delta = 0.7$, using $N(x^{(t)}, 0.01^2)$ as the proposal distribution. For each of three starting values, $x^{(0)} = 0, 7$, and 15, run the chain for 10,000 iterations. Plot the sample path of the output from each chain. If only one of the sample paths was available, what would you conclude about the chain? For each of the simulations, create a histogram of the realizations with the true density superimposed on the histogram. Based on your output from all three chains, what can you say about the behavior of the chain?

(b) Now change the proposal distribution to improve the convergence properties of the chain. Using the new proposal distribution, repeat part (a).

7.3   Consider a disk D of radius 1 inscribed within a square of perimeter 8 centered at the origin. Then the ratio of the area of the disk to that of the square is $\pi/4$. Let f represent the uniform distribution on the square. Then for a sample of points $(X_i, Y_i) \sim f(x, y)$ for $i = 1, \ldots, n$, $\hat{\pi} = \frac{4}{n} \sum_{i=1}^{n} 1_{\{(X_i,Y_i)\in D\}}$ is an estimator of π (where $1_{\{A\}}$ is 1 if A is true and 0 otherwise).

Consider the following strategy for estimating π. Start with $(x^{(0)}, y^{(0)}) = (0, 0)$. Thereafter, generate candidates as follows. First, generate $\epsilon_x^{(t)} \sim \text{Unif}(-h, h)$ and $\epsilon_y^{(t)} \sim \text{Unif}(-h, h)$. If $(x^{(t)} + \epsilon_x^{(t)}, y^{(t)} + \epsilon_y^{(t)})$ falls outside the square, regenerate $\epsilon_x^{(t)}$ and $\epsilon_y^{(t)}$ until the step taken remains within the square. Let $(X^{(t+1)}, Y^{(t+1)}) = (x^{(t)} + \epsilon_x^{(t)}, y^{(t)} + \epsilon_y^{(t)})$. Increment t. This generates a sample of points over the square. When $t = n$, stop and calculate $\hat{\pi}$ as given above.

(a) Implement this method for $h = 1$ and $n = 20,000$. Compute $\hat{\pi}$. What is the effect of increasing n? What is the effect of increasing and decreasing h? Comment.

(b) Explain why this method is flawed. Using the same method to generate candidates, develop the correct approach by referring to the Metropolis–Hastings ratio. Prove that your sampling approach has a stationary distribution that is uniform on the square.

(c) Implement your approach from part (b) and calculate $\hat{\pi}$. Experiment again with n and h. Comment.

Table 7.2 Breast cancer data.

	Hormone Treated						Control					
Recurrence	2	4	6	9	9	9	1	4	6	7	13	24
Times	13	14	18	23	31	32	25	35	35	39		
	33	34	43									
Censoring	10	14	14	16	17	18	1	1	3	4	5	8
Times	18	19	20	20	21	21	10	11	13	14	14	15
	23	24	29	29	30	30	17	19	20	22	24	24
	31	31	31	33	35	37	24	25	26	26	26	28
	40	41	42	42	44	46	29	29	32	35	38	39
	48	49	51	53	54	54	40	41	44	45	47	47
	55	56					47	50	50	51		

7.4 Derive the conditional distributions in equation (7.15).

7.5 A clinical trial was conducted to determine whether a hormone treatment benefits women who were treated previously for breast cancer. Each subject entered the clinical trial when she had a recurrence. She was then treated by irradiation and assigned to either a hormone therapy group or a control group. The observation of interest is the time until a second recurrence, which may be assumed to follow an exponential distribution with parameter $\tau\theta$ (hormone therapy group) or θ (control group). Many of the women did not have a second recurrence before the clinical trial was concluded, so that their recurrence times are censored.

In Table 7.2, a censoring time M means that a woman was observed for M months and did not have a recurrence during that time period, so that her recurrence time is known to exceed M months. For example, 15 women who received the hormone treatment suffered recurrences, and the total of their recurrence times is 280 months.

Let $y_i^H = (x_i^H, \delta_i^H)$ be the data for the ith person in the hormone group, where x_i^H is the time and δ_i^H equals 1 if x_i^H is a recurrence time and 0 if a censored time. The data for the control group can be written similarly.

The likelihood is then

$$L(\theta, \tau | \mathbf{y}) \propto \theta^{(\sum \delta_i^C + \sum \delta_i^H)} \tau^{(\sum \delta_i^H)} \exp\left\{ -\theta \sum x_i^C - \tau\theta \sum x_i^H \right\}.$$

You've been hired by the drug company to analyze their data. They want to know if the hormone treatment works, so the task is to find the marginal posterior distribution of τ using the Gibbs sampler. In a Bayesian analysis of these data, use the conjugate prior

$$f(\theta, \tau) \propto \theta^a \tau^b \exp\{-c\theta - d\theta\tau\}.$$

Physicians who have worked extensively with this hormone treatment have indicated that reasonable values for the hyperparameters are $(a, b, c, d) = (3, 1, 60, 120)$.

(a) Summarize and plot the data as appropriate.

(b) Derive the conditional distributions necessary to implement the Gibbs sampler.

(c) Program and run your Gibbs sampler. Use a suite of convergence diagnostics to evaluate the convergence and mixing of your sampler. Interpret the diagnostics.

(d) Compute summary statistics of the estimated joint posterior distribution, including marginal means, standard deviations, and 95% probability intervals for each parameter. Make a table of these results.

(e) Create a graph which shows the prior and estimated posterior for τ superimposed on the same scale.

(f) Interpret your results for the drug company. Specifically, what does your estimate of τ mean for the clinical trial? Are the recurrence times for the hormone group significantly different from those for the control group?

(g) A common criticism of Bayesian analyses is that the results are highly dependent on the priors. Investigate this issue by repeating your Gibbs sampler for values of the hyperparameters that are half and double the original hyperparameter values. Provide a table of summary statistics to compare your results. This is called a sensitivity analysis. Based on your results, what recommendations do you have for the drug company regarding the sensitivity of your results to hyperparameter values?

7.6   Problem 6.4 introduces data on coal-mining disasters from 1851 to 1962. For these data, assume the model

$$X_j \sim \begin{cases} \text{Poisson}(\lambda_1), & j = 1, \ldots, \theta, \\ \text{Poisson}(\lambda_2), & j = \theta + 1, \ldots, 112. \end{cases} \qquad (7.26)$$

Assume $\lambda_i | \alpha \sim \text{Gamma}(3, \alpha)$ for $i = 1, 2$, where $\alpha \sim \text{Gamma}(10, 10)$, and assume θ follows a discrete uniform distribution over $\{1, \ldots, 111\}$. The goal of this problem is to estimate the posterior distribution of the model parameters via a Gibbs sampler.

(a) Derive the conditional distributions necessary to carry out Gibbs sampling for the change-point model.

(b) Implement the Gibbs sampler. Use a suite of convergence diagnostics to evaluate the convergence and mixing of your sampler.

(c) Construct density histograms and a table of summary statistics for the approximate posterior distributions of θ, λ_1, and λ_2. Interpret the results in the context of the problem.

7.7   Consider a hierarchical nested model

$$Y_{ijk} = \mu + \alpha_i + \beta_{j(i)} + \epsilon_{ijk}, \qquad (7.27)$$

where $i = 1, \ldots, I$, $j = 1, \ldots, J_i$, and $k = 1, \ldots, K$. After averaging over k for each i and j, we can rewrite the model (7.27) as

$$Y_{ij} = \mu + \alpha_i + \beta_{j(i)} + \epsilon_{ij}, \qquad i = 1, \ldots, I, \quad j = 1, \ldots, J_i, \qquad (7.28)$$

where $Y_{ij} = \sum_{k=1}^{K} Y_{ijk}/K$. Assume that $\alpha_i \sim N(0, \sigma_\alpha^2)$, $\beta_{j(i)} \sim N(0, \sigma_\beta^2)$, and $\epsilon_{ij} \sim N(0, \sigma_\epsilon^2)$, where each set of parameters is independent a priori. Assume that σ_α^2, σ_β^2, and σ_ϵ^2 are known. To carry out Bayesian inference for this model, assume an improper flat prior for μ, so $f(\mu) \propto 1$. We consider two forms of the Gibbs sampler for this problem [463]:

(a) Let $n = \sum_i J_i$, $y_{..} = \sum_{ij} y_{ij}/n$, and $y_{i\cdot} = \sum_j y_{ij}/J_i$ hereafter. Show that at iteration t, the conditional distributions necessary to carry out Gibbs sampling for this model are given by

$$\mu^{(t+1)} \big| (\boldsymbol{\alpha}^{(t)}, \boldsymbol{\beta}^{(t)}, \mathbf{y}) \sim N\left(y_{..} - \frac{1}{n}\sum_i J_i \alpha_i^{(t)} - \frac{1}{n}\sum_{j(i)} \beta_{j(i)}^{(t)}, \frac{\sigma_\epsilon^2}{n} \right),$$

$$\alpha_i^{(t+1)} \big| (\mu^{(t+1)}, \boldsymbol{\beta}^{(t)}, \mathbf{y}) \sim N\left(\frac{J_i V_1}{\sigma_\epsilon^2}\left(y_{i\cdot} - \mu^{(t+1)} - \frac{1}{J_i}\sum_j \beta_{j(i)}^{(t)} \right), V_1 \right),$$

$$\beta_{j(i)}^{(t+1)} \big| (\mu^{(t+1)}, \boldsymbol{\alpha}^{(t+1)}, \mathbf{y}) \sim N\left(\frac{V_2}{\sigma_\epsilon^2}\left(y_{ij} - \mu^{(t+1)} - \alpha_i^{(t+1)} \right), V_2 \right),$$

where $V_1 = \left(\frac{J_i}{\sigma_\epsilon^2} + \frac{1}{\sigma_\alpha^2} \right)^{-1}$ and $V_2 = \left(\frac{1}{\sigma_\epsilon^2} + \frac{1}{\sigma_\beta^2} \right)^{-1}$.

(b) The convergence rate for a Gibbs sampler can sometimes be improved via reparameterization. One approach to reparameterization is called hierarchical centering. For this model, hierarchical centering can be described as follows. Let Y_{ij} follow (7.28), but now let $\eta_{ij} = \mu + \alpha_i + \beta_{j(i)}$ and $\epsilon_{ij} \sim N\left(0, \sigma_\epsilon^2\right)$. Then let $\gamma_i = \mu + \alpha_i$ with $\eta_{ij}|\gamma_i \sim N\left(\gamma_i, \sigma_\beta^2\right)$ and $\gamma_i|\mu \sim N\left(\mu, \sigma_\alpha^2\right)$. As above, assume σ_α^2, σ_β^2, and σ_ϵ^2 are known, and assume a flat prior for μ. Show that the conditional distributions necessary to carry out Gibbs sampling for this model are given by

$$\mu^{(t+1)} \big| (\boldsymbol{\gamma}^{(t)}, \boldsymbol{\eta}^{(t)}, \mathbf{y}) \sim N\left(\frac{1}{I}\sum_i \gamma_i^{(t)}, \frac{1}{I}\sigma_\alpha^2 \right),$$

$$\gamma_i^{(t+1)} \big| (\mu^{(t+1)}, \boldsymbol{\eta}^{(t)}, \mathbf{y}) \sim N\left(V_3\left(\frac{1}{\sigma_\beta^2}\sum_j \eta_{ij}^{(t)} + \frac{\mu^{(t+1)}}{\sigma_\alpha^2} \right), V_3 \right),$$

$$\eta_{ij}^{(t+1)} \big| (\mu^{(t+1)}, \boldsymbol{\gamma}^{(t+1)}, \mathbf{y}) \sim N\left(V_2\left(\frac{y_{ij}}{\sigma_\epsilon^2} + \frac{\gamma_i^{(t+1)}}{\sigma_\beta^2} \right), V_2 \right)$$

where $V_3 = \left(\frac{J_i}{\sigma_\beta^2} + \frac{1}{\sigma_\alpha^2} \right)^{-1}$.

7.8   In Problem 7.7, you were asked to derive Gibbs samplers under two model parameterizations. The goal of this problem is to compare the performance of the samplers.

The website for this book provides a dataset on the moisture content in the manufacture of pigment paste [52]. Batches of the pigment were produced, and the moisture content of each batch was tested analytically. Consider data from 15 randomly selected batches of pigment. For each batch, two independent samples were randomly selected and each of these samples was measured twice. For the analyses below, let $\sigma_\alpha^2 = 86$, $\sigma_\beta^2 = 58$, and $\sigma_\epsilon^2 = 1$.

Implement the two Gibbs samplers described below. To facilitate comparisons between the samplers, use the same number of iterations, random seed, starting values, and burn-in period for both implementations.

(a) Analyze these data by applying the Gibbs sampler from part (a) of Problem 7.7. Implement the sampler in blocks. For example, $\alpha = (\alpha_1, \ldots, \alpha_{15})$ is one block where all parameters can be updated simultaneously because their conditional distributions are independent. Update the blocks using a deterministic order within each cycle. For example, generate $\mu^{(0)}, \alpha^{(0)}, \beta^{(0)}$ in sequence, followed by $\mu^{(1)}, \alpha^{(1)}, \beta^{(1)}$, and so on.

(b) Analyze these data by applying the Gibbs sampler from part (b) of Problem 7.7. Implement the sampler update the blocks using a deterministic order within each cycle, updating $\mu^{(0)}, \gamma^{(0)}, \eta^{(0)}$ in sequence, followed by $\mu^{(1)}, \gamma^{(1)}, \eta^{(1)}$, and so on.

(c) Compare performance of the two algorithms by constructing the following diagnostics for each of the above implementations.

  i. After deleting the burn-in iterations, compute the pairwise correlations between all parameters.

  ii. Select several of the parameters in each implementation, and construct an autocorrelation plot for each parameter.

You may also wish to explore other diagnostics to facilitate these comparisons. For this problem, do you recommend the standard or the reparameterized model?

8

Advanced Topics in MCMC

Innovations in the theory and practice of MCMC continue at a rapid pace. In this chapter we survey a variety of advanced MCMC methods and explore some of the possible uses of MCMC to solve challenging statistical problems. Much recent effort has been focused on the development of methods for Bayesian inference. Sections 8.1–8.3 introduce auxiliary variable, reversible jump, and perfect sampling methods in Bayesian contexts, although these methods are also relevant in other contexts. Section 8.4 applies MCMC methods to facilitate Bayesian inference for spatial or image data. In Section 8.5 we discuss an application of MCMC to maximum likelihood estimation.

8.1 AUXILIARY VARIABLE METHODS

An important area of development in MCMC methods concerns auxiliary variable strategies. In many cases, such as Bayesian spatial lattice models, standard MCMC methods can take too long to mix properly to be of practical use. In such cases, one potential remedy is to augment the state space of the variable of interest. This approach can lead to chains which mix faster and require less tuning than the standard MCMC methods described in Chapter 7.

Continuing with the notation introduced in Chapter 7, we let $\mathbf{X}$ denote a random variable on whose state space we will simulate a Markov chain, usually for the purpose of estimating the expectation of a function of $\mathbf{X} \sim f(\mathbf{x})$. In Bayesian applications, it is important to remember that the random variables $\mathbf{X}^{(t)}$ simulated in a MCMC procedure are typically parameter vectors whose posterior distribution is of primary interest. Consider a target distribution f which can be evaluated but not easily sampled. To construct an auxiliary variable algorithm, the state space of $\mathbf{X}$ is augmented by the state space of a vector of auxiliary variables, $\mathbf{U}$. Then one constructs a Markov chain over the joint state space of $(\mathbf{X}, \mathbf{U})$ having stationary distribution $(\mathbf{X}, \mathbf{U}) \sim f(\mathbf{x}, \mathbf{u})$ which marginalizes to the target $f(\mathbf{x})$. When simulation has been completed, inference is based only on the marginal distribution of $\mathbf{X}$. For example, a Monte Carlo estimator of $\mu = \int h(\mathbf{x}) f(\mathbf{x}) \, d\mathbf{x}$ is $\hat{\mu} = \frac{1}{n} \sum_{t=1}^{n} h(\mathbf{X}^{(t)})$ where $(\mathbf{X}^{(t)}, \mathbf{U}^{(t)})$ are simulated in the augmented chain, but the $U^{(t)}$ are discarded.

Auxiliary variable MCMC methods were introduced in the statistical physics literature [151, 526]. Besag and Green noted the potential usefulness of this strategy, and a variety of refinements have subsequently been developed [35, 113, 286]. Augmenting the variables of interest to solve challenging statistical problems is effective in other areas, such as the EM algorithm described in Chapter 4 and the reversible jump algorithms described below in Section 8.2. The links between EM and auxiliary variable methods for MCMC algorithms are further explored in [542].

Below we describe simulated tempering as an example of an auxiliary variable strategy. Another important example is slice sampling, which is discussed in the next subsection. In Section 8.4.2 we present another application of auxiliary variable methods for the analysis of spatial or image data.

Example 8.1 (Simulated tempering) In problems with high dimensionality, multi-modality, or slow MCMC mixing, it may take extremely long chain runs to obtain good estimates of quantities of interest. The approach of *simulated tempering* provides a potential remedy [206, 371]. Simulations are based on a sequence of unnormalized densities f_i for $i = 1, \dots, m$, on a common sample space. These densities are viewed as ranging from cold $(i = 1)$ to hot $(i = m)$. Typically only the cold density is desired for inference, with the other densities being exploited to improve mixing. Indeed, the warmer densities should be designed so that MCMC mixing is faster for them than it is for f_1.

Consider the augmented variable $(\mathbf{X}, I)$ where the temperature I is now viewed as random with prior $I \sim p(i)$. From a starting value, $(\mathbf{x}^{(0)}, i^{(0)})$, we may construct a Metropolis–Hastings sampler in the augmented space as follows:

1. Use a Metropolis–Hastings or Gibbs update to draw $\mathbf{X}^{(t+1)} | i^{(t)}$ from a chain with stationary distribution $f_{i^{(t)}}$.

2. Generate I^* from a proposal density, $g\left(\cdot | i^{(t)}\right)$. A simple option is

$$
g\left(i^* | i^{(t)}\right) = \begin{cases} 1 & \text{if } \left(i^{(t)}, i^*\right) = (1, 2) \text{ or } \left(i^{(t)}, i^*\right) = (m, m-1), \\ 1/2 & \text{if } \left|i^* - i^{(t)}\right| = 1 \text{ and } i^{(t)} \in \{2, \dots, m-1\}, \\ 0 & \text{otherwise.} \end{cases}
$$

3. Accept or reject the candidate I^* as follows. Define the Metropolis–Hastings ratio to be $R_{\mathrm{ST}}\left(i^{(t)}, I^*, \mathbf{X}^{(t+1)}\right)$, where

$$
R_{\mathrm{ST}}(\mathbf{u}, \mathbf{v}, \mathbf{z}) = \frac{f_{\mathbf{v}}(\mathbf{z}) p(\mathbf{v}) g(\mathbf{u} | \mathbf{v})}{f_{\mathbf{u}}(\mathbf{z}) p(\mathbf{u}) g(\mathbf{v} | \mathbf{u})}, \tag{8.1}
$$

and accept $I^{(t+1)} = I^*$ with probability min $\{R_{ST}(i^{(t)}, I^*, \mathbf{X}^{(t+1)}), 1\}$. Otherwise, keep another copy of the current state, setting $I^{(t+1)} = i^{(t)}$.

4. Return to step 1.

The simplest way to estimate an expectation under the cold distribution is to average realizations generated from it, throwing away realizations generated from other f_i. To use more of the data, note that a state $(\mathbf{x}, i)$ drawn from the stationary distribution of the augmented chain has density proportional to $f_i(\mathbf{x})p(i)$. Therefore, importance weights $w^*(\mathbf{x}) = \frac{\tilde{f}(\mathbf{x})}{f_i(\mathbf{x})p(i)}$ can be used to estimate expectations with respect to a target density $\tilde{f}$; see Chapter 6.

The prior p is set by the user and ideally should be chosen so that the m tempering distributions (i.e., the m states for i) are visited roughly equally. In order for all the tempering distributions to be visited in a tolerable running time, m must be fairly small. On the other hand, each pair of adjacent tempering distributions must have sufficient overlap for the augmented chain to move easily between them. This requires a large m. To balance these competing concerns, choices for m that provide acceptance rates roughly in the range suggested in Section 7.3.1.1 are recommended. Improvements, extensions, and related techniques are discussed in [203, 206, 297, 357, 409]. Relationships between simulated tempering and other MCMC and importance sampling methods are explored in [367, 581].

Simulated tempering is reminiscent of the simulated annealing optimization algorithm described in Chapter 3. Suppose we run simulated tempering on the state space for $\boldsymbol{\theta}$. Let $L(\boldsymbol{\theta})$ and $q(\boldsymbol{\theta})$ be a likelihood and prior for $\boldsymbol{\theta}$, respectively. If we let $f_i(\boldsymbol{\theta}) = \exp\left\{\frac{1}{\tau_i}\log\{q(\boldsymbol{\theta})L(\boldsymbol{\theta})\}\right\}$ for $\tau_i = i$ and $i = 1, 2, \ldots$, then $i = 1$ makes the cold distribution match the posterior for $\boldsymbol{\theta}$, and $i > 1$ generates heated distributions that are increasingly flattened to improve mixing. Equation (8.1) then evokes step 2 of the simulated annealing algorithm described in Section 3.4 to minimize the negative log posterior. We have previously noted that simulated annealing produces a time-inhomogeneous Markov chain in its quest to find an optimum (Section 3.4.1.2). The output of simulated tempering is also a Markov chain, but simulated tempering does not systematically cool in the same sense as simulated annealing. The two procedures share the idea of using warmer distributions to facilitate exploration of the state space. $\qquad\Box$

8.1.1 Slice sampler

An important auxiliary variable MCMC technique is called the *slice sampler* [113, 286, 410]. Consider MCMC for a univariate variable $X \sim f(x)$, and suppose that it is impossible to sample directly from f. Introducing any univariate auxiliary variable U would allow us to consider a target density for $(X, U) \sim f(x, u)$. Writing $f(x, u) = f(x)f(u|x)$ suggests an auxiliary variable Gibbs sampling strategy that alternates between updates for X and U [286]. The trick is to choose a U variable that speeds MCMC mixing for X. At iteration $t + 1$ of the slice sampler we alternately generate

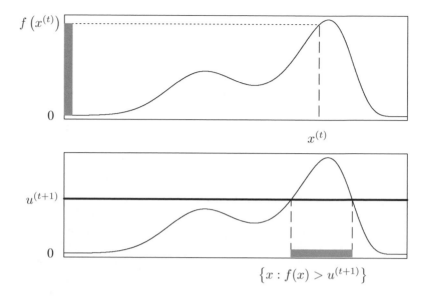

Fig. 8.1 Two steps of a univariate slice sampler for target distribution f.

$X^{(t+1)}$ and $U^{(t+1)}$ according to

$$U^{(t+1)}\big|x^{(t)} \sim \text{Unif}\left(0, f\left(x^{(t)}\right)\right),\tag{8.2}$$

$$X^{(t+1)}\big|u^{(t+1)} \sim \text{Unif}\left\{x : f(x) \geq u^{(t+1)}\right\}.\tag{8.3}$$

Figure 8.1 illustrates the approach. At iteration $t + 1$, the algorithm starts at $x^{(t)}$ shown in the upper panel. Then $U^{(t+1)}$ is drawn from $\text{Unif}\left(0, f\left(x^{(t)}\right)\right)$. In the top panel this corresponds to sampling along the vertical shaded strip. Now $X^{(t+1)}\big|\left(U^{(t+1)} = u^{(t+1)}\right)$ is drawn uniformly from the set of x values for which $f(x) \geq u^{(t+1)}$. In the lower panel this corresponds to sampling along the horizontal shaded strip.

While simulating from (8.3) is straightforward for this example, in other settings the set $\left\{x : f(x) \geq u^{(t+1)}\right\}$ may be more complicated. In particular, sampling $X^{(t+1)}\big|\left(U^{(t+1)} = u^{(t+1)}\right)$ in (8.3) can be challenging if f is not invertible. One approach to implementing equation (8.3) is to adopt a rejection sampling approach; see Section 6.2.3.

Example 8.2 (Moving between distant modes) When the target distribution is multimodal, one advantage of a slice sampler becomes more apparent. Figure 8.2 shows a univariate multimodal target distribution. If a standard Metropolis–Hastings algorithm is used to generate samples from the target distribution, then the algorithm may

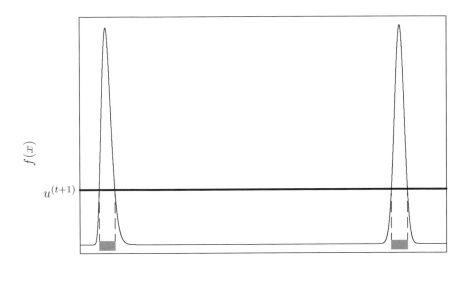

Fig. 8.2 The slice sampler for this multimodal target distribution draws $X^{(t+1)}|u^{(t+1)}$ uniformly from the set indicated by the two horizontal shaded strips.

find one mode of the distribution, but it may take many iterations to find the other mode unless the proposal distribution is very well tuned. Even if it finds both modes, it will almost never jump from one to the other. This problem will be exacerbated when the number of dimensions increases. In contrast, consider a slice sampler constructed to sample from the density shown in Figure 8.2. The horizontal shaded areas indicate the set defined in (8.3) from which $X^{(t+1)}|u^{(t+1)}$ is uniformly drawn. Hence the slice sampler will have about a 50% chance of switching modes each iteration. Therefore the slice sampler will mix much better with many fewer iterations required. □

Slice samplers have been shown to have attractive theoretical properties [398, 462], but can be challenging to implement in practice [410, 460]. The basic slice sampler described above can be generalized to include multiple auxiliary variables $U_1, \ldots, U_k$ and to accommodate multidimensional **X** [113, 286, 398, 462]. It is also possible to construct a slice sampler such that the algorithm is guaranteed to sample from the stationary distribution of the Markov chain [83, 397]. This is a variant of *perfect sampling*, which is discussed in Section 8.3.

8.2 REVERSIBLE JUMP MCMC

In Chapter 7 we considered MCMC methods for simulating $\mathbf{X}^{(t)}$ for $t = 1, 2, \ldots$ from a Markov chain with stationary distribution f. The methods described in Chapter 7 required that the dimensionality of $\mathbf{X}^{(t)}$ (i.e., of its state space) and the interpretation of the elements of $\mathbf{X}^{(t)}$ do not change with t. In many applications, it may be of interest to develop a chain that allows for changes in the dimension of the parameter space from one iteration to the next. Green's *reversible jump Markov chain Monte Carlo* (RJMCMC) method permits transdimensional Markov chain Monte Carlo simulation [243]. We discuss this approach below in the context of Bayesian model uncertainty. The full generality of RJMCMC is described in many of the references cited here.

Consider constructing a Markov chain to explore a space of candidate models, each of which might be used to fit observed data $\mathbf{y}$. Let $\mathcal{M}_1, \ldots, \mathcal{M}_K$ denote a countable collection of models under consideration. A parameter vector $\boldsymbol{\theta}_m$ denotes the parameters in the mth model. Different models may have different numbers of parameters, so we let p_m denote the number of parameters in the mth model. In the Bayesian paradigm, we may envision random variables $\mathbf{X} = (M, \boldsymbol{\theta}_M)$ which together index the model and parameterize inference for that model. We may assign prior distributions to these parameters, then seek to simulate from their posterior distribution using a MCMC method for which the tth random draw is $\mathbf{X}^{(t)} = \left(M^{(t)}, \boldsymbol{\theta}_{M^{(t)}}^{(t)} \right)$. Here $\boldsymbol{\theta}_{M^{(t)}}^{(t)}$, which denotes the parameters drawn for the model indexed by $M^{(t)}$, has dimension $p_{M^{(t)}}$ that can vary with t.

Thus, the goal of RJMCMC is to generate samples with joint posterior density $f(m, \boldsymbol{\theta}_m | \mathbf{y})$. This posterior arises from Bayes' theorem via

$$f(m, \boldsymbol{\theta}_m | \mathbf{y}) \propto f(\mathbf{y} | m, \boldsymbol{\theta}_m) f(\boldsymbol{\theta}_m | m) f(m), \tag{8.4}$$

where $f(\mathbf{y} | m, \boldsymbol{\theta}_m)$ denotes the density of the observed data under the mth model and its parameters, $f(\boldsymbol{\theta}_m | m)$ denotes the prior density for the parameters in the mth model, and $f(m)$ denotes the prior density of the mth model. A prior weight of $f(m)$ is assigned to the mth model so $\sum_{m=1}^{K} f(m) = 1$.

The posterior factorization

$$f(m, \boldsymbol{\theta}_m | \mathbf{y}) = f(m | \mathbf{y}) f(\boldsymbol{\theta}_m | m, \mathbf{y}) \tag{8.5}$$

suggests two important types of inference. First, $f(m | \mathbf{y})$ can be interpreted as the posterior probability for the mth model, normalized over all models under consideration. Second, $f(\boldsymbol{\theta}_m | m, \mathbf{y})$ is the posterior density of the parameters in the mth model.

RJMCMC enables the construction of an appropriate Markov chain for $\mathbf{X}$ that jumps between models with parameter spaces of different dimensions. Like simpler MCMC methods, RJMCMC proceeds with the generation of a proposed step from the current value $\mathbf{x}^{(t)}$ to $\mathbf{X}^*$, and then a decision whether to accept the proposal or to keep another copy of $\mathbf{x}^{(t)}$. The stationary distribution for our chain will be the posterior in (8.5) if the chain is constructed so that

$$f(m_1, \boldsymbol{\theta}_{m_1} | \mathbf{y}) a(m_2, \boldsymbol{\theta}_{m_2} | m_1, \boldsymbol{\theta}_{m_1}, \mathbf{y}) = f(m_2, \boldsymbol{\theta}_{m_2} | \mathbf{y}) a(m_1, \boldsymbol{\theta}_{m_1} | m_2, \boldsymbol{\theta}_{m_2}, \mathbf{y})$$

for all m_1 and m_2, where $a(\mathbf{x}_2|\mathbf{x}_1, \mathbf{Y})$ denotes the density for the chain moving to state $\mathbf{x}_2 = (m_2, \boldsymbol{\theta}_{m_2})$ at time $t + 1$, given that it was in state $\mathbf{x}_1 = (m_1, \boldsymbol{\theta}_{m_1})$ at time t. Chains that meet this *detailed balance* condition are termed *reversible*, because the direction of time does not matter in the dynamics of the chain.

The key to the RJMCMC algorithm is the introduction of auxiliary random variables at times t and $t + 1$ with dimensions chosen so that the augmented variables (namely $\mathbf{X}$ and the auxiliary variables) at times t and $t+1$ have equal dimensions. We can then construct a Markov transition for the augmented variable at time t that maintains dimensionality. This dimension-matching strategy enables the time-reversibility condition to be met by using a suitable acceptance probability, thereby ensuring that the Markov chain converges to the joint posterior for $\mathbf{X}$. Details of the limiting theory for these chains are given in [244, 243].

To understand dimension matching, it is simplest to begin by considering how one might propose parameters $\boldsymbol{\theta}_2$ corresponding to a proposed move from a model $\mathcal{M}_1$ with p_1 parameters to a model $\mathcal{M}_2$ with p_2 parameters when $p_2 > p_1$. A simple approach is to generate $\boldsymbol{\theta}_2$ from an invertible deterministic function of both $\boldsymbol{\theta}_1$ and an independent random component $\mathbf{U}_1$. We can write $\boldsymbol{\theta}_2 = \mathbf{q}_{1,2}(\boldsymbol{\theta}_1, \mathbf{U}_1)$. Proposing parameters for the reverse move can be carried out via the inverse transformation, $(\boldsymbol{\theta}_1, \mathbf{U}_1) = \mathbf{q}_{1,2}^{-1}(\boldsymbol{\theta}_2) = \mathbf{q}_{2,1}(\boldsymbol{\theta}_2)$. Note that $\mathbf{q}_{2,1}$ is an entirely deterministic way to propose $\boldsymbol{\theta}_1$ from a given $\boldsymbol{\theta}_2$.

Now generalize this idea to generate an augmented candidate parameter vector ($\boldsymbol{\theta}_{M^*}^*$ and auxiliary variables $\mathbf{U}^*$), given a proposed move to M^* from the current model, $m^{(t)}$. We can apply an invertible deterministic function $\mathbf{q}_{t,*}$ to $\boldsymbol{\theta}^{(t)}$ and some auxiliary random variables $\mathbf{U}$ to generate

$$(\boldsymbol{\theta}_{M^*}^*, \mathbf{U}^*) = \mathbf{q}_{t,*}(\boldsymbol{\theta}^{(t)}, \mathbf{U}), \tag{8.6}$$

where $\mathbf{U}$ is generated from proposal density $h(\cdot|m^{(t)}, \boldsymbol{\theta}^{(t)}, m^*)$. The auxiliary variables $\mathbf{U}^*$ and $\mathbf{U}$ are used so that $\mathbf{q}_{t,*}$ maintains dimensionality during the Markov chain transition at time t, but are discarded subsequently.

When $p_{M^*} = p_{M^{(t)}}$, the approach in (8.6) allows familiar proposal strategies. For example, a random walk could be obtained using $(\boldsymbol{\theta}_{M^*}^*, \mathbf{U}^*) = (\boldsymbol{\theta}^{(t)} + \mathbf{U}, \mathbf{U})$ with $\mathbf{U} \sim N(\mathbf{0}, \sigma^2 \mathbf{I})$ having dimension $p_{M^{(t)}}$. Alternatively, a Metropolis–Hastings chain can be constructed by using $\boldsymbol{\theta}_{M^*}^* = \mathbf{q}_{t,*}(\mathbf{U})$ when $p_U = p_{M^*}$, for an appropriate functional form of $\mathbf{q}_{t,*}$ and suitable $\mathbf{U}$. No $\mathbf{U}^*$ would be required to equalize dimensions. When $p_{M^{(t)}} < p_{M^*}$, the $\mathbf{U}$ can be used to expand parameter dimensionality; $\mathbf{U}^*$ may or may not be necessary to equalize dimensions, depending on the strategy employed. When $p_{M^{(t)}} > p_{M^*}$, both $\mathbf{U}$ and $\mathbf{U}^*$ may be unnecessary: for example, the simplest dimension reduction is deterministically to reassign some elements of $\boldsymbol{\theta}^{(t)}$ to $\mathbf{U}^*$ and retain the rest for $\boldsymbol{\theta}_{M^*}^*$. In all these cases, the reverse proposal is again obtained from the inverse of $\mathbf{q}_{t,*}$.

Assume that the chain is currently visiting model $m^{(t)}$, so the chain is in the state $\mathbf{x}^{(t)} = \left(m^{(t)}, \boldsymbol{\theta}_{m^{(t)}}^{(t)}\right)$. The next iteration of the RJMCMC algorithm can be summarized as follows:

1. Sample a candidate model $M^*|m^{(t)}$ from a proposal density with conditional density $g(\cdot|m^{(t)})$. The candidate model requires parameters $\boldsymbol{\theta}_{M^*}$ of dimension p_{M^*}.

2. Given $M^* = m^*$, generate an augmenting variable $\mathbf{U}|\left(m^{(t)}, \boldsymbol{\theta}_{m^{(t)}}^{(t)}, m^*\right)$ from a proposal distribution with density $h(\cdot|m^{(t)}, \boldsymbol{\theta}_{m^{(t)}}^{(t)}, m^*)$. Let

$$(\boldsymbol{\theta}_{m^*}^*, \mathbf{U}^*) = \mathbf{q}_{t,*}\left(\boldsymbol{\theta}_{m^{(t)}}^{(t)}, \mathbf{U}\right),$$

where $\mathbf{q}_{t,*}$ is an invertible mapping from $\left(\boldsymbol{\theta}_{m^{(t)}}^{(t)}, \mathbf{U}\right)$ to $(\boldsymbol{\theta}_{M^*}^*, \mathbf{U}^*)$ and the auxiliary variables have dimensions satisfying $p_{m^{(t)}} + p_\mathbf{U} = p_{m^*} + p_{\mathbf{U}^*}$.

3. For a proposed model, $M^* = m^*$, and the corresponding proposed parameter values $\boldsymbol{\theta}_{m^*}^*$, compute the Metropolis–Hastings ratio given by

$$\frac{f(m^*, \boldsymbol{\theta}_{m^*}^*|\mathbf{y})g(m^{(t)}|m^*)h(\mathbf{u}^*|m^*, \boldsymbol{\theta}_{m^*}^*, m^{(t)})}{f(m^{(t)}, \boldsymbol{\theta}_{m^{(t)}}^{(t)}|\mathbf{Y})g(m^*|m^{(t)})h(\mathbf{u}|m^{(t)}, \boldsymbol{\theta}_{m^{(t)}}^{(t)}, m^*)} \, |\mathbf{J}(t)|, \qquad (8.7)$$

where

$$\mathbf{J}(t) = \left.\frac{d\mathbf{q}_{t,*}(\boldsymbol{\theta}, \mathbf{u})}{d(\boldsymbol{\theta}, \mathbf{u})}\right|_{(\boldsymbol{\theta}, \mathbf{u}) = \left(\boldsymbol{\theta}_{m^{(t)}}^{(t)}, \mathbf{U}\right)}. \qquad (8.8)$$

Accept the move to the model M^* with probability equal to the minimum of 1 and the expression in (8.7). If the proposal is accepted, set $\mathbf{X}^{(t+1)} = (M^*, \boldsymbol{\theta}_{M^*}^*)$. Otherwise, reject the candidate draw and set $\mathbf{X}^{(t+1)} = \mathbf{x}^{(t)}$.

4. Discard $\mathbf{U}$ and $\mathbf{U}^*$. Return to step 1.

The last term in (8.7) is the absolute value of the determinant of the Jacobian matrix arising from the change of variables from $\left(\boldsymbol{\theta}_{m^{(t)}}^{(t)}, \mathbf{U}\right)$ to $(\boldsymbol{\theta}_{m^*}^*, \mathbf{U}^*)$. If $p_{M^{(t)}} = p_{M^*}$, then (8.7) simplifies to the standard Metropolis–Hastings ratio (7.1). Note that it is implicitly assumed here that the transformation $\mathbf{q}_{t,*}$ is differentiable.

Example 8.3 (Jumping between two simple models) An elementary example illustrates some of the details described above [243, 460]. Consider a problem with $K = 2$ possible models: the model $\mathcal{M}_1$ has a one-dimensional parameter space $\boldsymbol{\theta}_1 = \alpha$, and the model $\mathcal{M}_2$ has a two-dimensional parameter space $\boldsymbol{\theta}_2 = (\beta, \gamma)$. Thus $p_1 = 1$ and $p_2 = 2$. Let $m_1 = 1$ and $m_2 = 2$.

If the chain is currently in state $(1, \boldsymbol{\theta}_1)$ and the model $\mathcal{M}_2$ is proposed, then a random variable $U \sim h(u|1, \theta_1, 2)$ is generated from some proposal density h. Let $\beta = \alpha - U$ and $\gamma = \alpha + U$, so $\mathbf{q}_{1,2}(\alpha, u) = (\alpha - u, \alpha + u)$ and $\left|\frac{d\mathbf{q}_{1,2}(\alpha,u)}{d(\alpha,u)}\right| = 2$.

If the chain is currently in state $(2, \boldsymbol{\theta}_2)$ and $\mathcal{M}_1$ is proposed, then $(\alpha, u) = \mathbf{q}_{2,1}(\beta, \gamma) = \left(\frac{\beta+\gamma}{2}, \frac{\beta-\gamma}{2}\right)$ is the inverse mapping. Therefore $\left|\frac{d\mathbf{q}_{2,1}(\beta,\gamma)}{d(\beta,\gamma)}\right| = \frac{1}{2}$, and U^* is not required to match dimensions. This transition is entirely deterministic, so we replace $h(u^*|2, \theta_2, 1)$ in (8.7) with 1.

Thus for a proposed move from $\mathcal{M}_1$ to $\mathcal{M}_2$, the Metropolis–Hasting ratio (8.7) is equal to

$$\frac{f\left(2,\beta,\gamma|\mathbf{Y}\right)g\left(1|2\right)}{f\left(1,\alpha|\mathbf{Y}\right)g\left(2|1\right)h(u|1,\theta_1,2)} \times 2. \tag{8.9}$$

The Metropolis–Hastings ratio equals the reciprocal of (8.9) for a proposed move from $\mathcal{M}_2$ to $\mathcal{M}_1$. □

There are several significant challenges to implementing RJMCMC. Since the number of dimensions can be enormous, it can be critical to select an appropriate proposal distribution h and to construct efficient moves between the different dimensions of the model space. Another challenge is the diagnosis of convergence for RJMCMC algorithms. Research in these areas is ongoing [66, 67, 68].

RJMCMC is a very general method, and reversible jump methods have been developed for a myriad of application areas, including model selection and parameter estimation for linear regression [128], variable and link selection for generalized linear models [416], selection of the number of components in a mixture distribution [68, 453, 481], knot selection and other applications in nonparametric regression [42, 141, 292], and model determination for graphical models [127, 218]. There are many other areas for potential application of RJMCMC. One area of keen interest is genetic mapping [104, 547, 550].

RJMCMC unifies earlier MCMC methods to compare models with different numbers of parameters. For example, earlier methods for Bayesian model selection and model averaging for linear regression analysis, such as stochastic search variable selection [200] and MCMC model composition [445], can be shown to be special cases of RJMCMC [101].

8.2.1 RJMCMC for variable selection in regression

Consider a multiple linear regression problem with p potential predictor variables in addition to the intercept. A fundamental problem in regression is selection of a suitable model. Let m_k denote model k, which is defined by its inclusion of the i_1th through i_dth predictors, where the indices $\{i_1, \ldots, i_d\}$ are a subset of $\{1, \ldots, p\}$. For the consideration of all possible subsets of p predictor variables, there are therefore $K = 2^p$ models under consideration. Using standard regression notation, let $\mathbf{Y}$ denote the vector of n independent responses. For any model m_k, arrange the corresponding predictors in a design matrix $\mathbf{X}_{m_k} = (\mathbf{1} \ \mathbf{x}_{i_1} \ \cdots \ \mathbf{x}_{i_d})$, where $\mathbf{x}_{i_j}$ is the n-vector of observations of the i_jth predictor. The predictor data are assumed fixed. We seek the best ordinary least squares model of the form

$$\mathbf{Y} = \mathbf{X}_{m_k}\boldsymbol{\beta}_{m_k} + \boldsymbol{\epsilon} \tag{8.10}$$

among all m_k, where $\boldsymbol{\beta}_{m_k}$ is a parameter vector corresponding to the design matrix for m_k and the error variance is σ^2. In the remainder of this section, conditioning on the predictor data is assumed.

The notion of what model is best may have any of several meanings. In Example 3.2 the goal was to use the Akaike information criterion (AIC) to select the best model

[7, 75]. Here, we adopt a Bayesian approach for variable selection with priors on the regression coefficients and σ^2, and with the prior for the coefficients depending on σ^2. The immediate goal is to select the most promising subset of predictor variables, but we will also show how the output of an RJMCMC algorithm can be used to estimate a whole array of quantities of interest such as posterior model probabilities, the posterior distribution of the parameters of each model under consideration, and model-averaged estimates of various quantities of interest.

In our RJMCMC implementation, based on [101, 445], each iteration begins at a model $m^{(t)}$ which is described by a specific subset of predictor variables. To advance one iteration, the model M^* is proposed from among those models having either one predictor variable more or one predictor variable fewer than the current model. Thus the model proposal distribution is given by $g\left(\cdot|m^{(t)}\right)$, where

$$g\left(m^*|m^{(t)}\right) = \begin{cases} \frac{1}{p} & \text{if } M^* \text{ has one more or one fewer predictor than } m^{(t)}, \\ 0 & \text{otherwise.} \end{cases}$$

Given a proposed model $M^* = m^*$, step 2 of the RJMCMC algorithm requires us to sample $\mathbf{U}|\left(m^{(t)}, \boldsymbol{\beta}_{m^{(t)}}^{(t)}, m^*\right) \sim h\left(\cdot \mid m^{(t)}, \boldsymbol{\beta}_{m^{(t)}}^{(t)}, m^*\right)$. A simplifying approach is to let $\mathbf{U}$ become the next value for the parameter vector, in which case we may set the proposal distribution h equal to the posterior for $\boldsymbol{\beta}_m|(m, \mathbf{y})$, namely $f(\boldsymbol{\beta}_m|m, \mathbf{y})$. For appropriate conjugate priors, $\boldsymbol{\beta}_{m^*}^*|(m^*, \mathbf{y})$ has a noncentral t distribution [52]. We draw $\mathbf{U}$ from this proposal and set $\boldsymbol{\beta}_{m^*}^* = \mathbf{U}$ and $\mathbf{U}^* = \boldsymbol{\beta}_{m^{(t)}}^{(t)}$. Thus $\mathbf{q}_{t,*} = \left(\boldsymbol{\beta}_{m^{(t)}}^{(t)}, \mathbf{U}\right) = \left(\boldsymbol{\beta}_{m^*}^*, \mathbf{U}^*\right)$, yielding a Jacobian of 1. Since $g\left(m^{(t)}|m^*\right) = g\left(m^*|m^{(t)}\right) = 1/p$, the Metropolis–Hastings ratio in (8.7) can be written as

$$\frac{f\left(\mathbf{y}\big|m^*, \boldsymbol{\beta}_{m^*}^*\right) f\left(\boldsymbol{\beta}_{m^*}^*\big|m^*\right) f\left(m^*\right) f\left(\boldsymbol{\beta}_{m^{(t)}}^{(t)}\big|m^{(t)}, \mathbf{y}\right)}{f\left(\mathbf{y}\big|m^{(t)}, \boldsymbol{\beta}_{m^{(t)}}^{(t)}\right) f\left(\boldsymbol{\beta}_{m^{(t)}}^{(t)}\big|m^{(t)}\right) f\left(m^{(t)}\right) f\left(\boldsymbol{\beta}_{m^*}^*\big|m^*, \mathbf{y}\right)}$$

$$= \frac{f\left(\mathbf{y}|m^*\right) f\left(m^*\right)}{f\left(\mathbf{y}|m^{(t)}\right) f\left(m^{(t)}\right)} \quad (8.11)$$

after simplification. Here $f\left(\mathbf{y}|m^*\right)$ is the marginal likelihood, and $f\left(m^*\right)$ is the prior density, for the model m^*. Observe that this ratio does not depend on $\boldsymbol{\beta}_{m^*}^*$ or $\boldsymbol{\beta}_{m^{(t)}}$. Therefore, when implementing this approach with conjugate priors, one can treat the proposal and acceptance of $\boldsymbol{\beta}$ as a purely conceptual construct useful for placing the algorithm in the RJMCMC framework. In other words, we don't need to simulate $\boldsymbol{\beta}^{(t)}|m^{(t)}$, because $f\left(\boldsymbol{\beta}|m, \mathbf{y}\right)$ is available in closed form. The posterior model probabilities and $f\left(\boldsymbol{\beta}|m, \mathbf{y}\right)$ fully determine the joint posterior.

After running the RJMCMC algorithm, inference about many quantities of interest is possible. For example, from (8.5) the posterior model probabilities $f(m_k|\mathbf{y})$ can be approximated by the ratio of the number of times the chain visited the kth model to the number of iterations of the chain. These estimated posterior model probabilities can be used to select models. In addition, the output from the RJMCMC algorithm can be used to implement Bayesian model averaging. For example, if μ is some

Table 8.1 RJMCMC model selection results for baseball example: the five models with the highest posterior model probability (PMP). The bullets indicate inclusion of the corresponding predictor in the given model, using the predictor indices given in Table 3.2.

		Predictors					
3	4	8	10	13	14	24	PMP
•		•	•	•	•		0.22
	•	•	•	•	•		0.08
	•	•		•	•		0.05
•	•	•	•	•	•		0.04
•		•	•	•	•	•	0.03

quantity of interest such as a future observable, the utility of a course of action, or an effect size, then the posterior distribution of μ given the data is given by

$$f\left(\mu|\mathbf{y}\right) = \sum_{k=1}^{K} f\left(\mu|m_k, \mathbf{y}\right) f\left(m_k|\mathbf{y}\right). \tag{8.12}$$

This is the average of the posterior distribution for μ under each model, weighted by the posterior model probability. It has been shown that taking account of uncertainty about the form of the model can protect against underestimation of uncertainty [289].

Example 8.4 (Baseball salaries, continued) Recall Example 3.3, where we sought the best subset among 27 possible predictors to use in linear regression modeling of baseball players' salaries. Previously, the objective was to find the best subset as measured by the minimal AIC value. Here, we seek the best subset as measured by the model with the highest posterior model probability.

We adopt a uniform prior over model space, assigning $f(m_k) = 2^{-p}$ for each model. For the remaining parameters, we use a normal–gamma conjugate class of priors with $\boldsymbol{\beta}_{m_k}|m_k \sim N(\boldsymbol{\alpha}_{m_k}, \sigma^2 \mathbf{V}_{m_k})$ and $\nu\lambda/\sigma^2 \sim \chi_\nu^2$. For this construction, $f\left(\mathbf{y}|m_k\right)$ in (8.11) can be shown to be the noncentral t density (Problem 8.1). For the baseball data, the hyperparameters are set as follows. First, let $\nu = 2.58$ and $\lambda = 0.28$. Next, $\boldsymbol{\alpha}_{m_k} = \left(\hat{\beta}_0, 0, \ldots, 0\right)$ is a vector of length p_{m_k} whose first element equals the least squares estimated intercept from the full model. Finally, $\mathbf{V}_{m_k}$ is a diagonal matrix with entries $\left(s_{\mathbf{y}}^2, c^2/s_1^2, \ldots, c^2/s_p^2\right)$, where $s_{\mathbf{y}}^2$ is the sample variance of $\mathbf{y}$, s_i^2 is the sample variance of the ith predictor, and $c = 2.58$. Additional details are given in [445].

We ran 200,000 RJMCMC iterations. Table 8.1 shows the five models with the highest estimated posterior model probabilities. If the goal is to select the best model, then the model with the predictors 3, 8, 10, 13, and 14 should be chosen, where the predictor indices correspond to those in Table 3.2.

Table 8.2   RJMCMC results for baseball example: the estimated posterior effect probabilities $P(\beta_i \neq 0|\mathbf{y})$ exceeding 0.10. The predictor indices and labels correspond to those given in Table 3.2.

Index	Predictor	$P(\beta_i \neq 0|\mathbf{y})$
13	FA	1.00
14	Arb	1.00
8	RBIs	0.97
10	SOs	0.78
3	Runs	0.55
4	Hits	0.52
25	SBs×OBP	0.13
24	SOs×errors	0.12
9	Walks	0.11

The posterior effect probabilities, $P(\beta_i \neq 0|\mathbf{y})$, for those predictors with probabilities greater than 0.10 are given in Table 8.2. Each entry is a weighted average of an indicator variable that equals 1 only when the coefficient is in the model, where the weights correspond to the posterior model probabilities as in equation (8.12). These results indicate that free agency, arbitration status, and the number of runs batted in are strongly associated with baseball players' salaries.

Other quantities of interest based on variants of equation (8.12) can be computed, such as the model-averaged posterior expectation and variance for each regression parameter, or various posterior salary predictions.   □

Alternative approaches to transdimensional Markov chain simulation have been proposed. Stephens proposed a promising method based on the construction of a continuous-time Markov birth-and-death process [517]. In this approach, the parameters are modeled via a point process. Connections between Green's RJMCMC and Stephens's birth-and-death process have also been noted [78]. A general form of RJMCMC has been proposed, which also unifies many of the existing methods for assessing uncertainty about the dimension of the parameter space [230]. Continued interest and rapid development in these areas is likely.

8.3   PERFECT SAMPLING

MCMC is useful because at the tth iteration it generates a random draw $\mathbf{X}^{(t)}$ whose distribution approximates the target distribution f as $t \to \infty$. Since run lengths are finite in practice, much of the discussion in Chapter 7 pertained to assessing when the approximation becomes sufficiently good. For example, Section 7.3 presents methods to determine the run length and the number of iterations to discard (i.e., the burn-in). However, these convergence diagnostics all have various drawbacks.

Perfect sampling algorithms avoid all these concerns by generating a chain that has exactly reached the stationary distribution. This sounds wonderful, but there are challenges in implementation.

8.3.1   Coupling from the past

Propp and Wilson introduced a perfect sampling MCMC algorithm called *coupling from the past* (CFTP) [438]. Other expositions of the CFTP algorithm include [81, 144, 437]. The website maintained by Wilson surveys much of the early literature on CFTP and related methods [568].

CFTP is often motivated by saying that the chain is started at $t = -\infty$ and run forward to time $t = 0$. While this is true, convergence does not suddenly occur in the step from $t = -1$ to $t = 0$, and you are not required to somehow set $t = -\infty$ on your computer. Instead, we will identify a window of time from $t = \tau < 0$ to $t = 0$ for which whatever happens before τ is irrelevant, and the infinitely long progression of the chain prior to τ means that the chain is in its stationary distribution by time 0.

While this strategy might sound reasonable at the outset, in practice it is impossible to know what state the chain is in at time τ. Therefore, we must consider multiple chains: in fact, one chain started in every possible state at time τ. Each chain can be run forward from $t = \tau$ to $t = 0$. Because of the Markov nature of these chains, the chain outcomes at time $\tau + 1$ depend only on their status at time τ. Therefore, this collection of chains completely represents every possible chain that could have been run from infinitely long ago in the past.

The next problem is that we no longer have a single chain, and it seems that chain states at time 0 will differ. To remedy this multiplicity, we rely on the idea of *coupling*. Two chains on the same state space with the same transition probabilities have coupled (or *coalesced*) at time t if they share the same state at time t. At this point, the two chains will have identical probabilistic properties, due to the Markov property and the equal transition probabilities. A third such chain could couple with these two at time t or any time thereafter. Thus, to eliminate the multiple chains introduced above, we use an algorithm that ensures that once chains have coupled, they follow the same sample path thereafter. Further, we insist that all chains must have coalesced by time 0. This algorithm will therefore yield one chain from $t = 0$ onwards which is in the desired stationary distribution.

To simplify presentation, we assume that X is unidimensional and has finite state space with K states. Neither assumption is necessary for CFTP strategies more general than the one we describe below.

We consider an ergodic Markov chain with a deterministic transition rule q that updates the current state of the chain, $x^{(t)}$, as a function of some random variable $U^{(t+1)}$. Thus,

$$ X^{(t+1)} = q\left(x^{(t)}, U^{(t+1)}\right). \tag{8.13} $$

For example, a Metropolis–Hastings proposal from a distribution with cumulative distribution function F can be generated using $q(x, u) = F^{-1}(u)$, and a random walk proposal can be generated using $q(x, u) = x + u$. In (8.13) we used a univariate

$U^{(t+1)}$, but, more generally, chain transitions may be governed by a multivariate vector $\mathbf{U}^{(t+1)}$. We adopt the general case hereafter.

CFTP starts one chain from each state in the state space at some time $\tau < 0$ and transitions each chain forward using proposals generated by q. Proposals are accepted using the standard Metropolis–Hastings ratio. The goal is to find a starting time τ such that the chains have all coalesced by time $t = 0$ when run forwards in time from $t = \tau$. This approach provides $X^{(0)}$, which is a draw from the desired stationary distribution f.

The algorithm to find τ and thereby produce the desired chain is as follows. Let $X_k^{(t)}$ be the random state at time t of the Markov chain started in state k, with $k = 1, \ldots, K$.

1. Let $\tau = -1$. Generate $\mathbf{U}^{(0)}$. Start a chain in each state of the state space at time -1, namely $x_1^{(-1)}, \ldots, x_K^{(-1)}$, and run each chain forward to time 0 via the update $X_k^{(0)} = q\left(x_k^{(-1)}, \mathbf{U}^{(0)}\right)$ for $k = 1, \ldots, K$. If all K chains are in the same state at time 0, then the chains have coalesced and $X^{(0)}$ is a draw from f; the algorithm stops.

2. If the chains have not coalesced, then let $\tau = -2$. Generate $\mathbf{U}^{(-1)}$. Start a chain in each state of the state space at time -2, and run each chain forward to time 0. To do this, let $X_k^{(-1)} = q\left(x_k^{(-2)}, \mathbf{U}^{(-1)}\right)$. Next, you must reuse the $\mathbf{U}^{(0)}$ generated in step 1, so $X_k^{(0)} = q\left(X_k^{(-1)}, \mathbf{U}^{(0)}\right)$. If all K chains are in the same state at time 0, then the chains have coalesced and $X^{(0)}$ is a draw from f; the algorithm stops.

3. If the chains have not coalesced, move the starting time back to time $\tau = -3$ and update as above. We continue restarting the chains one step further back in time and running them forward to time 0 until we start at a τ for which all K chains have coalesced by time $t = 0$. At this point the algorithm stops. In every attempt, it is imperative that the random updating variables be reused. Specifically, when starting the chains at time τ, you must reuse the previously drawn random number updates $\mathbf{U}^{(\tau+1)}, \mathbf{U}^{(\tau+2)} \ldots, \mathbf{U}^{(0)}$. Also note that the same $\mathbf{U}^{(t)}$ vector is used to update all K chains at the tth iteration.

Propp and Wilson show that the value of $X^{(0)}$ returned from the CFTP algorithm for a suitable q is a realization of a random variable distributed according to the stationary distribution of the Markov chain and that this coalescent value will be produced in finite time [438]. Even if all chains coalesce before time 0, you must use $X^{(0)}$ as the perfect sampling draw; otherwise sampling bias is introduced.

Obtaining the perfect sampling draw $X^{(0)}$ from f is not sufficient for most uses. Typically we desire an i.i.d. n-sample from f, either for simulation or to use in a Monte Carlo estimate of some expectation, $\mu = \int h(x)f(x)\,dx$. A perfect i.i.d. sample from f can be obtained by running the CFTP algorithm n times to generate n individual values for $X^{(0)}$. If you only want to ensure that the algorithm is, indeed,

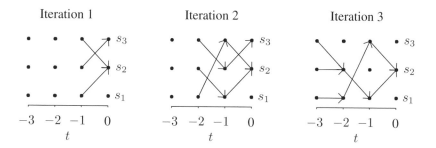

Fig. 8.3 Example of perfect sampling sample paths. See Example 8.5 for details.

sampling from f, but independence is not required, you can run CFTP once and continue to run the chain forward from its state at time $t = 0$. While the first option is probably preferable, the second option may be more reasonable in practice, especially for cases where the CFTP algorithm requires many iterations before coalescence is achieved. These are only the two simplest strategies available for using the output of a perfect sampling algorithm; see also [404] and the references in [568].

Example 8.5 (Sample paths in a small state space) In the example shown in Figure 8.3, there are three possible states, s_1, s_2, s_3. At iteration 1, a sample path is started from each of the three states at time $\tau = -1$. A random update $U^{(0)}$ is selected, and $X_k^{(0)} = q\left(s_k, U^{(0)}\right)$ for $k = 1, 2, 3$. The paths have not all coalesced at time $t = 0$, so the algorithm moves to iteration 2. In iteration 2, the algorithm begins at time $\tau = -2$. The transition rule for the moves from $t = -2$ to $t = -1$ is based on a newly sampled update variable, $U^{(-1)}$. The transition rule for the moves from $t = -1$ to $t = 0$ relies on the same $U^{(0)}$ value obtained previously in iteration 1. The paths have not all coalesced at time $t = 0$, so the algorithm moves to iteration 3. Here, the previous draws for $U^{(0)}$ and $U^{(-1)}$ are reused and a new $U^{(-2)}$ is selected. In iteration 3, all three sample paths visit state s_2 at time $t = 0$, thus the paths have coalesced, and $X^{(0)} = s_2$ is a draw from the stationary distribution f. □

Several finer details of CFTP implementation merit mention. First, note that CFTP requires reuse of previously generated variables $\mathbf{U}^{(t)}$ and the shared use of the same $\mathbf{U}^{(t)}$ realization to update all chains at time t. If the $\mathbf{U}^{(t)}$ are not reused, the samples will be biased. Propp and Wilson show an example where the regeneration of the $\mathbf{U}^{(t)}$ at each time biases the chain towards the extreme states in an ordered state space [438]. The reuse and sharing of past $\mathbf{U}^{(t)}$ ensures that all chains coalesce by $t = 0$ when started at any time $\tau' \leq \tau$, where τ is the starting time chosen by CFTP. Moreover, this practice ensures that the coalescent state at time 0 is the same for all such chains in a given run, which enables the proof that CFTP produces an exact draw from f.

Second, CFTP introduces a dependence between the τ and $X^{(0)}$ it chooses. Therefore, bias can be induced if a CFTP run is stopped prematurely before the coupling time has been determined. Suppose a CFTP algorithm is run for a long time during which coupling does not occur. If the computer crashes or an impatient user stops and then restarts the algorithm to find a coupling time, this will generally bias the sample towards those states in which coupling is easier. An alternative perfect sampling method known as Fill's algorithm was designed to avoid this problem [169].

Third, our description of the CFTP algorithm uses the sequence of starting times $\tau = -1, -2, \ldots$ for successive CFTP iterations. For many problems, this will be inefficient. It may be more efficient to use the sequence $\tau = -1, -2, -4, -8, -16, \ldots$, which minimizes the worst case number of simulation steps required and nearly minimizes the expected number of required steps [438].

Finally, it may seem that this coupling strategy should work if the chains were run forwards from time $t = 0$ instead of backwards; but that is not the case. To understand why, consider a Markov chain for which some state x' has a unique predecessor. It is impossible for x' to occur at the random time of first coalescence. If x' occurred, the chain must have already coalesced at the previous time, since all chains must have been in the predecessor state. Therefore the marginal distribution of the chain at the first coalescence time must assign zero probability to x', and hence cannot be the stationary distribution. Although this forward coupling approach fails, there is a clever way to modify the CFTP construct to produce a perfect sampling algorithm for a Markov chain that only runs forward in time [567].

8.3.1.1 *Stochastic monotonicity and sandwiching* When applying CFTP to a chain with a vast finite state space or an infinite (e.g., continuous) state space, it can be challenging to monitor whether sample paths started from all possible elements in the state space have coalesced by time zero. However, if the state space can be ordered in some way such that the deterministic transition rule q preserves the state space ordering, then only the sample paths started from the minimum state and maximum state in the ordering need to be monitored.

Let $\mathbf{x}, \mathbf{y} \in S$ denote any two possible states of a Markov chain exploring a possibly huge state space S. Formally, S is said to admit the natural *componentwise partial ordering*, $\mathbf{x} \leq \mathbf{y}$, if $x_i \leq y_i$ for $i = 1, \ldots, n$ and $\mathbf{x}, \mathbf{y} \in S$. The transition rule q is *monotone* with respect to this partial ordering if $q(\mathbf{x}, \mathbf{u}) \leq q(\mathbf{y}, \mathbf{u})$ for all $\mathbf{u}$ when $\mathbf{x} \leq \mathbf{y}$. Now, if there exist a minimum and a maximum element of the state space S, so $\mathbf{x}_{\min} \leq \mathbf{x} \leq \mathbf{x}_{\max}$ for all $\mathbf{x} \in S$ and the transition rule q is monotone, then a MCMC procedure that uses this q preserves the ordering of the states at each time step. Therefore, CFTP using a monotone transition rule can be carried out by simulating only two chains: one started at $\mathbf{x}_{\min}$ and the other at $\mathbf{x}_{\max}$. Sample paths for chains started at all other states will be sandwiched between the paths started in the maximum and minimum states. When the sample paths started in the minimum and maximum states have coalesced at time zero, coalescence of all the intermediate chains is also guaranteed. Therefore, CFTP samples from the stationary distribution at $t = 0$. Many problems satisfy these monotonicity properties; one example is given in Section 8.4.3.

In problems where this form of monotonicity isn't possible, other related strategies can be devised [399, 403, 567]. Considerable effort has been focused on developing methods to apply perfect sampling for specific problem classes, such as perfect Metropolis–Hastings independence chains [105], perfect slice samplers [397], and perfect sampling algorithms for Bayesian model selection [295, 486].

Perfect sampling is currently an area of active research, and many extensions of the ideas presented here have been proposed. While this idea is quite promising, perfect sampling algorithms have not been widely implemented for problems of realistic size. Challenges in implementation and long coalescence times have sometimes discouraged large-scale realistic applications. Nonetheless, the attractive properties of perfect sampling algorithms and continued research in this area will likely motivate new and innovative MCMC algorithms for practical problems.

8.4 EXAMPLE: MCMC FOR MARKOV RANDOM FIELDS

We offer here an introduction to Bayesian analysis of Markov random field models with emphasis on the analysis of spatial or image data. This topic provides interesting examples of many of the methods discussed in this chapter.

A *Markov random field* specifies a probability distribution for spatially referenced random variables. Markov random fields are quite general and can be used for many lattice-type structures such as regular rectangular, hexagonal, and irregular grid structures [110, 539]. There are a number of complex issues with Markov random field construction that we do not attempt to resolve here. Besag has published a number of key papers on Markov random fields for spatial statistics and image analysis, including his seminal 1974 paper [29, 30, 34, 35, 36, 37]. Additional comprehensive coverage of Markov random fields is given in [110, 329, 353, 569].

For simplicity, we focus here on the application of Markov random fields to a regular rectangular lattice. For example, we might overlay a rectangular grid on a map or image and label each pixel or cell in the lattice. The value for the ith pixel in the lattice is denoted by x_i for $i = 1, \ldots, n$, where n is finite. We will focus on binary random fields where x_i can take on only two values, 0 and 1, for $i = 1, ..., n$. It is generally straightforward to extend methods to the case where x_i is continuous or takes on more than two discrete values [110].

Let $\mathbf{x}_{\delta_i}$ define the set of x values for the pixels that are near pixel i. The pixels that define δ_i are called the *neighborhood* of pixel i. The pixel x_i is not in δ_i. A proper neighborhood definition must meet the condition that if pixel i is a neighbor of pixel j then pixel j is a neighbor of pixel i. In a rectangular lattice, a first-order neighborhood is the set of pixels that are vertically and horizontally adjacent to the pixel of interest see Figure 8.4). A second-order neighborhood also includes the pixels diagonally adjacent from the pixel of interest.

Imagine that the value x_i for the ith pixel is a realization of a random variable X_i. A *locally dependent Markov random field* specifies that the distribution of X_i given the remaining pixels, $\mathbf{X}_{-i}$, is dependent only on the neighboring pixels. Therefore, for $\mathbf{X}_{-i} = \mathbf{x}_{-i}$,

$$f\left(x_i | \mathbf{x}_{-i}\right) = f\left(x_i | \mathbf{x}_{\delta_i}\right) \tag{8.14}$$

First-order   Second-order

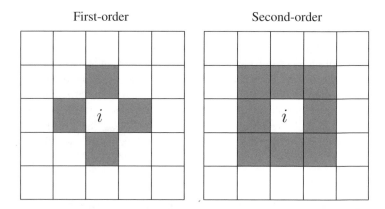

Fig. 8.4 The shaded pixels indicate a first-order and a second-order neighborhood of pixel i for a rectangular lattice.

for $i = 1, \ldots, n$. Assuming each pixel has a nonzero probability of equaling 0 or 1 means that the so-called *positivity condition* is satisfied: that the minimal state space of $\mathbf{X}$ equals the Cartesian product of the state spaces of its components. The positivity condition ensures that the conditional distributions considered later in this section are well defined.

The Hammersley–Clifford theorem shows that the conditional distributions in (8.14) together specify the joint distribution of $\mathbf{X}$ up to a normalizing constant [29]. For our discrete binary state space, this normalizing constant is the sum of $f(\mathbf{x})$ over all $\mathbf{x}$ in the state space. This sum is not usually available by direct calculation, because the number of terms is enormous. Even for an unrealistically small 40×40 pixel image where the pixels take on binary values, there are $2^{1600} = 4.4 \times 10^{481}$ terms in the summation. Bayesian MCMC methods provide a Monte Carlo basis for inference about images, despite such difficulties. We describe below several approaches for MCMC analysis of Markov random field models.

8.4.1 Gibbs sampling for Markov random fields

We begin by adopting a Bayesian model for analysis of a binary Markov random field. In the introduction to Markov random fields above, we used x_i to denote the value of the ith pixel. Here we let X_i denote the unknown true value of the ith pixel, where X_i is treated as a random variable in the Bayesian paradigm. Let y_i denote the observed value for the ith pixel. Thus $\mathbf{X}$ is a parameter vector and $\mathbf{y}$ is the data. In an image analysis application, $\mathbf{y}$ is the degraded image and $\mathbf{X}$ is the unknown true image. In a spatial statistics application of mapping of plant or animal species distributions, $y_i = 0$ might indicate that the species was not observed in pixel i during the sampling period and X_i might denote the true (unobserved) presence or absence of the species in pixel i.

Three assumptions are fundamental to the formulation of this model. First, we assume that observations are mutually independent given true pixel values. So the joint conditional density of $\mathbf{Y}$ given $\mathbf{X} = \mathbf{x}$ is

$$f(y_1, \ldots, y_n | x_1, \ldots, x_n) = \prod_{i=1}^{n} f(y_i | x_i), \tag{8.15}$$

where $f(y_i | x_i)$ is the density of the observed data in pixel i given the true value. Thus, viewed as a function of $\mathbf{x}$, (8.15) is the likelihood function. Second, we adopt a locally dependent Markov random field (8.14) to model the true image. Finally, we assume the positivity condition, defined above.

The parameters of the model are $x_1, \ldots, x_n$, and the goal of the analysis is to estimate these true values. To do this, adopt a Gibbs sampling approach. Assume the prior $\mathbf{X} \sim f(\mathbf{x})$ for the parameters. The goal in the Gibbs sampler, then, is to obtain a sample from the posterior of $\mathbf{X}$,

$$\cdots | \mathbf{x}) f(\mathbf{x}). \tag{8.16}$$

One class of prior densities for $\mathbf{X}$ is given by

$$f(\mathbf{x}) \propto \exp \left\{ \sum_{i \sim j}^{n} \phi(x_i - x_j) \right\}, \tag{8.17}$$

where $i \sim j$ denotes all pairs such that pixel i is a neighbor of pixel j, and ϕ is some function that is symmetric about 0 with $\phi(\mathbf{z})$ increasing as $|\mathbf{z}|$ increases. Equation (8.17) is called a *pairwise difference prior*. Adopting this prior distribution based on pairwise interactions simplifies computations but may not be realistic. Extensions to allow for higher-order interactions have been proposed [539].

The Gibbs sampler requires the derivation of the univariate conditional distributions whose form follows from (8.14)–(8.16) above. The Gibbs update at iteration t is therefore

$$X_i^{(t+1)} \Big| \left(\mathbf{x}_{-i}^{(t)}, \mathbf{y} \right) \sim f \left(x_i | \mathbf{x}_{-i}^{(t)}, \mathbf{y} \right). \tag{8.18}$$

A common strategy is to update each X_i in turn, but it can be more computationally efficient to update the pixels in independent blocks. The blocks are determined by the neighborhoods defined for a particular problem [34]. Other approaches to block updating for Markov random field models are given in [333, 474].

Example 8.6 (Utah serviceberry distribution map) An important problem in ecology is the mapping of species distributions over a landscape [251, 495]. These maps have a variety of uses, ranging from local land-use planning aimed at minimizing human development impacts on rare species to worldwide climate modeling. Here we consider the distribution of a deciduous small tree or shrub called the Utah serviceberry *Amelanchier utahensis* in the state of Colorado [355].

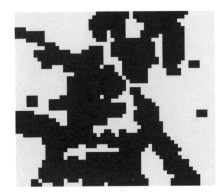

Fig. 8.5   Distribution of Utah serviceberry in western Colorado. The left panel is the true species distribution, and the right panel is the observed species distribution used in Example 8.6. Black pixels indicate presence.

We consider only the westernmost region of Colorado (west of approximately 104° W longitude), a region which includes the Rocky Mountains. We binned the presence–absence information into pixels that are approximately 8 km by 8 km. This grid consists of a lattice of 46 by 54 pixels, giving a total of $n = 2484$ pixels. The left panel in Figure 8.5 shows presence and absence, where each black pixel indicates that the species was observed in that pixel.

In typical applications of this model, the true image is not available. However, knowing the true image allows us to investigate various aspects of modeling binary spatially referenced data in what follows. Therefore, for purposes of illustration, we will use these pixelwise presence–absence data as a true image and consider estimating this truth from a degraded version of the image. A degraded image is shown in the right panel of Figure 8.5. We seek a map that reconstructs the true distribution of this species using this degraded image, which is treated as the observed data **y**. The observed data were generated by randomly selecting 30% of the pixels and switching their colors. Such errors might arise in satellite images or other error-prone approaches to species mapping.

Let $x_i = 1$ indicate that the species is truly present in pixel i. In a species mapping problem such as this one, such simple coding may not be completely appropriate. For example, a species may be present only in a portion of pixel i, or several sites may be included in one pixel, and thus we might consider modeling the proportion of sites in each pixel where the species was observed to be present. For simplicity, we assume that this application of Markov random fields is more akin to an image analysis problem where $x_i = 1$ indicates that the pixel is black.

We consider the simple likelihood function arising from the data density

$$f(\mathbf{y}|\mathbf{x}) \propto \exp\left\{\alpha \sum_{i=1}^{n} 1_{\{y_i = x_i\}}\right\} \tag{8.19}$$

for $x_i \in \{0, 1\}$. The parameter α can be specified as a user-selected constant or estimated by adopting a prior for it. We adopt the former approach here, setting $\alpha = 1$.

We assume the pairwise difference prior density for $\mathbf{X}$ given by

$$f(\mathbf{x}) \propto \exp\left\{\beta \sum_{i \sim j}^{n} 1_{\{x_i = x_j\}}\right\} \tag{8.20}$$

for $\mathbf{x} \in \mathcal{S} = \{0, 1\}^{46 \times 54}$. We consider a first-order neighborhood, so summation over $i \sim j$ in (8.20) indicates summation over the horizontally and vertically adjacent pixels of pixel i, for $i = 1, \ldots, n$. Equation (8.20) introduces the hyperparameter β, which can be assigned a hyperprior or specified as a constant. Usually β is restricted to be positive to encourage clustering of similar-colored pixels. Here we set $\beta = 0.8$. Sensitivity analysis to determine the effect of chosen values for α and β is recommended.

Assuming (8.19) and (8.20), the univariate conditional distribution for $X_i|\mathbf{x}_{-i}, \mathbf{y}$ is Bernoulli. Thus during the $(t+1)$th cycle of the Gibbs sampler, the ith pixel is set to 1 with probability

$$P\left(X_i^{(t+1)} = 1 \big| \mathbf{x}_{-i}^{(t)}, \mathbf{y}\right)$$
$$= \left(1 + \exp\left\{\alpha\left(1_{\{y_i=0\}} - 1_{\{y_i=1\}}\right) + \beta \sum_{i \sim j}\left(1_{\{x_j^{(t)}=0\}} - 1_{\{x_j^{(t)}=1\}}\right)\right\}\right)^{-1}$$
$$\tag{8.21}$$

for $i = 1, \ldots, n$. Recall that

$$\mathbf{x}_{-i}^{(t)} = \left(x_1^{(t+1)}, \ldots, x_{i-1}^{(t+1)}, x_{i+1}^{(t)}, \ldots, x_p^{(t)}\right),$$

so neighbors are always assigned their most recent values as soon as they become available within the Gibbs cycle.

Figure 8.6 gives the posterior mean probability of presence for the Utah service-berry in western Colorado as estimated using the Gibbs sampler described above. Figure 8.7 shows that the mean posterior estimates from the Gibbs sampler successfully discriminate between true presence and absence. Indeed, if pixels with posterior mean of 0.5 or larger are converted to black and pixels with posterior mean smaller than 0.5 are converted to white, then 86% of the pixels are labeled correctly. □

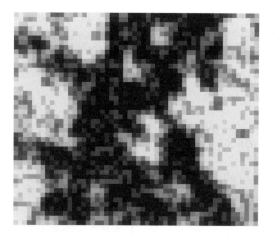

Fig. 8.6   Estimated posterior mean of **X** for the Gibbs sampler analysis in Example 8.6.

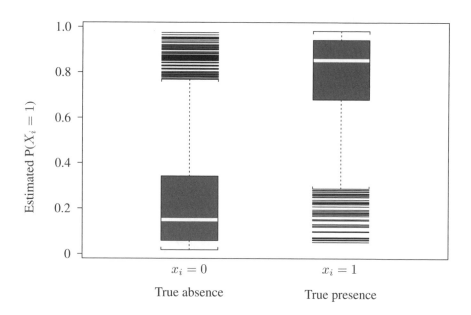

Fig. 8.7   Boxplots of posterior mean estimates of $P[X_i = 1]$ for Example 8.6. Averaging pixel-specific sample paths from the Gibbs sampler provides an estimate of $P[X_i = 1]$ for each i. The boxplots show these estimates split into two groups corresponding to pixels where the serviceberry was truly present and pixels where it wasn't.

The model used in Example 8.6 is elementary, ignoring many of the important issues that may arise in the analysis of such spatial lattice data. For example, when the pixels are created by binning spatially referenced data, it is unclear how to code the observed response for pixel i if the species was observed to be present in some portions of it and not in other portions.

A model that addresses this problem uses a latent binary spatial process over the region of interest [110, 192]. Let $\lambda(\mathbf{s})$ denote a binary process over the image region, where $\mathbf{s}$ denotes coordinates. Then the proportion of pixel i that is occupied by the species of interest is given by

$$p_i = \frac{1}{|A_i|} \int_{\mathbf{s} \text{ within pixel } i} 1_{\{\lambda(\mathbf{s})=1\}} \, d\mathbf{s}, \tag{8.22}$$

where $|A_i|$ denotes the area of pixel i. The $Y_i|x_i$ are assumed to be conditionally independent Bernoulli trials with probability of detecting presence given by p_i, so $P[Y_i = 1|X_i = 1] = p_i$. This formalization allows for direct modeling when pixels may contain a number of sites that were sampled. A more complex form of this model is described in [192]. We may also wish to incorporate covariate data to improve our estimates of species distributions. For example, the Bernoulli trials may be modeled as having parameters p_i for which

$$\log\left\{\frac{p_i}{1 - p_i}\right\} = \mathbf{w}_i^T \boldsymbol{\beta} + \gamma_i, \tag{8.23}$$

where $\mathbf{w}_i$ is a vector of covariates for the ith pixel, $\boldsymbol{\beta}$ is the vector of coefficients associated with the covariates, and γ_i is a spatially dependent random effect. These models are popular in the field of spatial epidemiology; see [38, 39, 351, 428].

8.4.2 Auxiliary variable methods for Markov random fields

Although convenient, the Gibbs sampler implemented as described in Section 8.4.1 can have poor convergence properties. In Section 8.1 we introduced the idea of incorporating auxiliary variables to improve convergence and mixing of Markov chain algorithms. For binary Markov random field models, the improvement can be profound.

One notable auxiliary variable technique is the Swendsen–Wang algorithm [151, 526]. As applied to binary Markov random fields, this approach creates a coarser version of an image by clustering neighboring pixels that are colored similarly. Each cluster is then updated with an appropriate Metropolis–Hastings step. This coarsening of the image allows for faster exploration of the parameter space in some applications [286].

In the Swendsen–Wang algorithm, clusters are created via the introduction of *bond variables*, U_{ij}, for each adjacency $i \sim j$ in the image. Clusters consist of bonded pixels. Adjacent like-colored pixels may or may not be bonded, depending on U_{ij}. Let $U_{ij} = 1$ indicate that pixels i and j are bonded, and $U_{ij} = 0$ indicate that they are not bonded. The bond variables U_{ij} are assumed to be conditionally independent given $\mathbf{X} = \mathbf{x}$. Let $\mathbf{U}$ denote the vector of all the U_{ij}.

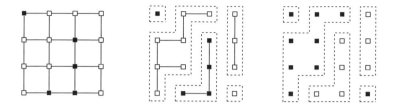

Fig. 8.8   An illustration of the Swendsen–Wang algorithm.

Loosely speaking, the Swendsen–Wang algorithm alternates between growing clusters and coloring them. Figure 8.8 shows one cycle of the algorithm applied to a 4×4 pixel image. The left panel in Figure 8.8 shows the current image and the set of all possible bonds for a 4×4 image. The middle panel shows the bonds that were generated at the start of the next iteration of the Swendsen–Wang algorithm. We will see below that bonds between like-colored pixels are generated with probability $1 - \exp\{-\beta\}$, so like-colored neighbors are not forced to be bonded. Connected sets of bonded pixels form clusters. We've drawn boxes around the five clusters in the middle panel of Figure 8.8. This shows the coarsening of the image allowed by the Swendsen–Wang algorithm. At the end of each iteration, the color of each cluster is updated: Clusters are randomly recolored in a way that depends upon the posterior distribution for the image. The updating produces the new image in the right panel in Figure 8.8. The observed data $\mathbf{y}$ are not shown here.

Formally, the Swendsen–Wang algorithm is a special case of a Gibbs sampler that alternates between updates to $\mathbf{X}|\mathbf{u}$ and $\mathbf{U}|\mathbf{x}$. It proceeds as follows:

1. Draw independent bond variables

$$U_{ij}^{(t+1)}\Big|\, x^{(t)} \sim \text{Unif}\left(0, \exp\left\{\beta 1_{\left\{x_i^{(t)}=x_j^{(t)}\right\}}\right\}\right)$$

   for all $i \sim j$ adjacencies. Note that $U_{ij}^{(t+1)}$ can exceed 1 only if $x_i^{(t)} = x_j^{(t)}$, and in this case $U_{ij}^{(t+1)} > 1$ with probability $1 - \exp\{-\beta\}$. When $U_{ij}^{(t+1)} > 1$, we declare the ith and jth pixels to be bonded for iteration $t+1$.

2. Sample $\mathbf{X}^{(t+1)}\big|\, \mathbf{u}^{(t+1)} \sim f\big(\cdot\,|u^{(t+1)}\big)$, where

$$f\left(\mathbf{x}|\mathbf{u}^{(t+1)}\right) \propto \exp\left\{\alpha \sum_{i=1}^{n} 1_{\{y_i=x_i\}}\right\}$$

$$\times \prod_{i\sim j} 1_{\left\{0 \leq u_{ij}^{(t+1)} \leq \exp\left\{\beta 1_{\{x_i=x_j\}}\right\}\right\}}. \tag{8.24}$$

   Note that (8.24) forces the color of each cluster to be updated as a single unit.

3. Increment t and return to step 1.

Gibbs

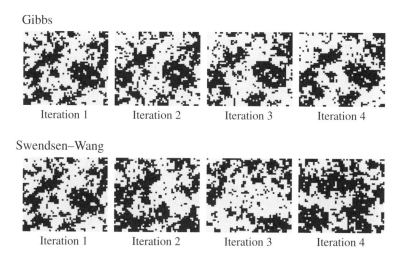

| Iteration 1 | Iteration 2 | Iteration 3 | Iteration 4 |

Swendsen–Wang

| Iteration 1 | Iteration 2 | Iteration 3 | Iteration 4 |

Fig. 8.9 A comparison between Gibbs sampling and the Swendsen–Wang algorithm simulating a Markov random field. Iteration 1 is the same for both algorithms. See Example 8.7 for details.

Thus for our simple model, pixel pairs with the same color are bonded with probability $1 - \exp\{-\beta\}$. The bond variables define clusters of pixels, with each cluster consisting of a set of pixels that are interlinked with at least one bond. Each cluster is updated independently with all pixels in the cluster taking on the same color. Updates in (8.24) are implemented by simulating from a Bernoulli distribution where the probability of coloring a cluster of pixels, C, black is

$$\frac{\exp\left\{\alpha \sum_{i \in C} 1_{\{y_i=1\}}\right\}}{\exp\left\{\alpha \sum_{i \in C} 1_{\{y_i=0\}}\right\} + \exp\left\{\alpha \sum_{i \in C} 1_{\{y_i=1\}}\right\}}. \tag{8.25}$$

The local dependence structure of the Markov random field is decoupled from the coloring decision given in (8.25), thereby potentially enabling faster mixing.

Example 8.7 (Utah serviceberry distributions, continued) To compare the performance of the Gibbs sampler and the Swendsen–Wang algorithm, we return to Example 8.6. For this problem the likelihood has a dominant influence on the posterior. Thus to highlight the differences between the algorithms, we set $\alpha = 0$ to understand what sort of mixing can be enabled by the Swendsen–Wang algorithm. In Figure 8.9, both algorithms were started in the same image in iteration 1, and the three subsequent iterations are shown. The Swendsen–Wang algorithm produces images that vary greatly over iterations, while the Gibbs sampler produces images that are quite similar. In the Swendsen–Wang iterations, large clusters of pixels switch colors abruptly, thereby providing faster mixing.

When the likelihood is included, there are fewer advantages to the Swendsen–Wang algorithm when analyzing the data from Example 8.6. For the chosen α and β, clusters grow large and change less frequently than in Figure 8.9. In this application, sequential images from a Swendsen–Wang algorithm look quite similar to those for a Gibbs sampler, and the differences between results produced by the Swendsen–Wang algorithm and the Gibbs sampler are small. $\qquad\square$

Exploiting a property called *decoupling*, the Swendsen–Wang algorithm grows clusters without regard to the likelihood, conditional on $\mathbf{X}^{(t)}$. The likelihood and image prior terms are separated in steps 1 and 2 of the algorithm. This feature is appealing because it can improve mixing rates in MCMC algorithms. Unless α and β are carefully chosen, however, decoupling may not be helpful. If clusters tend to grow large but change color very infrequently, the sample path will consist of rare drastic image changes. This constitutes poor mixing. Further, when the posterior distribution is highly multimodal, both the Gibbs sampler and the Swendsen–Wang algorithm can miss potential modes if the chain is not run long enough. A partial decoupling method has been proposed to address such problems, and offers some potential advantages for challenging imaging problems [285, 286].

8.4.3 Perfect sampling for Markov random fields

Implementing standard perfect sampling for a binary image problem would require monitoring sample paths that start from all possible images. This is clearly impossible even in a binary image problem of moderate size. In Section 8.3.1.1 we introduced the idea of stochastic monotonicity to cope with large state spaces. We can apply this strategy to implement perfect sampling for the Bayesian analysis of Markov random fields.

To exploit the stochastic monotonicity strategy, the states must be partially ordered, so $\mathbf{x} \leq \mathbf{y}$ if $x_i \leq y_i$ for $i = 1, \ldots, n$ and for $\mathbf{x}, \mathbf{y} \in \mathcal{S}$. In the binary image problem, such an ordering is straightforward. If $\mathcal{S} = \{0, 1\}^n$, define $\mathbf{x} \leq \mathbf{y}$ if $y_i = 1$ whenever $x_i = 1$ for all $i = 1, \ldots, n$. If the deterministic transition rule q maintains this partial ordering of states, then only the sample paths that start from all-black and all-white images need to be monitored for coalescence.

Example 8.8 (Sandwiching binary images) Figure 8.10 shows five iterations of a Gibbs sampler CFTP algorithm for a 4×4 binary image with order-preserving pixelwise updates. The sample path in the top row starts at iteration $\tau = -1000$, where the image is all black. In other words, $x_i^{(-1000)} = 1$ for $i = 1, \ldots, 16$. The sample path in the bottom row starts at all white. The path starting from all black is the upper bound and the path starting from all white is the lower bound used for sandwiching.

After some initial iterations, we examine the paths around $t = -400$. In the lower sample path, the circled pixel at iteration $t = -400$ changed from white to black at $t = -399$. Monotonicity requires that this pixel change to black in the upper path too. This requirement is implemented directly via the monotone update function q. Note, however, that changes from white to black in the upper image do not compel

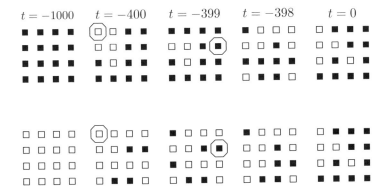

Fig. 8.10 A sequence of images from a perfect sampling algorithm for a binary image problem. See Example 8.8 for details.

the same change in the lower image; see, for example, the pixel to the right of the circled pixel.

Changes from black to white in the upper image compel the same change in the lower image. For example, the circled pixel in the upper sample path at $t = -399$ has changed from black to white at $t = -398$, thereby forcing the corresponding pixel in the lower sample path to change to white. A pixel change from black to white in the lower image does not necessitate a like change to the upper image.

Examination of the pixels in these sequences of images shows that pixelwise image ordering is maintained over the simulation. At iteration $t = 0$, the two sample paths have coalesced. Therefore a chain started at any image at $\tau = -1000$ must also have coalesced to the same image by iteration $t = 0$. The image shown at $t = 0$ is a realization from the stationary distribution of the chain.

$\square$

Example 8.9 (Utah serviceberry distribution, continued) The setup for the CFTP algorithm for the species distribution mapping problem closely follows the development of the Gibbs sampler described in Example 8.6. To update the ith pixel at iteration $t + 1$, generate $U^{(t+1)}$ from Unif$(0, 1)$. Then the update is given by

$$
\begin{aligned}
X_i^{(t+1)} &= q\left(\mathbf{x}_{-i}^{(t)}, U^{(t+1)}\right) \\
&= \begin{cases} 1 & \text{if } U^{(t+1)} < \mathrm{P}\left[X_i^{(t+1)} = 1 | \mathbf{x}_{-i}^{(t)}, \mathbf{y}\right], \\ 0 & \text{otherwise,} \end{cases}
\end{aligned} \tag{8.26}
$$

where $P\left[X_i^{(t+1)} = 1 \middle| \mathbf{x}_{-i}^{(t)}, \mathbf{y}\right]$ is given in (8.21). Such updates maintain the partial ordering of the state space. Therefore, the CFTP algorithm can be implemented by starting at two initial images: all black and all white. These images are monitored, and the CFTP algorithm proceeds until they coalesce by iteration $t = 0$. The CFTP algorithm has been implemented for similar binary image problems in [144, 145]. □

8.5 MARKOV CHAIN MAXIMUM LIKELIHOOD

We have presented Markov chain Monte Carlo in the context of Monte Carlo integration, with many Bayesian examples. However, MCMC techniques can also be useful for maximum likelihood estimation, particularly in exponential families [205, 429]. Consider data generated from an exponential family model $\mathbf{X} \sim f(\cdot|\boldsymbol{\theta})$ where

$$f(\mathbf{x}|\boldsymbol{\theta}) = c_1(\mathbf{x})c_2(\boldsymbol{\theta})\exp\{\boldsymbol{\theta}^T\mathbf{s}(\mathbf{x})\}. \tag{8.27}$$

Here $\boldsymbol{\theta} = (\theta_1, \ldots, \theta_p)$ and $\mathbf{s}(\mathbf{x}) = (s_1(\mathbf{x}), \ldots, s_p(\mathbf{x}))$ are vectors of canonical parameters and sufficient statistics, respectively. For many problems, $c_2(\boldsymbol{\theta})$ cannot be determined analytically, so the likelihood cannot be directly maximized.

Suppose that we generate $\mathbf{X}^{(1)}, \ldots, \mathbf{X}^{(n)}$ from a MCMC approach having $f(\cdot|\boldsymbol{\psi})$ as the stationary density, where $\boldsymbol{\psi}$ is any particular choice for $\boldsymbol{\theta}$ and $f(\cdot|\boldsymbol{\psi})$ is in the same exponential family as the data density. Then it is easy to show that

$$c_2(\boldsymbol{\theta})^{-1} = c_2(\boldsymbol{\psi})^{-1} \int \exp\{(\boldsymbol{\theta} - \boldsymbol{\psi})^T\mathbf{s}(\mathbf{x})\}f(\mathbf{x}|\boldsymbol{\psi})\,d\mathbf{x}. \tag{8.28}$$

Although the MCMC draws are dependent and not exactly from $f(\cdot|\boldsymbol{\psi})$,

$$\hat{k}(\boldsymbol{\theta}) = \frac{1}{n}\sum_{t=1}^{n}\exp\left\{(\boldsymbol{\theta} - \boldsymbol{\psi})^T\mathbf{s}(\mathbf{X}^{(t)})\right\} \rightarrow \frac{c_2(\boldsymbol{\psi})}{c_2(\boldsymbol{\theta})} \tag{8.29}$$

as $n \rightarrow \infty$ by the strong law of large numbers (1.46). Therefore, a Monte Carlo estimator of the log likelihood given data $\mathbf{x}$ is

$$\hat{l}(\boldsymbol{\theta}|\mathbf{x}) = \boldsymbol{\theta}^T\mathbf{s}(\mathbf{x}) - \log\hat{k}(\boldsymbol{\theta}), \tag{8.30}$$

up to an additive constant. The maximizer of $\hat{l}(\boldsymbol{\theta}|\mathbf{x})$ converges to the maximizer of the true log likelihood as $n \rightarrow \infty$. Therefore, we take the *Monte Carlo maximum likelihood estimate* of $\boldsymbol{\theta}$ to be the maximizer of (8.30), which we denote $\hat{\boldsymbol{\theta}}_{\boldsymbol{\psi}}$.

Hence we can approximate the MLE $\hat{\boldsymbol{\theta}}$ using simulations from $f(\cdot|\boldsymbol{\psi})$ generated via MCMC. Of course, the quality of $\hat{\boldsymbol{\theta}}_{\boldsymbol{\psi}}$ will depend greatly on $\boldsymbol{\psi}$. Analogously to importance sampling, $\boldsymbol{\psi} = \hat{\boldsymbol{\theta}}$ is best. In practice, however, we must choose one or more values wisely, perhaps through adaptation or empirical estimation [205].

Problems

8.1   One approach to Bayesian variable selection for linear regression models is described in Section 8.2.1 and further examined in Example 8.4. For a Bayesian analysis for the model in equation (8.10), we might adopt the normal–gamma conjugate class of priors $\beta|m_k \sim N(\alpha_{m_k}, \sigma^2 \mathbf{V}_{m_k})$ and $\nu\lambda/\sigma^2 \sim \chi^2_\nu$. Show that the marginal density of $\mathbf{Y}|m_k$ is given by

$$
\frac{\Gamma\left(\frac{\nu+n}{2}\right)(\nu\lambda)^{\nu/2}}{\pi^{n/2}\Gamma(\frac{\nu}{2})|\mathbf{I}+\mathbf{X}_{m_k}\mathbf{V}_{m_k}\mathbf{X}^T_{m_k}|^{1/2}}
$$
$$
\times\left[\lambda\nu + (\mathbf{Y}-\mathbf{X}_{m_k}\alpha_{m_k})^T\left(\mathbf{I}+\mathbf{X}_{m_k}\mathbf{V}_{m_k}\mathbf{X}^T_{m_k}\right)^{-1}(\mathbf{Y}-\mathbf{X}_{m_k}\alpha_{m_k})\right]^{-\frac{\nu+n}{2}},
$$

where $\mathbf{X}_{m_k}$ is the design matrix, α_{m_k} is the mean vector, and $\mathbf{V}_{m_k}$ is the covariance matrix for β_{m_k} for the model m_k.

8.2   Consider the CFTP algorithm described in Section 8.3.

(a) Construct an example with a finite state space to which both the Metropolis–Hastings algorithm and the CFTP algorithm can be applied to simulate from some multivariate stationary distribution f. For your example, define both the Metropolis–Hastings ratio (7.1) and the deterministic transition rule (8.13), and show how these quantities are related.

(b) Construct an example with a state space with two elements so that the CFTP algorithm can be applied to simulate from some stationary distribution f. Define two deterministic transition rules of the form in (8.13). One transition rule, q_1, should allow for coalescence in one iteration, and the other transition rule, q_2, should be defined so that coalescence is impossible. Which assumption of the CFTP algorithm is violated for q_2?

(c) Construct an example with a state space with two elements that shows why the CFTP algorithm cannot be started at $\tau = 0$ and run forward to coalescence. This should illustrate the argument mentioned on page 234.

8.3   Suppose we desire a draw from the marginal distribution of X that is determined by the assumptions that $\theta \sim \text{Beta}(\alpha, \beta)$ and $X|\theta \sim \text{Bin}(n, \theta)$ [81].

(a) Show that $\theta|x \sim \text{Beta}(\alpha + x, \beta + n - x)$.

(b) What is the marginal expected value of X?

(c) Implement a Gibbs sampler to obtain a joint sample of (θ, X), using $x^{(0)} = 0$, $\alpha = 10$, $\beta = 5$, and $n = 10$.

(d) Let $U^{(t+1)}$ and $V^{(t+1)}$ be independent Unif(0,1) random variables. Then the transition rule from $X^{(t)} = x^{(t)}$ to $X^{(t+1)}$ can be written as

$$X^{(t+1)} = q(x^{(t)}, U^{(t+1)}, V^{(t+1)})$$
$$= F_{\text{Bin}}^{-1}\left(V^{(t+1)}; n, F_{\text{Beta}}^{-1}\left(U^{(t+1)}; \alpha + x^{(t)}, \beta + n - x^{(t)}\right)\right), \quad (8.31)$$

where $F_d^{-1}(p; \mu_1, \mu_2)$ is the inverse cumulative distribution function of the distribution d with parameters μ_1 and μ_2, evaluated at p. Implement the CFTP algorithm from Section 8.3.1, using the transition rule given in (8.31), to draw a perfect sample for this problem. Decrement τ by one unit each time the sample paths do not coalesce by time 0. Run the function 100 times to produce 100 draws from the stationary distribution for $\alpha = 10$, $\beta = 5$, and $n = 10$. Make a histogram of the 100 starting times (the finishing times are all $t = 0$, by construction). Make a histogram of the 100 realizations of $X^{(0)}$. Discuss your results.

(e) Run the function from part (d) several times for $\alpha = 1.001$, $\beta = 1$, and $n = 10$. Pick a run where the chains were required to start at $\tau = -15$ or earlier. Graph the sample paths (from each of the 11 starting values) from their starting time to $t = 0$, connecting sequential states with lines. The goal is to observe the coalescence as in the right panel in Figure 8.3. Comment on any interesting features of your graph.

(f) Run the algorithm from part (d) several times. For each run, collect a perfect chain of length 20 (i.e., once you have achieved coalescence, don't stop the algorithm at $t = 0$, but continue the chain from $t = 0$ through $t = 19$). Pick one such chain having $x^{(0)} = 0$, and graph its sample path for $t = 0, \ldots, 19$. Next, run the Gibbs sampler from part (c) through $t = 19$ starting with $x^{(0)} = 0$. Superimpose the sample path of this chain on your existing graph, using a dashed line.

    i. Is $t = 2$ sufficient burn-in for the Gibbs sampler? Why or why not?

    ii. Of the two chains (CFTP conditional on $x^{(0)} = 0$ and Gibbs starting from $x^{(0)} = 0$), which should produce subsequent variates $X^{(t)}$ for $t = 1, 2, \ldots$ whose distribution more closely resembles the target? Why does this conditional CFTP chain fail to produce a perfect sample?

8.4 Consider the one-dimensional black-and-white image represented by a vector of zeros and ones. The data (observed image) are

$$10101111010000101000010110101001101$$

for the 35 pixels $\mathbf{y} = (y_1, \ldots, y_{35})$. Suppose the posterior density for the true image $\mathbf{x}$ is given by

$$f(\mathbf{x}|\mathbf{y}) \propto \exp\left\{\sum_{i=1}^{35} \alpha(x_i, y_i)\right\} \exp\left\{\sum_{i\sim j} \beta 1_{\{x_i = x_j\}}\right\},$$

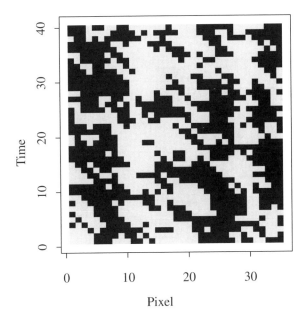

Fig. 8.11   Forty Gibbs sampler iterates for Problem 8.4, with $\beta = 1$.

where
$$\alpha(x_i, y_i) = \begin{cases} \log\{2/3\} & \text{if } x_i = y_i, \\ \log\{1/3\} & \text{if } x_i \neq y_i. \end{cases}$$

Consider the Swendsen–Wang algorithm for this problem where the bond variable is drawn according to $U_{ij}|\mathbf{x} \sim \text{Unif}\left(0, \exp\left\{\beta 1_{\{x_i = x_j\}}\right\}\right)$.

(a) Implement the Swendsen–Wang algorithm described above with $\beta = 1$. Create a chain of length 40, starting from the initial image $\mathbf{x}^{(0)}$ equal to the observed data.

Note that the entire sequence of images can be displayed in a two-dimensional graph as shown in Figure 8.11. This figure was created using a Gibbs sampler. Using your output from your implementation of the Swendsen–Wang algorithm, create a graph analogous to Figure 8.11 for your Swendsen–Wang iterations. Comment on the differences between your graph and Figure 8.11.

(b) Investigate the effect of β by repeating part (a) for $\beta = 0.5$ and $\beta = 2$. Comment on the differences between your graphs and the results in part (a).

(c) Investigate the effect of the starting value by repeating part (a) for three different starting values: first with $\mathbf{x}^{(0)} = (0, \ldots, 0)$, second with $\mathbf{x}^{(0)} = (1, \ldots, 1)$, and third with $x_i^{(0)} = 0$ for $i = 1, \ldots, 17$ and $x_i^{(0)} = 1$ for $i = 18, \ldots, 35$. Compare the results of these trials with the results from part (a).

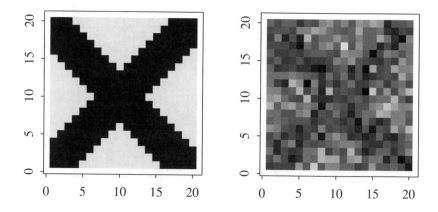

Fig. 8.12 Images for Problem 8.5. The left panel is the true image, and the right panel is an observed image.

(d) What would be a good way to produce a single best image to represent your estimate of the truth?

8.5 Data corresponding to the true image and observed images given in Figure 8.12 are available on the website for this book. The true image is a binary 20×20 pixel image with prior density given by

$$f\left(x_i | \mathbf{x}_{\delta_i}\right) = N\left(\bar{x}_{\delta_i}, \ \sigma^2/\nu_i\right)$$

for $i = 1, \ldots, n$, where ν_i is the number of neighbors in the neighborhood δ_i of x_i and $\bar{x}_{\delta_i}$ is the mean value of the neighbors of the ith pixel. This density promotes local dependence. The observed image is a gray scale degraded version of the true image with noise that can be modeled via a normal distribution. Suppose the likelihood is given by

$$f\left(y_i | x_i\right) = N\left(x_i, \ \sigma^2\right)$$

for $i = 1, \ldots, n$.

(a) Prove that univariate conditional posterior distribution used for Gibbs sampling for this problem is given by

$$f\left(x_i | \mathbf{x}_{-i}, \mathbf{y}\right) = N\left(\frac{1}{\nu_i + 1} y_i + \frac{\nu_i}{\nu_i + 1} \bar{x}_{\delta_i}, \ \frac{\sigma^2}{\nu_i + 1}\right).$$

(b) Starting with the initial image $\mathbf{x}^{(0)}$ equal to the observed data image, and using $\sigma = 5$ and a second-order neighborhood, use the Gibbs sampler (with no burn-in period or subsampling) to generate a collection of 100 images from the posterior. Do not count an image as new until an update has been proposed for each of its pixels (i.e., a full cycle). Record the data necessary to make the following plots: the data image, the first image sampled from the posterior distribution $(\mathbf{X}^{(1)})$, the last image sampled from the posterior distribution $(\mathbf{X}^{(100)})$, and the mean image.

Hints:

- Dealing with the edges is tricky because the neighborhood size varies. You may find it convenient to create a matrix with 42 rows and 42 columns consisting of the observed data surrounded on each of the four sides by a row or column of zeros. If you use this strategy, be sure that this margin area does not affect the analysis.

- Plot $\mathbf{X}^{(t)}$ at the end of each full cycle so that you can better understand the behavior of your chain.

(c) Fill in the rest of a 2×3 factorial design with runs analogous to (b), crossing the following factors and levels:

- Neighborhood structure chosen to be (i) first-order neighborhoods or (ii) second-order neighborhoods.

- Pixelwise error chosen to have variability given by (i) $\sigma = 2$, (ii) $\sigma = 5$, or (iii) $\sigma = 15$.

Provide plots and detailed comments comparing the results from each design point in this experiment.

(d) Repeat a run analogous to part (b) once more, but this time using the initial starting image $\mathbf{x}^{(0)}$ equal to 57.5 (the true posterior mean pixel color) everywhere, for $\sigma = 5$ and a first-order neighborhood. Discuss your results and their implications for the behavior of the chain.

9

Bootstrapping

9.1 THE BOOTSTRAP PRINCIPLE

Let $\theta = T(F)$ be an interesting feature of a distribution function, F, expressed as a functional of F. For example, $T(F) = \int z \, dF(z)$ is the mean of the distribution. Let $x_1, \ldots, x_n$ be data observed as a realization of the random variables $X_1, \ldots, X_n \sim$ i.i.d. F. In this chapter, we use $X \sim F$ to denote that X is distributed with density function f having corresponding cumulative distribution function F. Let $\mathcal{X} = \{X_1, \ldots, X_n\}$ denote the entire dataset.

If $\widehat{F}$ is the empirical distribution function of the observed data, then an estimate of θ is $\widehat{\theta} = T(\widehat{F})$. For example, when θ is a univariate population mean, the estimator is the sample mean, $\widehat{\theta} = \int z \, d\widehat{F}(z) = \sum_{i=1}^{n} X_i/n$.

Statistical inference questions are usually posed in terms of $T(\widehat{F})$ or some $R(\mathcal{X}, F)$, a statistical function of the data and their unknown distribution function F. For example, a general test statistic might be $R(\mathcal{X}, F) = \left[T(\widehat{F}) - T(F)\right]/S(\widehat{F})$, where S is a functional that estimates the standard deviation of $T(\widehat{F})$.

The distribution of the random variable $R(\mathcal{X}, F)$ may be intractable or altogether unknown. This distribution also may depend on the unknown distribution F. The *bootstrap* provides an approximation to the distribution of $R(\mathcal{X}, F)$ derived from the empirical distribution function of the observed data (itself an estimate of F) [152, 154]. Several thorough reviews of bootstrap methods have been published since its introduction [122, 157, 159].

Let $\mathcal{X}^*$ denote a *bootstrap sample* of *pseudo-data*, which we will call a *pseudo-dataset*. The elements of $\mathcal{X}^* = \{X_1^*, \ldots, X_n^*\}$ are i.i.d. random variables with distribution $\widehat{F}$. The bootstrap strategy is to examine the distribution of $R(\mathcal{X}^*, \widehat{F})$,

Table 9.1 Possible bootstrap pseudo-datasets from $\{1, 2, 6\}$ (ignoring order), the resulting values of $\widehat{\theta}^* = T(\widehat{F}^*)$, the probability of each outcome in the bootstrapping experiment $(P^*[\widehat{\theta}^*])$, and the observed relative frequency in 1000 bootstrap iterations.

$\mathcal{X}^*$	$\widehat{\theta}^*$	$P^*[\widehat{\theta}^*]$	Observed Frequency
1 1 1	3/3	1/27	36/1000
1 1 2	4/3	3/27	101/1000
1 2 2	5/3	3/27	123/1000
2 2 2	6/3	1/27	25/1000
1 1 6	8/3	3/27	104/1000
1 2 6	9/3	6/27	227/1000
2 2 6	10/3	3/27	131/1000
1 6 6	13/3	3/27	111/1000
2 6 6	14/3	3/27	102/1000
6 6 6	18/3	1/27	40/1000

i.e., the random variable formed by applying R to $\mathcal{X}^*$. In some special cases it is possible to derive or estimate the distribution of $R(\mathcal{X}^*, \widehat{F})$ through analytical means (see Example 9.1 and Problems 9.1 and 9.2). However, the usual approach is via simulation, as described in Section 9.2.1.

Example 9.1 (Simple illustration) Suppose $n = 3$ univariate data points, namely $\{x_1, x_2, x_3\} = \{1, 2, 6\}$, are observed as an i.i.d. sample from a distribution F that has mean θ. At each observed data value, $\widehat{F}$ places mass 1/3. Suppose the estimator to be bootstrapped is the sample mean $\widehat{\theta}$, which we may write as $T(\widehat{F})$ or $R(\mathcal{X}, F)$, where R does not depend on F in this case.

Let $\mathcal{X}^* = \{X_1^*, X_2^*, X_3^*\}$ consist of elements drawn i.i.d. from $\widehat{F}$. There are $3^3 = 27$ possible outcomes for $\mathcal{X}^*$. Let $\widehat{F}^*$ denote the empirical distribution function of such a sample, with corresponding estimate $\widehat{\theta}^* = T(\widehat{F}^*)$. Since $\widehat{\theta}^*$ does not depend on the ordering of the data, it has only 10 distinct possible outcomes. Table 9.1 lists these.

In Table 9.1, $P^*[\widehat{\theta}^*]$ represents the probability distribution for $\widehat{\theta}^*$ with respect to the bootstrap experiment of drawing $\mathcal{X}^*$ conditional on the original observations. To distinguish this distribution from F, we will use an asterisk when referring to such conditional probabilities or moments, as when writing $P^*[\widehat{\theta}^* \leq 6/3] = 8/27$.

The bootstrap principle is to equate the distributions of $R(\mathcal{X}, F)$ and $R(\mathcal{X}^*, \widehat{F})$. In this example, that means we base inference on the distribution of $\widehat{\theta}^*$. This distribution is summarized in the columns of Table 9.1 labeled $\widehat{\theta}^*$ and $P^*[\widehat{\theta}^*]$. So, for example, a simple bootstrap 25/27 (roughly 93%) confidence interval for θ is (4/3, 14/3) using quantiles of the distribution of $\widehat{\theta}^*$. The point estimate is still calculated from the observed data as $\widehat{\theta} = 9/3$. $\square$

9.2 BASIC METHODS

9.2.1 Nonparametric bootstrap

For realistic sample sizes the number of potential bootstrap pseudo-datasets is very large, so complete enumeration of the possibilities is not practical. Instead, B independent random bootstrap pseudo-datasets are drawn from the empirical distribution function of the observed data, namely $\widehat{F}$. Denote these $\mathcal{X}_i^* = \{\mathbf{X}_{i1}^*, \ldots, \mathbf{X}_{in}^*\}$ for $i = 1, \ldots, B$. The empirical distribution of the $R(\mathcal{X}_i^*, \widehat{F})$ for $i = 1, \ldots, B$ is used to approximate the distribution of $R(\mathcal{X}, F)$, allowing inference. The simulation error introduced by avoiding complete enumeration of all possible pseudo-datasets can be made arbitrarily small by increasing B. Using the bootstrap frees the analyst from making parametric assumptions to carry out inference, provides answers to problems for which analytic solutions are impossible, and can yield more accurate answers than given by routine application of standard parametric theory.

Example 9.2 (Simple illustration, continued) Continuing with the dataset in Example 9.1, recall that the empirical distribution function of the observed data, $\widehat{F}$, places mass 1/3 on 1, 2, and 6. A nonparametric bootstrap would generate $\mathcal{X}_i^*$ by sampling X_{i1}^*, X_{i2}^*, and X_{i3}^* i.i.d. from $\widehat{F}$. In other words, draw the X_{ij}^* with replacement from $\{1, 2, 6\}$ with equal probability. Each bootstrap pseudo-dataset yields a corresponding estimate $\widehat{\theta}^*$. Table 9.1 shows the observed relative frequencies of the possible values for $\widehat{\theta}^*$ resulting from $B = 1000$ randomly drawn pseudo-datasets, $\mathcal{X}_i^*$. These relative frequencies approximate $P^*[\widehat{\theta}^*]$. The bootstrap principle asserts that $P^*[\widehat{\theta}^*]$ in turn approximates the sampling distribution of $\widehat{\theta}$.

For this simple illustration, the space of all possible bootstrap pseudo-datasets can be completely enumerated and the $P^*[\widehat{\theta}^*]$ exactly derived. Therefore there is no need to resort to simulation. In realistic applications, however, the sample size is too large to enumerate the bootstrap sample space. Thus, in real applications (e.g., Section 9.2.3), only a small proportion of possible pseudo-datasets will ever be drawn, often yielding only a subset of possible values for the estimator. $\square$

A fundamental requirement of bootstrapping is that the data to be resampled must have originated as an i.i.d. sample. If the sample is not i.i.d., the distributional approximation of $R(\mathcal{X}, F)$ by $R(\mathcal{X}^*, \widehat{F})$ will not hold. Section 9.2.3 illustrates that the user must carefully consider the relationship between the stochastic mechanism generating the observed data and the bootstrap resampling strategy employed. Methods for bootstrapping with dependent data are described in [122, 159, 344, 352, 518].

9.2.2 Parametric bootstrap

The ordinary nonparametric bootstrap described above generates each pseudo-dataset $\mathcal{X}^*$ by drawing $\mathbf{X}_1^*, \ldots, \mathbf{X}_n^*$ i.i.d. from $\widehat{F}$. When the data are modeled to originate from a parametric distribution, so $\mathbf{X}_1, \ldots, \mathbf{X}_n \sim$ i.i.d. $F(\mathbf{x}, \boldsymbol{\theta})$, another estimate of F may be employed. Suppose that the observed data are used to estimate $\boldsymbol{\theta}$ by $\widehat{\boldsymbol{\theta}}$. Then each *parametric bootstrap* pseudo-dataset $\mathcal{X}^*$ can be generated by drawing $\mathbf{X}_1^*, \ldots, \mathbf{X}_n^* \sim$ i.i.d. $F(\mathbf{x}, \widehat{\boldsymbol{\theta}})$. When the model is known or believed to be a good representation of reality, the parametric bootstrap can be a powerful tool, allowing inference in otherwise intractable situations and producing confidence intervals that are much more accurate than those produced by standard asymptotic theory.

In some cases, however, the model upon which bootstrapping is based is almost an afterthought. For example, a deterministic biological population model might predict changes in population abundance over time, based on biological parameters and initial population size. Suppose animals are counted at various times using various methodologies. The observed counts are compared with the model predictions to find model parameter values that yield a good fit. One might fashion a second model asserting that the observations are, say, lognormally distributed with mean equal to the prediction from the biological model and with a predetermined coefficient of variation. This provides a convenient—if weakly justified—link between the parameters and the observations. A parametric bootstrap from the second model can then be applied by drawing bootstrap pseudo-datasets from this lognormal distribution. In this case, the sampling distribution of the observed data can hardly be viewed as arising from the lognormal model.

Such an analysis, relying on an ad hoc error model, should be a last resort. It is tempting to use a convenient but inappropriate model. If the model is not a good fit to the mechanism generating the data, the parametric bootstrap can lead inference badly astray. There are occasions, however, when few other inferential tools seem feasible.

9.2.3 Bootstrapping regression

Consider the ordinary multiple regression model, $Y_i = \mathbf{x}_i^T \boldsymbol{\beta} + \epsilon_i$, for $i = 1, \ldots, n$, where the ϵ_i are assumed to be i.i.d. mean zero random variables with constant variance. Here, $\mathbf{x}_i$ and $\boldsymbol{\beta}$ are p-vectors of predictors and parameters, respectively. A naive bootstrapping mistake would be to resample from the collection of response values a new pseudo-response, say Y_i^*, for each observed $\mathbf{x}_i$, thereby generating a new regression dataset. Then a bootstrap parameter vector estimate, $\widehat{\boldsymbol{\beta}}^*$, would be calculated from these pseudo-data. After repeating the sampling and estimation steps many times, the empirical distribution of $\widehat{\boldsymbol{\beta}}^*$ would be used for inference about $\boldsymbol{\beta}$. The mistake is that the $Y_i \mid \mathbf{x}_i$ are not i.i.d.—they have different conditional means. Therefore, it is not appropriate to generate bootstrap regression datasets in the manner described.

We must ask what variables are i.i.d. in order to determine a correct bootstrapping approach. The ϵ_i are i.i.d. given the model. Thus a more appropriate strategy would be to *bootstrap the residuals* as follows.

Table 9.2 Copper–nickel alloy data for illustrating methods of obtaining a bootstrap confidence interval for β_1/β_0.

x_i	0.01	0.48	0.71	0.95	1.19	0.01	0.48
y_i	127.6	124.0	110.8	103.9	101.5	130.1	122.0
x_i	1.44	0.71	1.96	0.01	1.44	1.96	
y_i	92.3	113.1	83.7	128.0	91.4	86.2	

Start by fitting the regression model to the observed data and obtaining the fitted responses $\widehat{y}_i$ and residuals $\hat{\epsilon}_i$. Sample a bootstrap set of residuals, $\{\hat{\epsilon}_1^*, \ldots, \hat{\epsilon}_n^*\}$, from the set of fitted residuals, completely at random with replacement. (Note that the $\hat{\epsilon}_i$ are actually not independent, though they are usually roughly so.) Create a bootstrap set of pseudo-responses, $Y_i^* = \widehat{y}_i + \hat{\epsilon}_i^*$, for $i = 1, \ldots, n$. Regress Y^* on $\mathbf{x}$ to obtain a bootstrap parameter estimate $\widehat{\boldsymbol{\beta}}^*$. Repeat this process many times to build an empirical distribution for $\widehat{\boldsymbol{\beta}}^*$ that can be used for inference.

This approach is most appropriate for designed experiments or other data where the $\mathbf{x}_i$ values are fixed in advance. The strategy of bootstrapping residuals is at the core of simple bootstrapping methods for other models such as AR(1), nonparametric regression, and generalized linear models.

Bootstrapping the residuals is reliant on the chosen model providing an appropriate fit to the observed data, and on the assumption that the residuals have constant variance. Without confidence that these conditions hold, a different bootstrapping method is probably more appropriate.

Suppose that the data arose from an observational study, where both response and predictors are measured from a collection of individuals selected at random. In this case, the data pairs $\mathbf{z}_i = (\mathbf{x}_i, y_i)$ can be viewed as values observed for i.i.d. random variables $\mathbf{Z}_i = (\mathbf{X}_i, Y_i)$ drawn from a joint response–predictor distribution. To bootstrap, sample $\mathbf{Z}_1^*, \ldots, \mathbf{Z}_n^*$ completely at random with replacement from the set of observed data pairs, $\{\mathbf{z}_1, \ldots, \mathbf{z}_n\}$. Apply the regression model to the resulting pseudo-dataset to obtain a bootstrap parameter estimate $\widehat{\boldsymbol{\beta}}^*$. Repeat these steps many times, then proceed to inference as in the first approach. This approach of *bootstrapping the cases* is sometimes called the *paired bootstrap*.

If you have doubts about the adequacy of the regression model, the constancy of the residual variance, or other regression assumptions, the paired bootstrap will be less sensitive to violations in the assumptions than will bootstrapping the residuals. The paired bootstrap sampling more directly mirrors the original data generation mechanism in cases where the predictors are not considered fixed.

There are other, more complex methods for bootstrapping regression problems [122, 156, 159, 288].

Example 9.3 (Copper–nickel alloy) Table 9.2 gives thirteen measurements of corrosion loss (y_i) in copper–nickel alloys, each with a specific iron content (x_i) [147].

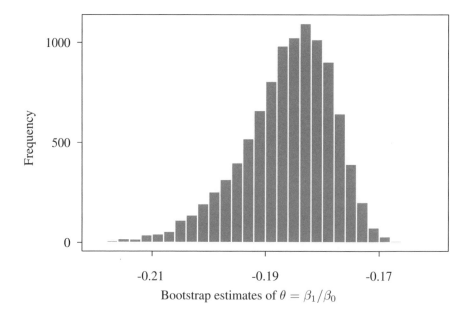

Fig. 9.1 Histogram of 10,000 bootstrap estimates of β_1/β_0 from the nonparametric paired bootstrap analysis with the copper–nickel alloy data.

Of interest is the change in corrosion loss in the alloys as the iron content increases, relative to the corrosion loss when there is no iron. Thus, consider the estimation of $\theta = \beta_1/\beta_0$ in a simple linear regression.

Letting $\mathbf{z}_i = (x_i, y_i)$ for $i = 1, \ldots, 13$, suppose we adopt the paired bootstrapping approach. The observed data yield the estimate $\widehat{\theta} = \widehat{\beta}_1/\widehat{\beta}_0 = -0.185$. For $i = 2, \ldots, 10,000$, we draw a bootstrap dataset $\{\mathbf{Z}_1^*, \ldots, \mathbf{Z}_{13}^*\}$ by resampling 13 data pairs from the set $\{\mathbf{z}_1, \ldots, \mathbf{z}_{13}\}$ completely at random with replacement. Figure 9.1 shows a histogram of the estimates obtained from regressions of the bootstrap datasets. The histogram summarizes the sampling variability of $\widehat{\theta}$ as an estimator of θ. □

9.2.4 Bootstrap bias correction

A particularly interesting choice for bootstrap analysis when $T(F) = \theta$ is the quantity $R(\mathcal{X}, F) = T(\widehat{F}) - T(F)$. This represents the bias of $T(\widehat{F}) = \widehat{\theta}$, and it has mean equal to $\mathrm{E}\{\widehat{\theta}\} - \theta$. The bootstrap estimate of the bias is $\mathrm{E}^*\{\widehat{\theta}^*\} - \widehat{\theta} = \overline{\theta}^* - \widehat{\theta}$, where $\overline{\theta}^* = \sum_{j=1}^{B} \widehat{\theta}_j^*/B$.

Example 9.4 (Copper–nickel alloy, continued) For the copper–nickel alloy regression data introduced in Example 9.3, the mean value of $\widehat{\theta}^* - \widehat{\theta}$ among the bootstrap pseudo-datasets is -0.00125, indicating a small degree of negative bias. Thus, the bias-corrected bootstrap estimate of β_1/β_0 is $-0.18507 - (-0.00125) = -0.184$.

The bias estimate can naturally be incorporated into confidence interval estimates via the nested bootstrap of Section 9.3.2.4. □

An improved bias estimate requires little additional effort. Let $\widehat{F}_j^*$ denote the empirical distribution of the jth bootstrap pseudo-dataset, and define $\overline{F}^*(\mathbf{x}) = \sum_{j=1}^{B} \widehat{F}_j^*(\mathbf{x})/B$. Then $\overline{\theta}^* - T(\overline{F}^*)$ is a better estimate of bias. Compare this strategy with bootstrap bagging, discussed in Section 9.5. Study of the merits of these and other bias corrections has shown that $\overline{\theta}^* - T(\overline{F}^*)$ has superior performance and convergence rate [159].

9.3 BOOTSTRAP INFERENCE

9.3.1 Percentile method

The simplest method for drawing inference about a univariate parameter θ using bootstrap simulations is to construct a confidence interval using the percentile method. This amounts to reading percentiles off the histogram of $\widehat{\theta}^*$ values produced by bootstrapping. It has been the approach implicit in the preceding discussion.

Example 9.5 (Copper–nickel alloy, continued) Returning to the estimation of $\theta = \beta_1/\beta_0$ for the copper–nickel alloy regression data introduced in Example 9.3, recall that Figure 9.1 summarizes the sampling variability of $\widehat{\theta}$ as an estimator of θ. A bootstrap $1 - \alpha$ confidence interval based on the percentile method could be constructed by finding the $((1 - \alpha/2)100)$th and $((\alpha/2)100)$th empirical percentiles in the histogram. The 95% confidence interval for β_1/β_0 using the simple bootstrap percentile method is $(-0.205, -0.174)$. □

Conducting a hypothesis test is closely related to estimating a confidence interval. The simplest approach for bootstrap hypothesis testing is to base the p-value on a bootstrap confidence interval. Specifically, consider a null hypothesis expressed in terms of a parameter whose estimate can be bootstrapped. If the $(1 - \alpha)100\%$ bootstrap confidence interval for the parameter does not cover the null value, then the null hypothesis is rejected with a p-value no greater than α. The confidence interval itself may be obtained from the percentile method or one of the superior approaches discussed later.

Using a bootstrap confidence interval to conduct a hypothesis test often sacrifices statistical power. Greater power is possible if the bootstrap simulations are carried out using a sampling distribution that is consistent with the null hypothesis [501]. Use of the null hypothesis sampling distribution of a test statistic is a fundamental tenet of hypothesis testing. Unfortunately, there will usually be many different bootstrap sampling strategies that are consistent with a given null hypotheses, with each imposing various extra restrictions in addition to those imposed by the null hypothesis. These different sampling models will yield hypothesis tests of different quality. More empirical and theoretical research is needed to develop bootstrap hypothesis testing methods, particularly methods for appropriate bootstrap sampling under the null hypothesis. Strategies for specific situations are illustrated by [122, 159].

Although simple, the percentile method is prone to bias and inaccurate coverage probabilities. The bootstrap works better when θ is essentially a location parameter. This is particularly important when using the percentile method. To ensure best bootstrap performance, the bootstrapped statistic should be approximately pivotal: Its distribution should not depend on the true value of θ. Since a variance-stabilizing transformation g naturally renders the variance of $g(\widehat{\theta})$ independent of θ, it frequently provides a good pivot. Section 9.3.2 discusses several approaches that rely on pivoting to improve bootstrap performance.

9.3.1.1 *Justification for the percentile method*   The percentile method can be justified by a consideration of a continuous, strictly increasing transformation ϕ and a distribution function H that is continuous and symmetric (i.e., $H(z) = 1 - H(-z)$), with the property that

$$P\left[h_{\alpha/2} \le \phi(\widehat{\theta}) - \phi(\theta) \le h_{1-\alpha/2}\right] = 1 - \alpha, \tag{9.1}$$

where h_α is the α quantile of H. For instance, if ϕ is a normalizing, variance-stabilizing transformation, then H is the standard normal distribution. In principle, when F is continuous we may transform any random variable $X \sim F$ to have any desired distribution G, using the monotone transformation $G^{-1}(F(X))$. There is therefore nothing special about normalization. In fact, the remarkable aspect of the percentile approach is that we are never actually required to specify explicitly ϕ or H.

Applying the bootstrap principle to (9.1), we have

$$\begin{aligned}
1 - \alpha &\approx P^*\left[h_{\alpha/2} \le \phi(\widehat{\theta}^*) - \phi(\widehat{\theta}) \le h_{1-\alpha/2}\right] \\
&= P^*\left[h_{\alpha/2} + \phi(\widehat{\theta}) \le \phi(\widehat{\theta}^*) \le h_{1-\alpha/2} + \phi(\widehat{\theta})\right] \\
&= P^*\left[\phi^{-1}\left(h_{\alpha/2} + \phi(\widehat{\theta})\right) \le \widehat{\theta}^* \le \phi^{-1}\left(h_{1-\alpha/2} + \phi(\widehat{\theta})\right)\right]. \tag{9.2}
\end{aligned}$$

Since the bootstrap distribution is observed by us, its percentiles are known quantities (aside from Monte Carlo variability which can be made arbitrarily small by increasing the number of pseudo-datasets, B). Let ξ_α denote the α quantile of the empirical distribution of $\widehat{\theta}^*$. Then $\phi^{-1}\left(h_{\alpha/2} + \phi(\widehat{\theta})\right) \approx \xi_{\alpha/2}$ and $\phi^{-1}\left(h_{1-\alpha/2} + \phi(\widehat{\theta})\right) \approx \xi_{1-\alpha/2}$.

Next, the original probability statement (9.1) from which we hope to build a confidence interval is reexpressed to isolate θ. Exploiting symmetry by noting that $h_{\alpha/2} = -h_{1-\alpha/2}$ yields

$$P\left[\phi^{-1}\left(h_{\alpha/2} + \phi(\widehat{\theta})\right) \le \theta \le \phi^{-1}\left(h_{1-\alpha/2} + \phi(\widehat{\theta})\right)\right] = 1 - \alpha. \tag{9.3}$$

The confidence limits in this equation happily coincide with the limits in (9.2), for which we already have estimates $\xi_{\alpha/2}$ and $\xi_{1-\alpha/2}$. Hence we may simply read off the quantiles for $\widehat{\theta}^*$ from the bootstrap distribution and use these as the confidence limits for θ. Note that the percentile method is *transformation-respecting* in the sense that the percentile method confidence interval for a monotone transformation of θ is the same as the transformation of the interval for θ itself [159].

9.3.2 Pivoting

9.3.2.1 *Accelerated bias-corrected percentile method, BC$_a$*   The *accelerated bias-corrected percentile method*, BC_a, usually offers substantial improvement over the simple percentile approach [142, 155]. For the basic percentile method to work well, it is necessary for the transformed estimator $\phi(\widehat{\theta})$ to be unbiased with variance that does not depend on θ. BC_a augments ϕ with two parameters to better meet these conditions, thereby ensuring an approximate pivot.

Assume there exists a monotonically increasing function ϕ and constants a and b such that

$$U = \frac{\phi(\widehat{\theta}) - \phi(\theta)}{1 + a\phi(\theta)} + b \tag{9.4}$$

has a $N(0, 1)$ distribution, with $1 + a\phi(\theta) > 0$. Note that if $a = b = 0$, this transformation leads us back to the simple percentile method.

By the bootstrap principle,

$$U^* = \frac{\phi(\widehat{\theta}^*) - \phi(\widehat{\theta})}{1 + a\phi(\widehat{\theta})} + b \tag{9.5}$$

has approximately a standard normal distribution. For any quantile of a standard normal distribution, say z_α,

$$\alpha \approx P^*[U^* \le z_\alpha]$$
$$= P^* \left[\widehat{\theta}^* \le \phi^{-1} \left(\phi(\widehat{\theta}) + (z_\alpha - b)[1 + a\phi(\widehat{\theta})] \right) \right]. \tag{9.6}$$

However, the α quantile of the empirical distribution of $\widehat{\theta}^*$, denoted ξ_α, is observable from the bootstrap distribution. Therefore

$$\phi^{-1} \left(\phi(\widehat{\theta}) + (z_\alpha - b)[1 + a\phi(\widehat{\theta})] \right) \approx \xi_\alpha. \tag{9.7}$$

In order to use (9.7), consider U itself:

$$1 - \alpha = P[U > z_\alpha]$$
$$= P \left[\theta < \phi^{-1} \left(\phi(\widehat{\theta}) + u(a, b, \alpha)[1 + a\phi(\widehat{\theta})] \right) \right] \tag{9.8}$$

where $u(a, b, \alpha) = \frac{b - z_\alpha}{1 - a(b - z_\alpha)}$. Notice the similarity between (9.6) and (9.8). If we can find a β such that $u(a, b, \alpha) = z_\beta - b$, then the bootstrap principle can be applied to conclude that $\theta < \xi_\beta$ will approximate a $1 - \alpha$ upper confidence limit. A straightforward inversion of this requirement yields

$$\beta = \Phi\left(b + u(a, b, \alpha)\right) = \Phi\left(b + \frac{b + z_{1-\alpha}}{1 - a(b + z_{1-\alpha})}\right), \qquad (9.9)$$

where Φ is the standard normal cumulative distribution function and the last equality follows from symmetry. Thus, if we knew a suitable a and b, then to find a $1 - \alpha$ upper confidence limit we would first compute β and then find the βth quantile of the empirical distribution of $\widehat{\theta}^*$, namely ξ_β, using the bootstrap pseudo-datasets.

For a two-sided $1 - \alpha$ confidence interval, this approach yields $P\left[\xi_{\beta_1} \leq \theta \leq \xi_{\beta_2}\right] \approx 1 - \alpha$, where

$$\beta_1 = \Phi\left(b + \frac{b + z_{\alpha/2}}{1 - a(b + z_{\alpha/2})}\right), \qquad (9.10)$$

$$\beta_2 = \Phi\left(b + \frac{b + z_{1-\alpha/2}}{1 - a(b + z_{1-\alpha/2})}\right), \qquad (9.11)$$

and ξ_{β_1} and ξ_{β_2} are the corresponding quantiles from the bootstrapped values of $\widehat{\theta}^*$.

As with the percentile method, the beauty of the above justification for BC_a is that explicit specification of the transformation ϕ is not necessary. Further, since the BC_a approach merely corrects the percentile levels determining the confidence interval endpoints to be read from the bootstrap distribution, it shares the transformation-respecting property of the simple percentile method.

The remaining question is the choice of a and b. The simplest nonparametric choices are $b = \Phi^{-1}\left(\widehat{F}^*(\widehat{\theta})\right)$ and

$$a = \frac{1}{6} \sum_{i=1}^{n} \psi_i^3 \bigg/ \left(\sum_{i=1}^{n} \psi_i^2\right)^{3/2}, \qquad (9.12)$$

where

$$\psi_i = \widehat{\theta}_{(\cdot)} - \widehat{\theta}_{(-i)} \qquad (9.13)$$

with $\widehat{\theta}_{(-i)}$ denoting the statistic computed omitting the ith observation, and $\widehat{\theta}_{(\cdot)} = \frac{1}{n} \sum_{i=1}^{n} \widehat{\theta}_{(-i)}$. A related alternative is to let

$$\psi_i = \lim_{\epsilon \to 0} \frac{1}{\epsilon} \left(T\left((1 - \epsilon)\widehat{F} + \epsilon\delta_i\right) - T(\widehat{F})\right), \qquad (9.14)$$

where δ_i represents the distribution function that steps from zero to one at the observation x_i (i.e., unit mass on x_i). The ψ_i in (9.14) can be approximated using finite differences. Shao and Tu discuss the motivation for these quantities and give additional alternatives for a and b [501].

Example 9.6 (Copper–nickel alloy, continued) Continuing the copper–nickel alloy regression problem introduced in Example 9.3, we have $a = 0.0486$ (using (9.13)) and $b = 0.00802$. The adjusted quantiles are therefore $\beta_1 = 0.038$ and $\beta_2 = 0.986$. The main effect of BC_a was therefore to shift the confidence interval slightly to the right. The resulting interval is $(-0.203, -0.172)$. □

9.3.2.2 *The bootstrap* t

Another approximate pivot that is quite easy to implement is provided by the *bootstrap* t method, also called the *studentized bootstrap* [153, 159]. Suppose $\theta = T(F)$ is to be estimated using $\hat{\theta} = T(\hat{F})$, with $V(\hat{F})$ estimating the variance of $\hat{\theta}$. Then it is reasonable to hope that $R(\mathcal{X}, F) = \frac{T(\hat{F}) - T(F)}{\sqrt{V(\hat{F})}}$ will be roughly pivotal. Bootstrapping $R(\mathcal{X}, F)$ yields a collection of $R(\mathcal{X}^*, \hat{F})$.

Denote by $\hat{G}$ and $\hat{G}^*$ the distributions of $R(\mathcal{X}, F)$ and $R(\mathcal{X}^*, \hat{F})$, respectively. By definition, a $1 - \alpha$ confidence interval for θ is obtained from the relation

$$P\left[\xi_{\alpha/2}(\hat{G}) \leq R(\mathcal{X}, F) \leq \xi_{1-\alpha/2}(\hat{G})\right]$$
$$= P\left[\hat{\theta} - \sqrt{V(\hat{F})}\xi_{1-\alpha/2}(\hat{G}) \leq \theta \leq \hat{\theta} - \sqrt{V(\hat{F})}\xi_{\alpha/2}(\hat{G})\right]$$
$$= 1 - \alpha,$$

where $\xi_\alpha(\hat{G})$ is the α quantile of $\hat{G}$. These quantiles are unknown, because F (and hence $\hat{G}$) is unknown. However, the bootstrap principle implies that the distributions $\hat{G}$ and $\hat{G}^*$ should be roughly equal, so $\xi_\alpha(\hat{G}) \approx \xi_\alpha(\hat{G}^*)$ for any α. Thus, a bootstrap confidence interval can be constructed as

$$\left(T(\hat{F}) - \sqrt{V(\hat{F})}\xi_{1-\alpha/2}(\hat{G}^*), \quad T(\hat{F}) - \sqrt{V(\hat{F})}\xi_{\alpha/2}(\hat{G}^*)\right), \quad (9.15)$$

where the percentiles of $\hat{G}^*$ are taken from the histogram of bootstrap values of $R(\mathcal{X}^*, \hat{F})$. Since these are percentiles in the tail of the distribution, at least several thousand bootstrap pseudo-datasets are needed for adequate precision.

Example 9.7 (Copper–nickel alloy, continued) Continuing the copper–nickel alloy regression problem introduced in Example 9.3, an estimator $V(\hat{F})$ of the variance of $\hat{\beta}_1/\hat{\beta}_0$ based on the delta method is

$$\left(\frac{\hat{\beta}_1}{\hat{\beta}_0}\right)^2 \left(\frac{\widehat{\text{var}}\{\hat{\beta}_1\}}{\hat{\beta}_1^2} + \frac{\widehat{\text{var}}\{\hat{\beta}_0\}}{\hat{\beta}_0^2} - \frac{2\,\widehat{\text{cov}}\{\hat{\beta}_0, \hat{\beta}_1\}}{\hat{\beta}_0\hat{\beta}_1}\right), \quad (9.16)$$

where the estimated variances and covariance can be obtained from basic regression results. Carrying out the bootstrap t method then yields the histogram shown in Figure 9.2, which corresponds to $\hat{G}^*$. The 0.025 and 0.975 quantiles of $\hat{G}^*$ are -5.77 and 4.44, respectively, and $\sqrt{V(\hat{F})} = 0.00273$. Thus, the 95% bootstrap t confidence interval is $(-0.197, -0.169)$. □

This method requires an estimator of the variance of $\hat{\theta}$, namely $V(\hat{F})$. If no such estimator is readily available, a delta method approximation may be used [122].

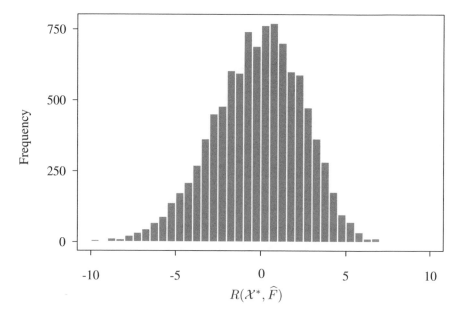

Fig. 9.2   Histogram of 10,000 values of $R(\mathcal{X}^*, \widehat{F})$ from a studentized bootstrap analysis with the copper–nickel alloy data.

The bootstrap t usually provides confidence interval coverage rates that closely approximate the nominal confidence level. Confidence intervals from the bootstrap t are most reliable when $T(\widehat{F})$ is approximately a location statistic in the sense that a constant shift in all the data values will induce the same shift in $T(\widehat{F})$. They are also more reliable for variance-stabilized estimators. Coverage rates for bootstrap t intervals can be sensitive to the presence of outliers in the dataset, and should be used with caution in such cases. The bootstrap t does not share the transformation-respecting property of the percentile-based methods above.

9.3.2.3 *Empirical variance stabilization*
A variance-stabilizing transformation is often the basis for a good pivot. A variance-stabilizing transformation of the estimator $\widehat{\theta}$ is one for which the sampling variance of the transformed estimator does not depend on θ. Usually a variance-stabilizing transformation of the statistic to be bootstrapped is unknown, but it can be estimated using the bootstrap.

Start by drawing B_1 bootstrap pseudo-datasets $\mathcal{X}_j^*$ for $j = 1, \ldots, B_1$. Calculate $\widehat{\theta}_j^*$ for each bootstrap pseudo-dataset, and let $\widehat{F}_j^*$ be the empirical distribution function of the jth bootstrap pseudo-dataset.

For each $\mathcal{X}_j^*$, next draw B_2 bootstrap pseudo-datasets $\mathcal{X}_{j1}^{**}, \ldots, \mathcal{X}_{jB_2}^{**}$ from $\widehat{F}_j^*$. For each j, let $\widehat{\theta}_{jk}^{**}$ denote the parameter estimate from the kth subsample, and let $\overline{\theta}_j^{**}$

be the mean of the $\widehat{\theta}_{jk}^{**}$. Then

$$\widehat{s}(\widehat{\theta}_j^*) = \frac{1}{B_2 - 1} \sum_{k=1}^{B_2} \left(\widehat{\theta}_{jk}^{**} - \overline{\theta}_j^{**} \right)^2 \tag{9.17}$$

is an estimate of the standard error of $\widehat{\theta}$ given $\theta = \widehat{\theta}_j^*$.

Fit a curve to the set of points $\{\widehat{\theta}_j^*, \widehat{s}(\widehat{\theta}_j^*)\}$, $j = 1, \ldots, B_1$. For a flexible, non-parametric fit, Chapter 11 reviews many suitable approaches. The fitted curve is an estimate of the relationship between the standard error of the estimator and θ. We seek a variance-stabilizing transformation to neutralize this relationship.

Recall that if Z is a random variable with mean θ and standard deviation $s(\theta)$, then Taylor series expansion (i.e., the delta method) yields $\text{var}\{g(Z)\} \approx g'(\theta)^2 s^2(\theta)$. For the variance of $g(Z)$ to be constant, we require

$$g(z) = \int_a^z \frac{1}{s(u)} du, \tag{9.18}$$

where a is any convenient constant for which $\frac{1}{s(u)}$ is continuous on $[a, z]$. Therefore, an approximately variance-stabilizing transformation for $\widehat{\theta}$ may be obtained from our bootstrap data by applying (9.18) to the fitted curve from the previous step. The integral can be approximated using a numerical integration technique from Chapter 5. Let $\widehat{g}(\theta)$ denote the result.

Now that an approximate variance-stabilizing transformation has been estimated, the bootstrap t may be carried out on the transformed scale. Draw B_3 new bootstrap pseudo-datasets from $\widehat{F}$, and apply the bootstrap t method to find an interval for $\widehat{g}(\theta)$. Note, however, that the standard error of $\widehat{g}(\theta)$ is roughly constant, so we can use $R(\mathcal{X}^*, \widehat{F}) = \widehat{g}(\widehat{\theta}^*) - \widehat{g}(\widehat{\theta})$ for computing the bootstrap t confidence interval. Finally, the endpoints of the resulting interval can be converted back to the scale of θ by applying the transformation $\widehat{g}^{-1}$.

The strategy of drawing iterated bootstrap pseudo-datasets from each original pseudo-dataset sample can be quite useful in a variety of settings. In fact, it is the basis for the confidence interval approach described below.

9.3.2.4 *Nested bootstrap and prepivoting* Another style of pivoting is provided by the nested bootstrap [23, 24]. This approach is sometimes also called the iterated or double bootstrap.

Consider constructing a confidence interval or conducting a hypothesis test based on a test statistic $R_0(\mathcal{X}, F)$, given observed data values $\mathbf{x}_1, \ldots, \mathbf{x}_n$ from the model $\mathbf{X}_1, \ldots, \mathbf{X}_n \sim$ i.i.d. F. Let $F_0(q, F) = P[R_0(\mathcal{X}, F) \leq q]$. The notation for F_0 makes explicit the dependence of the distribution of R_0 on the distribution of the data used in R_0. Then a two-sided confidence interval can be fashioned after the statement

$$P\left[F_0^{-1}(\alpha/2, F) \leq R_0(\mathcal{X}, F) \leq F_0^{-1}(1 - \alpha/2, F) \right] = 1 - \alpha, \tag{9.19}$$

and a hypothesis test based on the statement

$$P\left[R_0(\mathcal{X}, F) \leq F_0^{-1}(1 - \alpha, F) \right] = 1 - \alpha. \tag{9.20}$$

Of course, these probability statements depend on the quantiles of F_0, which are not known. In the estimation case, F is not known; for hypothesis testing, the null value for F is hypothesized. In both cases, the distribution of R_0 is not known. We can use the bootstrap to approximate F_0 and its quantiles.

The bootstrap begins by drawing B bootstrap pseudo-datasets, $\mathcal{X}_1^*, \ldots, \mathcal{X}_B^*$, from the empirical distribution $\widehat{F}$. For the jth bootstrap pseudo-dataset, compute the statistic $R_0(\mathcal{X}_j^*, \widehat{F})$. Let $\widehat{F}_0(q, \widehat{F}) = \frac{1}{B} \sum_{j=1}^{B} 1_{\{R_0(\mathcal{X}_j^*, \widehat{F}) \leq q\}}$, where $1_{\{A\}} = 1$ if A is true and zero otherwise. Thus $\widehat{F}_0$ estimates $P^* \left[R_0(\mathcal{X}^*, \widehat{F}) \leq q \right]$, which itself estimates $P[R_0(\mathcal{X}, F) \leq q] = F_0(q, F)$ according to the bootstrap principle. Thus, the upper limit of the confidence interval would be estimated as $\widehat{F}_0^{-1}(1 - \alpha/2, \widehat{F})$, or we would reject the null hypothesis if $R_0(\{\mathbf{x}_1, \ldots, \mathbf{x}_n\}, F) > \widehat{F}_0^{-1}(1 - \alpha, \widehat{F})$. This is the ordinary nonparametric bootstrap.

Note, however, that a confidence interval constructed in this manner will not have coverage probability exactly equal to $1 - \alpha$, because $\widehat{F}_0$ is only a bootstrap approximation to the distribution of $R_0(\mathcal{X}, F)$. Similarly, the size of the hypothesis test is $P \left[R_0(\mathcal{X}, F) > \widehat{F}_0^{-1}(1 - \alpha, \widehat{F}) \right] \neq \alpha$, since $F_0(q, F) \neq F_0(q, \widehat{F})$.

Not knowing the distribution F_0 also deprives us of a perfect pivot: the random variable $R_1(\mathcal{X}, F) = F_0(R_0(\mathcal{X}, F), F)$ has a standard uniform distribution independent of F. The bootstrap principle asserts the approximation of F_0 by $\widehat{F}_0$, and hence the approximation of $R_1(\mathcal{X}, F)$ by $\widehat{R}_1(\mathcal{X}, F) = \widehat{F}_0(R_0(\mathcal{X}, F), \widehat{F})$. This allows bootstrap inference based on a comparison of $\widehat{R}_1(\mathcal{X}, F)$ to the quantiles of a uniform distribution. For hypothesis testing, this amounts to accepting or rejecting the null hypothesis based on the bootstrap p-value.

However, we could instead proceed by acknowledging that $\widehat{R}_1(\mathcal{X}, F) \sim F_1$, for some nonuniform distribution F_1. Let $F_1(q, F) = P[\widehat{R}_1(\mathcal{X}, F) \leq q]$. Then the correct size test rejects the null hypothesis if $\widehat{R}_1 > F_1^{-1}(1 - \alpha, F)$. A confidence interval with the correct coverage probability is motivated by the statement $P \left[F_1^{-1}(\alpha/2, F) \leq \widehat{R}_1(\mathcal{X}, F) \leq F_1^{-1}(1 - \alpha/2, F) \right] = 1 - \alpha$. As before, F_1 is unknown but may be approximated using the bootstrap. Now the randomness $\widehat{R}_1$ comes from two sources: (1) the observed data were random observations from F, and (2) given the observed data (and hence $\widehat{F}$), $\widehat{R}_1$ is calculated from random resamplings from $\widehat{F}$. To capture both sources of randomness, we use the following nested bootstrapping algorithm:

1. Generate bootstrap pseudo-datasets $\mathcal{X}_1^*, \ldots, \mathcal{X}_{B_0}^*$, each as an i.i.d. random sample from the original data with replacement.

2. Compute $R_0(\mathcal{X}_j^*, \widehat{F})$ for $j = 1, \ldots, B_0$.

3. For $j = 1, \ldots, B_0$:

   (a) Let $\widehat{F}_j$ denote the empirical distribution function of $\mathcal{X}_j^*$. Draw B_1 iterated bootstrap pseudo-datasets, $\mathcal{X}_{j1}^{**}, \ldots, \mathcal{X}_{jB_1}^{**}$, each as an i.i.d. random sample from $\widehat{F}_j$.

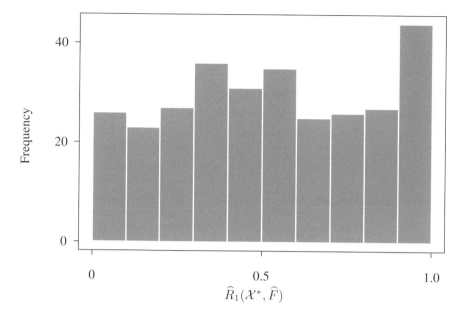

Fig. 9.3 Histogram of 300 values of $\widehat{R}_1(\mathcal{X}^*, \widehat{F})$ from a nested bootstrap analysis with the copper–nickel alloy data.

(b) Compute the $R_0(\mathcal{X}^{**}_{jk}, \widehat{F}_j)$ for $k = 1, \ldots, B_1$.

(c) Compute

$$\widehat{R}_1(\mathcal{X}^*_j, \widehat{F}) = \widehat{F}_0(R_0(\mathcal{X}^*_j, \widehat{F}), \widehat{F})$$

$$= \frac{1}{B_1} \sum_{k=1}^{B_1} 1_{\left\{R_0(\mathcal{X}^{**}_{jk}, \widehat{F}_j) \leq R_0(\mathcal{X}^*_j, \widehat{F})\right\}}. \tag{9.21}$$

4. Denote the empirical distribution function of the resulting sample of $\widehat{R}_1(\mathcal{X}^*_j, \widehat{F})$ as $\widehat{F}_1$.

5. Use $\widehat{R}_1(\{\mathbf{x}_1, \ldots, \mathbf{x}_n\}, F)$ and the quantiles of $\widehat{F}_1$ to construct the confidence interval or hypothesis test.

Steps 1 and 2 capture the first source of randomness by applying the bootstrap principle to approximate F by $\widehat{F}$. Step 3 captures the second source of randomness introduced in $\widehat{R}_1$ when R_0 is bootstrapped conditional on $\widehat{F}$.

Example 9.8 (Copper–nickel alloy, continued) Returning to the regression problem introduced in Example 9.3, let $R_0(\{\mathbf{x}_1, \ldots, \mathbf{x}_{13}\}, F) = \frac{\widehat{\beta}_1}{\widehat{\beta}_0} - \frac{\beta_1}{\beta_0}$. Figure 9.3 shows a histogram of $\widehat{R}_1$ values obtained by the nested bootstrap with $B_0 = B_1 = $

300. This distribution shows that $\widehat{F}_1$ differs noticeably from uniform. Indeed, the nested bootstrap gave 0.025 and 0.975 quantiles of $\widehat{R}_1$ as 0.0316 and 0.990, respectively. The 3.16% and 99.0% percentiles of $R_0(\mathcal{X}^*, \widehat{F})$ are then found and used to construct a confidence interval for β_1/β_0, namely $(-0.197, -0.168)$. $\square$

With its nested looping, the double bootstrap can be much slower than other pivoting methods: in this case nine times more bootstrap draws were used than for the preceding methods. There are reweighting methods such as bootstrap recycling that allow reuse of the initial sample, thereby reducing the computational burden [121, 413].

9.3.3 Hypothesis testing

The preceding discussion about bootstrap construction of confidence intervals is relevant for hypothesis testing, too. A hypothesized parameter value outside a $(1 - \alpha)100\%$ confidence interval can be rejected at a p-value of α. Hall and Wilson offer some additional advice to improve the statistical power and accuracy of bootstrap hypothesis tests [263].

First, bootstrap resampling should be done in a manner that reflects the null hypothesis. To understand what this means, consider a null hypothesis about a univariate parameter θ with null value θ_0. Let the test statistic be $R(\mathcal{X}, F) = \hat{\theta} - \theta_0$. The null hypothesis would be rejected in favor of a simple two-sided alternative when $|\hat{\theta} - \theta_0|$ is large compared to a reference distribution. To generate the reference distribution, it may be tempting to resample values $R(\mathcal{X}^*, F) = \hat{\theta}^* - \theta_0$ via the bootstrap. However, if the null is false, this statistic does not have the correct reference distribution. If θ_0 is far from the true value of θ, then $|\hat{\theta} - \theta_0|$ will not seem unusually large compared to the bootstrap distribution of $|\hat{\theta}^* - \theta_0|$. A better approach is to use values of $R(\mathcal{X}^*, \widehat{F}) = \hat{\theta}^* - \hat{\theta}$ to generate a bootstrap estimate of the null distribution of $R(\mathcal{X}, F)$. When θ_0 is far from the true value of θ, the bootstrap values of $|\hat{\theta}^* - \hat{\theta}|$ will seem quite small compared to $|\hat{\theta} - \theta_0|$. Thus, comparing $\hat{\theta} - \theta_0$ to the bootstrap distribution of $\hat{\theta}^* - \hat{\theta}$ yields greater statistical power.

Second, we should reemphasize the importance of using a suitable pivot. It is often best to base the hypothesis test on the bootstrap distribution of $(\hat{\theta}^* - \hat{\theta})/\hat{\sigma}^*$, where $\hat{\sigma}^*$ is the value of a good estimator of the standard deviation of $\hat{\theta}^*$ computed from a bootstrap pseudo-dataset. This pivoting approach is usually superior to basing the test on the bootstrap distribution of $(\hat{\theta}^* - \hat{\theta})/\hat{\sigma}$, $(\hat{\theta}^* - \theta_0)/\hat{\sigma}$, $\hat{\theta}^* - \hat{\theta}$, or $\hat{\theta}^* - \theta_0$, where $\hat{\sigma}$ estimates the standard deviation of $\hat{\theta}$ from the original dataset.

9.4 REDUCING MONTE CARLO ERROR

9.4.1 Balanced bootstrap

Consider a bootstrap bias correction of the sample mean. The bias correction should equal zero, because $\overline{\mathbf{X}}$ is unbiased for the true mean μ. Now, $R(\mathcal{X}, F) = \overline{\mathbf{X}} - \mu$, and the corresponding bootstrap values are $R(\mathcal{X}_j^*, \widehat{F}) = \overline{\mathbf{X}}_j^* - \overline{\mathbf{X}}$ for $j = 1, \ldots, B$. Even though $\overline{\mathbf{X}}$ is unbiased, random selection of pseudo-datasets is unlikely to produce a set of $R(\mathcal{X}^*, \widehat{F})$ values whose mean is exactly zero. The ordinary bootstrap exhibits unnecessary Monte Carlo variation in this case.

However, if each data value occurs in the combined collection of bootstrap pseudo-datasets with the same relative frequency as it does in the observed data, then the bootstrap bias estimate $\frac{1}{B}\sum_{j=1}^{B} R(\mathcal{X}_j^*, \widehat{F})$ must equal zero. By balancing the bootstrap data in this manner, a source of potential Monte Carlo error is eliminated.

The simplest way to achieve this balance is to concatenate B copies of the observed data values, randomly permute this series, and then read off B blocks of size n sequentially. The jth block becomes $\mathcal{X}_j^*$. This is the *balanced bootstrap*—sometimes called the *permutation bootstrap* [123]. More elaborate balancing algorithms have been proposed [223], but other methods of reducing Monte Carlo error may be easier or more effective [159].

9.4.2 Antithetic bootstrap

For a sample of univariate data, $x_1, \ldots, x_n$, denote the ordered data as $x_{(1)}, \ldots, x_{(n)}$, where $x_{(i)}$ is the value of the ith order statistic (i.e., the ith smallest data value). Let $\pi(i) = n - i + 1$ be a permutation operator that reverses the order statistics. Then for each bootstrap dataset $\mathcal{X}^* = \{X_1^*, \ldots, X_n^*\}$, let $\mathcal{X}^{**} = \{X_1^{**}, \ldots, X_n^{**}\}$ denote the dataset obtained by substituting $X_{(\pi(i))}$ for every instance of $X_{(i)}$ in $\mathcal{X}^*$. Thus, for example, if $\mathcal{X}^*$ has an unrepresentative predominance of the larger observed data values, then the smaller observed data values will predominate in $\mathcal{X}^{**}$.

Using this strategy, each bootstrap draw provides two estimators: $R(\mathcal{X}^*, \widehat{F})$ and $R(\mathcal{X}^{**}, \widehat{F})$. These two estimators will often be negatively correlated. For example, if R is a statistic that is monotone in the sample mean, then negative correlation is likely [349].

Let $R_a(\mathcal{X}^*, \widehat{F}) = \frac{1}{2}\left(R(\mathcal{X}^*, \widehat{F}) + R(\mathcal{X}^{**}, \widehat{F})\right)$. Then R_a has the desirable property that it estimates the quantity of interest with variance

$$\text{var}\{R_a(\mathcal{X}^*, \widehat{F})\} = \frac{1}{4}\left(\text{var}\{R(\mathcal{X}^*, \widehat{F})\} + \text{var}\{R(\mathcal{X}^{**}, \widehat{F})\}\right.$$
$$\left. + 2\,\text{cov}\{R(\mathcal{X}^*, \widehat{F}), R(\mathcal{X}^{**}, \widehat{F})\}\right)$$
$$\leq \text{var}\{R(\mathcal{X}^*, \widehat{F})\} \tag{9.22}$$

if the covariance is negative.

There are clever ways of establishing orderings of multivariate data, too, to permit an antithetic bootstrap strategy [257].

9.5 OTHER USES OF THE BOOTSTRAP

By viewing $\mathcal{X}^*$ as a random sample from a distribution $\widehat{F}$ with known parameter $\widehat{\theta}$, the bootstrap principle can be seen as a tool used to approximate the likelihood function itself. *Bootstrap likelihood* [121] is one such approach, which has connections to empirical likelihood methods. By ascribing random weights to likelihood components, a *Bayesian bootstrap* can be developed [469]. A generalization of this is the *weighted likelihood bootstrap*, which is a powerful tool for approximating likelihood surfaces in some difficult circumstances [414].

The bootstrap is generally used for assessing the statistical accuracy and precision of an estimator. *Bootstrap aggregating*, or *bagging*, uses the bootstrap to improve the estimator itself [57]. Suppose that the bootstrapped quantity, $R(\mathcal{X}, F)$, depends on F only though $\boldsymbol{\theta}$. Thus, the bootstrap values of $R(\mathcal{X}, \boldsymbol{\theta})$ are $R(\mathcal{X}^*, \widehat{\boldsymbol{\theta}})$. In some cases, $\boldsymbol{\theta}$ is the result of a model-fitting exercise where the form of the model is uncertain or unstable. For example, classification and regression trees, neural nets, and linear regression subset selection are all based on models whose form may change substantially with small changes to the data.

In these cases, a dominant source of variability in predictions or estimates may be the model form. Bagging consists of replacing $\widehat{\boldsymbol{\theta}}$ with $\overline{\boldsymbol{\theta}}^* = \frac{1}{B} \sum_{j=1}^{B} \widehat{\boldsymbol{\theta}}_j^*$, where $\widehat{\boldsymbol{\theta}}_j^*$ is the parameter estimate arising from the jth bootstrap pseudo-dataset. Since each bootstrap pseudo-dataset represents a perturbed version of the original data, the models fit to each pseudo-dataset can vary substantially in form. Thus $\overline{\boldsymbol{\theta}}^*$ provides a sort of model averaging that can reduce mean squared estimation error in cases where perturbing the data can cause significant changes to $\widehat{\boldsymbol{\theta}}$. A review of the model-averaging philosophy is provided in [289].

A related strategy is the *bootstrap umbrella of model parameters*, or *bumping* approach [536]. For problems suitable for bagging, notice that the bagged average is not always an estimate from a model of the same class as those being fit to the data. For example, the average of classification trees is not a classification tree. Bumping avoids this problem.

Suppose that $h(\boldsymbol{\theta}, \mathcal{X})$ is some objective function relevant to estimation in the sense that high values of h correspond to $\boldsymbol{\theta}$ that are very consistent with $\mathcal{X}$. For example, h could be the log likelihood function. The bumping strategy generates bootstrap pseudo-values via $\widehat{\boldsymbol{\theta}}_j^* = \arg\max_{\boldsymbol{\theta}} h(\boldsymbol{\theta}, \mathcal{X}_j^*)$. The original dataset is included among the bootstrap pseudo-datasets, and the final estimate of $\boldsymbol{\theta}$ is taken to be the $\widehat{\boldsymbol{\theta}}_j$ that maximizes $h(\boldsymbol{\theta}, \mathcal{X})$ with respect to $\boldsymbol{\theta}$. Thus, bumping is really a method for searching through a space of models (or parameterizations thereof) for a model that yields a good estimator.

9.6 DEGREE OF BOOTSTRAP APPROXIMATION

All the bootstrap methods described in this chapter rely on the principle that the bootstrap distribution should approximate the true distribution for a quantity of interest. Standard parametric approaches such as a t-test and the comparison of a log likelihood ratio to a χ^2 distribution also rely on distributional approximation.

We have already discussed one situation where the bootstrap approximation fails: for dependent data. For example, consider the bootstrapping the sample mean as an estimator of the mean of some stationary time series having autocorrelation ρ_ℓ at lag

ℓ. Then the variance of $\overline{X}$ is $\frac{\sigma^2}{n}\left[1 + 2\sum_{\ell=1}^{n-1}\left(1 - \frac{\ell}{n}\right)\rho_\ell\right]$, whereas the variance of the ordinary bootstrap mean $\overline{X}^*$ is about σ^2/n. These two quantities can often differ substantially. The reason that the ordinary bootstrap fails in this case is that resampling destroys the dependence structure in the original data. Bootstrap approaches for dependent data are discussed in [122, 159, 344, 352, 518]. The bootstrap also fails for estimation of extremes. For example, bootstrapping the sample maximum can be catastrophic; see [122] for details. Finally, the bootstrap can fail for heavy-tailed distributions. In these circumstances, the bootstrap samples outliers too frequently.

There is a substantial asymptotic theory for the consistency and rate of convergence of bootstrap methods, thereby formalizing the degree of approximation provided by the bootstrap. These results are mostly beyond the scope of this book, but we mention a few main ideas below.

The Glivenko–Cantelli theorem [401] gives $P\left[\sup_\mathbf{x}\left|\widehat{F}(\mathbf{x}) - F(\mathbf{x})\right| \to 0\right] = 1$ as $n \to \infty$. Thus it is intuitively clear that $T(\widehat{F}) \to T(F)$ when T is in some sense smooth. A general, formal theory is provided by [497]. In this chapter we have focused on plug-in estimators, where $T(\widehat{F})$ estimates $T(F)$. Applications of the bootstrap usually work even if the estimator is of a slightly different form, say $T_n(\widehat{F})$, where $T_n \to T$ at a suitable rate. For example, the usual sample variance statistic differs from the plug-in estimator by a factor of $n/(n-1)$. Nevertheless, it is a sensible estimator to bootstrap, since it is unbiased and consistent for the population variance.

More generally, consider a suitable a space of distribution functions containing F, and let $\mathcal{N}_F$ denote a neighborhood of F into which $\widehat{F}$ eventually falls with probability 1. If the distribution of a standardized $R(\mathcal{X}, G)$ is uniformly weakly convergent when the elements of $\mathcal{X}$ are drawn from any $G \in \mathcal{N}_F$, and if the mapping from G to the corresponding limiting distribution of R is continuous, then the bootstrap is consistent [122]. Consistency means that $P^*\left[\left|P[R(\mathcal{X}^*, \widehat{F}) \leq q] - P[R(\mathcal{X}, F) \leq q]\right| > \epsilon\right] \to 0$ for any ϵ and any q as $n \to \infty$.

Edgeworth expansions can be used to assess the rate of convergence [258]. When $R(\mathcal{X}, F)$ is constructed to be asymptotically pivotal, the usual rate of convergence for the bootstrap is given by $P^*[R(\mathcal{X}^*, \widehat{F}) \leq q] - P[R(\mathcal{X}, F) \leq q] = \mathcal{O}_p(n^{-1})$. Without pivoting, the rate is typically only $\mathcal{O}_p(n^{-1/2})$. In other words, coverage probabilities for one-sided confidence intervals are $\mathcal{O}(n^{-1/2})$ accurate for the basic, unpivoted percentile method, but $\mathcal{O}(n^{-1})$ accurate for BC_a and the bootstrap t. For two-sided intervals, all three methods are $\mathcal{O}(n^{-1})$ accurate. The improvement offered by the nested bootstrap depends on the accuracy of the original interval and the type of interval: For a two-sided, equi-tailed interval, nested bootstrapping can reduce the coverage probability error from $\mathcal{O}(n^{-1})$ to $\mathcal{O}(n^{-2})$. Most common inferential problems are covered by these convergence results, including estimation of smooth functions of sample moments and solutions to smooth maximum likelihood problems. Accessible discussions of the increases in convergence rates provided by BC_a, the nested bootstrap, and other bootstrap improvements are given in [122, 159]. More advanced theoretical discussion is also available [41, 258, 501].

9.7 PERMUTATION TESTS

There are other important techniques aside from the bootstrap that share the underlying strategy of basing inference on "experiments" within the observed dataset. Perhaps the most important of these is the classic permutation test that dates back to the era of Fisher [170] and Pitman [433, 434]. Comprehensive introductions to this field include [150, 240, 372]. The basic approach is most easily explained through a hypothetical example.

Example 9.9 (Comparison of independent group means)   Consider a medical experiment where rats are randomly assigned to treatment and control groups. The outcome X_i is then measured for the ith rat. Under the null hypothesis, the outcome does not depend on whether a rat was labeled as treatment or control. Under the alternative hypothesis, outcomes tend to be larger for rats labeled as treatment.

A test statistic T measures the difference in outcomes observed for the two groups. For example, T might be the difference between group mean outcomes, having value t_1 for the observed dataset.

Under the null hypothesis, the individual labels "treatment" and "control" are meaningless, because they have no influence on the outcome. Since they are meaningless, the labels could be randomly shuffled among rats without changing the joint null distribution of the data. Shuffling the labels creates a new dataset: Although one instance of each original outcome is still seen, the outcomes appear to have arisen from a different assignment of treatment and control. Each of these permuted datasets is as likely to have been observed as the actual dataset, since the experiment relied on random assignment.

Let t_2 be the value of the test statistic computed from the dataset with this first permutation of labels. Suppose all M possible permutations (or a large number of randomly chosen permutations) of the labels are examined, thereby obtaining $t_2, \ldots, t_M$.

Under the null hypothesis, $t_2, \ldots, t_M$ were generated from the same distribution that yielded t_1. Therefore, t_1 can be compared to the empirical quantiles of $t_1, \ldots, t_M$ to test a hypothesis or construct confidence limits.   □

To pose this strategy more formally, suppose that we observe a value t for a test statistic T having density f under the null hypothesis. Suppose large values of T indicate that the null hypothesis is false. Monte Carlo hypothesis testing proceeds by generating a random sample of $M - 1$ values of T drawn from f. If the observed value t is the kth largest among all M values, then the null hypothesis is rejected at a significance level of k/M. If the distribution of the test statistic is highly discrete, then ties found when ranking t can be dealt with naturally by reporting a range of p-values. Barnard [20] posed the approach in this manner; interesting extensions are offered in [32, 33].

There are a variety of approaches for sampling from the null distribution of the test statistic. The permutation approach described in Example 9.9 works because "treatment" and "control" are meaningless labels assigned completely at random and independent of outcome, under the null hypothesis. This simple permutation approach can be broadened for application to a variety of more complicated situations. In all cases, the permutation test relies heavily on the condition of exchangeability. The data are exchangeable if the probability of any particular joint outcome is the same regardless of the order in which the observations are considered.

There are two advantages to the permutation test over the bootstrap. First, if the basis for permuting the data is random assignment, then the resulting p-value is exact (if all possible permutations are considered). For such experiments, the approach is usually called a *randomization test*. In contrast, standard parametric approaches and the bootstrap are founded on asymptotic theory that is relevant for large sample sizes. Second, permutation tests are often more powerful than their bootstrap counterparts. However, the permutation test is a specialized tool for making a comparison between distributions, whereas a bootstrap tests hypotheses about parameters, thereby requiring less stringent assumptions and providing greater flexibility. The bootstrap can also provide a reliable confidence interval and standard error, beyond the mere p-value given by the permutation test. The standard deviation observed in the permutation distribution is not a reliable standard error estimate. Additional guidance on choosing between a permutation test and a bootstrap is offered in [159, 240, 241].

Problems

9.1   Let $X_1, \ldots, X_n \sim$ i.i.d. Bernoulli(θ). Define $R(\mathcal{X}, F) = \overline{X} - \theta$ and $R^* = R(\mathcal{X}^*, \widehat{F})$, where $\mathcal{X}^*$ is a bootstrap pseudo-dataset and $\widehat{F}$ is the empirical distribution of the data. Derive the exact $E^*\{R^*\}$ and $\text{var}^*\{R^*\}$ analytically.

9.2   Suppose $\theta = g(\mu)$, where g is a smooth function and μ is the mean of the distribution from which the data arise. Consider bootstrapping $R(\mathcal{X}, F) = g(\overline{X}) - g(\mu)$.

(a)   Show that $E^*\{\overline{X}^*\} = \overline{x}$ and $\text{var}^*\{\overline{X}^*\} = \widehat{\mu}_2/n$, where $\widehat{\mu}_k = \sum_{i=1}^{n} (x_i - \overline{x})^k$.

(b)   Use Taylor series to show that

$$E^*\{R(\mathcal{X}^*, \widehat{F})\} = \frac{g''(\overline{x})\widehat{\mu}_2}{2n} + \frac{g'''(\overline{x})\widehat{\mu}_3}{6n^2} + \cdots$$

and

$$\text{var}^*\{R(\mathcal{X}^*, \widehat{F})\} = \frac{g'(\overline{x})^2\widehat{\mu}_2}{n} - \frac{g''(\overline{x})^2}{4n^2}\left(\widehat{\mu}_2 - \frac{\widehat{\mu}_4}{n}\right) + \cdots .$$

9.3   Justify the choice of b for BC$_a$ given in Section 9.3.2.1.

9.4   Table 9.3 contains 40 annual counts of the numbers of recruits and spawners in a salmon population. The units are thousands of fish. Recruits are fish that enter the catchable population. Spawners are fish that are laying eggs. Spawners die after laying eggs.

Table 9.3 Forty years of fishery data: numbers of recruits (R) and spawners (S).

R	S	R	S	R	S	R	S
68	56	222	351	311	412	244	265
77	62	205	282	166	176	222	301
299	445	233	310	248	313	195	234
220	279	228	266	161	162	203	229
142	138	188	256	226	368	210	270
287	428	132	144	67	54	275	478
276	319	285	447	201	214	286	419
115	102	188	186	267	429	275	490
64	51	224	389	121	115	304	430
206	289	121	113	301	407	214	235

The classic Beverton–Holt model for the relationship between spawners and recruits is

$$R = \frac{1}{\beta_1 + \beta_2/S}, \qquad \beta_1 \geq 0 \text{ and } \beta_2 \geq 0,$$

where R and S are the numbers of recruits and spawners, respectively [40]. This model may be fit using linear regression with the transformed variables $1/R$ and $1/S$.

Consider the problem of maintaining a sustainable fishery. The total population abundance will only stabilize if $R = S$. The total population will decline if fewer recruits are produced than the number of spawners who died producing them. If too many recruits are produced, the population will also decline eventually because there is not enough food for them all. Thus, only some middle level of recruits can be sustained indefinitely in a stable population. This stable population level is the point where the 45° line intersects the curve relating R and S.

(a) Fit the Beverton–Holt model, and find a point estimate for the stable population level where $R = S$. Use the bootstrap to obtain a corresponding 95% confidence interval and a standard error for your estimate, from two methods: bootstrapping the residuals and bootstrapping the cases. Histogram each bootstrap distribution, and comment on the differences in your results.

(b) Provide a bias-corrected estimate and a corresponding standard error for the corrected estimator.

(c) Use the nested bootstrap with prepivoting to find a 95% confidence interval for the stabilization point.

9.5 Patients with advanced terminal cancer of the stomach and breast were treated with ascorbate in an attempt to prolong survival [76]. Table 9.4 shows survival times (days). Work with the data on the log scale.

Table 9.4 Survival times (days) for patients with two types of terminal cancer.

Stomach	25	42	45	46	51	103	124
	146	340	396	412	876	1112	
Breast	24	40	719	727	791	1166	1235
	1581	1804	3460	3808			

(a) Use the bootstrap t and BC_a methods to construct 95% confidence intervals for the mean survival time of each group.

(b) Use a permutation test to examine the hypothesis that there is no difference in mean survival times between groups.

(c) Having computed a reliable confidence interval in (a), let us explore some possible missteps. Construct a 95% confidence interval for the mean breast cancer survival time by applying the simple bootstrap to the logged data and exponentiating the resulting interval boundaries. Construct another such confidence interval by applying the simple bootstrap to the data on the original scale. Compare with (a).

9.6 Use the problem of estimating the mean of a standard Cauchy distribution to illustrate how the bootstrap can fail for heavy-tailed distributions. Use the problem of estimating θ for the $\mathrm{Unif}(0, \theta)$ distribution to illustrate how the bootstrap can fail for extremes.

9.7 Perform a simulation experiment on an artificial problem of your design, to compare the accuracy of coverage probabilities and the widths of 95% bootstrap confidence intervals constructed using the percentile method, the BC_a method, and the bootstrap t. Discuss your findings.

10

Nonparametric Density Estimation

This chapter concerns estimation of a density function f using observations of random variables $\mathbf{X}_1, \ldots, \mathbf{X}_n$ sampled independently from f. Initially, this chapter focuses on univariate density estimation. Section 10.4 introduces some methods for estimating a multivariate density function.

In exploratory data analysis, an estimate of the density function can be used to assess multimodality, skew, tail behavior, and so forth. For inference, density estimates are useful for decision making, classification, and summarizing Bayesian posteriors. Density estimation is also a useful presentational tool, since it provides a simple, attractive summary of a distribution. Finally, density estimation can serve as a tool in other computational methods, including some simulation algorithms and Markov chain Monte Carlo approaches. Comprehensive monographs on density estimation include [492, 507, 553].

The parametric solution to a density estimation problem begins by assuming a parametric model, $\mathbf{X}_1, \ldots, \mathbf{X}_n \sim$ i.i.d. $f_{\mathbf{X}|\theta}$, where θ is a very low-dimensional parameter vector. Parameter estimates $\widehat{\theta}$ are found using some estimation paradigm, such as maximum likelihood, Bayesian, or method-of-moments estimation. The resulting density estimate at $\mathbf{x}$ is $f_{\mathbf{X}|\theta}(\mathbf{x}|\widehat{\theta})$. The danger with this approach lies at the start: Relying on an incorrect model $f_{\mathbf{X}|\theta}$ can lead to serious inferential errors, regardless of the estimation strategy used to generate $\widehat{\theta}$ from the chosen model.

In this chapter, we focus on nonparametric approaches to density estimation which assume very little about the form of f. These approaches use predominantly local information to estimate f at a point $\mathbf{x}$. More precise viewpoints on what makes an estimator nonparametric are offered in [492, 532].

One familiar nonparametric density estimator is a histogram, which is a piecewise constant density estimator. Histograms are produced automatically by most software packages and are used so routinely that one rarely considers their underlying complexity. Optimal choice of the locations, widths, and number of bins is based on sophisticated theoretical analysis.

Another elementary density estimator can be motivated by considering how density functions assign probability to intervals. If we observe a data point $X_i = x_i$, we assume that f assigns some density not only at x_i but also in a region around x_i, if f is smooth enough. Therefore, to estimate f from $X_1, \ldots, X_n \sim$ i.i.d. f, it makes sense to accumulate localized probability density contributions in regions around each X_i.

Specifically, to estimate the density at a point x, suppose we consider a region of width $dx = 2h$, centered at x, where h is some fixed number. Then the proportion of the observations that fall in the interval $\gamma = [x - h, x + h]$ gives an indication of the density at x. More precisely, we may take $\widehat{f(x)dx} = \frac{1}{n} \sum_{i=1}^{n} 1_{\{|x-X_i|<h\}}$, that is,

$$\hat{f}(x) = \frac{1}{2hn} \sum_{i=1}^{n} 1_{\{|x-X_i|<h\}}, \tag{10.1}$$

where $1_{\{A\}} = 1$ if A is true, and 0 otherwise.

Let $N_\gamma(h, n) = \sum_{i=1}^{n} 1_{\{|x-X_i|<h\}}$ denote the number of sample points falling in the interval γ. Then N_γ is a $\text{Bin}(n, p(\gamma))$ random variable, where $p(\gamma) = \int_{x-h}^{x+h} f(t)\, dt$. Thus $\text{E}\{N_\gamma/n\} = p(\gamma)$ and $\text{var}\{N_\gamma/n\} = p(\gamma)(1-p(\gamma))/n$. Clearly nh must increase as N_γ increases in order for (10.1) to provide a reasonable estimator, yet we can be more precise about separate requirements for n and h. The proportion of points falling in the interval γ estimates the probability assigned to γ by f. In order to approximate the density at x, we must shrink γ by letting $h \to 0$. Then $\lim_{h\to 0} \text{E}\{\hat{f}(x)\} = \lim_{h\to 0} \frac{p(\gamma)}{2h} = f(x)$. Simultaneously, however, we want to increase the total sample size since $\text{var}\{\hat{f}(x)\} \to 0$ as $n \to \infty$. Thus, a fundamental requirement for the pointwise consistency of the estimator $\hat{f}$ in (10.1) is that $nh \to \infty$ and $h \to 0$ as $n \to \infty$. We will see later that these requirements hold in far greater generality.

10.1 MEASURES OF PERFORMANCE

To better understand what makes a density estimator perform well, we must first consider how to assess the quality of a density estimator. Let $\hat{f}$ denote an estimator of f that is based on some fixed number h that controls how localized the probability density contributions used to construct $\hat{f}$ should be. A small h will indicate that $\hat{f}(x)$ should depend more heavily on data points observed near x, whereas a larger h will indicate that distant data should be weighted nearly equally to observations near x.

To evaluate $\hat{f}$ as an estimator of f over the entire range of support, one could use the *integrated squared error*,

$$\text{ISE}(h) = \int_{-\infty}^{\infty} \left(\hat{f}(x) - f(x) \right)^2 dx. \tag{10.2}$$

Note that $\text{ISE}(h)$ is a function of the observed data, through its dependence on $\hat{f}(x)$. Thus it summarizes the performance of $\hat{f}$ conditional on the observed sample. If we want to discuss the generic properties of an estimator without reference to a particular observed sample, it seems more sensible to further average $\text{ISE}(h)$ over all samples that might be observed. The *mean integrated squared error* is

$$\text{MISE}(h) = \text{E}\{\text{ISE}(h)\}, \tag{10.3}$$

where the expectation is taken with respect to the distribution f. Thus $\text{MISE}(h)$ may be viewed as the average value of a global measure of error (namely $\text{ISE}(h)$) with respect to the sampling density. Moreover, with an interchange of expectation and integration,

$$\text{MISE}(h) = \int \text{MSE}_h(\hat{f}(x)) \, dx, \tag{10.4}$$

where

$$\text{MSE}_h(\hat{f}(x)) = \text{E}\left\{ \left(\hat{f}(x) - f(x) \right)^2 \right\} = \text{var}\{\hat{f}(x)\} + \left(\text{bias}\{\hat{f}(x)\} \right)^2 \tag{10.5}$$

and $\text{bias}\{\hat{f}(x)\} = \text{E}\{\hat{f}(x)\} - f(x)$. Equation (10.4) suggests that $\text{MISE}(h)$ can also be viewed as accumulating the local mean squared error at every x.

For multivariate density estimation, $\text{ISE}(h)$ and $\text{MISE}(h)$ are defined analogously. Specifically, $\text{ISE}(h) = \int [\hat{f}(\mathbf{x}) - f(\mathbf{x})]^2 d\mathbf{x}$ and $\text{MISE}(h) = \text{E}\{\text{ISE}(h)\}$.

$\text{MISE}(h)$ and $\text{ISE}(h)$ both measure the quality of the estimator $\hat{f}$, and each can be used to develop criteria for selecting a good value for h. Preference between these two measures is a topic of some debate [249, 260, 313]. The distinction is essentially one between the statistical concepts of loss and risk. Using $\text{ISE}(h)$ is conceptually appealing because it assesses the estimator's performance with the observed data. However, focusing on $\text{MISE}(h)$ is an effective way to approximate ISE-based evaluation while reflecting the sensible goal of seeking optimal performance on average over many data sets. We will encounter both measures in the following sections.

Although we limit attention to performance criteria based on squared error for the sake of simplicity and familiarity, squared error is not the only reasonable option. For example, there are several potentially appealing reasons to replace integrated squared error with the L_1 norm $\int |\hat{f}(x) - f(x)| \, dx$, and $\text{MISE}(h)$ with the corresponding expectation. Notably among these, the L_1 norm is unchanged under any monotone continuous change of scale. This dimensionless character of L_1 makes it a sort of universal measure of how near $\hat{f}$ is to f. Devroye and Györfi study the theory of density estimation using L_1 and present other advantages of this approach [138, 139]. In principle, the optimality of an estimator depends on the metric by which

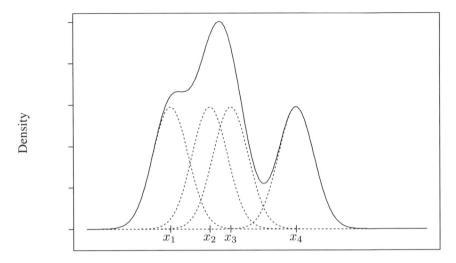

Fig. 10.1   Normal kernel density estimate (solid) and kernel contributions (dotted) for the sample $x_1, \ldots, x_4$. The kernel density estimate at any x is the sum of the kernel contributions centered at each x_i.

performance is assessed, so the adoption of different metrics may favor different types of estimators. In practice, however, many other factors generally affect the quality of a density estimator more than the metric that might have been used to motivate it.

10.2   KERNEL DENSITY ESTIMATION

The density estimator given in equation (10.1) weights all points within h of x equally. A univariate *kernel density estimator* allows a more flexible weighting scheme, fitting

$$\hat{f}(x) = \frac{1}{nh} \sum_{i=1}^{n} K\left(\frac{x - X_i}{h}\right),   \tag{10.6}$$

where K is a *kernel function* and h is a fixed number usually termed the *bandwidth*.

A kernel function assigns weights to the contributions given by each X_i to the kernel density estimator $\hat{f}(x)$, depending on the proximity of X_i to x. Typically, kernel functions are positive everywhere and symmetric about zero. K is usually a density, such as a normal or Student's t density. Other popular choices include the triweight and Epanechnikov kernels (see Section 10.2.2), which don't correspond to familiar densities. Note that the univariate uniform kernel, namely $K(z) = \frac{1}{2}1_{\{|z|<1\}}$, yields the estimator given by (10.1). Constraining K so that $\int z^2 K(z)\,dz = 1$ allows h to play the role of the scale parameter of the density K, but is not required.

Figure 10.1 illustrates how a kernel density estimate is constructed from a sample of four univariate observations, $x_1, \ldots, x_4$. Centered at each observed data point is a scaled kernel: in this case, a normal density function divided by 4. These contributions are shown with the dotted lines. Summing the contributions yields the estimate $\hat{f}$ shown with the solid line.

The estimator in (10.6) is more precisely termed a *fixed-bandwidth kernel density estimator*, because h is constant. The value chosen for the bandwidth exerts a strong influence on the estimator $\hat{f}$. If h is too small, the density estimator will tend to assign probability density too locally near observed data, resulting in a very wiggly estimated density function with many false modes. When h is too large, the density estimator will spread probability density contributions too diffusely. Averaging over neighborhoods that are too large smooths away important features of f.

Notice that computing a kernel density estimate at every observed sample point based on a sample of size n requires $n(n-1)$ evaluations of K. Thus, the computational burden of calculating $\hat{f}$ grows quickly with n. However, for most practical purposes such as graphing the density, the estimate need not be computed at each X_i. A practical strategy is to calculate $\hat{f}(x)$ over a grid of values for x, then linearly interpolate between grid points. A grid of a few hundred values is usually sufficient to provide a graph of $\hat{f}$ that appears smooth. An even faster, approximate method of calculating the kernel density estimate relies on binning the data and rounding each value to the nearest bin center [274]. Then, the kernel need only be evaluated at each nonempty bin center, with density contributions weighted by bin counts. Drastic reductions in computing time can thereby be obtained in situations where n is so large as to prevent calculating individual contributions to $\hat{f}$ centered at every X_i.

10.2.1 Choice of bandwidth

The bandwidth parameter controls the smoothness of the density estimate. Recall from (10.4) and (10.5) that MISE(h) equals the integrated mean squared error. This emphasizes that the bandwidth determines the tradeoff between the bias and variance of $\hat{f}$. Such a tradeoff is a pervasive theme in nearly all kinds of model selection, including regression, density estimation, and smoothing (see Chapters 11 and 12). A small bandwidth produces a density estimator with wiggles indicative of high variability caused by undersmoothing. A large bandwidth causes important features of f to be smoothed away, thereby causing bias.

Example 10.1 (Bimodal density) The effect of bandwidth is shown in Figure 10.2. This histogram shows a sample of 100 points from an equally weighted mixture of $N(4, 1^2)$ and $N(9, 2^2)$ densities. Three density estimates that use a standard normal kernel are superimposed, with $h = 1.875$ (dashed), $h = 0.625$ (heavy), and $h = 0.3$ (solid). The bandwidth $h = 1.875$ is clearly too large, because it leads to an oversmooth density estimate that fails to reveal the bimodality of f. On the other hand, $h = 0.3$ is too small a bandwidth, leading to undersmoothing. The density estimate is too wiggly, exhibiting many false modes. The bandwidth $h = 0.625$ is adequate, correctly representing the main features of f while suppressing most effects of sampling variability. □

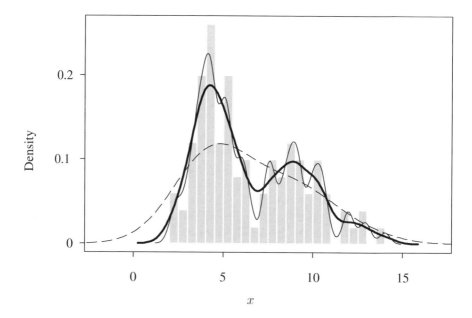

Fig. 10.2 Histogram of 100 data points drawn from the bimodal distribution in Example 10.1, and three normal kernel density estimates. The estimates correspond to bandwidths $h = 1.875$ (dashed line), $h = 0.625$ (heavy), and $h = 0.3$ (solid).

In the following subsections we discuss a variety of ways to choose h. When density estimation is used mainly for exploratory data analysis, span choice based on visual inspection is defensible, and the trial-and-error process leading to your choice might itself provide useful insight into the stability of features observed in the density estimate. In practice, one may simply try a sequence of values for h, and choose a value that is large enough to avoid the threshold where smaller bandwidths cause features of the density estimate to become unstable or the density estimate exhibits obvious wiggles so localized that they are unlikely to represent modes of f. Although the density estimate is sensitive to the choice of bandwidth, we stress that there is no single correct choice in any application. Indeed, bandwidths within 10%–20% of each other will often produce qualitatively similar results.

There are situations when a more formal bandwidth selection procedure might be desired: for use in automatic algorithms, for use by a novice data analyst, or for use when a greater degree of objectivity or formalism is desired. A comprehensive review of approaches is given in [316]; other good reviews include [27, 77, 315, 426, 492, 502, 507].

To understand bandwidth selection, it is necessary to further analyze MISE(h). Suppose that K is a symmetric, continuous probability density function with mean zero and variance $0 < \sigma_K^2 < \infty$. Let $R(g)$ denote a measure of the roughness of a given function g, defined by

$$R(g) = \int g^2(z)\, dz. \tag{10.7}$$

Assume hereafter that $R(K) < \infty$ and that f is sufficiently smooth. In this section, this means that f must have two bounded continuous derivatives and $R(f'') < \infty$; higher-order smooth derivatives are required for some methods discussed later. Recall that

$$\text{MISE}(h) = \int \text{MSE}_h(\hat{f}(x))\, dx = \int \text{var}\{\hat{f}(x)\} + \left(\text{bias}\{\hat{f}(x)\}\right)^2\, dx. \tag{10.8}$$

We further analyze this expression, allowing $h \to 0$ and $nh \to \infty$ as $n \to \infty$.

To compute the bias term in (10.8), note that

$$\begin{aligned} \text{E}\{\hat{f}(x)\} &= \frac{1}{h}\int K\left(\frac{x-u}{h}\right) f(u)\, du \\ &= \int K(t) f(x - ht)\, dt \end{aligned} \tag{10.9}$$

by applying a change of variable. Next, substituting the Taylor series expansion

$$f(x - ht) = f(x) - ht f'(x) + h^2 t^2 f''(x)/2 + o(h^2) \tag{10.10}$$

in (10.9) and noting that K is symmetric about zero leads to

$$\text{E}\{\hat{f}(x)\} = f(x) + h^2 \sigma_K^2 f''(x)/2 + o(h^2), \tag{10.11}$$

where $o(h^2)$ is a quantity that converges to zero faster than h^2 does as $h \to 0$. Thus,

$$\left(\text{bias}\{\hat{f}(x)\}\right)^2 = h^4 \sigma_K^4 [f''(x)]^2/4 + o(h^4), \tag{10.12}$$

and integrating the square of this quantity over x gives

$$\int \left(\text{bias}\{\hat{f}(x)\}\right)^2\, dx = h^4 \sigma_K^4 R(f'')/4 + o(h^4). \tag{10.13}$$

To compute the variance term in (10.8), a similar strategy is employed:

$$\begin{aligned} \text{var}\{\hat{f}(x)\} &= \frac{1}{n}\text{var}\left\{\frac{1}{h}K\left(\frac{x - X_i}{h}\right)\right\} \\ &= \frac{1}{nh}\int K(t)^2 f(x - ht)\, dt - \frac{1}{n}\left[\text{E}\left\{\frac{1}{h}K\left(\frac{x - X_i}{h}\right)\right\}\right]^2 \\ &= \frac{1}{nh}\int K(t)^2 \left[f(x) + o(1)\right]\, dt - \frac{1}{n}\left[f(x) + o(1)\right]^2 \\ &= \frac{1}{nh}f(x)R(K) + o\left(\frac{1}{nh}\right). \end{aligned} \tag{10.14}$$

Integrating this quantity over x gives

$$\int \text{var}\{\hat{f}(x)\} \, dx = \frac{R(K)}{nh} + o\left(\frac{1}{nh}\right). \tag{10.15}$$

Thus,

$$\text{MISE}(h) = \text{AMISE}(h) + o\left(\frac{1}{nh} + h^4\right), \tag{10.16}$$

where

$$\text{AMISE}(h) = \frac{R(K)}{nh} + \frac{h^4 \sigma_K^4 R(f'')}{4} \tag{10.17}$$

is termed the *asymptotic mean integrated squared error*. If $nh \to \infty$ and $h \to 0$ as $n \to \infty$, then $\text{MISE}(h) \to 0$, confirming our intuition from the uniform kernel estimator discussed in the chapter introduction. The error term in (10.16) can be shown to equal $\mathcal{O}(n^{-1} + h^5)$ with a more delicate analysis of the squared bias as in [491], but it is the AMISE which interests us most.

To minimize $\text{AMISE}(h)$ with respect to h, we must set h at an intermediate value that avoids excessive bias and excessive variability in $\hat{f}$. Minimizing $\text{AMISE}(h)$ with respect to h shows that exactly balancing the orders of the bias and variance terms in (10.17) is best. The optimal bandwidth is

$$h = \left(\frac{R(K)}{n \sigma_K^4 R(f'')}\right)^{1/5}, \tag{10.18}$$

but this result is not immediately helpful, since it depends on the unknown density f.

Note that optimal bandwidths have $h = \mathcal{O}(n^{-1/5})$, in which case $\text{MISE} = \mathcal{O}(n^{-4/5})$. This result reveals how quickly the bandwidth should shrink with increasing sample size, but it says little about what bandwidth would be appropriate for density estimation with a given dataset. A variety of automated bandwidth selection strategies have been proposed; see the following subsections. Their relative performance in real applications varies with the nature of f and the observed data. There is no universally best way to proceed.

Many bandwidth selection methods rely on optimizing or finding the root of a function of h—for example, minimizing an approximation to $\text{AMISE}(h)$. In these cases, a search may be conducted over a logarithmic-spaced grid of 50 or more values, linearly interpolating between grid points. When there are multiple roots or local minima, a grid search permits a better understanding of the bandwidth selection problem than would an automated optimization or root-finding algorithm.

10.2.1.1 *Cross-validation*

Many bandwidth selection strategies begin by relating h to some measure of the quality of $\hat{f}$ as an estimator of f. The quality is quantified by some $Q(h)$, whose estimate, $\hat{Q}(h)$, is optimized to find h.

If $\hat{Q}(h)$ evaluates the quality of $\hat{f}$ based on how well it fits the observed data in some sense, then the observed data are being used twice: once to calculate $\hat{f}$ from the data and a second time to evaluate the quality of $\hat{f}$ as an estimator of f. Such double use of the data provides an overoptimistic view of the quality of the estimator. When misled in this way, the chosen estimator tends to be overfitted (i.e., undersmoothed), with too many wiggles or false modes.

Cross-validation provides a remedy to this problem. To evaluate the quality of $\hat{f}$ at the ith data point, the model is fitted using all the data except the ith point. Let

$$\hat{f}_{-i}(X_i) = \frac{1}{h(n-1)} \sum_{j \neq i} K\left(\frac{X_i - X_j}{h}\right) \qquad (10.19)$$

denote the estimated density at X_i using a kernel density estimator with all the observations except X_i. Choosing $\hat{Q}$ to be a function of the $\hat{f}_{-i}(X_i)$ separates the tasks of fitting and evaluating $\hat{f}$ to select h.

Although cross-validation enjoys great success as a span selection strategy for scatterplot smoothing (see Chapter 11), it is not always effective for bandwidth selection in density estimation. The h estimated by cross-validation approaches can be highly sensitive to sampling variability. Despite the persistence of these methods in general practice and in some software, a sophisticated plug-in method like the Sheather–Jones approach (Section 10.2.1.2) is a much more reliable choice. Nevertheless, cross-validation methods introduce some ideas that are useful in a variety of contexts.

One easy cross-validation option is to let $\hat{Q}(h)$ be the *pseudo-likelihood*

$$\text{PL}(h) = \prod_{i=1}^{n} \hat{f}_{-i}(X_i), \qquad (10.20)$$

as proposed in [148, 252]. The bandwidth is chosen to maximize the pseudo-likelihood. Although simple and intuitively appealing, this approach frequently produces kernel density estimates that are too wiggly and too sensitive to outliers [493]. The theoretical limiting performance of kernel density estimators with span chosen to minimize $\text{PL}(h)$ is also poor: In many cases the estimator is not consistent [489].

Another approach is motivated by reexpressing the integrated squared error as

$$\text{ISE}(h) = \int \hat{f}^2(x)\, dx - 2\text{E}\{\hat{f}(x)\} + \int f(x)^2\, dx$$
$$= R(\hat{f}) - 2\text{E}\{\hat{f}(x)\} + R(f). \qquad (10.21)$$

The final term in this expression is constant, and the middle term can be estimated by $\frac{2}{n}\sum_{i=1}^{n} \hat{f}_{-i}(X_i)$. Thus, minimizing

$$\text{UCV}(h) = R(\hat{f}) - \frac{2}{n}\sum_{i=1}^{n} \hat{f}_{-i}(X_i) \qquad (10.22)$$

with respect to h should provide a good bandwidth [50, 472]. $\text{UCV}(h)$ is called the *unbiased cross-validation* criterion, because $\text{E}\{\text{UCV}(h) + R(f)\} = \text{MISE}(h)$. The approach is also called *least squares cross-validation*, because choosing h to minimize $\text{UCV}(h)$ minimizes the integrated squared error between $\hat{f}$ and f.

If analytic evaluation of $R(\hat{f})$ is not possible, the best way to evaluate (10.22) is probably to use a different kernel which permits an analytic simplification. For a normal kernel ϕ it can be shown that

$$
\begin{aligned}
\text{UCV}(h) = {} & \frac{R(\phi)}{nh} \\
& + \frac{1}{n(n-1)h} \sum_{i=1}^{n} \sum_{j \neq i} \left[\frac{1}{(8\pi)^{1/4}} \phi^{1/2} \left(\frac{X_i - X_j}{h} \right) - 2\phi \left(\frac{X_i - X_j}{h} \right) \right], \quad (10.23)
\end{aligned}
$$

following the steps outlined in Problem 10.3. This expression can be computed efficiently without numerical approximation.

Although the bandwidth identified by minimizing $\text{UCV}(h)$ with respect to h is asymptotically as good as the best possible bandwidth [256, 519], its convergence to the optimum is extremely slow [259, 494]. In practical settings, using unbiased cross-validation is risky because the resulting bandwidth tends to exhibit a strong dependence on the observed data. In other words, when applied to different datasets drawn from the same distribution, unbiased cross-validation can yield very different answers. Its performance is erratic in application, and undersmoothing is frequently seen.

The high sampling variability of unbiased cross-validation is mainly attributable to the fact that the target performance criterion, $Q(h) = \text{ISE}(h)$, is itself random, unlike $\text{MISE}(h)$. Scott and Terrell have proposed a *biased cross-validation* criterion $(\text{BCV}(h))$ that seeks to minimize an estimate of $\text{AMISE}(h)$ [494]. In practice, this approach is generally outperformed by the best plug-in methods (Section 10.2.1.2) and can yield excessively large bandwidths and oversmooth density estimates.

Example 10.2 (Whale migration) Figure 10.3 shows a histogram of times of 121 bowhead whale calf sightings during the spring 2001 visual census conducted at the ice edge near Point Barrow, Alaska. This census is the central component of international efforts to manage this endangered whale population while allowing a small traditional subsistence hunt by coastal Inupiat communities [135, 219, 446].

The timing of the northeasterly spring migration is surprisingly regular, and it is important to characterize the migration pattern for planning of future scientific efforts to study these animals. There is speculation that the migration may occur in loosely defined pulses. If so, this is important to discover, because it may lead to new insights about bowhead whale biology and stock structure.

Figure 10.4 shows the results of kernel density estimates for these data using the normal kernel. Three different cross-validation criteria were used to select h. Maximizing cross-validated $\text{PL}(h)$ with respect to h yields $h = 9.75$ and the density estimate shown with the dashed curve. This density estimate is barely adequate, exhibiting likely false modes in several regions. The result from minimizing $\text{UCV}(h)$ with respect to h is even worse in this application, giving $h = 5.08$ and the density estimate shown with the dotted curve. This bandwidth is clearly too small. Finally, minimizing $\text{BCV}(h)$ with respect to h yields $h = 26.52$ and the density estimate shown with the solid line. Clearly the best of the three options, this density estimate

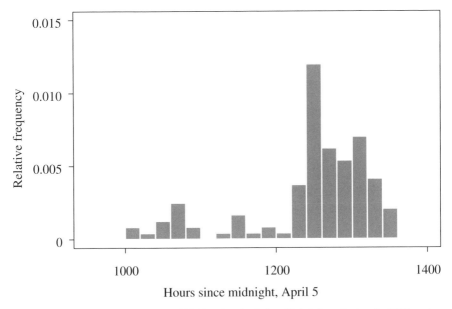

Fig. 10.3 Histogram of the times of 121 bowhead whale calf sightings during the 2001 spring migration discussed in Example 10.2. The date of each sighting is expressed as the number of hours since midnight, April 5, when the first adult whale was sighted.

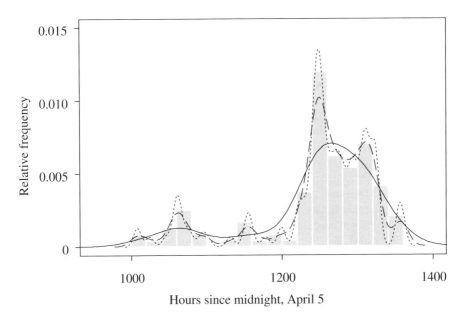

Fig. 10.4 Kernel density estimates for the whale calf migration data in Example 10.2 using normal kernels and bandwidths chosen by three different cross-validation criteria. The bandwidths are 9.75 using $PL(h)$ (dashed), 5.08 using $UCV(h)$ (dotted), and 26.52 using $BCV(h)$ (solid).

emphasizes only the most prominent features of the data distribution but may seem oversmooth. Perhaps a bandwidth between 10 and 26 would be preferable. □

10.2.1.2 Plug-in methods

Plug-in methods apply a pilot bandwidth to estimate one or more important features of f. The bandwidth for estimating f itself is then estimated at a second stage using a criterion that depends on the estimated features. The best plug-in methods have proven to be very effective in diverse applications and are more popular than cross-validation approaches. However, Loader offers arguments against the uncritical rejection of cross-validation approaches [361].

For unidimensional kernel density estimation, recall that the bandwidth that minimizes AMISE is given by

$$h = \left(\frac{R(K)}{n\sigma_K^4 R(f'')} \right)^{1/5}, \tag{10.24}$$

where σ_K^2 is the variance of K, viewing K as a density. At first glance, (10.24) seems unhelpful because the optimal bandwidth depends on the unknown density f through the roughness of its second derivative. A variety of methods have been proposed to estimate $R(f'')$.

Silverman suggests an elementary approach: replacing f by a normal density with variance set to match the sample variance [507]. This amounts to estimating $R(f'')$ by $R(\phi'')/\hat{\sigma}^5$, where ϕ is the standard normal density function. Silverman's rule of thumb therefore gives

$$h = \left(\frac{4}{3n} \right)^{1/5} \hat{\sigma}. \tag{10.25}$$

If f is multimodal, the ratio of $R(f'')$ to $\hat{\sigma}$ may be larger than it would be for normally distributed data. This would result in oversmoothing. A better bandwidth can be obtained by considering the interquartile range (IQR), which is a more robust measure of spread than is $\hat{\sigma}$. Thus, Silverman suggests replacing $\hat{\sigma}$ in (10.25) by $\tilde{\sigma} = \min\{\hat{\sigma}, \text{IQR}/(\Phi^{-1}(0.75) - \Phi^{-1}(0.25))\} \approx \min\{\hat{\sigma}, \text{IQR}/1.35\}$, where Φ is the standard normal cumulative distribution function. Although simple, this approach cannot be recommended for general use, because it has a strong tendency to oversmooth. Silverman's rule of thumb is valuable, however, as a method for producing approximate bandwidths effective for pilot estimation of quantities used in sophisticated plug-in methods.

Empirical estimation of $R(f'')$ in (10.24) is a better option than Silverman's rule of thumb. The kernel-based estimator is

$$\hat{f}''(x) = \frac{d^2}{dx^2} \left\{ \frac{1}{nh_0} \sum_{i=1}^{n} L\left(\frac{x - X_i}{h_0} \right) \right\}$$

$$= \frac{1}{nh_0^3} \sum_{i=1}^{n} L''\left(\frac{x - X_i}{h_0} \right), \tag{10.26}$$

where h_0 is the bandwidth and L is a sufficiently differentiable kernel used to estimate f''. Estimation of $R(f'')$ follows from (10.26).

It is important to recognize, however, that the best bandwidth for estimating f will differ from the best bandwidth for estimating f'' or $R(f'')$. This is because $\text{var}\{\hat{f}''\}$ contributes a proportionally greater share to the mean squared error for estimating f'' than $\text{var}\{\hat{f}\}$ does for estimating f. Therefore, a larger bandwidth is required for estimating f''. We therefore anticipate $h_0 > h$.

Suppose we use bandwidth h_0 with kernel L to estimate $R(f'')$, and bandwidth h with kernel K to estimate f. Then the asymptotic mean squared error for estimation of $R(f'')$ using kernel L is minimized when $h_0 \propto n^{-1/7}$. To determine how h_0 should be related to h, recall that optimal bandwidths for estimating f have $h \propto n^{-1/5}$. Solving this expression for n and replacing n in the equation $h_0 \propto n^{-1/7}$, one can show that

$$h_0 = C_1(R(f''), R(f'''))C_2(L)h^{5/7}, \tag{10.27}$$

where C_1 and C_2 are functionals that depend on derivatives of f and on the kernel L, respectively. Equation (10.27) still depends on the unknown f, but the quality of the estimate of $R(f'')$ produced using h_0 and L is not excessively deteriorated if h_0 is set using relatively simple estimates to find C_1 and C_2. In fact, we may estimate C_1 and C_2 using a bandwidth chosen by Silverman's rule of thumb.

The result is a two-stage process for finding the bandwidth, known as the *Sheather–Jones method* [315, 503]. At the first stage, a simple rule of thumb is used to calculate the bandwidth h_0. This bandwidth is used to estimate $R(f'')$, which is the only unknown in expression (10.24) for the optimal bandwidth. Then the bandwidth h is computed via (10.24) and is used to produce the final kernel density estimate.

For univariate kernel density estimation with pilot kernel $L = \phi$, the Sheather–Jones bandwidth is the value of h that solves the equation

$$\left(\frac{R(K)}{n\sigma_K^4 \hat{R}_{\hat{\alpha}(h)}(f'')} \right)^{1/5} - h = 0, \tag{10.28}$$

where

$$\hat{R}_{\hat{\alpha}(h)}(f'') = \frac{1}{n(n-1)\alpha^5} \sum_{i=1}^{n} \sum_{j=1}^{n} \phi^{(4)}\left(\frac{X_i - X_j}{\alpha} \right),$$

$$\hat{\alpha}(h) = \left(\frac{6\sqrt{2}h^5 \hat{R}_a(f'')}{\hat{R}_b(f''')} \right)^{1/7},$$

$$\hat{R}_a(f'') = \frac{1}{n(n-1)a^5} \sum_{i=1}^{n} \sum_{j=1}^{n} \phi^{(4)}\left(\frac{X_i - X_j}{a} \right),$$

$$\hat{R}_b(f''') = \frac{1}{n(n-1)b^7} \sum_{i=1}^{n} \sum_{j=1}^{n} \phi^{(6)} \left(\frac{X_i - X_j}{b} \right),$$

$$a = 0.920(\text{IQR})/n^{1/7},$$

$$b = 0.912(\text{IQR})/n^{1/9},$$

$\phi^{(i)}$ is the ith derivative of the normal density function, and IQR is the interquartile range of the data. The solution to (10.28) can be found using grid search or a root-finding technique from Chapter 2, such as Newton's method.

The Sheather–Jones method generally performs extremely well [315, 316, 427, 502]. There are a variety of other good methods based on carefully chosen approximations to MISE(h) or its minimizer [77, 261, 262, 314, 426]. In each case, careful pilot estimation of various quantities plays a critical role in ensuring that the final bandwidth performs well. Some of these approaches give bandwidths that asymptotically converge more quickly to the optimal bandwidth than does the Sheather–Jones method; all can be useful options in some circumstances. However, none of these offer substantially easier practical implementation or broadly better performance than the Sheather–Jones approach.

Example 10.3 (Whale migration, continued) Figure 10.5 illustrates the use of Silverman's rule of thumb and the Sheather–Jones method on the bowhead whale migration data introduced in Example 10.2. The bandwidth given by the Sheather–Jones approach is 10.22, yielding the density estimate shown with the solid line. This bandwidth seems a bit too narrow, yielding a density estimate that is too wiggly. Silverman's rule of thumb gives a bandwidth of 32.96, larger than the bandwidth given by any previous method. The resulting density estimate is probably too smooth, hiding important features of the distribution.   □

10.2.1.3 *Maximal smoothing principle*   Recall again that the minimal AMISE is obtained when

$$h = \left(\frac{R(K)}{n\sigma_K^4 R(f'')} \right)^{1/5}, \tag{10.29}$$

but f is unknown. Silverman's rule of thumb replaces $R(f'')$ by $R(\phi'')$. The Sheather–Jones method estimates $R(f'')$. Terrell's maximal smoothing approach replaces $R(f'')$ with the most conservative (i.e., smallest) possible value [531].

Specifically, Terrell considered the collection of all h that would minimize (10.29) for various f, and recommended that the largest such bandwidth be chosen. In other words, the right hand side of (10.29) should be maximized with respect to f. This will bias bandwidth selection against undersmoothing. Since $R(f'')$ vanishes as the variance of f shrinks, the maximization is carried out subject to the constraint that the variance of f matches the sample variance $\hat{\sigma}^2$.

Constrained maximization of (10.29) with respect to f is an exercise in the calculus of variations. The f that maximizes (10.29) is a polynomial. Substituting its

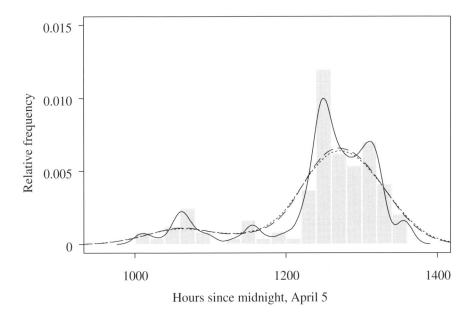

Fig. 10.5 Kernel density estimates for the whale calf migration data using normal kernels with bandwidths chosen by three different criteria. The bandwidths are 10.22 using the Sheather–Jones approach (solid), 32.96 using Silverman's rule of thumb (dashed), and 35.60 using the maximal smoothing span of Terrell (dotted).

roughness for $R(f'')$ in (10.29) yields

$$h = 3 \left(\frac{R(K)}{35n} \right)^{1/5} \hat{\sigma} \qquad (10.30)$$

as the chosen bandwidth. Table 10.1 provides the values of $R(K)$ for some common kernels.

Terrell proposed the *maximal smoothing principle* to motivate this choice of bandwidth. When interpreting a density estimate, the analyst's eye is naturally drawn to modes. Further, modes usually have important scientific implications. Therefore, the bandwidth should be selected to discourage false modes, producing an estimate that shows modes only where the data indisputably require them.

The maximal smoothing approach is appealing because it is quick and simple to calculate. In practice, the resulting kernel density estimate is often too smooth. We would be reluctant to use a maximal smoothing bandwidth when the density estimate is being used for inference. For exploratory analyses, the maximal smoothing bandwidth can be quite helpful, allowing the analyst to focus on the dominant features of the density without being misled by highly variable indications of possibly false modes.

Table 10.1   Some kernel choices and related quantities discussed in the text. The kernels are listed in increasing order of roughness, $R(K)$. $K(z)$ should be multiplied by $1_{\{|z|<1\}}$ in all cases except the normal kernel, which has positive support over the entire real line. R.E. is the asymptotic relative efficiency, as described in Section 10.2.2.1.

Name	$K(z)$	$R(K)$	$\delta(K)$	R.E.
Normal	$\exp\{-z^2/2\}/\sqrt{2\pi}$	$1/(2\sqrt{\pi})$	$(1/(2\sqrt{\pi}))^{1/5}$	1.051
Uniform	$1/2$	$1/2$	$(9/2)^{1/5}$	1.076
Epanechnikov	$(3/4)(1-z^2)$	$3/5$	$15^{1/5}$	1.000
Triangle	$1-|z|$	$2/3$	$24^{1/5}$	1.014
Biweight	$(15/16)(1-z^2)^2$	$5/7$	$35^{1/5}$	1.006
Triweight	$(35/32)(1-z^2)^3$	$350/429$	$(9450/143)^{1/5}$	1.013

Example 10.4 (Whale migration, continued)   The dotted line in Figure 10.5 shows the density estimate obtained using the maximal smoothing bandwidth of 35.60. Even larger than Silverman's bandwidth, this choice appears too large for the whale data. Generally, both Silverman's rule of thumb and Terrell's maximal smoothing principle tend to produce oversmoothed density estimates.   □

10.2.2   Choice of kernel

Kernel density estimation requires specification of two components: the kernel and the bandwidth. It turns out that the shape of the kernel has much less influence on the results than does the bandwidth. Table 10.1 lists a few choices for kernel functions.

10.2.2.1   *Epanechnikov kernel*   Suppose K is limited to bounded, symmetric densities with finite moments and variance equal to 1. Then Epanechnikov showed that minimizing AMISE with respect to K amounts to minimizing $R(K)$ with respect to K subject to these constraints [162]. The solution to this variational calculus problem is the kernel assigning density $\frac{1}{\sqrt{5}}K^*(z/\sqrt{5})$, where K^* is the Epanechnikov kernel

$$K^*(z) = \begin{cases} \frac{3}{4}(1-z^2) & \text{if } |z| < 1, \\ 0 & \text{otherwise.} \end{cases} \tag{10.31}$$

This is a symmetric quadratic function, centered at zero, where its mode is reached, and decreasing to zero at the limits of its support.

From (10.17) and (10.18) we see that the minimal AMISE for a kernel density estimator with positive kernel K is $\frac{5}{4}[\sigma_K R(K)/n]^{4/5} R(f'')^{1/5}$. Switching to a K that doubles $\sigma_K R(K)$ therefore requires doubling n to maintain the same minimal AMISE. Thus, $\sigma_{K_2} R(K_2)/(\sigma_{K_1} R(K_1))$ measures the asymptotic relative efficiency of K_2 compared to K_1. The relative efficiencies of a variety of kernels compared to the Epanechnikov kernel are given in Table 10.1. Notice that the relative efficiencies are all quite close to 1, reinforcing the point that kernel choice is fairly unimportant.

10.2.2.2 Canonical kernels and rescalings Unfortunately, a particular value of h corresponds to a different amount of smoothing depending on which kernel is being used. For example, $h = 1$ corresponds to a kernel standard deviation nine times larger for the normal kernel than for the triweight kernel.

Let h_K and h_L denote the bandwidths that minimize AMISE(h) when using symmetric kernel densities K and L, respectively, which have mean zero and finite positive variance. Then from (10.29) it is clear that

$$\frac{h_K}{h_L} = \frac{\delta(K)}{\delta(L)}, \tag{10.32}$$

where for any kernel we have $\delta(K) = \left(R(K)/\sigma_K^4\right)^{1/5}$. Thus, to change from bandwidth h for kernel K to a bandwidth that gives an equivalent amount of smoothing for kernel L, use the bandwidth $h\delta(L)/\delta(K)$. Table 10.1 lists values for $\delta(K)$ for some common kernels.

Suppose further that we rescale each kernel shape in Table 10.1 so that $h = 1$ corresponds to a bandwidth of $\delta(K)$. The kernel density estimator can then be written as $\hat{f}_X(x) = \frac{1}{n}\sum_{i=1}^n K_{h\delta(K)}(x - X_i)$, where $K_{h\delta(K)}(z) = \frac{1}{h\delta(K)}K\left(\frac{z}{h\delta(K)}\right)$, and K represents one of the original kernel shapes and scalings shown in Table 10.1. Scaling kernels in this way provides a *canonical kernel* $K_{\delta(K)}$ of each shape [373]. A key benefit of this viewpoint is that a single value of h can be used interchangeably for each canonical kernel without affecting the amount of smoothing in the density estimate.

Note that

$$\text{AMISE}(h) = C(K_{\delta(K)})\left(\frac{1}{nh} + \frac{h^4 R(f'')}{4}\right) \tag{10.33}$$

for an estimator using a canonical kernel with bandwidth h (i.e., a kernel from Table 10.1 with bandwidth $h\delta(K)$) and with $C(K_{\delta(K)}) = (\sigma_K R(K))^{4/5}$. This means that the balance between variance and squared bias determined by the factor $(nh)^{-1} + h^4 R(f'')/4$ is no longer confounded with the chosen kernel. It also means that the contributions made by the kernel to the variance and squared bias terms in AMISE(h) are equal. It follows that the optimal kernel shape does not depend on the bandwidth: The Epanechnikov kernel shape is optimal for any desired degree of smoothing [373].

Example 10.5 (Bimodal density, continued) Figure 10.6 shows kernel density estimates for the data from Example 10.1, which originated from the equally weighted mixture of $N(4, 1^2)$ and $N(9, 2^2)$ densities. All the bandwidths were set at 0.69 for the canonical kernels of each shape, that being the Sheather–Jones bandwidth for the normal kernel. The uniform kernel produces a noticeably rougher result due to its discontinuity. The Epanechnikov and uniform kernels provide a slight (false) suggestion that the lower mode contains two small local modes. Aside from these small differences, the results for all the kernels are qualitatively the same. This example illustrates that even quite different kernels can be scaled to produce such similar results that the choice of kernel is unimportant.   □

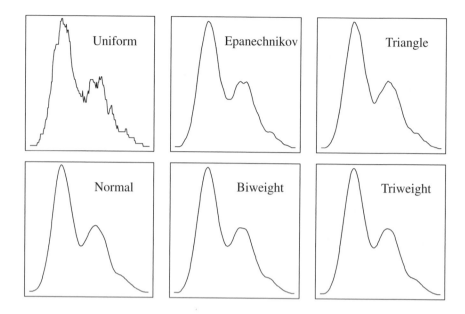

Fig. 10.6 Kernel density estimates for the data in Example 10.1 using the canonical version of each of the six kernels from Table 10.1, with $h = 0.69$ (dotted).

10.3 NONKERNEL METHODS

10.3.1 Logspline

A *cubic spline* is a piecewise cubic function that is everywhere twice continuously differentiable but whose third derivative may be discontinuous at a finite number of prespecified *knots*. One may view a cubic spline as a function created from cubic polynomials on each between-knot interval by pasting them together twice continuously differentiably at the knots. Kooperberg and Stone's logspline density estimation approach estimates the log of f by a cubic spline of a certain form [339, 520].

This method provides univariate density estimation on an interval (L, U), where each endpoint may be infinite. Suppose there are $M \geq 3$ knots, t_j for $j = 1, \ldots, M$, with $L < t_1 < t_2 < \cdots < t_M < U$. Knot selection will be discussed later.

Let S denote the M-dimensional space consisting of cubic splines with knots at $t_1, \ldots, t_M$ and which are linear on $(L, t_1]$ and $[t_M, U)$. Let a basis for S be denoted by the functions $\{1, B_1, \ldots, B_{M-1}\}$. There are numerical advantages to certain types of bases; books on splines and the other references in this section provide additional detail [124, 488]. It is possible to choose the basis functions so that on $(L, t_1]$ the function B_1 is linear with a negative slope and all other B_i are constant, and so that on $[t_M, U)$ the function B_{M-1} is linear with a positive slope and all other B_i are constant.

Now consider modeling f with a parameterized density, $f_{X|\theta}$, defined by

$$\log f_{X|\theta}(x|\theta) = \theta_1 B_1(x) + \cdots + \theta_{M-1} B_{M-1}(x) - c(\theta), \qquad (10.34)$$

where

$$\exp\{c(\theta)\} = \int_L^U \exp\{\theta_1 B_1(x) + \cdots + \theta_{M-1} B_{M-1}(x)\} \, dx \qquad (10.35)$$

and $\theta = (\theta_1, \ldots, \theta_{M-1})$. For this to be a reasonable model for a density, we require $c(\theta)$ to be finite, which is ensured if (i) $L > -\infty$ or $\theta_1 < 0$ and (ii) $U < \infty$ or $\theta_{M-1} < 0$. Under this model, the log likelihood of θ is

$$l(\theta|x_1, \ldots, x_n) = \sum_{i=1}^{n} \log f_{X|\theta}(x_i|\theta), \qquad (10.36)$$

given observed data values $x_1, \ldots, x_n$. As long as the knots are positioned so that each interval contains sufficiently many observations for estimation, maximizing (10.36) subject to the constraint that $c(\theta)$ is finite provides the maximum likelihood estimate, $\hat{\theta}$. This estimate is unique because $l(\theta|x_1, \ldots, x_n)$ is concave. Having estimated the model parameters, take

$$\hat{f}(x) = f_{X|\theta}(x|\hat{\theta}) \qquad (10.37)$$

as the maximum likelihood logspline density estimate of $f(x)$.

The maximum likelihood estimation of θ is conditional on the number of knots and their placement. Kooperberg and Stone suggest an automated strategy for placement of a prespecified number of knots [340]. Their strategy places knots at the smallest and largest observed data point and at other positions symmetrically distributed about the median but not equally spaced.

To place a prespecified number of knots, let $x_{(i)}$ denote the ith order statistic of the data, for $i = 1, \ldots, n$, so $x_{(1)}$ is the minimum observed value. Define an approximate quantile function $q\left(\frac{i-1}{n-1}\right) = x_{(i)}$ for $1 \leq i \leq n$, where the value of q is obtained by linear interpolation for noninteger i.

The M knots will be placed at $x_{(1)}$, at $x_{(n)}$, and at the positions of the order statistics indexed by $q(r_2), \ldots, q(r_{M-1})$ for a sequence of numbers $0 < r_2 < r_3 < \cdots < r_{M-1} < 1$.

When $(L, U) = (-\infty, \infty)$, placement of the interior knots is governed by the following constraint on between-knot distances:

$$n(r_{i+1} - r_i) = 4 \cdot \max\{4 - \epsilon, 1\} \cdot \max\{4 - 2\epsilon, 1\} \cdots \max\{4 - (i-1)\epsilon, 1\}$$

for $1 \leq i \leq M/2$, where $r_1 = 0$ and ϵ is chosen to satisfy $r_{(M+1)/2} = 1/2$ if M is odd, or $r_{M/2} + r_{M/2+1} = 1$ if M is even. The remaining knots are placed to maintain quantile symmetry, so that

$$r_{M+1-i} - r_{M-i} = r_{i+1} - r_i \qquad (10.38)$$

for $M/2 \leq i < M - 1$, where $r_M = 1$.

When (L, U) is not a doubly infinite interval, similar knot placement rules have been suggested. In particular, if (L, U) is a interval of finite length, then $r_2, \ldots, r_{M-1}$ are chosen equally spaced, so $r_i = \frac{i-1}{M-1}$.

The preceding paragraphs assumed that M, the number of knots, was prespecified. A variety of methods for choosing M are possible, but methods for choosing the number of knots have evolved to the point where a complete description of the recommended strategy is beyond the scope of our discussion. Roughly, the process is as follows. Begin by placing a small number of knots in the positions given above. The suggested minimum number is the first integer exceeding $\min\{2.5n^{1/5}, n/4, n^*, 25\}$, where n^* is the number of distinct data points. Additional knots are then added to the existing set, one at time. At each iteration, a single knot is added in a position that gives the largest value of the Rao statistic for testing that the model without that knot suffices [341, 520]. Without examining significance levels, this process continues until the total number of knots reaches $\min\{4n^{1/5}, n/4, n^*, 30\}$, or until no new candidate knots can be placed due to constraints on the positions or nearness of knots.

Next, single knots are sequentially deleted. The deletion of a single knot corresponds to the removal of one basis function. Let $\hat{\boldsymbol{\theta}} = (\hat{\theta}_1, \ldots, \hat{\theta}_{M-1})$ denote the maximum likelihood estimate of the parameters of the current model. Then the Wald statistic for testing the significance of the contribution of the ith basis function is $\hat{\theta}_i / \mathrm{SE}\{\hat{\theta}_i\}$, where $\mathrm{SE}\{\hat{\theta}_i\}$ is the square root of the ith diagonal entry of $-\mathbf{l}''(\hat{\boldsymbol{\theta}})^{-1}$, the inverted observed information matrix [341, 520]. The knot whose removal would yield the smallest Wald statistic value is deleted. Sequential deletion is continued until only about three knots remain.

Sequential knot addition followed by sequential knot deletion generates a sequence of S models, with varying numbers of knots. Denote the number of knots in the sth model by m_s, for $s = 1, \ldots, S$. To choose the best model in the sequence, let

$$\mathrm{BIC}(s) = -2l(\hat{\boldsymbol{\theta}}_s | x_1, \ldots, x_n) + (m_s - 1) \log n \qquad (10.39)$$

measure the quality of the sth model having corresponding MLE parameter vector $\hat{\boldsymbol{\theta}}_s$. The quantity $\mathrm{BIC}(s)$ is a *Bayes information criterion* for model comparison [321, 490]; other measures of model quality can also be motivated. Selection of the model with the minimal value of $\mathrm{BIC}(s)$ among those in the model sequence provides the chosen number of knots.

Additional details of the knot selection process are given in [341, 520]. Software to carry out logspline density estimation in the S-Plus and R languages is available in [97, 338]. Stepwise addition and deletion of knots is a greedy search strategy that is not guaranteed to find the best collection of knots. Other search strategies are also effective, including MCMC strategies [265, 520].

The logspline approach is one of several effective methods for density estimation based on spline approximation; another is given in [250].

Example 10.6 (Whale migration, continued) Figure 10.7 shows the logspline density estimate (solid line) for the whale calf migration data from Example 10.2. Using the procedure outlined above, a model with seven knots was selected. The locations

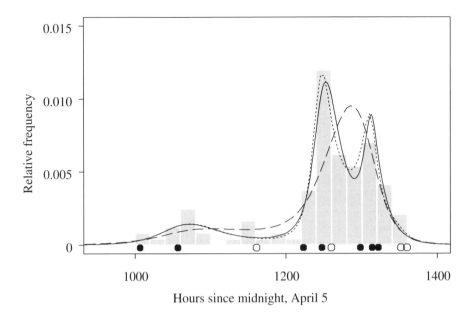

Fig. 10.7 Logspline density estimate (solid line) from bowhead whale calf migration data in Example 10.6. Below the histogram are dots indicating where knots were used (solid) and where knots were considered but rejected (hollow). Two other logspline density estimates for other knot choices are shown with the dotted and dashed lines; see the text for details.

of these seven knots are shown with the solid dots in the figure. During initial knot placement, stepwise knot addition, and stepwise knot deletion, four other knots were considered at various stages but not used in the model finally selected according to the BIC criterion. These discarded knots are shown with hollow dots in the figure. The degree of smoothness seen in Figure 10.7 is typical of logspline estimates since splines are piecewise cubic and twice continuously differentiable.

Estimation of local modes can sometimes be a problem if the knots are insufficient in number or poorly placed. The other lines in Figure 10.7 show the logspline density estimates with two other choices for the knots. The very poor estimate (dashed line) was obtained using six knots. The other estimate (dotted line) was obtained using all eleven knots shown in the figure with either hollow or solid dots.   □

10.4 MULTIVARIATE METHODS

Multivariate density estimation of a density function f is based on i.i.d. random variables sampled from f. A p-dimensional variable is denoted $\mathbf{X}_i = (X_{i1}, \ldots, X_{ip})$.

10.4.1 The nature of the problem

Multivariate density estimation is a significantly different task than univariate density estimation. It is very difficult to visualize any resulting density estimate when the region of support spans more than two or three dimensions. As an exploratory data analysis tool, multivariate density estimation therefore has diminishing usefulness unless some dimension-reduction strategy is employed. However, multivariate density estimation is a useful component in many more elaborate statistical computing algorithms, where visualization of the estimate is not required.

Multivariate density estimation is also hindered by the *curse of dimensionality*. High-dimensional space is very different than 1, 2, or 3-dimensional space. Loosely speaking, high-dimensional space is vast, and points lying in such a space have very few near neighbors. To illustrate, Scott defined the tail region of a standard p-dimensional normal density to comprise all points at which the probability density is less than one percent of the density at the mode [492]. While only 0.2% of the probability density falls in this tail region when $p = 1$, more than half the density falls in it when $p = 10$, and 98% falls in it when $p = 20$.

The curse of dimensionality has important implications for density estimation. For example, consider a kernel density estimator based on a random sample of n points whose distribution is p-dimensional standard normal. Below we mention several ways to construct such an estimator; our choice here is the so-called product kernel approach with normal kernels sharing a common bandwidth, but it is not necessary to understand this technique yet to follow our argument. Define the optimal relative root mean squared error at the origin to be

$$\text{ORRMSE}(p, n) = \frac{\sqrt{\min_h \left\{ \text{MSE}_h \left(\hat{f}(\mathbf{0}) \right) \right\}}}{f(\mathbf{0})},$$

where $\hat{f}$ estimates f from a sample of n points using the best possible bandwidth. This quantity measures the quality of the multivariate density estimator at the true mode. When $p = 1$ and $n = 30$, $\text{ORRMSE}(1, 30) = 0.0289$. Table 10.2 shows the sample sizes required to achieve as low a value of $\text{ORRMSE}(p, n)$ for different values of p. The sample sizes are shown to about three significant digits. A different bandwidth minimizes $\text{ORRMSE}(p, n)$ for each different p and n, so the table entries were computed by fixing p and searching over n, with each trial value of n requiring an optimization over h. This table confirms that desirable sample sizes grow very rapidly with p. In practice, things are not as hopeless as Table 10.2 might suggest. Adequate estimates can be sometimes obtained with a variety of techniques, especially those that attempt to simplify the problem via dimension reduction.

10.4.2 Multivariate kernel estimators

The most literal generalization of the univariate kernel density estimator in (10.6) to the case of p-dimensional density estimation is the *general multivariate kernel*

Table 10.2 Sample sizes required to match the optimal relative root mean squared error at the origin achieved for one-dimensional data when $n = 30$. These results pertain to estimation of a p-variate normal density using a normal product kernel density estimator with bandwidths that minimize the relative root mean squared error at the origin in each instance.

p	n
1	30
2	180
3	806
5	17,400
10	112,000,000
15	2,190,000,000,000
30	806,000,000,000,000,000,000,000,000

estimator

$$\hat{f}(\mathbf{x}) = \frac{1}{n|\mathbf{H}|} \sum_{i=1}^{n} K\left(\mathbf{H}^{-1}(\mathbf{x} - \mathbf{X}_i)\right), \tag{10.40}$$

where $\mathbf{H}$ is a $p \times p$ nonsingular matrix of constants, whose absolute determinant is denoted $|\mathbf{H}|$. The function K is a real-valued multivariate kernel function for which $\int K(\mathbf{z})\, d\mathbf{z} = 1$, $\int \mathbf{z}K(\mathbf{z})\, d\mathbf{z} = \mathbf{0}$, and $\int \mathbf{z}\mathbf{z}^T K(\mathbf{z})\, d\mathbf{z} = \mathbf{I}_p$, where $\mathbf{I}_p$ is the $p \times p$ identity matrix.

This estimator is quite a bit more flexible than usually is required. It allows both a p-dimensional kernel of arbitrary shape and an arbitrary linear rotation and scaling via $\mathbf{H}$. It can be quite inconvenient to try to specify large number of bandwidth parameters contained in $\mathbf{H}$ and to specify a kernel shape over p-dimensional space. It is more practical to resort to more specialized forms of $\mathbf{H}$ and K that have far fewer parameters.

The *product kernel* approach provides a great deal of simplification. The density estimator is

$$\hat{f}(\mathbf{x}) = \frac{1}{n} \sum_{i=1}^{n} \prod_{j=1}^{p} \frac{1}{h_j} K\left(\frac{x_j - X_{ij}}{h_j}\right) \tag{10.41}$$

where $K(z)$ is a univariate kernel function, $\mathbf{x} = (x_1, \ldots, x_p)$, $\mathbf{X_i} = (X_{i1}, \ldots, X_{ip})$, and the h_j are fixed bandwidths for each coordinate, $j = 1, \ldots, p$.

Another simplifying approach would be to allow K to be a radially symmetric, unimodal density function in p dimensions, and to set

$$\hat{f}(\mathbf{x}) = \frac{1}{nh^p} \sum_{i=1}^{n} K\left(\frac{\mathbf{x} - \mathbf{X}_i}{h}\right). \tag{10.42}$$

In this case, the multivariate Epanechnikov kernel shape

$$K(\mathbf{z}) = \begin{cases} \frac{(p+2)\Gamma(1+p/2)}{2\pi^{p/2}}(1 - \mathbf{z}^T\mathbf{z}) & \text{if } \mathbf{z}^T\mathbf{z} \leq 1, \\ 0 & \text{otherwise} \end{cases} \tag{10.43}$$

is optimal with respect to the asymptotic mean integrated squared error. However, as with univariate kernel density estimation, many other kernels provide nearly equivalent results.

The single fixed bandwidth in (10.42) means that probability contributions associated with each observed data point will diffuse in all directions equally. When the data have very different variability in different directions, or when the data nearly lie on a lower-dimensional manifold, treating all dimensions as if they were on the same scale can lead to poor estimates. Fukunaga [186] suggested linearly transforming the data so that they have identity covariance matrix, then estimating the density of the transformed data using (10.42) with a radially symmetric kernel, and then back-transforming to obtain the final estimate. To carry out the transformation, compute an eigenvalue–eigenvector decomposition of the sample covariance matrix so $\hat{\Sigma} = \mathbf{P}\mathbf{\Lambda}\mathbf{P}^T$, where $\mathbf{\Lambda}$ is a $p \times p$ diagonal matrix with the eigenvalues in descending order and $\mathbf{P}$ is an orthonormal $p \times p$ matrix whose columns consist of the eigenvectors corresponding to the eigenvalues in $\mathbf{\Lambda}$. Let $\overline{\mathbf{X}}$ be the sample mean. Then setting $\mathbf{Z}_i = \mathbf{\Lambda}^{-1/2}\mathbf{P}^T(\mathbf{X}_i - \overline{\mathbf{X}})$ for $i = 1, \ldots, n$ provides the transformed data. This process is commonly called *whitening* or *sphering* the data. Using the kernel density estimator in (10.42) on the transformed data is equivalent to using the density estimator

$$\frac{|\hat{\Sigma}|^{-1/2}}{nh^p} \sum_{i=1}^{n} K\left(\frac{(\mathbf{x} - \mathbf{X}_i)^T \hat{\Sigma}^{-1} (\mathbf{x} - \mathbf{X}_i)}{h} \right) \qquad (10.44)$$

on the original data, for a symmetric kernel K.

Within the range of complexity presented by the choices above, the product kernel approach in (10.41) is usually preferred to (10.42) and (10.44), in view of its performance and flexibility. Using a product kernel also simplifies the numerical calculation and scaling of kernels.

As with the univariate case, it is possible to derive an expression for the asymptotic mean integrated squared error for a product kernel density estimator. The minimizing bandwidths $h_1, \ldots, h_p$ are the solutions to a set of p nonlinear equations. The optimal h_i are all $\mathcal{O}(n^{-1/(p+4)})$, and $\text{AMISE}(h_1, \ldots, h_p) = \mathcal{O}(n^{-1/(p+4)})$ for these optimal h_i. Bandwidth selection for product kernel density estimators and other multivariate approaches is far less well studied than in the univariate case.

Perhaps the simplest approach to bandwidth selection in this case is to assume that f is normal, thereby simplifying the minimization of $\text{AMISE}(h_1, \ldots, h_p)$ with respect to $h_1, \ldots, h_p$. This provides a bandwidth selection rationale akin to Silverman's rule of thumb in the univariate case. The resulting bandwidths for the normal product kernel approach are

$$h_i = \left(\frac{4}{n(p+2)} \right)^{1/(p+4)} \hat{\sigma}_i \qquad (10.45)$$

for $i = 1, \ldots, p$, where $\hat{\sigma}$ is an estimate of the standard deviation along the ith coordinate. As with the univariate case, using a robust scale estimator can improve performance. When nonnormal kernels are being used, the bandwidth for the normal kernel can be rescaled using (10.32) and Table 10.1 to provide an analogous bandwidth for the chosen kernel.

Terrell's maximal smoothing principle can also be applied to p-dimensional problems. Suppose we apply the general kernel density estimator given by (10.40) with a kernel function that is a density with identity covariance matrix. Then the maximal smoothing principle indicates choosing a bandwidth matrix **H** that satisfies

$$\mathbf{H}\mathbf{H}^T = \left[\frac{(p+8)^{(p+6)/2}\pi^{p/2}R(K)}{16n(p+2)\Gamma((p+8)/2)} \right]^{2/(p+4)} \widehat{\Sigma}, \qquad (10.46)$$

where $\widehat{\Sigma}$ is the sample covariance matrix. One could apply this result to find the maximal smoothing bandwidths for a normal product kernel, then rescale the coordinatewise bandwidths using (10.32) and Table 10.1 if another product kernel shape was desired.

Cross-validation methods can also be generalized to the multivariate case, as can some other automatic bandwidth selection procedures. However, the overall performance of such methods in general p-dimensional problems is not thoroughly documented.

10.4.3 Adaptive kernels and nearest neighbors

With ordinary fixed-kernel density estimation, the shape of K and the bandwidth are fixed. These determine an unchanging notion of proximity. Weighted contributions from nearby $\mathbf{X}_i$ determine $\hat{f}(\mathbf{x})$, where the weights are based on the proximity of $\mathbf{X}_i$ to $\mathbf{x}$. For example, with a uniform kernel, the estimate is based on variable numbers of observations within a sliding window of fixed shape.

It is worthwhile to consider the opposite viewpoint: allowing regions to vary in size, but requiring them (in some sense) to have a fixed number of observations falling in them. Then regions of larger size correspond to areas of low density, and regions of small size correspond to areas of high density.

It turns out that estimators derived from this principle can be written as kernel estimators with a changing bandwidth that adapts to the local density of observed data points. Such approaches are variously termed *adaptive kernel estimators*, *variable-bandwidth kernel estimators*, or *variable-kernel estimators*. Three particular strategies are reviewed below.

The motivation for adaptive methods is that a fixed bandwidth may not be equally suitable everywhere. In regions where data are sparse, wider bandwidths can help prevent excessive local sensitivity to outliers. Conversely, where data are abundant, narrower bandwidths can prevent bias introduced by oversmoothing. Consider again the kernel density estimate of bowhead whale calf migration times given in Figure 10.5 using the fixed Sheather–Jones bandwidth. For migration times below 1200 and above 1270 hours, the estimate exhibits a number of modes, yet it is unclear how many of these are true and how many are artifacts of sampling variability. It is not possible to increase the bandwidth sufficiently to smooth away some of the small local modes in the tails without also smoothing away the prominent bimodality between 1200 and 1270. Only local changes to the bandwidth will permit such improvements.

In theory, when $p = 1$ there is little to recommend adaptive methods over simpler approaches, but in practice some adaptive methods have been demonstrated to be quite effective in some examples. For moderate or large p, theoretical analysis suggests that the performance of adaptive methods can be excellent compared to standard kernel estimators, but the practical performance of adaptive approaches in such cases is not thoroughly understood. Some performance comparisons for adaptive methods can be found in [312, 492, 532].

10.4.3.1   *Nearest neighbor approaches*   The k*th nearest neighbor density estimator,*

$$\hat{f}(\mathbf{x}) = \frac{k}{nV_p d_k(\mathbf{x})^p}, \tag{10.47}$$

was the first approach to explicitly adopt a variable-bandwidth viewpoint [362]. For this estimator, $d_k(\mathbf{x})$ is the Euclidean distance from $\mathbf{x}$ to the kth nearest observed data point, and V_p is the volume of the unit sphere in p dimensions, where p is the dimensionality of the data. Since $V_p = \pi^{p/2}/\Gamma(p/2 + 1)$, note that $d_k(\mathbf{x})$ is the only random quantity in (10.47), as it depends on $\mathbf{X}_1, \ldots, \mathbf{X}_n$. Conceptually, the kth nearest neighbor estimate of the density at $\mathbf{x}$ is k/n divided by the volume of the smallest sphere centered at $\mathbf{x}$ that contains k of the n observed data values. The number of nearest neighbors, k, plays a role analogous to that of bandwidth: Large values of k yield smooth estimates, and small values of k yield wiggly estimates.

The estimator (10.47) may be viewed as a kernel estimator with a bandwidth that varies with $\mathbf{x}$ and a kernel equal to the density function that is uniform on the unit sphere in p dimensions. For an arbitrary kernel, the nearest neighbor estimator can be written as

$$\hat{f}(\mathbf{x}) = \frac{1}{nd_k(\mathbf{x})^p} \sum_{i=1}^{n} K\left(\frac{\mathbf{x} - \mathbf{X}_i}{d_k(\mathbf{x})}\right). \tag{10.48}$$

If $d_k(\mathbf{x})$ is replaced with an arbitrary function $h_k(x)$, which may not explicitly represent distance, the name *balloon estimator* has been suggested, because the bandwidth inflates or deflates through a function that depends on $\mathbf{x}$ [532]. The nearest neighbor estimator is asymptotically of this type: For example, using $d_k(\mathbf{x})$ as the bandwidth for the uniform-kernel nearest neighbor estimator is asymptotically equivalent to using a balloon estimator bandwidth of $h_k(\mathbf{x}) = \left(\frac{k}{nV_p f(\mathbf{x})}\right)^{1/p}$, since $\frac{k}{nV_p d_k(\mathbf{x})^p} \to f(\mathbf{x})$ as $n \to \infty$, $k \to \infty$, and $k/n \to 0$.

Nearest neighbor and balloon estimators exhibit a number of surprising attributes. First, choosing K to be a density does not ensure that $\hat{f}$ is a density; for instance, the estimator in (10.47) does not have a finite integral. Second, when $p = 1$ and K is a density with zero mean and unit variance, choosing $h_k(x) = \frac{k}{2nf(x)}$ does not offer any asymptotic improvement relative to a standard kernel estimator, regardless of the choice of k [492]. Finally, one can show that the pointwise asymptotic mean squared error of a univariate balloon estimator is minimized when $h_k(x) = h(x) = \left(\frac{f(x)R(K)}{nf''(x)}\right)^{1/5}$. Even with this optimal pointwise adaptive bandwidth, however, the asymptotic efficiency of univariate balloon estimators does not greatly exceed that of ordinary fixed-bandwidth kernel estimators when f is roughly symmetric and unimodal [532]. Thus, nearest neighbor and balloon estimators seem a poor choice when $p = 1$.

On the other hand, for multivariate data, balloon estimators offer much more promise. The asymptotic efficiency of the balloon estimator can greatly surpass that of a standard multivariate kernel estimator, even for fairly small p and symmetric, unimodal data [532]. If we further generalize (10.48) to

$$\hat{f}(\mathbf{x}) = \frac{1}{n|\mathbf{H}(\mathbf{x})|} \sum_{i=1}^{n} K\left(\mathbf{H}(\mathbf{x})^{-1}(\mathbf{x} - \mathbf{X}_i)\right), \tag{10.49}$$

where $\mathbf{H}(\mathbf{x})$ is a bandwidth matrix that varies with $\mathbf{x}$, then we have effectively allowed the shape of kernel contributions to vary with $\mathbf{x}$. When $\mathbf{H}(\mathbf{x}) = h_k(\mathbf{x})\mathbf{I}$, the general form reverts to the balloon estimator. Further, setting $h_k(\mathbf{x}) = d_k(\mathbf{x})$ yields the nearest neighbor estimator in (10.48). More general choices for $\mathbf{H}(\mathbf{x})$ are mentioned in [532].

10.4.3.2 *Variable-kernel approaches and transformations* A *variable-kernel* or *sample point adaptive estimator* can be written as

$$\hat{f}(\mathbf{x}) = \frac{1}{n} \sum_{i=1}^{n} \frac{1}{h_i^p} K\left(\frac{\mathbf{x} - \mathbf{X}_i}{h_i}\right), \tag{10.50}$$

where K is a multivariate kernel and h_i is a bandwidth individualized for the kernel contribution centered at $\mathbf{X}_i$ [60]. For example, h_i might be set equal to the distance from $\mathbf{X}_i$ to the kth nearest other observed data point, so $h_i = d_k(\mathbf{X}_i)$. A more general variable kernel estimator with bandwidth matrix $\mathbf{H}_i$ that depends on the ith sampled point is also possible (cf. (10.49)), but we focus on the simpler form here.

The variable-kernel estimator in (10.50) is a mixture of kernels with identical shape but different scales, centered at each observation. Letting the bandwidth vary as a function of $\mathbf{X}_i$ rather than of $\mathbf{x}$ guarantees that $\hat{f}$ is a density whenever K is a density.

Optimal bandwidths for variable-kernel approaches depend on f. Pilot estimation of f can be used to guide bandwidth adaptation. Consider the following general strategy:

1. Construct a pilot estimator $\tilde{f}(\mathbf{x})$ which is strictly positive for all observed $\mathbf{x}_i$. Pilot estimation might employ, for example, a normal product kernel density estimator with bandwidth chosen according to (10.45). If $\tilde{f}$ is based on an estimator that may equal or approach zero at some x_i, then let $\tilde{f}(\mathbf{x})$ equal the estimated density whenever the estimate exceeds ϵ, and let $\tilde{f}(\mathbf{x}) = \epsilon$ otherwise. The choice of an arbitrary small constant $\epsilon > 0$ improves performance by providing an upper bound for adaptively chosen bandwidths.

2. Let the adaptive bandwidth be $h_i = h/\tilde{f}(\mathbf{X}_i)^\alpha$, for a sensitivity parameter $0 \leq \alpha \leq 1$. The parameter h assumes the role of a bandwidth parameter that can be adjusted to control the overall smoothness of the final estimate.

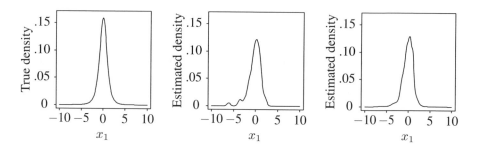

Fig. 10.8 Results from Example 10.7. From left to right, the three panels show the bivariate density value along the one-dimensional slice for which $x_2 = 0$ for: the true bivariate t distribution with two degrees of freedom, the bivariate estimate using a fixed-bandwidth product kernel approach, and the bivariate estimate using Abramson's adaptive approach as described in the text.

3. Apply the variable-kernel estimator in (10.50) with the bandwidths h_i found in step 2 to produce the final estimate.

The parameter α affects the degree of local adaptation by controlling how quickly the bandwidth changes in response to suspected changes in f. Asymptotic arguments and practical experience support setting $\alpha = 1/2$, which yields the approach of Abramson [3]. Several investigators have found this approach to perform well in practice [507, 575].

An alternative proposal is $\alpha = 1/p$, which yields an approach that is asymptotically equivalent to the adaptive kernel estimator of Breiman, Meisel, and Purcell [60]. This choice ensures that the number of observed data points captured by the scaled kernel will be roughly equal everywhere [507]. In their algorithm, these authors used a nearest neighbor approach for $\tilde{f}$ and set $h_i = h d_k(\mathbf{X}_i)$ for a smoothness parameter h that may depend on k.

Example 10.7 (Bivariate t distribution) To illustrate the potential benefit of adaptive approaches, consider estimating the bivariate t distribution (with two degrees of freedom) from a sample of size $n = 500$. In the nonadaptive approach, we use a normal product kernel with individual bandwidths chosen using the Sheather–Jones approach. As an adaptive alternative, we use Abramson's variable-kernel approach ($\alpha = 1/2$) with a normal product kernel, the pilot estimate taken to be the result of the nonadaptive approach, $\epsilon = 0.005$, and h set equal to the mean of the coordinate-wise bandwidths used in the nonadaptive approach times the geometric mean of the $\tilde{f}(\mathbf{X}_i)^{1/2}$.

The left panel of Figure 10.8 shows the true values of the bivariate t distribution with two degrees of freedom, f, along the line $x_2 = 0$. In other words, this shows a slice from the true density. The center panel of Figure 10.8 shows the result of the nonadaptive approach. The tails of the estimate exhibit undesirable wiggliness caused by an inappropriately narrow bandwidth in the tail regions, where a few outliers fall. The right panel of Figure 10.8 shows the result from Abramson's approach. Bandwidths are substantially wider in the tail areas, thereby producing smoother estimates in these regions than were obtained from the fixed-bandwidth approach. Abramson's method also uses much narrower bandwidths near the estimated mode. There is a slight indication of this for our random sample, but the effect can sometimes be pronounced.   □

Having discussed the variable-kernel approach, emphasizing its application in higher dimensions, we next consider a related approach primarily used for univariate data. This method illustrates the potential advantage of data transformation for density estimation.

Wand, Marron, and Ruppert noted that conducting fixed-bandwidth kernel density estimation on data that have been nonlinearly transformed is equivalent to using a variable-bandwidth kernel estimator on the original data [554]. The transformation induces separate bandwidths h_i at each data point.

Suppose univariate data $X_1, \ldots, X_n$ are observed from a density f_X. Let

$$y = t_\lambda(x) = \sigma_X t_\lambda^*(x) / \sigma_{t_\lambda^*(X)} \tag{10.51}$$

denote a transformation, where t_λ^* is a monotonic increasing mapping of the support of f to the real line parameterized by λ, and σ_X^2 and $\sigma_{t_\lambda^*(X)}^2$ are the variances of X and $Y = t_\lambda^*(X)$ respectively. Then t_λ is a scale-preserving transformation that maps the random variable $X \sim f_X$ to Y having density

$$g_\lambda(y) = f_X(t_\lambda^{-1}(y)) \left| \frac{d}{dy} t_\lambda^{-1}(y) \right|. \tag{10.52}$$

For example, if X is a standard normal random variable and $t_\lambda^*(X) = \exp\{X\}$, then Y has the same variance as X. However, a window of fixed width 0.3 on the Y scale centered at any value y has variable width when back-transformed to the X scale: the width is roughly 2.76 when $x = -1$ but only 0.24 when $x = 1$. In practice, sample standard deviations or robust measures of spread may be used in t_λ to preserve scale.

Suppose we transform the data using t_λ to obtain $Y_1, \ldots, Y_n$, then construct a fixed-bandwidth kernel density estimate for these transformed data, and then back-transform the resulting estimate to the original scale to produce an estimate of f_X. From (10.18) we know that the bandwidth that minimizes AMISE(h) for a kernel estimate of g_λ is

$$h_\lambda = \left(\frac{R(K)}{n \sigma_K^4 R(g_\lambda'')} \right)^{1/5} \tag{10.53}$$

for a given choice of λ.

Since h_λ depends on the unknown density g_λ, a plug-in method is suggested to estimate $R(g_\lambda'')$ by $\hat{R}(g_\lambda'') = R(\hat{g}_\lambda'')$, where $\hat{g}$ is a kernel estimator using pilot bandwidth h_0. Wand, Marron, and Ruppert suggest using a normal kernel with

Silverman's rule of thumb to determine h_0, thereby yielding the estimator

$$\hat{R}(g''_\lambda) = \frac{1}{n^2 h_0^5} \sum_{i \neq j} \sum \phi^{(4)} \left(\frac{Y_i - Y_j}{h_0} \right), \tag{10.54}$$

where $h_0 = \sqrt{2}\hat{\sigma}_X \left(\frac{84\sqrt{\pi}}{5n^2} \right)^{1/13}$ and $\phi^{(4)}$ is the fourth derivative of the standard normal density [554]. Since t_λ is scale-preserving, the sample standard deviation of $X_1, \ldots, X_n$, say $\hat{\sigma}_X$, provides an estimate of the standard deviation of Y to use in the expression for h_0. Related derivative estimation ideas are discussed in [259, 492].

The familiar Box–Cox transformation [51],

$$t_\lambda(x) = \begin{cases} (x^\lambda - 1)/\lambda & \text{if } \lambda \neq 0, \\ \log x & \text{if } \lambda = 0, \end{cases} \tag{10.55}$$

is among the parameterized transformation families available for (10.51). When any good transformation will suffice, or in multivariate settings, it can be useful to rely upon the notion that the transformation should make the data more nearly symmetric and unimodal, because fixed-bandwidth kernel density estimation is known to perform well in this case.

This transformation approach to variable-kernel density estimation can work well for univariate skewed unimodal densities. Extensions to multivariate data are challenging, and applications to multimodal data can result in poor estimates. Without all the formalism outlined above, data analysts routinely transform variables to convenient scales using functions such as the log, often retaining this transformation thereafter for displaying results and even making inferences. When inferences on the original scale are preferred, one could pursue a transformation strategy based in graphical or quantitative assessments of the symmetry and unimodality achieved, rather than optimizing the transformation within a class of functions as described above.

10.4.4 Exploratory projection pursuit

Exploratory projection pursuit focuses on discovering low-dimensional structure in a high-dimensional density. The final density estimate is constructed by modifying a standard multivariate normal distribution to reflect the structure found. The approach described below follows Friedman [181], which extends previous work [185, 296].

In this subsection, reference will be made to a variety of density functions with assorted arguments. For notational clarity, we therefore assign a subscript to the density function to identify the random variable whose density is being discussed.

Let the data consist of n observations of p-dimensional variables, $\mathbf{X}_1, \ldots, \mathbf{X}_n \sim$ i.i.d. $f_{\mathbf{X}}$. Before beginning exploratory projection pursuit, the data are transformed to have mean $\mathbf{0}$ and variance–covariance matrix $\mathbf{I}_p$. This is accomplished using the *whitening* or *sphering* transformation described in Section 10.4.2. Let $f_{\mathbf{Z}}$ denote the density corresponding to the transformed variables, $\mathbf{Z}_1, \ldots, \mathbf{Z}_n$. Both $f_{\mathbf{Z}}$ and $f_{\mathbf{X}}$ are unknown. To estimate $f_{\mathbf{X}}$ it suffices to estimate $f_{\mathbf{Z}}$ and then reverse the transformation to obtain an estimate of $f_{\mathbf{X}}$. Thus our primary concern will be the estimation of $f_{\mathbf{Z}}$.

Several steps in the process rely on another density estimation technique, based on Legendre polynomial expansion. The Legendre polynomials are a sequence of orthogonal polynomials on $[-1, 1]$ defined by $P_0(u) = 1$, $P_1(u) = u$, and $P_j(u) = \left[(2j - 1)uP_{j-1}(u) - (j - 1)P_{j-2}(u) \right] / j$ for $j \geq 2$, having the property that the L_2 norm $\int_{-1}^{1} P_j^2(u)\, du = 2/(2j + 1)$ for all j [2, 479]. These polynomials can be used as a basis for representing functions on $[-1, 1]$. In particular, we can represent a univariate density f that has support only on $[-1, 1]$ by its Legendre polynomial expansion

$$f(x) = \sum_{j=0}^{\infty} a_j P_j(x), \tag{10.56}$$

where

$$a_j = \frac{2j + 1}{2} \mathrm{E}\{P_j(X)\} \tag{10.57}$$

and the expectation in (10.57) is taken with respect to f. Equation (10.57) can be confirmed by noting the orthogonality and L_2 norm of the P_j. If we observe $X_1, \ldots, X_n \sim$ i.i.d. f, then $\frac{1}{n} \sum_{i=1}^{n} P_j(X_i)$ is an estimator of $\mathrm{E}\{P_j(X)\}$. Therefore

$$\hat{a}_j = \frac{2j + 1}{2n} \sum_{i=1}^{n} P_j(X_i) \tag{10.58}$$

may be used as estimates of the coefficients in the Legendre expansion of f. Truncating the sum in (10.56) after $J + 1$ terms suggests the estimator

$$\hat{f}(x) = \sum_{j=0}^{J} \hat{a}_j P_j(x). \tag{10.59}$$

Having described this Legendre expansion approach, we can now move on to study exploratory projection pursuit.

The first step of exploratory projection pursuit is a projection step. If $Y_i = \boldsymbol{\alpha}^T \mathbf{Z}_i$, then we say that Y_i is the one-dimensional projection of $\mathbf{Z}_i$ in the direction $\boldsymbol{\alpha}$. The goal of the first step is to project the multivariate observations onto the one-dimensional line for which the distribution of the projected data has the most structure.

The degree of structure in the projected data is measured as the amount of departure from normality. Let $U(y) = 2\Phi(y) - 1$, where Φ is the standard normal cumulative distribution function. If $Y \sim N(0, 1)$, then $U(Y) \sim \mathrm{Unif}(-1, 1)$. To measure the structure in the distribution of Y it suffices to measure the degree to which the density of $U(Y)$ differs from $\mathrm{Unif}(-1, 1)$.

Define a *structure index* as

$$S(\boldsymbol{\alpha}) = \int_{-1}^{1} \left[f_U(u) - \tfrac{1}{2} \right]^2 du = R(f_U) - \tfrac{1}{2}, \tag{10.60}$$

where f_U is the probability density function of $U(\boldsymbol{\alpha}^T \mathbf{Z})$ when $\mathbf{Z} \sim f_{\mathbf{Z}}$. When $S(\boldsymbol{\alpha})$ is large, a large amount of nonnormal structure is present in the projected data. When

$S(\boldsymbol{\alpha})$ is nearly zero, the projected data are nearly normal. Note that $S(\boldsymbol{\alpha})$ depends on f_U, which must be estimated.

To estimate $S(\boldsymbol{\alpha})$ from the observed data, use the Legendre expansion for f_U to reexpress $R(f_U)$ in (10.60) as

$$R(f_U) = \sum_{j=0}^{\infty} \frac{2j+1}{2} \left[\mathrm{E}\{P_j(U)\} \right]^2, \tag{10.61}$$

where the expectations are taken with respect to f_U. Since $U(\boldsymbol{\alpha}^T\mathbf{Z}_1), \ldots, U(\boldsymbol{\alpha}^T\mathbf{Z}_n)$ represent draws from f_U, the expectations in (10.61) can be estimated by sample moments. If we also truncate the sum in (10.61) at $J+1$ terms, we obtain

$$\hat{S}(\boldsymbol{\alpha}) = \sum_{j=0}^{J} \frac{2j+1}{2} \left(\frac{1}{n} \sum_{i=1}^{n} P_j(2\Phi(\boldsymbol{\alpha}^T\mathbf{Z}_i) - 1) \right)^2 - \frac{1}{2} \tag{10.62}$$

as an estimator of $S(\boldsymbol{\alpha})$.

Thus, to estimate the projection direction yielding the greatest nonnormal structure, we maximize $\hat{S}(\boldsymbol{\alpha})$ with respect to $\boldsymbol{\alpha}$, subject to the constraint that $\boldsymbol{\alpha}^T\boldsymbol{\alpha} = 1$. Denote the resulting direction by $\hat{\boldsymbol{\alpha}}_1$. Although $\hat{\boldsymbol{\alpha}}_1$ is estimated from the data, we treat it as a fixed quantity when discussing distributions of projections of random vectors onto it. For example, let $f_{\hat{\boldsymbol{\alpha}}_1^T\mathbf{Z}}$ denote the univariate marginal density of $\hat{\boldsymbol{\alpha}}_1^T\mathbf{Z}$ when $\mathbf{Z} \sim f_{\mathbf{Z}}$, treating $\mathbf{Z}$ as random and $\hat{\boldsymbol{\alpha}}_1$ as fixed.

The second step of exploratory projection pursuit is a structure removal step. The goal is to apply a transformation to $\mathbf{Z}_1, \ldots, \mathbf{Z}_n$ which makes the density of the projection of $f_{\mathbf{Z}}$ on $\hat{\boldsymbol{\alpha}}_1$ a standard normal density, while leaving the distribution of a projection along any orthogonal direction unchanged. To do this, let $\mathbf{A}_1$ be an orthonormal matrix with first row equal to $\hat{\boldsymbol{\alpha}}_1^T$. Also, for observations from a random vector $\mathbf{V} = (V_1, \ldots, V_p)$, define the vector transformation $\mathbf{T}(\mathbf{v}) = \left(\Phi^{-1}(F_{V_1}(v_1)), v_2, \ldots, v_p \right)$, where F_{V_1} is the cumulative distribution function of the first element of $\mathbf{V}$. Then letting

$$\mathbf{Z}_i^{(1)} = \mathbf{A}_1^T \mathbf{T}(\mathbf{A}_1 \mathbf{Z}_i) \tag{10.63}$$

for $i = 1, \ldots, n$ would achieve the desired transformation. The transformation in (10.63) cannot be used directly to achieve the structure removal goal, because it depends on the cumulative distribution function corresponding to $f_{\hat{\boldsymbol{\alpha}}_1^T\mathbf{Z}}$. To get around this problem, simply replace the cumulative distribution function with the corresponding empirical distribution function of $\hat{\boldsymbol{\alpha}}_1^T\mathbf{Z}_1, \ldots, \hat{\boldsymbol{\alpha}}_1^T\mathbf{Z}_n$. An alternative replacement is suggested in [298].

The $\mathbf{Z}_i^{(1)}$ for $i = 1, \ldots, n$ may be viewed as a new dataset consisting of the observed values of random variables $\mathbf{Z}_1^{(1)}, \ldots, \mathbf{Z}_n^{(1)}$ whose unknown distribution $f_{\mathbf{Z}^{(1)}}$ depends on $f_{\mathbf{Z}}$. There is an important relationship between the conditionals determined by $f_{\mathbf{Z}^{(1)}}$ and $f_{\mathbf{Z}}$ given a projection onto $\hat{\boldsymbol{\alpha}}_1$. Specifically, the conditional distribution of $\mathbf{Z}_i^{(1)}$ given $\hat{\boldsymbol{\alpha}}_1^T\mathbf{Z}_i^{(1)}$ equals the conditional distribution of $\mathbf{Z}_i$ given $\hat{\boldsymbol{\alpha}}_1^T\mathbf{Z}_i$ because the structure removal step creating $\mathbf{Z}_i^{(1)}$ leaves all coordinates of $\mathbf{Z}_i$ except the first unchanged. Therefore

$$f_{\mathbf{Z}^{(1)}}(\mathbf{z}) = \frac{f_{\mathbf{Z}}(\mathbf{z})\phi(\hat{\alpha}_1^T\mathbf{z})}{f_{\hat{\alpha}_1^T\mathbf{Z}}(\hat{\alpha}_1^T\mathbf{z})}. \tag{10.64}$$

Equation (10.64) provides no immediate way to estimate $f_{\mathbf{Z}}$, but iterating the entire process described above will eventually prove fruitful.

Suppose a second projection step is conducted. A new direction to project the working variables $\mathbf{Z}_1^{(1)}, \ldots, \mathbf{Z}_n^{(1)}$ is sought to isolate the greatest amount of one-dimensional structure. Finding this direction requires the calculation of a new structure index based on the transformed sample $\mathbf{Z}_1^{(1)}, \ldots, \mathbf{Z}_n^{(1)}$, leading to the estimation of $\hat{\alpha}_2$ as the projection direction revealing greatest structure.

Taking a second structure removal step requires the reapplication of equation (10.63) with a suitable matrix $\mathbf{A}_2$, yielding new working variables $\mathbf{Z}_1^{(2)}, \ldots, \mathbf{Z}_n^{(2)}$.

Iterating the same conditional distribution argument as expressed in (10.64) allows us to write the density from which the new working data arise as

$$f_{\mathbf{Z}^{(2)}}(\mathbf{z}) = f_{\mathbf{Z}}(\mathbf{z}) \frac{\phi(\hat{\alpha}_1^T\mathbf{z})\phi(\hat{\alpha}_2^T\mathbf{z})}{f_{\hat{\alpha}_1^T\mathbf{Z}}(\hat{\alpha}_1^T\mathbf{z})f_{\hat{\alpha}_2^T\mathbf{Z}^{(1)}}(\hat{\alpha}_2^T\mathbf{z})}, \tag{10.65}$$

where $f_{\hat{\alpha}_2^T\mathbf{Z}^{(1)}}$ is the marginal density of $\hat{\alpha}_2^T\mathbf{Z}^{(1)}$ when $\mathbf{Z}^{(1)} \sim f_{\mathbf{Z}^{(1)}}$.

Suppose the projection and structure removal steps are iterated several additional times. At some point, the identification and removal of structure will lead to new variables whose distribution has little or no remaining structure. In other words, their distribution will be approximately normal along any possible univariate projection. At this point, iterations are stopped. Suppose that a total of M iterations were taken. Then (10.65) extends to give

$$f_{\mathbf{Z}^{(M)}}(\mathbf{z}) = f_{\mathbf{Z}}(\mathbf{z}) \prod_{m=1}^{M} \frac{\phi(\hat{\alpha}_m^T\mathbf{z})}{f_{\hat{\alpha}_m^T\mathbf{Z}^{(m-1)}}(\hat{\alpha}_m^T\mathbf{z})}, \tag{10.66}$$

where $f_{\hat{\alpha}_m^T\mathbf{Z}^{(m-1)}}$ is the marginal density of $\hat{\alpha}_m^T\mathbf{Z}^{(m-1)}$ when $\mathbf{Z}^{(m-1)} \sim f_{\mathbf{Z}^{(m-1)}}$, and $\mathbf{Z}^{(0)} \sim f_{\mathbf{Z}}$.

Now, equation (10.66) can be used to estimate $f_{\mathbf{Z}}$, because—having eliminated all structure from the distribution of our working variables $\mathbf{Z}_i^{(M)}$—we may set $f_{\mathbf{Z}^{(M)}}$ equal to a p-dimensional multivariate normal density, denoted ϕ_p. Solving for $f_{\mathbf{Z}}$ gives

$$f_{\mathbf{Z}}(\mathbf{z}) = \phi_p(\mathbf{z}) \prod_{m=1}^{M} \frac{f_{\hat{\alpha}_m^T\mathbf{Z}^{(m-1)}}(\hat{\alpha}_m^T\mathbf{z})}{\phi(\hat{\alpha}_m^T\mathbf{z})}. \tag{10.67}$$

Although this equation still depends on the unknown densities $f_{\hat{\alpha}_m^T\mathbf{Z}^{(m-1)}}$, these can be estimated using the Legendre approximation strategy. Note that if $U^{(m-1)} = 2\Phi(\hat{\alpha}_m^T\mathbf{Z}^{(m-1)}) - 1$ for $\mathbf{Z}^{(m-1)} \sim f_{\mathbf{Z}^{(m-1)}}$ then

$$f_{U^{(m-1)}}(u) = \frac{f_{\hat{\alpha}_m^T\mathbf{Z}^{(m-1)}}\left(\Phi^{-1}\left((u+1)/2\right)\right)}{2\phi\left(\Phi^{-1}\left((u+1)/2\right)\right)}. \tag{10.68}$$

Use the Legendre expansion of $f_{U^{(m-1)}}$ and sample moments to estimate

$$\hat{f}_{U^{(m-1)}}(u) = \sum_{j=0}^{J} \left\{ \frac{2j+1}{2} P_j(u) \sum_{i=1}^{n} P_j(U_i^{(m-1)})/n \right\} \qquad (10.69)$$

from $U_1^{(m-1)}, \ldots, U_n^{(m-1)}$ derived from $\mathbf{Z}_1^{(m-1)}, \ldots, \mathbf{Z}_n^{(m-1)}$. After substituting $\hat{f}_{U^{(m-1)}}$ for $f_{U^{(m-1)}}$ in (10.68) and isolating $f_{\hat{\boldsymbol{\alpha}}_m^T \mathbf{Z}^{(m-1)}}$, we obtain

$$\hat{f}_{\hat{\boldsymbol{\alpha}}_m^T \mathbf{Z}^{(m-1)}}(\hat{\boldsymbol{\alpha}}_m^T \mathbf{z}) = 2\hat{f}_{U^{(m-1)}}\left(2\Phi(\hat{\boldsymbol{\alpha}}_m^T \mathbf{z}) - 1\right)\phi(\hat{\boldsymbol{\alpha}}_m^T \mathbf{z}). \qquad (10.70)$$

Thus, from (10.67) the estimate for $f_{\mathbf{Z}}(\mathbf{z})$ is

$$\hat{f}(\mathbf{z}) = \phi_p(\mathbf{z}) \prod_{m=1}^{M} \left\{ \sum_{j=0}^{J} (2j+1) P_j \left(2\Phi(\hat{\boldsymbol{\alpha}}_m^T \mathbf{z}) - 1\right) \bar{P}_{jm} \right\}, \qquad (10.71)$$

where the

$$\bar{P}_{jm} = \frac{1}{n} \sum_{i=1}^{n} P_j \left(2\Phi(\hat{\boldsymbol{\alpha}}_m^T \mathbf{Z}_i^{(m-1)}) - 1\right) \qquad (10.72)$$

are estimated using the working variables accumulated during the structure removal process, and $\mathbf{Z}_i^{(0)} = \mathbf{Z}_i$. Reversing the sphering by applying the change of variable $\mathbf{X} = \mathbf{P}\boldsymbol{\Lambda}^{1/2}\mathbf{Z} + \bar{\mathbf{x}}$ to $\hat{f}_{\mathbf{Z}}$ provides the estimate $\hat{f}_{\mathbf{X}}$.

The estimate $\hat{f}_{\mathbf{Z}}$ is most strongly influenced by the central portion of the data, because the transformation U compresses information about the tails of $f_{\mathbf{Z}}$ into the extreme portions of the interval $[-1, 1]$. Low-degree Legendre polynomial expansion has only limited capacity to capture substantial features of f_U in these narrow margins of the interval. Furthermore, the structure index driving the choice of each $\hat{\boldsymbol{\alpha}}_m$ will not assign high structure to directions for which only the tail behavior of the projection is nonnormal. Therefore, exploratory projection pursuit should be viewed foremost as a way to extract key low-dimensional features of the density which are exhibited by the bulk of the data, and to reconstruct a density estimate that reflects these key features.

Example 10.8 (Bivariate rotation) To illustrate exploratory projection pursuit, we will attempt to reconstruct the density of some bivariate data. Let $\mathbf{W} = (W_1, W_2)$, where $W_1 \sim \text{Gamma}(4, 2)$ and $W_2 \sim N(0, 1)$ independently. Then $\text{E}\{\mathbf{W}\} = (2, 0)$ and $\text{var}\{\mathbf{W}\} = \mathbf{I}$. Use

$$\mathbf{R} = \begin{pmatrix} -0.581 & -0.814 \\ -0.814 & 0.581 \end{pmatrix}$$

to rotate $\mathbf{W}$ to produce data via $\mathbf{X} = \mathbf{R}\mathbf{W}$. Let $f_{\mathbf{X}}$ denote the density of $\mathbf{X}$, which we will try to estimate from a sample of $n = 500$ points drawn from $f_{\mathbf{X}}$. Since $\text{var}\{\mathbf{X}\} = \mathbf{R}\mathbf{R}^T = \mathbf{I}$, the whitening transformation is nearly just a translation (aside

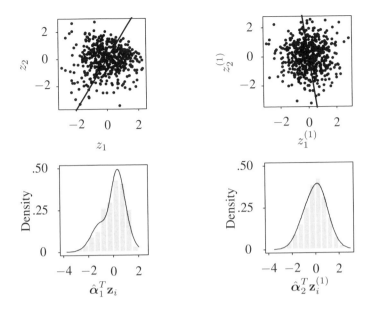

Fig. 10.9 The first two projection and structure removal steps for Example 10.8, as described in the text.

from the fact that the theoretical and sample variance–covariance matrices differ slightly).

The whitened data values, $z_1, \ldots, z_{500}$, are plotted in the top left panel of Figure 10.9. The underlying gamma structure is detectable in this graph as an abruptly declining frequency of points in the top right region of the plot: Z and X are rotated about 135 degrees counterclockwise with respect to W.

The direction $\hat{\alpha}_1$ that reveals the most univariate projected structure is shown with the line in the top left panel of Figure 10.9. Clearly this direction corresponds roughly to the original gamma-distributed coordinate. The bottom left panel of Figure 10.9 shows a histogram of the z_i values projected onto $\hat{\alpha}_1$, revealing a rather nonnormal distribution. The curve superimposed on this histogram is the univariate density estimate for $\hat{\alpha}_1^T Z$ obtained using the Legendre expansion strategy. Throughout this example, the number of Legendre polynomials was set to $J + 1 = 4$.

Removing the structure revealed by the projection on $\hat{\alpha}_1$ yields new working data values, $z_1^{(1)}, \ldots, z_{500}^{(1)}$, graphed in the top right panel of Figure 10.9. The projection direction, $\hat{\alpha}_2$, showing the most nonnormal structure is again shown with a line. The bottom right panel shows a histogram of $\hat{\alpha}_2^T z^{(1)}$ values and the corresponding Legendre density estimate.

At this point, there is little need to proceed with additional projection and structure removal steps: The working data are already nearly multivariate normal. Employing (10.71) to reconstruct an estimate of f_Z yields the density estimate shown in Fig-

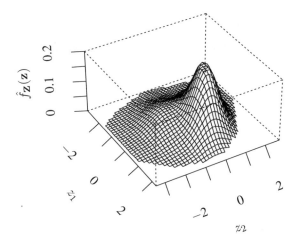

Fig. 10.10 The exploratory projection pursuit density estimate $\hat{f}_Z$ for Example 10.8.

ure 10.10. The rotated gamma–normal structure is clearly seen in the figure, with the heavier gamma tail extending leftward and the abrupt tail terminating on the right. The final step in application would be to reexpress this result in terms of a density for **X** rather than **Z**. $\qquad\square$

Problems

10.1 Sanders et al. provide a comprehensive dataset of infrared emissions and other characteristics of objects beyond our galaxy [478]. These data are available from the website for our book. Let X denote the log of the variable labeled F12, which is the total 12-μm-band flux measurement on each object.

(a) Fit a normal kernel density estimate for X, using bandwidths derived from the $UCV(h)$ criterion, Silverman's rule of thumb, the Sheather–Jones approach, Terrell's maximal smoothing principle, and any other approaches you wish. Comment on the apparent suitability of each bandwidth for these data.

(b) Fit kernel density estimates for X using uniform, normal, Epanechnikov, and triweight kernels, each with bandwidth equivalent to the Sheather–Jones bandwidth for a normal kernel. Comment.

(c) Fit a nearest neighbor density estimate for X as in (10.48) with the uniform and normal kernels. Next fit an Abramson adaptive estimate for X using a normal kernel and setting h equal to the Sheather–Jones bandwidth for a fixed-bandwidth estimator times the geometric mean of the $\tilde{f}_X(x_i)^{1/2}$ values.

(d) If code for logspline density estimation is available, experiment with this approach for estimating the density of X.

(e) Let $\hat{f}_X$ denote the normal kernel density estimate for X computed using the Sheather–Jones bandwidth. Note the ratio of this bandwidth to the bandwidth given by Silverman's rule of thumb. Transform the data back to the original scale (i.e., $Z = \exp\{X\}$), and fit a normal kernel density estimate $\hat{f}_Z$, using a bandwidth equal to Silverman's rule of thumb scaled down by the ratio noted previously. (This is an instance where the robust scale measure is far superior to the sample standard deviation.) Next, transform $\hat{f}_X$ back to the original scale using the change-of-variable formula for densities, and compare the two resulting estimates of density for Z on the region between 0 and 8. Experiment further to investigate the relationship between density estimation and nonlinear scale transformations. Comment.

10.2   This problem continues using the infrared data on extragalactic objects and the variable X (the log of the 12-μm-band flux measurement) from Problem 10.1. The dataset also includes F100 data: the total 100-μm-band flux measurements for each object. Denote the log of this variable by Y. Construct bivariate density estimates for the joint density of X and Y using the following approaches.

(a) Use a standard bivariate normal kernel with bandwidth matrix $h\mathbf{I}_2$. Describe how you chose h.

(b) Use a bivariate normal kernel with bandwidth matrix $\mathbf{H}$ chosen by Terrell's maximal smoothing principle. Find a constant c for which the bandwidth matrix $c\mathbf{H}$ provides a superior density estimate.

(c) Use a normal product kernel with the bandwidth for each coordinate chosen using the Sheather–Jones approach.

(d) Use a nearest neighbor estimator (10.48) with a normal kernel. Describe how you chose k.

(e) Use an Abramson adaptive estimator with the normal product kernel and bandwidths chosen in the same manner as Example 10.7.

10.3   Starting from equation (10.22), derive a simplification for UCV(h) when $K(z) = \phi(z) = \frac{\exp\{-z^2/2\}}{\sqrt{2\pi}}$, pursuing the following steps:

(a) Show that

$$\mathrm{UCV}(h) = \frac{1}{n^2 h^2} \sum_{i=1}^{n} \int K^2 \left(\frac{x - X_i}{h} \right) dx$$

$$+ \frac{1}{n(n-1)h^2} \sum_{i=1}^{n} \sum_{j \neq i} \int K \left(\frac{x - X_i}{h} \right) K \left(\frac{x - X_j}{h} \right) dx$$

$$- \frac{2}{n(n-1)h} \sum_{i=1}^{n} \sum_{j \neq i} K \left(\frac{X_i - X_j}{h} \right)$$

$$= A + B + C,$$

where A, B, and C denote to the three terms given above.

(b) Show that $A = \frac{1}{2nh\sqrt{\pi}}$.

(c) Show that

$$B = \frac{1}{2n(n-1)h\sqrt{\pi}} \sum_{i=1}^{n} \sum_{j \neq i} \exp \left\{ \frac{-1}{4h^2} (X_i - X_j)^2 \right\}. \qquad (10.73)$$

(d) Finish by showing (10.23).

10.4   Replicate the first four rows of Table 10.2. Assume $\hat{f}$ is a product kernel estimator. You may find it helpful to begin with the expression $\mathrm{MSE}_h(\hat{f}(\mathbf{x})) = \mathrm{var}\{\hat{f}(\mathbf{x})\} + \left(\mathrm{bias}\{\hat{f}(x)\}\right)^2$, and to use the result

$$\phi(x; \mu, \sigma^2)\phi(x; \nu, \tau^2) = \phi \left(x; \frac{\mu\tau^2 + \nu\sigma^2}{\sigma^2 + \tau^2}, \frac{\sigma^2\tau^2}{\sigma^2 + \tau^2} \right) \left(\frac{\exp\left\{ -\frac{(\mu-\nu)^2}{2(\sigma^2+\tau^2)} \right\}}{\sqrt{2\pi(\sigma^2 + \tau^2)}} \right),$$

where $\phi(x; \alpha, \beta^2)$ denotes a univariate normal density function with mean α and variance β^2.

10.5   Available from the website for this book are some manifold data exhibiting some strong structure. Specifically, these four-dimensional data come from a mixture distribution, with a low weighting of a density that lies nearly on a three-dimensional manifold and a high weighting of a heavy-tailed density that fills four-dimensional space.

(a) Estimate the direction of the least normal univariate projection of these data. Use a sequence of graphs to guess a nonnormal projection direction, or follow the method described for the projection step of exploratory projection pursuit.

(b) Estimate the univariate density of the data projected in the direction found in part (a), using any means you wish.

(c) Use the ideas in this chapter to estimate and/or describe the density of these data via any productive means. Discuss the difficulties you encounter.

11

Bivariate Smoothing

Consider the bivariate data shown in Figure 11.1. If asked, virtually anyone could draw a smooth curve that fits the data well, yet most would find it surprisingly difficult to describe precisely how they had done it. We focus here on a variety of methods for this task, called *scatterplot smoothing*.

Effective smoothing methods for bivariate data are usually much simpler than for higher-dimensional problems; therefore we initially limit consideration to the case of n bivariate data points (x_i, y_i), $i = 1, \ldots, n$. Chapter 12 covers smoothing multivariate data.

The goal of smoothing is different for predictor–response data than for general bivariate data. With predictor–response data, the random response variable Y is assumed to be a function (probably stochastic) of the value of a predictor variable X. For example, a model commonly assumed for predictor–response data is $Y_i = s(x_i) + \epsilon_i$, where the ϵ_i are zero-mean stochastic noise and s is a smooth function. In this case, the conditional distribution of $Y|x$ describes how Y depends on $X = x$. One sensible smooth curve through the data would connect the conditional means of $Y|x$ for the range of predictor values observed.

In contrast to predictor–response data, general bivariate data have the characteristic that neither X or Y is distinguished as the response. In this case, it is sensible to summarize the joint distribution of (X, Y). One smooth curve that would capture a primary aspect of the relationship between X and Y would correspond to the ridge top of their joint density; there are other reasonable choices, too. Estimating such relationships can be considerably more challenging than smoothing predictor–response data; see Sections 11.6 and 12.2.1.

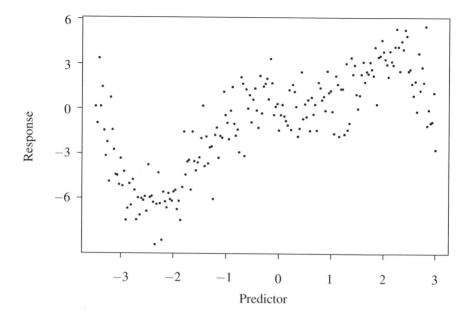

Fig. 11.1   Predictor–response data. A smooth curve sketched through these data would likely exhibit several peaks and troughs.

Detailed discussion of smoothing techniques includes [86, 164, 268, 269, 273, 280, 484, 508, 544, 553].

11.1   PREDICTOR–RESPONSE DATA

Suppose that $E\{Y|x\} = s(x)$ for a smooth function s. Because smoothing predictor–response data usually focuses on estimation of the conditional mean function s, smoothing is often called *nonparametric regression*.

For a given point x, let $\hat{s}(x)$ be an estimator of $s(x)$. What estimator is best? One natural approach is to assess the quality of $\hat{s}(x)$ as an estimator of $s(x)$ at x using the mean squared error (of estimation) at x, namely $\mathrm{MSE}(\hat{s}(x)) = E\{[\hat{s}(x) - s(x)]^2\}$, where the expectation is taken with respect to the joint distribution of the responses. By adding and subtracting $E\{\hat{s}(x)|x\}$ inside the squared term in this expression, it is straightforward to obtain the familiar result that

$$\mathrm{MSE}(\hat{s}(x)) = (\mathrm{bias}\,\{\hat{s}(x)\})^2 + \mathrm{var}\,\{\hat{s}(x)\}, \qquad (11.1)$$

where $\mathrm{bias}\{\hat{s}(x)\} = E\{\hat{s}(x)\} - s(x)$.

Although we motivate smoothing by considering estimation of conditional means under squared error loss, alternative viewpoints are reasonable. For example, using absolute error loss shifts focus to the median$\{Y|x\}$. Thus, smoothing may be seen more generally as an attempt to describe how the center of the distribution of $Y|x$ varies with x, for some notion of what constitutes the center.

The smoother $\hat{s}(x)$ is usually based not only on the observed data (x_i, y_i), for $i = 1, \ldots, n$, but also on a user-specified *smoothing parameter* λ, whose value is chosen to control the overall behavior of the smoother. Thus, we often write $\hat{s}_\lambda$ and $\text{MSE}_\lambda(\hat{s}_\lambda(x))$ hereafter.

Consider prediction of the response at a new point x^*, using the smoother $\hat{s}_\lambda$. We introduced $\text{MSE}_\lambda(\hat{s}_\lambda(x^*))$ to assess the quality of $\hat{s}_\lambda(x^*)$ as an estimator of the true conditional mean, $s(x^*) = \text{E}\{Y|X = x^*\}$. Now, to assess the quality of the smoother as a predictor of a single response at $X = x^*$, we use the *mean squared prediction error* at x^*, namely

$$\begin{aligned}
\text{MSPE}_\lambda(\hat{s}_\lambda(x^*)) &= \text{E}\{(Y - \hat{s}_\lambda(x^*))^2 \mid X = x^*\} \\
&= \text{var}\{Y|X = x^*\} + \text{MSE}_\lambda(\hat{s}_\lambda(x^*)).
\end{aligned} \tag{11.2}$$

More should be required of $\hat{s}_\lambda$ beyond good prediction at a single x^*. If $\hat{s}_\lambda$ is a good smoother, it should limit $\text{MSPE}_\lambda(\hat{s}_\lambda(x))$ over a range of x. For the observed dataset, a good global measure of the quality of $\hat{\mathbf{s}}_\lambda = (\hat{s}_\lambda(x_1), \ldots, \hat{s}_\lambda(x_n))$ would be $\overline{\text{MSPE}}_\lambda(\hat{\mathbf{s}}_\lambda) = \frac{1}{n}\sum_{i=1}^n \text{MSPE}_\lambda(\hat{s}_\lambda(x_i))$, namely the average mean squared prediction error. There are other good global measures of the quality of a smooth, but in many cases the choice is asymptotically unimportant in the sense that they provide equivalent asymptotic guidance about optimal smoothing [272].

Having discussed theoretical measures of performance of smoothers, we now turn our focus to practical methods for constructing smoothers that perform well. For predictor–response data, it's difficult to resist the notion that a smoother should summarize the conditional distribution of Y_i given $X_i = x_i$ by some measure of location like the conditional mean, even if the model $Y_i = s(x_i) + \epsilon_i$ is not assumed explicitly. In fact, regardless of the type of data, nearly all smoothers rely on the concept of *local averaging*: the Y_i whose corresponding x_i are near x should be averaged in some way to glean information about the appropriate value of the smooth at x.

A generic local-averaging smoother can be written as

$$\hat{s}(x) = \text{ave}\{Y_i|x_i \in \mathcal{N}(x)\} \tag{11.3}$$

for some generalized average function "ave" and some neighborhood of x, say $\mathcal{N}(x)$. Different smoothers result from different choices for the averaging function (e.g., mean, weighted mean, median, or M-estimate) and the neighborhood (e.g., the nearest few neighboring points, or all points within some distance). In general, the form of $\mathcal{N}(x)$ may vary with x so that different neighborhood sizes or shapes may be used in different regions of the dataset.

The most important characteristic of a neighborhood is its *span*, which is represented by the smoothing parameter λ. In a general sense, the span of a neighborhood measures its inclusiveness: neighborhoods with small span are strongly local, including only very nearby points, whereas neighborhoods with large span have wider membership. There are many ways to measure a neighborhood's inclusiveness, including its *size* (number of points), *span* (proportion of sample points that are members), *bandwidth* (physical length or volume of the neighborhood), and other concepts discussed later. We use λ to denote whichever concept is most natural for each smoother.

The smoothing parameter controls the wiggliness of $\hat{s}_\lambda$. Smoothers with small spans tend to reproduce local patterns very well, but draw little information from more distant data. A smoother that ignores distant data containing useful information about the local response will have higher variability than could otherwise be achieved. In contrast, smoothers with large spans draw lots of information from distant data when making local predictions. When these data are of questionable relevance, potential bias is introduced. Adjusting λ controls this tradeoff between bias and variance.

Below we introduce some strategies for constructing local-averaging smoothers. This chapter focuses on smoothing methods for predictor–response data, but Section 11.6 briefly addresses issues regarding smoothing general bivariate data, which are further considered in Chapter 12.

11.2 LINEAR SMOOTHERS

An important class of smoothers are the *linear smoothers*. For such smoothers, the prediction at any point x is a linear combination of the response values. Linear smoothers are faster to compute and easier to analyze than nonlinear smoothers.

Frequently it suffices to consider estimation of the smooth at only the observed x_i points. For a vector of predictor values, $\mathbf{x} = (x_1 \ \ldots \ x_n)^T$, denote the vector of corresponding response variables as $\mathbf{Y} = (Y_1 \ \ldots \ Y_n)^T$, and define $\hat{\mathbf{s}} = (\hat{s}(x_1) \ \ldots \ \hat{s}(x_n))^T$. Then a linear smoother can be expressed as $\hat{\mathbf{s}} = \mathbf{S}\mathbf{Y}$ for an $n \times n$ *smoothing matrix* $\mathbf{S}$ whose entries do not depend on $\mathbf{Y}$. A variety of linear smoothers are introduced below.

11.2.1 Constant-span running mean

A very simple smoother takes the sample mean of k nearby points:

$$\hat{s}_k(x_i) = \sum_{\{j:\, x_j \in \mathcal{N}(x_i)\}} Y_j/k. \tag{11.4}$$

Insisting on odd k, we define $\mathcal{N}(x_i)$ as x_i itself, the $(k-1)/2$ points whose predictor values are nearest below x_i, and the $(k-1)/2$ points whose predictor values are nearest above x_i. This $\mathcal{N}(x_i)$ is termed the *symmetric nearest neighborhood*, and the smoother is sometimes called a *moving average*.

Without loss of generality, assume hereafter that the data pairs have been sorted so that the x_i are in increasing order. Then the constant-span running-mean smoother can be written as

$$\hat{s}_k(x_i) = \text{mean}\left\{ Y_j \text{ for } \max\left(i - \frac{k-1}{2}, 1\right) \leq j \leq \min\left(i + \frac{k-1}{2}, n\right) \right\}. \tag{11.5}$$

For the purposes of graphing or prediction, one can compute $\hat{s}$ at each of the x_i and interpolate linearly in between. Note that by stepping through i in order, we can efficiently compute $\hat{s}_k$ at x_{i+1} with the recursive update

$$\hat{s}_k(x_{i+1}) = \hat{s}_k(x_i) - \frac{Y_{i-(k-1)/2}}{k} + \frac{Y_{i+(k+1)/2}}{k}. \tag{11.6}$$

This avoids recalculating the mean at each point. An analogous update holds for points whose predictor values lie near the edges of the data.

The constant-span running-mean smoother is a linear smoother. The middle rows of the smoothing matrix $\mathbf{S}$ resemble $\left(0 \ \ldots \ 0 \ \frac{1}{k} \ \ldots \ \frac{1}{k} \ 0 \ \ldots \ 0\right)$. An important detail in most smoothing problems is how to compute $\hat{s}_k(x_i)$ near the edges of the data. For example, x_1 does not have $(k-1)/2$ neighbors to its left. Some adjustment must be made to the top and bottom $(k-1)/2$ rows of $\mathbf{S}$. Three possible choices (e.g., for $k=5$) are to shrink symmetric neighborhoods by using

$$\mathbf{S} = \begin{pmatrix} 1 & 0 & 0 & 0 & 0 & 0 & \cdots & 0 \\ 1/3 & 1/3 & 1/3 & 0 & 0 & 0 & \cdots & 0 \\ 1/5 & 1/5 & 1/5 & 1/5 & 1/5 & 0 & \cdots & 0 \\ 0 & 1/5 & 1/5 & 1/5 & 1/5 & 1/5 & \cdots & 0 \\ \vdots & \vdots & \vdots & \vdots & \vdots & \vdots & & \vdots \end{pmatrix}; \tag{11.7}$$

to truncate neighborhoods by using

$$\mathbf{S} = \begin{pmatrix} 1/3 & 1/3 & 1/3 & 0 & 0 & 0 & \cdots & 0 \\ 1/4 & 1/4 & 1/4 & 1/4 & 0 & 0 & \cdots & 0 \\ 1/5 & 1/5 & 1/5 & 1/5 & 1/5 & 0 & \cdots & 0 \\ 0 & 1/5 & 1/5 & 1/5 & 1/5 & 1/5 & \cdots & 0 \\ \vdots & \vdots & \vdots & \vdots & \vdots & \vdots & & \vdots \end{pmatrix}; \tag{11.8}$$

or—in the case of circular data only—to wrap neighborhoods by using

$$\mathbf{S} = \begin{pmatrix} 1/5 & 1/5 & 1/5 & 0 & 0 & 0 & \cdots & 0 & 1/5 & 1/5 \\ 1/5 & 1/5 & 1/5 & 1/5 & 0 & 0 & 0 & \cdots & 0 & 1/5 \\ 1/5 & 1/5 & 1/5 & 1/5 & 1/5 & 0 & 0 & 0 & \cdots & 0 \\ 0 & 1/5 & 1/5 & 1/5 & 1/5 & 1/5 & 0 & 0 & \cdots & 0 \\ \vdots & \vdots & \vdots & \vdots & \vdots & \vdots & \vdots & \vdots & & \vdots \end{pmatrix}. \tag{11.9}$$

The truncation option is usually preferred, and is implicit in (11.5). Since k is intended to be a rather small fraction of n, the overall picture presented by the smooth is not greatly affected by the treatment of the edges, but regardless of how this detail is addressed, readers should be aware of the reduced reliability of $\hat{s}$ at the edges of the data.

Example 11.1 (Easy data) Figure 11.2 shows a constant-span running-mean smooth of the data introduced at the start of this chapter, which are easy to smooth well using a variety of methods we will discuss. These data are $n = 200$ equally spaced points from the model $Y_i = s(x_i) + \epsilon_i$, where the errors are mean-zero i.i.d. normal noise

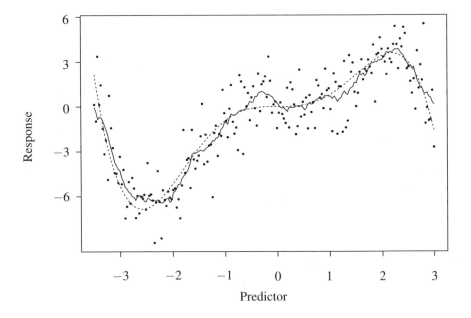

Fig. 11.2 Results from a constant-span running-mean smoother with $k = 13$ (solid line), compared with the true underlying curve (dotted line).

with a standard deviation of 1.5. The data are available from the website for this book. The true relationship, $s(x) = x^3 \sin\{(x + 3.4)/2\}$, is shown with a dotted line; the estimate $\hat{s}_k(x)$ is shown with the solid line. We used a smoothing matrix equivalent to (11.8) for $k = 13$. The result is not visually appealing: perhaps this emphasizes the surprising sophistication of whatever methods people employ when they sketch in a smooth curve by hand. □

11.2.1.1 *Effect of span*

A natural smoothing parameter for the constant-span running-mean smoother is $\lambda = k$. As for all smoothers, this parameter controls wiggliness, here by directly controlling the number of data points contained in any neighborhood. For sorted data and an interior point x_i whose neighborhood is not affected by the data edges, the span-k running-mean smoother given by (11.5) has

$$
\mathrm{MSE}_k(\hat{s}_k(x_i)) = \mathrm{E}\left\{ \left(s(x_i) - \frac{1}{k} \sum_{j=i-(k-1)/2}^{i+(k-1)/2} Y_j \right)^2 \right\}, \tag{11.10}
$$

where, recall, $s(x_i) = \mathrm{E}\{Y|X = x_i\}$. It is straightforward to reexpress this as

$$
\mathrm{MSE}_k(\hat{s}_k(x_i)) = \left(\mathrm{bias}\{\hat{s}_k(x_i)\}\right)^2 + \frac{1}{k^2} \sum_{j=i-(k-1)/2}^{i+(k-1)/2} \mathrm{var}\{Y|X = x_j\}, \tag{11.11}
$$

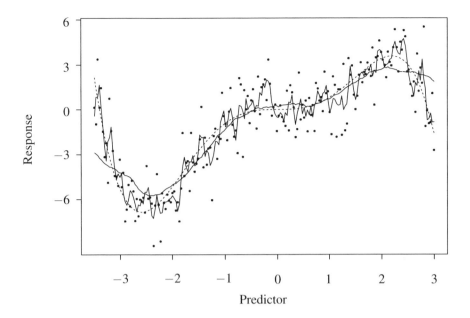

Fig. 11.3 Results from a constant-span running-mean smoother with $k = 3$ (wigglier solid line) and $k = 43$ (smoother solid line). The underlying true curve is shown with a dotted line.

where

$$\text{bias}\{\hat{s}_k(x_i)\} = s(x_i) - \frac{1}{k} \sum_{j=i-(k-1)/2}^{i+(k-1)/2} s(x_j). \tag{11.12}$$

To understand how the mean squared prediction error depends on the smoothing span, we can use (11.11) and make the simplifying assumption that $\text{var}\{Y|X = x_j\} = \sigma^2$ for all $x_j \in \mathcal{N}(x_i)$. Then

$$\text{MSPE}_k(\hat{s}_k(x_i)) = \text{var}\{Y|X = x_i\} + \text{MSE}_k(\hat{s}_k(x_i))$$
$$= (1 + 1/k)\sigma^2 + \left(\text{bias}\{\hat{s}_k(x_i)\}\right)^2. \tag{11.13}$$

Therefore, as the neighborhood size k is increased, the variance term in (11.13) decreases, but the bias term will typically increase, because $s(x_i)$ will not likely be similar to $s(x_j)$ for distant j. Likewise, if k is decreased, the variance term will increase, but the bias term will usually be smaller.

Example 11.2 (Easy data, continued) Figure 11.3 illustrates how k influences $\hat{s}_k$. In this graph, $k = 3$ leads to a result that is far too wiggly. In contrast, $k = 43$ leads to a result that is quite smooth, but systematically biased. The bias arises when a neighborhood is so wide that the response values at the fringes of the neighborhood are not representative of the response at the middle. This tends to erode peaks, fill in troughs, and flatten trends near the edges of the range of the predictor. □

11.2.1.2 Span selection for linear smoothers

The best choice for k clearly must balance a tradeoff between bias and variance. For small k, the estimated curve will be wiggly but exhibit more fidelity to the data. For large k, the estimated curve will be smooth, but exhibit substantial bias in some regions. For all smoothers, the role of the smoothing parameter is to control this tradeoff between bias and variance.

An expression for $\overline{\mathrm{MSPE}}_k(\hat{\mathbf{s}}_k)$ can be obtained by averaging values from (11.13) over all x_i, but this expression cannot be minimized to choose k, because it depends on unknown expected values. Furthermore, it may be more reasonable to choose the span that is best for the observed data, rather than the span that is best on average for datasets that might have been observed but weren't. Therefore, we might consider choosing the k that minimizes the residual mean squared error

$$\mathrm{RSS}_k(\hat{\mathbf{s}}_k)/n = \frac{1}{n}\sum_{i=1}^{n}(Y_i - \hat{s}_k(x_i))^2. \tag{11.14}$$

However,

$$\mathrm{E}\{\mathrm{RSS}_k(\hat{\mathbf{s}}_k)/n\} = \overline{\mathrm{MSPE}}_k(\hat{\mathbf{s}}_k) - \frac{1}{n}\sum_{i\neq j}\mathrm{cov}\{Y_i, \hat{s}_k(x_j)\}. \tag{11.15}$$

For constant-span running means, $\mathrm{cov}\{Y_i, \hat{s}_k(x_j)\} = \mathrm{var}\{Y|X = x_j\}/k$ for interior x_j. Therefore, $\mathrm{RSS}_k(\hat{\mathbf{s}}_k)/n$ is a downward-biased estimator of $\overline{\mathrm{MSPE}}_k(\hat{\mathbf{s}}_k)$.

To eliminate the correlation between Y_i and $\hat{s}_k(x_i)$, we may omit the ith point when calculating the smooth at x_i. This process is known as *cross-validation* [521]; it is used only for assessing the performance of the smooth, not for fitting the smooth itself. Denote by $\hat{s}_k^{(-i)}(x_i)$ the value of the smooth at x_i when it is fitted using the dataset that omits the ith data pair. A better (indeed, pessimistic) estimator of $\overline{\mathrm{MSPE}}_k(\hat{\mathbf{s}}_k)$ is

$$\mathrm{CVRSS}_k(\hat{\mathbf{s}}_k)/n = \frac{1}{n}\sum_{i=1}^{n}\left(Y_i - \hat{s}_k^{(-i)}(x_i)\right)^2, \tag{11.16}$$

where $\mathrm{CVRSS}_k(\hat{\mathbf{s}}_k)$ is called the *cross-validated residual sum of squares*. Typically, $\mathrm{CVRSS}_k(\hat{\mathbf{s}}_k)$ is plotted against k.

Example 11.3 (Easy data, continued) Figure 11.4 shows a plot of $\mathrm{CVRSS}_k(\hat{\mathbf{s}}_k)$ against k for smoothing the data introduced in Example 11.1. This plot usually shows a steep increase in $\mathrm{CVRSS}_k(\hat{\mathbf{s}}_k)$ for small k due to increasing variance, and a gradual increase in $\mathrm{CVRSS}_k(\hat{\mathbf{s}}_k)$ for large k due to increasing bias. The region of best performance is where the curve is lowest; this region is often quite broad and rather flat. In this example, good choices of k range between 11 and 23, with $k = 13$ being optimal. Minimizing $\mathrm{CVRSS}_k(\hat{\mathbf{s}}_k)$ with respect to k often produces a final smooth that is somewhat too wiggly. Undersmoothing can be reduced by choosing a larger

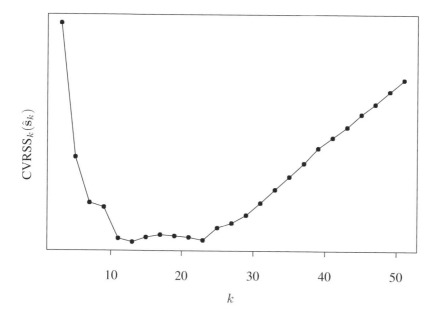

Fig. 11.4 Plot of $CVRSS_k(\hat{s}_k)$ versus k for the constant-span running-mean smoother applied to the data in Figure 11.1. Good choices for k range between about 11 and 23. The smaller values in this range would be especially good at bias reduction, whereas the larger ones would produce smoother fits.

k within the low $CVRSS_k(\hat{s}_k)$ valley in the cross-validation plot which corresponds to good performance. In this example, $k = 23$ would be worth trying. □

This approach of leave-one-out cross-validation is time-consuming, even for linear smoothers, since it seems to require computing n separate smooths of slightly different datasets. Two shortcuts are worth mentioning.

First, consider a linear smoother with smoothing matrix $\mathbf{S}$. The proper fit at x_i when the ith data pair is omitted from the dataset is a somewhat imprecise concept, even for a constant-span running-mean smoother, because smooths are typically calculated only at the x_i values in the dataset. Should smooths be fitted at the two data points adjacent to the omitted x_i, with linear interpolation used in between, or should some other approach be tried? The most unambiguous way to proceed is to define

$$\hat{s}_k^{(-i)}(x_i) = \sum_{\substack{j=1 \\ j \neq i}}^{n} \frac{Y_j S_{ij}}{1 - S_{ii}}, \tag{11.17}$$

where S_{ij} is the (i, j)th element of $\mathbf{S}$. In other words, the ith row of $\mathbf{S}$ is altered by replacing the (i, j)th element of $\mathbf{S}$ with zero and rescaling the remainder of the row so it sums to 1. In this case, to compute $CVRSS_k(\hat{s}_k)$ there is no need to actually delete the ith observation and recompute the smooth for each i. Following from (11.17), it can be shown that for linear smoothers, (11.16) can be reexpressed as

$$\text{CVRSS}_k(\hat{s}_k)/n = \frac{1}{n} \sum_{i=1}^{n} \left(\frac{y_i - \hat{s}_k(x_i)}{1 - S_{ii}} \right)^2. \tag{11.18}$$

This approach is analogous to the well-known shortcut for calculating deleted residuals in linear regression [412] and is further justified in [280].

Second, one may wish to reduce the number of cross-validation computations by generating fewer partial datasets, each with a greater number of points omitted. For example, one could randomly partition the observed dataset into 10 portions, then leave out one portion at a time. The cross-validated residual sum of squares would then be accumulated from the residuals of the points omitted in each portion. This approach tends to overestimate the true prediction error, while leaving-one-out is less biased but more variable; five- or tenfold cross-validation (i.e., 5 to 10 portions) has been recommended [281].

We mentioned above that different smoothers employ different smoothing parameters to control wiggliness. So far, we have focused on the number (k) or fraction (k/n) of nearest neighbors. Another reasonable choice, $\mathcal{N}(x) = \{x_i : |x_i - x| < h\}$, uses the positive real-valued distance h as a smoothing parameter. There are also schemes for weighting points based on their proximity to x, in which case the smoothing parameter may relate to these weights. Usually, the number of points in a neighborhood is smaller near the boundaries of the data, meaning that any fixed span chosen by cross-validation or another method may provide a poorer fit near the boundaries than in the middle of the data. The span may also be allowed to vary locally. For such alternative parameterizations of neighborhoods, plotting the cross-validated residual sum of squares and drawing conclusions about the bias–variance tradeoff proceed in a fashion analogous to the preceding discussion.

Cross-validated span selection is not limited to the constant-span running-mean smoother. The same strategy is effective for most other smoothers discussed in this chapter. The tradeoff between bias and variance is a fundamental principle in many areas of statistics: It arose previously for density estimation (Chapter 10), and it is certainly a major consideration for all types of smoothing.

There are a wide variety of other methods for choosing the span for a scatterplot smoother, resulting in different bias–variance tradeoffs [269, 270, 273, 280, 281]. One straightforward approach is to replace CVRSS with another criterion such as C_p, AIC, or BIC [281]. Two other popular alternatives are *generalized cross-validation* and plug-in methods [271, 475, 508]. In generalized cross-validation, (11.16) is replaced by

$$\text{GCVRSS}_k(\hat{s}_k) = \frac{\text{RSS}_k(\hat{s}_k)}{(1 - \text{tr}\{\mathbf{S}\}/n)^2}, \tag{11.19}$$

where $\text{tr}\{\mathbf{S}\}$ denotes the sum of the diagonal elements of $\mathbf{S}$. For equally spaced x_i, CVRSS and GCVRSS give similar results. When the data are not equally spaced, span selection based on GCVRSS is less affected by observations that exert strong influence on the fit. Notwithstanding this potential benefit of generalized cross-validation, reliance on GCVRSS often results in significant undersmoothing. Plug-in methods generally derive an expression for the expected mean squared prediction error or some other fitting criterion, whose theoretical minimum is found to depend on the type of smoother, the wiggliness of the true curve, and the conditional variance of $Y|x$. A preliminary smooth is completed using a span chosen informally (or by cross-validation). Then this smooth is used to estimate the unknown quantities in the expression for the optimal span, and the result is used in a final smooth.

It is tempting to select the span selection method that yields the picture most pleasing to your eye. That is fine, but it is worthwhile admitting up front that scatterplot smoothing is often an exercise in descriptive—not inferential—statistics, so selecting your favorite span from trial and error or a simple plot of CVRSS is as reasonable as the opportunistic favoring of any technical method. Since spans chosen by cross-validation vary with the random dataset observed and sometimes undersmooth, it is important for practitioners to develop their own expertise based on hands-on analysis and experience.

11.2.2   Running lines and running polynomials

The constant-span running-mean smoother exhibits visually unappealing wiggliness for any reasonable k. It also can have strong bias at the edges because it fails to recognize the local trend in the data. The running-line smoother can mitigate both problems.

Consider fitting a linear regression model to the k data points in $\mathcal{N}(x_i)$. Then the least squares linear regression prediction at x is

$$\ell_i(x) = \overline{Y}_i + \hat{\beta}_i(x - \bar{x}_i), \tag{11.20}$$

where $\overline{Y}_i$, $\bar{x}_i$, and $\hat{\beta}_i$ are the mean response, the mean predictor, and the estimated slope of the regression line, respectively, for the data in $\mathcal{N}(x_i)$. The running-line smooth at x_i is $\hat{s}_k(x_i) = \ell_i(x_i)$.

Let $\mathbf{X}_i = (\mathbf{1} \ \mathbf{x}_i)$, where $\mathbf{1}$ is a column of ones and $\mathbf{x}_i$ is the column vector of predictor data in $\mathcal{N}(x_i)$; and let $\mathbf{Y}_i$ be the corresponding column vector of response data. Then note that $\ell_i(x_i)$—and hence the smooth at x_i—is obtained by multiplying $\mathbf{Y}_i$ by one row of $\mathbf{H}_i = \mathbf{X}_i \left(\mathbf{X}_i^T \mathbf{X}_i\right)^{-1} \mathbf{X}_i^T$. (Usually $\mathbf{H}_i$ is called the ith *hat matrix*.) Therefore, this smoother is linear, with a banded smoothing matrix $\mathbf{S}$ whose nonzero entries are drawn from an appropriate row of each $\mathbf{H}_i$. Computing the smooth directly from $\mathbf{S}$ is not very efficient. For data ordered by x_i, it is faster to sequentially update the sufficient statistics for regression, analogously to the approach discussed for running means.

Example 11.4 (Easy data, continued) Figure 11.5 shows a running-line smooth of the data introduced in Example 11.1, for the span $k = 23$ chosen by cross-validation. The edge effects are much smaller and the smooth is less jagged than with the constant-span running-mean smoother. Since the true curve is usually well approximated by a

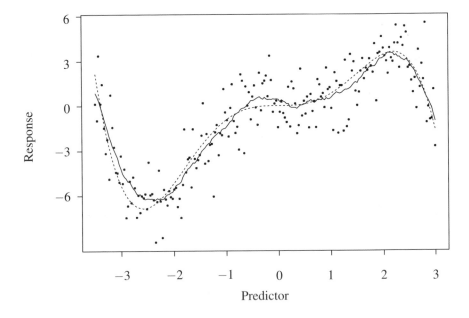

Fig. 11.5 Plot of the running-line smooth for $k = 23$ (solid line) and the true underlying curve (dotted line).

line even for fairly wide neighborhoods, k may be increased from the optimal value for the constant-span running-mean smoother. This reduces variance without seriously increasing bias. □

Nothing in this discussion limits the local fitting to simple linear regression. A running polynomial smoother could be produced by setting $\hat{s}_k(x_i)$ to the value at x_i of a least squares polynomial regression fit to the data in $\mathcal{N}(x_i)$. Such smoothers are sometimes called *local regression smoothers* (see Section 11.2.4). Odd-order polynomials are preferred [168, 508]. Since smooth functions are roughly locally linear, higher-order local polynomial regression often offers scant advantage over the simpler linear fits unless the true curve has very sharp wiggles.

11.2.3 Kernel smoothers

For the smoothers mentioned so far, there is a discontinuous change to the fit each time the neighborhood membership changes. Therefore, they tend to fit well statistically but exhibit visually unappealing jitters or wiggles.

One approach to increasing smoothness is to redefine the neighborhood so that points only gradually gain or lose membership in it. Let K be a symmetric kernel centered at 0. A kernel is essentially a weighting function—in this case it weights neighborhood membership. One reasonable kernel choice would be the standard normal density, $K(z) = \frac{1}{\sqrt{2\pi}} \exp\{-z^2/2\}$. Then let

$$\hat{s}_h(x) = \sum_{i=1}^{n} Y_i \frac{K\left(\frac{x-x_i}{h}\right)}{\sum_{j=1}^{n} K\left(\frac{x-x_j}{h}\right)}, \tag{11.21}$$

where the smoothing parameter h is called the *bandwidth*. Notice that for many common kernels such as the normal kernel, all data points are used to calculate the smooth at each point, but very distant data points receive very little weight. Proximity increases a point's influence on the local fit; in this sense the concept of local averaging remains. A large bandwidth yields a quite smooth result because the weightings of the data points change little across the range of the smooth. A small bandwidth ensures a much greater dominance of nearby points, thus producing more wiggles.

The choice of smoothing kernel is much less important than the choice of bandwidth. A similar smooth will be produced from diverse kernel shapes. Although kernels need not be densities, a smooth, symmetric, nonnegative function with tails tending continuously toward zero is generally best in practice. Thus, there are few reasons to look beyond a normal kernel, despite a variety of asymptotic arguments supporting more exotic choices.

Kernel smoothers are clearly linear smoothers. However, the computation of the smooth cannot be sequentially updated in the manner of the previous efficient approaches, because the weights for all points change each time x changes. Fast Fourier transform methods are helpful in the special case of equally spaced data [267, 505]. Further background on kernel smoothing is given by [484, 492, 508, 553].

Example 11.5 (Easy data, continued) Figure 11.6 shows a kernel smooth of the data introduced in Example 11.1, using a normal kernel with $h = 0.16$ chosen by cross-validation. Since neighborhood entries and exits are gradual, the result exhibits characteristically rounded features. However, note that the kernel smoother does not eliminate systematic bias at the edges, as the running-line smoother does. □

11.2.4 Local regression smoothing

Running polynomial smoothers and kernel smoothers share some important links [10, 268, 508]. Suppose that the data originated from a random design, so they are a random sample from the model $(X_i, Y_i) \sim$ i.i.d. $f(x, y)$. (A nonrandom design would have prespecified the x_i values.) We may write

$$s(x) = E\{Y|x\} = \int y f(y|x)\, dy = \int y \frac{f(x,y)}{f(x)}\, dy, \tag{11.22}$$

where, marginally, $X \sim f(x)$. Using the kernel density estimation approach described in Chapter 10 (and a product kernel for estimating $f(x, y)$), we may estimate

$$\widehat{f}(x, y) = \frac{1}{n h_x h_y} \sum_{i=1}^{n} K_x\left(\frac{x - X_i}{h_x}\right) K_y\left(\frac{y - Y_i}{h_y}\right) \tag{11.23}$$

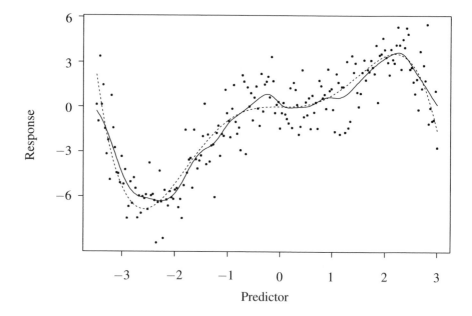

Fig. 11.6 Plot of kernel smooth using a normal kernel with $h = 0.16$ chosen by cross-validation (solid line) and the true underlying curve (dotted line).

and

$$\widehat{f}(x) = \frac{1}{nh_x} \sum_{i=1}^{n} K_x \left(\frac{x - X_i}{h_x} \right) \tag{11.24}$$

for suitable kernels K_x and K_y and corresponding bandwidths h_x and h_y. The Nadaraya–Watson estimator [406, 556] of $s(x)$ is obtained by substituting $\widehat{f}(x,y)$ and $\widehat{f}(x)$ in (11.22), yielding

$$\widehat{s}_{h_x}(x) = \sum_{i=1}^{n} Y_i \frac{K_x \left(\frac{x - X_i}{h_x} \right)}{\sum_{i=1}^{n} K_x \left(\frac{x - X_i}{h_x} \right)}. \tag{11.25}$$

Note that this matches the form of a kernel smoother (see (11.21)).

It is easy to show that the Nadaraya–Watson estimator minimizes

$$\sum_{i=1}^{n} (Y_i - \beta_0)^2 K_x \left(\frac{x - X_i}{h_x} \right) \tag{11.26}$$

with respect to β_0. This is a least squares problem that locally approximates $s(x)$ with a constant. Naturally, this locally constant model could be replaced with a local higher-order polynomial model. Fitting a local polynomial using weighted regression, with weights set according to some kernel function, yields a *locally weighted regression smooth*, often simply called a *local regression smooth* [100, 168, 553]. The *p*th-order local polynomial regression smoother minimizes the weighted least squares criterion

$$\sum_{i=1}^{n} \left[Y_i - \beta_0 - \beta_1(x - X_i) - \cdots - \beta_p(x - X_i)^p\right]^2 K_x\left(\frac{x - X_i}{h_x}\right) \quad (11.27)$$

and can be fitted using a weighted polynomial regression at each x, with weights determined by the kernel K_x according to the proximity to x. This is still a linear smoother, with a smoothing matrix composed of one row from the hat matrix used in each weighted polynomial regression.

The least squares criterion may be replaced by other choices. See Section 11.4.1 for an extension of this technique that relies on a robust fitting method.

11.2.5 Spline smoothing

Perhaps you have found the graphs of smooths presented so far in this chapter to be somewhat unsatisfying visually because they are more wiggly than you would have drawn by hand. They exhibit small-scale variations that your eye easily attributes to random noise rather than to signal. Then smoothing splines may better suit your taste.

Assume that the data have been sorted in increasing order of the predictor, so x_1 is the smallest predictor value and x_n is the largest. Define

$$Q_\lambda(\hat{s}) = \sum_{i=1}^{n} (Y_i - \hat{s}(x_i))^2 + \lambda \int_{x_1}^{x_n} \hat{s}''(x)^2 \, dx, \quad (11.28)$$

where $\hat{s}''(x)$ is the second derivative of $\hat{s}(x)$. Then the summation constitutes a penalty for misfit, and the integral is a penalty for wiggliness. The parameter λ controls the relative weighting of these two penalties.

It is an exercise in the calculus of variations to minimize $Q_\lambda(\hat{s})$ over all twice differentiable functions $\hat{s}$ for fixed λ. The result is a *cubic smoothing spline*, $\hat{s}_\lambda(x)$. This function is a cubic polynomial in each interval $[x_i, x_{i+1}]$ for $i = 1, \ldots, n-1$, with these polynomial pieces pasted together twice continuously differentiably at each x_i. Although usually inadvisable in practice, smoothing splines can be defined on ranges extending beyond the edges of the data. In this case, the extrapolative portions of the smooth are linear.

It turns out that cubic splines are linear smoothers, so $\hat{\mathbf{s}}_\lambda = \mathbf{SY}$. This result is presented clearly in [280], and efficient computation methods are covered in [124, 506]. Other useful references about smoothing splines include [143, 164, 245, 551].

The ith row of $\mathbf{S}$ consists of weights $S_{i1}, \ldots, S_{in}$, whose relationship to x_i is sketched in Figure 11.8 (discussed in Section 11.3). Such weights are reminiscent of kernel smoothing with a kernel that is not always positive, but in this case the kernel does not retain the same shape when centered at different points.

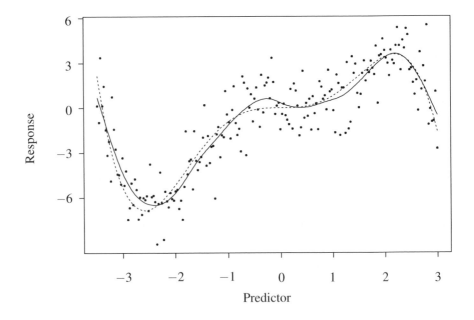

Fig. 11.7 Plot of a cubic smoothing spline using $\lambda = 0.066$ chosen by cross-validation (solid line) and the true underlying curve (dotted line).

Example 11.6 (Easy data, continued) Figure 11.7 shows a spline smooth of the data introduced in Example 11.1, using $\lambda = 0.066$ chosen by cross-validation. The result is a curve very similar to what you might have sketched by hand. □

11.2.5.1 Choice of penalty Smoothing splines depend on a smoothing parameter λ that relates to neighborhood size less directly than for smoothers we have discussed previously. We have already noted that λ controls the bias–variance trade-off, with large values of λ favoring variance reduction and small values favoring bias reduction. As $\lambda \to \infty$, $\hat{s}_\lambda$ approaches the least squares line. When $\lambda = 0$, $\hat{s}_\lambda$ is an interpolating spline that simply connects the data points.

Since smoothing splines are linear smoothers, the span selection methods discussed in Section 11.2.1 still apply. Calculating $\text{CVRSS}_\lambda(\hat{s}_\lambda)$ via (11.18) requires the S_{ii}, which can be calculated efficiently using the method in [424]. Calculating $\text{GCVRSS}_\lambda(\hat{s}_\lambda)$ requires $\text{tr}\{\mathbf{S}\}$, which also can be calculated efficiently [125].

11.3 COMPARISON OF LINEAR SMOOTHERS

Although the smoothers described so far may seem very different, they all rely on the principal of local averaging. The fit of each depends on a smoothing matrix $\mathbf{S}$, whose rows determine the weights used in a local average of the response values.

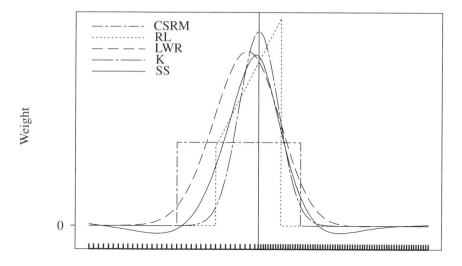

Fig. 11.8 Equivalent kernels for five different linear smoothing methods for which tr$\{\mathbf{S}\} =$ 7. The methods are: constant-span running mean (CSRM) with symmetric neighborhoods, running lines (RL) with symmetric neighborhoods, locally weighted regression (LWR), Gaussian kernel smoothing (K), and cubic smoothing spline (SS). The smoothing weights for an interior point (indicated by the vertical line) correspond to the 36th row of $\mathbf{S}$. The entire collection of 105 values of x_i is shown by hashes on the horizontal axis: they are equally spaced on each side, but twice as dense on the right.

Comparison of a typical row of $\mathbf{S}$ for different smoothers is a useful approach to understanding the differences between techniques.

Of course, the weights in a typical row of $\mathbf{S}$ depend on the smoothing parameter. In general, values of λ that favor greater smoothing will produce rows of $\mathbf{S}$ with a more diffuse allocation of weight, rather than high weights concentrated in just a few entries. Therefore, to enable a fair comparison, it is necessary to find a common link between the diverse smoothing parameters used by different techniques. The common basis for comparison is the number of *equivalent degrees of freedom* of the smooth, which can most simply be defined as df $=$ tr$\{\mathbf{S}\}$ for linear smoothers. Several alternative definitions and extensions for nonlinear smoothers are discussed in [280].

For fixed degrees of freedom, the entries in a row of $\mathbf{S}$ are functions of the x_i, their spacing, and their proximity to the edge of the data. If the weights in a row of $\mathbf{S}$ are plotted against the predictor values, we may view the result as an *equivalent kernel*, whose weights are analogous to the explicit weighting used in a kernel smoother. Figure 11.8 compares the equivalent kernel for various smoothers with seven degrees of freedom. The kernel is shown for the 36th of 105 ordered predictor values, of which 35 are equally spaced to the left and 69 are equally spaced twice as densely to the right. Notice that these kernels can be skewed, depending on the spacing of the x_i. Also, kernels need not be everywhere positive. In Figure 11.8, the equivalent kernel for the smoothing spline assigns negative weight in some regions. Although not shown in this figure, the shape of the kernels is markedly different for a point near the edge of the data. For such a point, weights generally increase towards the edge and decrease away from it.

11.4 NONLINEAR SMOOTHERS

Nonlinear smoothers can be much slower to calculate, and in ordinary cases they offer little improvement over simpler approaches. However, the simpler methods can exhibit very poor performance for some types of data. The *loess* smoother provides improved robustness to outliers that would introduce substantial noise in an ordinary smoother. We also examine the *supersmoother*, which allows the smoothing span to vary as best suits the local needs of the smoother. Such smoothing can be very helpful when var$\{Y|x\}$ varies with x.

11.4.1 Loess

The *loess* (for "locally weighted scatterplot smoother") smoother is a widely used method with good robustness properties [98, 99]. It is essentially a weighted running-line smoother, except that each local line is fitted using a robust method rather than least squares. As a result, the smoother is nonlinear.

Loess is fitted iteratively; let t index the iteration number. To start at $t = 0$, we let $d_k(x_i)$ denote the distance from x_i to its kth nearest neighbor, where k (or k/n) is a smoothing parameter. The kernel used for the local weighting around point x_i is

$$K_i(x) = K\left(\frac{x - x_i}{d_k(x_i)}\right), \tag{11.29}$$

where

$$K(z) = \begin{cases} (1 - |z|^3)^3 & \text{for } |z| \leq 1, \\ 0 & \text{otherwise} \end{cases} \tag{11.30}$$

is the *tricube kernel*.

The estimated parameters of the locally weighted regression for the ith point at iteration t are found by minimizing the weighted sum of squares

$$\sum_{j=1}^{n} \left(Y_j - (\beta_{0,i}^{(t)} + \beta_{1,i}^{(t)} x_j)\right)^2 K_i(x_j). \tag{11.31}$$

We denote these estimates as $\hat{\beta}_{m,i}^{(t)}$ for $m = 0, 1$ and $i = 1, \ldots, n$. Linear—rather than polynomial—regression is recommended, but the extension to polynomials would require only a straightforward change to (11.31). Note that the fitted value for the response variable given by the local regression is $\hat{Y}_i^{(t)} = \hat{\beta}_{0,i}^{(t)} + \hat{\beta}_{1,i}^{(t)} x_i$. This completes iteration t.

To prepare for the next iteration, observations are assigned new weights based on the size of their residual, in order to downweight apparent outliers. If $e_i^{(t)} = Y_i - \hat{Y}_i^{(t)}$, then define robustness weights as

$$r_i^{(t+1)} = B\left(\frac{e_i^{(t)}}{6 \times \text{median } |e_i^{(t)}|}\right), \tag{11.32}$$

where $B(z)$ is the *biweight kernel* given by

$$B(z) = \begin{cases} (1 - z^2)^2 & \text{for } |z| \leq 1, \\ 0 & \text{otherwise.} \end{cases} \tag{11.33}$$

Then the weights $K_i(x_j)$ in (11.31) are replaced with $r_i^{(t+1)} K_i(x_j)$, and new locally weighted fits are obtained. The resulting estimates for each i provide $\hat{Y}_i^{(t+1)}$. The process stops after $t = 3$ by default [98, 99].

Example 11.7 (Easy data, continued) Figure 11.9 shows a loess smooth of the data introduced in Example 11.1, using $k = 30$ chosen by cross-validation. The results are very similar to the running-line smooth.

Figure 11.10 shows the effect of outliers. The dotted line in each panel is the original smooth for loess and running lines; the solid lines are the result when three additional data points at $(1, -8)$ are inserted in the dataset. The spans for each smoother were left unchanged. Loess was so robust to these outliers that the two curves are nearly superimposed. The running-line smoother shows greater sensitivity to the outliers. □

11.4.2 Supersmoother

All of the previous methods employ a fixed span. There are cases, however, where a variable span may be more appropriate.

Example 11.8 (Difficult data) Consider the curve and data shown in Figure 11.11. These data are available from the website for this book. Suppose that the true conditional mean function for these data is given by the curve; thus the goal of smoothing is to estimate the curve using the observed data. The curve is very wiggly on the right side of the figure, but these wiggles could reasonably be detected by a smoother with a suitably small span, because the variability in the data is very low. On the left, the curve is very smooth, but the data have much larger variance. Therefore, a large span would be needed in this region to smooth the noisy data adequately. Thus a small span is needed to minimize bias in one region, and a large span is needed to control

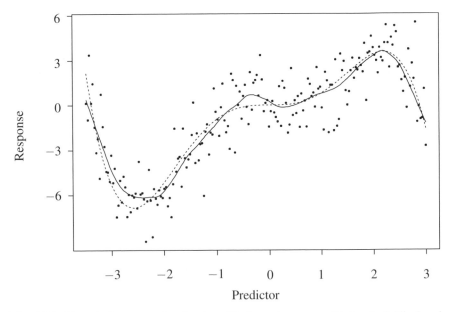

Fig. 11.9 Plot of a loess smooth using $k = 30$ chosen by cross-validation (solid line) and the true underlying curve (dotted line).

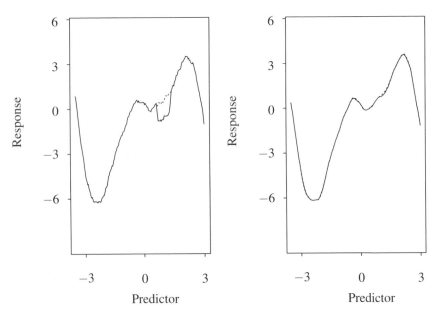

Fig. 11.10 Plot of running-line smooths (left) using $k = 23$, and loess smooths (right) using $k = 30$. In each panel, the dotted line is the smooth of the original data, and the solid line is the smooth after inserting in the dataset three new outlier points at $(1, -8)$.

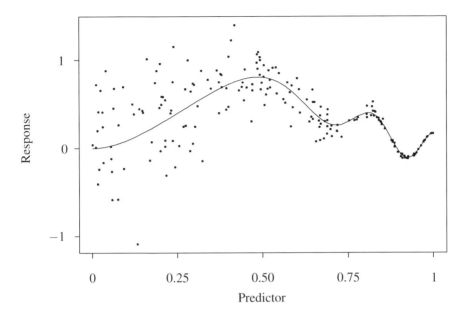

Fig. 11.11 These bivariate data with nonconstant variance and wiggles with changing frequency and amplitude would be fitted terribly by most fixed-span smoothers. The true $E\{Y|x\}$ is shown with the solid line.

variance in another region. The *supersmoother* [180, 183] was designed for this sort of problem. □

The supersmoothing approach begins with the calculation of m different smooths, which we denote $\hat{s}_1(x), \ldots, \hat{s}_m(x)$, using different spans, say $h_1, \ldots, h_m$. For $m = 3$, spans of $h_1 = 0.05n$, $h_2 = 0.2n$, and $h_3 = 0.5n$ are recommended. Each smooth should be computed over the full range of the data. For simplicity, use the running-line smoother to generate the $\hat{s}_j(x)$ for $j = 1, 2$, and 3. Figure 11.12 shows the three smooths.

Next, define $p(h_j, x)$ to be a measure of performance of the jth smooth at point x, for $j = 1, \ldots, m$. Ideally, we would like to assess the performance at point x_i according to $E\left\{g\big(Y - \hat{s}_j^{(i)}(x_i)\big) \mid X = x_i\right\}$, where g is a symmetric function that penalizes large deviations, and $\hat{s}_j^{(i)}(x_i)$ is the jth smooth at x_i estimated using the cross-validation dataset that omits x_i. This expected value is of course unknown, so, following the local-averaging paradigm, we estimate it as

$$\hat{p}(h_j, x_i) = \hat{s}^* \left(g\big(Y_i - \hat{s}_j^{(i)}(x_i)\big)\right), \tag{11.34}$$

where $\hat{s}^*$ is some fixed-span smoother. For the implementation suggested in [180], $\hat{s}^* = \hat{s}_2$ and $g(z) = |z|$. Figure 11.13 shows the smoothed absolute cross-validated

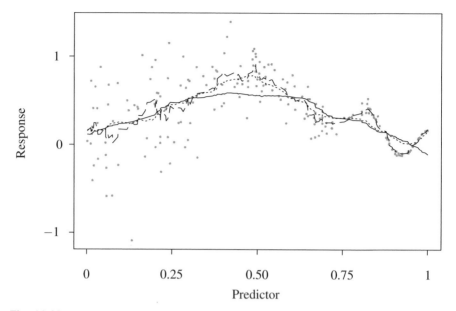

Fig. 11.12   The three preliminary fixed-span smooths employed by the supersmoother. The spans are $0.05n$ (dashed), $0.2n$ (dotted), and $0.5n$ (solid). The data points have been faded to enable a clearer view of the smooths.

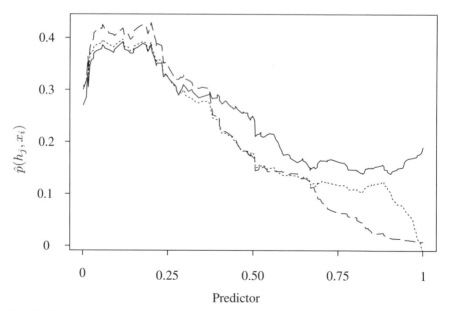

Fig. 11.13   The $\hat{p}(h_j, x_i)$ for $j = 1$ (dashed), 2 (dotted), and 3 (solid). For each j, the curve is a smooth of the absolute cross-validated residuals.

residuals $\left|Y_i - \hat{s}_j^{(i)}(x_i)\right|$ for the three different smooths. The curves in this figure represent $\hat{p}(h_j, x_i)$ for $j = 1, 2$, and 3. The data used in each smooth originate as residuals from alternative smooths using spans of $0.05n$ (dashed), $0.2n$ (dotted), and $0.5n$ (solid), but each set of absolute residuals is smoothed with a span of $0.2n$ to generate the curves shown.

At each x_i, the performances of the three smooths can be assessed using $\hat{p}(h_j, x_i)$ for $j = 1, 2$, and 3. Denote by $\hat{h}_i$ the best of these spans at x_i, that is, the particular span among h_1, h_2, and h_3 that provides the lowest $\hat{p}(h_j, x_i)$. Figure 11.14 plots $\hat{h}_i$ against x_i for our example. The best span can vary abruptly even for adjacent x_i, so next the data in Figure 11.14 are passed through a fixed-span smoother (say, $\hat{s}_2$) to estimate the optimal span as a function of x. Denote this smooth as $\hat{h}(x)$. Figure 11.14 also shows $\hat{h}(x)$.

Now we have the original data and a notion of the best span to use for any given x: namely $\hat{h}(x)$. What remains is to create a final, overall smooth. Among several strategies that might be employed at this point, [180] recommends setting $\hat{s}(x_i)$ equal to a linear interpolation between $\hat{s}_{h^-(x_i)}(x_i)$ and $\hat{s}_{h^+(x_i)}(x_i)$, where among the m fixed spans tried, $h^-(x_i)$ is the largest span less than $\hat{h}(x_i)$ and $h^+(x_i)$ is the smallest span that exceeds $\hat{h}(x_i)$. Thus,

$$\hat{s}(x_i) = \frac{\hat{h}(x_i) - h^-(x_i)}{h^+(x_i) - h^-(x_i)}\hat{s}_{h^+(x_i)}(x_i) + \frac{h^+(x_i) - \hat{h}(x_i)}{h^+(x_i) - h^-(x_i)}\hat{s}_{h^-(x_i)}(x_i). \qquad (11.35)$$

Figure 11.15 shows the final result. The supersmoother adjusted the span wisely, based on the local variability of the data. In comparison, the spline smooth shown in this figure undersmoothed the left side and oversmoothed the right side, for a fixed λ chosen by cross-validation.

Although the supersmoother is a nonlinear smoother, it is very fast compared to most other nonlinear smoothers, including loess.

11.5 CONFIDENCE BANDS

Producing reliable confidence bands for smooths is not straightforward. Intuitively, what is desired is an image that portrays the range and variety of smooth curves that might plausibly be obtained from data like what we observed. Bootstrapping (Chapter 9) provides a method for avoiding parametric assumptions, but it does not help clarify exactly what sort of region should be graphed.

Consider first the notion of a *pointwise confidence band*. Bootstrapping the residuals would proceed as follows. Let $\mathbf{e}$ denote the vector of residuals (so $\mathbf{e} = (\mathbf{I} - \mathbf{S})\mathbf{Y}$ for a linear smoother). Sample the elements of $\mathbf{e}$ with replacement to generate bootstrapped residuals $\mathbf{e}^*$. Add these to the fitted values to obtain bootstrapped responses $\mathbf{Y}^* = \hat{\mathbf{Y}} + \mathbf{e}^*$. Smooth $\mathbf{Y}^*$ over $\mathbf{x}$ to generate a bootstrapped fitted smooth, $\hat{\mathbf{s}}^*$. Start anew and repeat the bootstrapping many times. Then, for each x in the dataset, a bootstrap confidence interval for $\hat{s}(x)$ can be generated using the percentile method (Section 9.3.1) by deleting the few largest and smallest bootstrap fits at that point. If

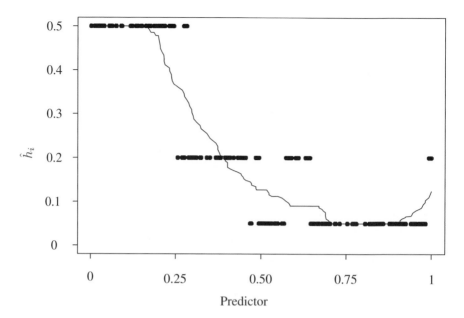

Fig. 11.14   The supersmoother estimate of the optimal span as a function of x. The points correspond to $(x_i, \hat{h}_i)$. A smooth of these points, namely $\hat{h}(x)$, is shown with the curve.

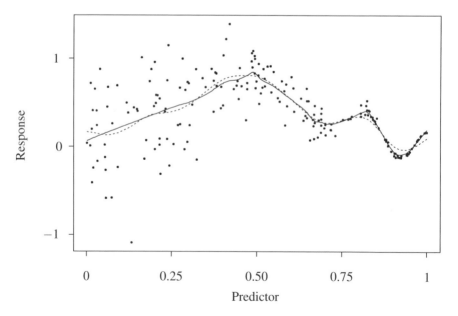

Fig. 11.15   A supersmoother fit (solid line). A spline smoother fit (with λ chosen by cross-validation) is also shown (dotted line).

the upper bounds of these pointwise confidence intervals are connected for each x, the result is a band lying above $\hat{s}(x)$. A plot showing this upper band along with the corresponding lower band provides a visually appealing confidence region.

Although this approach is appealing, it can be quite misleading. First, the confidence bands are composed of pointwise confidence intervals with no adjustment made for simultaneous inference. To correct the joint coverage probability to 95%, each of the individual intervals would need to represent much more than 95% confidence. The result would be a substantial widening of the pointwise bands.

Second, the pointwise confidence bands are not informative about features shared by all smooths supported by the data. For example, all the smooths could have an important bend at the same point, but the pointwise confidence region would not necessarily enforce this. It would be possible to sketch smooth curves lying entirely within the pointwise region that do not have such a bend, or even ones that have a reverse bend at that point. Similarly, suppose that all the smooths show generally the same curved shape and a linear fit is significantly inferior. If the confidence bands are wide or if the curve is not too severe, it will be possible to sketch a linear fit that lies entirely within the bands. In this case, the pointwise bands are failing to convey an important aspect of the inference: that a linear fit should be rejected.

Example 11.9 (Comparing the smooth with the null model) The shortcomings of pointwise confidence bands are illustrated in Figure 11.16, for some data for which the true conditional mean function is $E\{Y|x\} = x^2$. The smoothing span for a running-line smoother was chosen by cross-validation, and a pointwise 95% confidence band is shown by the shaded region. Unfortunately, the null model $E\{Y|x\} = 0$ lies entirely within the pointwise confidence band. Below we introduce an alternative method that convincingly rejects the null model. Figure 11.16 also shows that the band widens appropriately near the edges of the data to reflect the increased uncertainty of smoothing in these regions with fewer neighboring observations. □

The failure of pointwise confidence bands to capture the correct joint coverage probability can be remedied with a post hoc adjustment. Write the ordinary pointwise confidence bands as $(\hat{s}(x) - \hat{L}(x), \ \hat{s}(x) + \hat{U}(x))$, where $\hat{L}(x)$ and $\hat{U}(x)$ denote how far the lower and upper pointwise confidence bounds deviate from $\hat{s}(x)$ at the point x. (Alternatively, $\hat{s}(x)$ can be replaced by the pointwise median bootstrap curve; for hypothesis testing use the pointwise median null band.) Then the confidence bands can be stretched by finding the smallest w for which the bands given by $(\hat{s}(x) - w\hat{L}(x), \ \hat{s}(x) + w\hat{U}(x))$ contain at least $(1 - \alpha)100\%$ of the bootstrap curves in their entirety, where $(1 - \alpha)100\%$ is the desired confidence level. This method improves the joint coverage probability, but does not change the shape of the confidence bands.

The failure of pointwise confidence bands to represent accurately the shape of the bootstrap confidence set cannot be blamed on the pointwise nature of the bands; rather it is caused by the attempt to reduce a n-dimensional confidence set to a two-dimensional picture. Even if a band with correct joint coverage were used, the same problems would remain. For that reason, it may be more reasonable to superimpose a number of smooth curves that are known to belong to the joint confidence set, rather

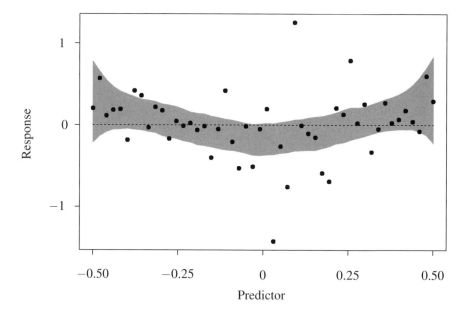

Fig. 11.16   Running-line smooth of some data for which $E\{Y|x\} = x^2$, with span chosen by cross-validation. The shaded region represents a pointwise 95% confidence band as described in the text. Note that the line $Y = 0$ is contained entirely within the band.

than trying to plot the boundaries of the set itself. With this in mind, we now describe a second bootstrapping approach suitable for linear smoothers.

Assume the response variable has constant variance. Among the estimators of this variance, Hastie and Tibshirani [280] recommend

$$\hat{\sigma}^2 = \frac{\text{RSS}_\lambda(\hat{\mathbf{s}}_\lambda)}{n - 2\text{tr}\{\mathbf{S}\} + \text{tr}\{\mathbf{SS}^T\}}. \tag{11.36}$$

The quantity

$$V = (\hat{\mathbf{s}}_\lambda - \mathbf{s})^T \left(\hat{\sigma}^2 \mathbf{SS}^T\right)^{-1} (\hat{\mathbf{s}}_\lambda - \mathbf{s}) \tag{11.37}$$

is approximately pivotal, so its distribution is roughly independent of the true underlying curve. Bootstrap the residuals as above, each time computing the vector of bootstrap fits, $\hat{\mathbf{s}}^*$, and the corresponding value

$$V^* = (\hat{\mathbf{s}}_\lambda^* - \hat{\mathbf{s}}_\lambda)^T \left(\hat{\sigma}^{*2} \mathbf{SS}^T\right)^{-1} (\hat{\mathbf{s}}_\lambda^* - \hat{\mathbf{s}}_\lambda). \tag{11.38}$$

Use the collection of V^* values to construct the empirical distribution of V^*. Eliminate those bootstrap fits whose V^* values are in the upper tail of this empirical distribution. Graph the remaining smooths—or a subset of them—superimposed. This provides a useful picture of the uncertainty of a smooth.

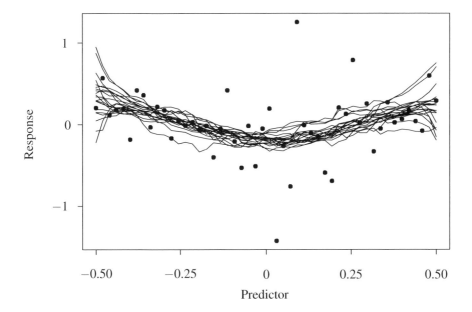

Fig. 11.17 Twenty bootstrap smooths of the data in Figure 11.16 for which V^* was within the central 95% region of its bootstrap distribution; see Example 11.10.

Example 11.10 (Comparing the smooth with the null model, continued) Applying the above method to the data described in Example 11.9 with a running-line smoother yields Figure 11.17. Roughly the same pointwise spread is indicated by this figure as is shown in the pointwise bands of Figure 11.16, but Figure 11.17 confirms that the smooth is curved like $y = x^2$. In fact, from 1000 bootstrap iterations, only three smooths resembled functions with a nonpositive second derivative. Thus, this bootstrap approach strongly rejects the null relationship $Y = 0$, while the pointwise confidence bands fail to rule it out. □

A variety of other bootstrapping and parametric approaches for assessing the uncertainty of results from smoothers are given in [168, 269, 280, 370].

11.6 GENERAL BIVARIATE DATA

For general bivariate data, there is no clear distinction of predictor and response variables, even though the two variables may exhibit a strong relationship. It is therefore more reasonable to label the variables as X_1 and X_2. As an example of such data, consider the two variables whose scatterplot is shown in Figure 11.18. For this example, the curve to be estimated corresponds to a curvilinear ridgetop of the joint distribution of X_1 and X_2.

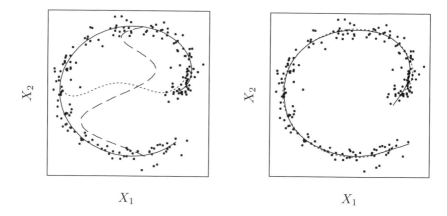

Fig. 11.18 The left panel shows data scattered around the time-parameterized curve given by $(x(\tau), y(\tau)) = ((1 - \cos \tau) \cos \tau, \ (1 - \cos \tau) \sin \tau)$ for $\tau \in [0, 3\pi/2]$, which is shown by the solid line. The dotted line shows the result of a fifth-order polynomial regression of X_2 on X_1, and the dashed line shows the result of a fifth-order polynomial regression of X_1 on X_2. The right panel shows a principal curve smooth (solid) for these data, along with the true curve (dotted). These are nearly superimposed.

Arbitrary designation of one variable as predictor and the other as response is counterproductive for such problems. For example, the left panel of Figure 11.18 shows the two fits obtained by ordinary fifth-order polynomial regression. Each of these lines is fitted by minimizing a set of residuals that measure the distances between points and the fitted curve, parallel to the axis of response. In one case X_1 was treated as the response, and in the other case X_2 was. Very different answers result, and in this case they are both terrible approximations to the true relationship.

The right panel of Figure 11.18 shows another curve fit to these data. Here, the curve was chosen to minimize the orthogonal distances between the points and the curve, without designation of either variable as the response. This approach corresponds to the local-averaging notion that points in any local neighborhood should fall near the curve. One approach to formalizing this notion is presented in Section 12.2.1, where we discuss the *principal curves* method for smoothing general p-dimensional data when there is no clear distinction of predictor and response. Setting $p = 2$ accommodates the bivariate case shown here.

Problems

11.1   Generate 100 random points from the following model: $X \sim \mathrm{Unif}(0, \pi)$ and $Y = g(X) + \epsilon$ with independent $\epsilon | x \sim \mathrm{N}(0, g(x)^2/64)$, where $g(x) = 1 + \frac{\sin\{x^2\}}{x^2}$. Smooth your data with a constant-span (symmetric nearest neighbor) running-mean smoother. Select a span of $2k + 1$ for $1 \leq k \leq 11$ chosen by cross-validation. Does a running-median smooth with the same span seem very different?

11.2 Use the data from Problem 11.1 to investigate kernel smoothers as described below:

(a) Smooth the data using a normal kernel smoother. Select the optimal standard deviation of the kernel using cross-validation.

(b) Define the symmetric triangle distribution as

$$
f(x; \mu, h) = \begin{cases} 0 & \text{if } |x - \mu| > h, \\ (x - \mu + h)/a^2 & \text{if } \mu - h \le x < \mu, \\ (\mu + h - x)/a^2 & \text{if } \mu \le x \le \mu + h. \end{cases}
$$

The standard deviation of this distribution is $a/\sqrt{6}$. Smooth the data using a symmetric triangle kernel smoother. Use cross-validation to search the same set of standard deviations used in the first case, and select the optimum.

(c) Let
$$
f(x; \mu, h) = c\big(1 + \cos\{2\pi z \log\{|z| + 1\}\}\big) \exp\{-z^2/2\},
$$

where $z = (x - \mu)/h$ and c is a constant. Plot this density function. The standard deviation of this density is about $0.90h$. Smooth the data using a kernel smoother with this kernel. Use cross-validation to search the same set of standard deviations used previously, and select the optimum.

(d) Compare the smooths produced using the three kernels. Compare their CVRSS values at the optimal spans. Compare the optimal spans themselves. For kernel smoothers, what can be said about the relative importance of the kernel and the span?

11.3 Use the data from Problem 11.1 to investigate running lines and running polynomial smoothers as described below:

(a) Smooth the data using a running-line smoother with symmetric nearest neighbors. Select a span of $2k + 1$ for $1 \le k \le 11$ chosen by cross-validation.

(b) Repeat this process for running local polynomial smoothers of degree 3 and 5; each time choose the optimal span using cross-validation over a suitable range for k. (Hints: You may need to orthogonalize polynomial terms; also reduce the polynomial degree as necessary for large spans near the edges of the data.)

(c) Comment on the quality and characteristics of the three smooths (local linear, cubic, and quintic).

(d) Does there appear to be a relationship between polynomial degree and optimal span?

(e) Comment on the three plots of CVRSS.

11.4   The book website provides data on the temperature-pressure profile of the Martian atmosphere, as measured by the Mars Global Surveyor spacefcraft in 2003 using a radio occultation technique [540]. Temperatures generally cool with increasing planetocentric radius (altitude).

(a) Smooth temperature as a function of radius using a smoothing spline, loess, and at least one other technique. Justify the choice of span for each procedure.

(b) The dataset also includes standard errors for the temperature measurements. Apply reasonable weighting schemes to produce weighted smooths using each smoother considered in part (a). Compare these results with the previous results. Discuss.

(c) Construct confidence bands for your smooths. Discuss.

(d) These data originate from seven separate orbits of the spacecraft. These orbits pass over somewhat different regions of Mars. A more complete dataset including orbit number, atmospheric pressure, longitude, lattitude, and other variables is available in the file 'mars-all.dat' at the book website. Introductory students may smooth some other interesting pairs of variables. Advanced students may seek to improve the previous analyses, for example by adjusting for orbit number or longitude and latitude. Such an analysis might include both parametric and non-parametric model components.

11.5   Reproduce Figure 11.8. (Hint: The kernel for a spline smoother can be reverse-engineered from the fit produced by any software package, using a suitable vector of response data.)

(a) Create a graph analogous to Figure 11.8 for smoothing at the second smallest predictor value. Compare this with the first graph.

(b) Graphically compare the equivalent kernels for cubic smoothing splines for different x_i and λ.

11.6   Figure 11.19 shows the pressure difference between two sensors on a steel plate exposed to a powerful air blast [299]. There are 161 observations during a period just before and after the blast. The noise in Figure 11.19 is attributable to inadequate temporal resolution and to error in the sensors and recording equipment; the underlying physical shock waves that generate these data are smooth. These data are available from the website for this book.

(a) Construct a running-line smooth of these data, with span chosen by eye.

(b) Make a plot of $\text{CVRSS}_k(\hat{s}_k)$ versus k for $k \in \{3, 5, 7, 11, 15, 20, 30, 50\}$. Comment.

(c) Produce the most appealing smooth you can for these data, using any smoother and span you wish. Why do you like it?

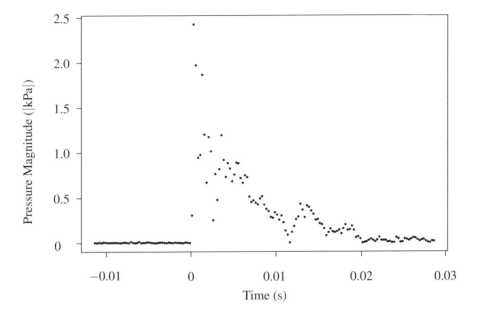

Fig. 11.19 Data on air blast pressure difference for Problem 11.6.

(d) Comment on difficulties of smoothing and span selection for these data.

11.7   Using the data from Problem 11.6 and your favorite linear smoothing method for these data, construct confidence bands for the smooth using each method described in Section 11.5. Discuss. (Using a spline smoother is particularly interesting.)

12

Multivariate Smoothing

12.1 PREDICTOR–RESPONSE DATA

Multivariate predictor–response smoothing methods fit smooth surfaces to obser-
vations $(\mathbf{x}_i, y_i)$, where $\mathbf{x}_i$ is a vector of p predictors and y_i is the corresponding
response value. The $y_1, \ldots, y_n$ values are viewed as observations of the random variables $Y_1, \ldots, Y_n$, where the distribution of Y_i depends on the ith vector of predictor
variables.

Many of the bivariate smoothing methods discussed in Chapter 11 can be generalized to the case of several predictors. Running lines can be replaced by running
planes. Univariate kernels can be replaced by multivariate kernels. One generalization of spline smoothing is thin plate splines [245, 382]. In addition to the significant
complexities of actually implementing some of these approaches, there is a fundamental change in the nature of the smoothing problem when using more than one
predictor.

The _curse of dimensionality_ is that high-dimensional space is vast, and points
have few near neighbors. This same problem was discussed in Section 10.4.1 as it
applied to multivariate density estimation. Consider a unit sphere in p dimensions
with volume $\pi^{p/2}/\Gamma(\frac{p}{2}+1)$. Suppose that several p-dimensional predictor points are
distributed uniformly within the ball of radius 4. In one dimension, 25% of predictors
are expected within the unit ball; hence unit balls might be reasonable neighborhoods
for smoothing. Table 12.1 shows that this proportion vanishes rapidly as p increases.
In order to retain 25% of points in a neighborhood when the full set of points lies in
a ball of radius 4, the neighborhood ball would need to have radius 3.73 if $p = 20$.
Thus, the concept of local neighborhoods is effectively lost.

p	Ratio
1	0.25
2	0.063
3	0.016
4	0.0039
5	0.00098
10	9.5×10^{-7}
20	9.1×10^{-13}
100	6.2×10^{-61}

Table 12.1 Ratio of the volume of the unit sphere in p dimensions to the volume of the sphere with radius 4.

The curse of dimensionality raises concerns about the effectiveness of smoothers for multivariate data. Effective local averaging will require a large number of points in each neighborhood, but to find such points, the neighborhoods must stretch over most of predictor space. A variety of effective multivariate surface smoothing methods are described in [280, 281, 484].

There is a rich set of smoothing methods developed for geostatistics and spatial statistics that are suitable for two and three dimensions. In particular, kriging methods offer a more principled foundation for inference than many of the generic smoothers considered here. We do not consider such methods further, but refer readers to books on spatial statistics such as [110, 254].

12.1.1 Additive models

Simple linear regression is based on the model $E\{Y|x\} = \beta_0 + \beta_1 x$. Nonparametric smoothing of bivariate predictor–response data generalizes this to $E\{Y|x\} = s(x)$ for a smooth function s. Now we seek to extend the analogy to the case with p predictors. Multiple regression uses the model $E\{Y|\mathbf{x}\} = \beta_0 + \sum_{k=1}^{p} \beta_k x_k$ where $\mathbf{x} = (x_1, \ldots, x_p)^T$. The generalization for smoothing is the *additive model*

$$E\{Y|\mathbf{x}\} = \alpha + \sum_{k=1}^{p} s_k(x_k), \qquad (12.1)$$

where s_k is a smooth function of the kth predictor variable. Thus, the overall model is composed of univariate effects whose influence on the mean response is additive.

Fitting such a model relies on the relationship

$$s_k(x_k) = E\left\{ Y - \alpha - \sum_{j \neq k} s_j(x_j) \middle| \mathbf{x} \right\}, \qquad (12.2)$$

where x_k is the kth component in $\mathbf{x}$. Suppose that we wished to estimate s_k at x_k^*, and that many replicate values of the kth predictor were observed at exactly this x_k^*.

Suppose further that all the s_j ($j \neq k$) were known except s_k. Then the expected value on the right side of (12.2) could be estimated as the mean of the values of $Y_i - \alpha - \sum_{j \neq k} s_j(x_{ij})$ corresponding to indices i for which the ith observation of the kth predictor satisfies $x_{ik} = x_k^*$. For actual data, however, there will likely be no such replicates. This problem can be overcome by smoothing: the average is taken over points whose kth coordinate is in a neighborhood of x_k^*. A second problem— that none of the s_j are actually known—is overcome by iteratively cycling through smoothing steps based on isolations like (12.2) updating s_k using the best current guesses for all s_j for $j \neq k$.

The iterative strategy is called the *backfitting algorithm*. Let $\mathbf{Y} = (Y_1, \ldots, Y_n)^T$, and for each k, let $\hat{\mathbf{s}}_k^{(t)}$ denote the vector of estimated values of $s_k(x_{ik})$ at iteration t for $i = 1, \ldots, n$. These n-vectors of estimated smooths at each observation are updated as follows:

1. Let $\hat{\alpha}$ be the n-vector $(\overline{Y}, \ldots, \overline{Y})^T$. Some other generalized average of the response values may replace the sample mean $\overline{Y}$. Let $t = 0$, where t indexes the iteration number.

2. Let $\hat{\mathbf{s}}_k^{(0)}$ represent initial guesses for coordinatewise smooths evaluated at the observed data. A reasonable initial guess is to let $\hat{\mathbf{s}}_k^{(0)} = (\hat{\beta}_k x_{1k}, \ldots, \hat{\beta}_k x_{nk})^T$ for $k = 1, \ldots, p$, where the $\hat{\beta}_k$ are the linear regression coefficients found when Y is regressed on the predictors.

3. For $k = 1, \ldots, p$ in turn, let

$$\hat{\mathbf{s}}_k^{(t+1)} = \text{smooth}_k(\mathbf{r}_k), \tag{12.3}$$

where

$$\mathbf{r}_k = \mathbf{Y} - \alpha - \sum_{j<k} \hat{\mathbf{s}}_j^{(t+1)} - \sum_{j>k} \hat{\mathbf{s}}_j^{(t)} \tag{12.4}$$

and $\text{smooth}_k(\mathbf{r}_k)$ denotes the vector obtained by smoothing the elements of $\mathbf{r}_k$ against the kth coordinate values of predictors, namely $x_{1k}, \ldots, x_{nk}$, and evaluating the smooth at each x_{ik}. The smoothing technique used for the kth smooth may vary with k.

4. Increment t and go to step 3.

The algorithm can be stopped when none of the $\hat{\mathbf{s}}_k^{(t)}$ change very much—perhaps when

$$\sum_{k=1}^{p} \left(\hat{\mathbf{s}}_k^{(t+1)} - \hat{\mathbf{s}}_k^{(t)} \right)^T \left(\hat{\mathbf{s}}_k^{(t+1)} - \hat{\mathbf{s}}_k^{(t)} \right) \Big/ \sum_{k=1}^{p} \left(\hat{\mathbf{s}}_k^{(t)} \right)^T \hat{\mathbf{s}}_k^{(t)}$$

is very small.

To understand why this algorithm works, recall the Gauss–Seidel algorithm for solving a linear system of the form $\mathbf{Az} = \mathbf{b}$ for $\mathbf{z}$ given a matrix $\mathbf{A}$ and a vector $\mathbf{b}$ of known constants (see Section 2.2.4). The Gauss–Seidel procedure is initialized with starting value $\mathbf{z}_0$. Then each component of $\mathbf{z}$ is solved for in turn, given the current values for the other components. This process is iterated until convergence.

Suppose only linear smoothers are used to fit an additive model, and let $\mathbf{S}_k$ be the $n \times n$ smoothing matrix for the kth component smoother. Then the backfitting algorithm solves the set of equations given by $\hat{\mathbf{s}}_k = \mathbf{S}_k \left(\mathbf{Y} - \sum_{j \neq k} \hat{\mathbf{s}}_j \right)$. Writing this set of equations in matrix form yields

$$
\begin{pmatrix}
\mathbf{I} & \mathbf{S}_1 & \mathbf{S}_1 & \cdots & \mathbf{S}_1 \\
\mathbf{S}_2 & \mathbf{I} & \mathbf{S}_2 & \cdots & \mathbf{S}_2 \\
\vdots & \vdots & \vdots & & \vdots \\
\mathbf{S}_p & \mathbf{S}_p & \mathbf{S}_p & \cdots & \mathbf{I}
\end{pmatrix}
\begin{pmatrix}
\hat{\mathbf{s}}_1 \\
\hat{\mathbf{s}}_2 \\
\vdots \\
\hat{\mathbf{s}}_p
\end{pmatrix}
=
\begin{pmatrix}
\mathbf{S}_1 \mathbf{Y} \\
\mathbf{S}_2 \mathbf{Y} \\
\vdots \\
\mathbf{S}_p \mathbf{Y}
\end{pmatrix},
\tag{12.5}
$$

which is of the form $\mathbf{A}\mathbf{z} = \mathbf{b}$ where $\mathbf{z} = (\hat{\mathbf{s}}_1, \hat{\mathbf{s}}_2, \ldots, \hat{\mathbf{s}}_p)^T = \hat{\mathbf{s}}$. Note that $\mathbf{b} = \mathbf{\Lambda}\mathbf{Y}$ where $\mathbf{\Lambda}$ is a block-diagonal matrix with the individual $\mathbf{S}_k$ matrices along the diagonal. Since the backfitting algorithm sequentially updates each vector $\hat{\mathbf{s}}_k$ as a single block, it is more formally a *block Gauss–Seidel algorithm*. The iterative backfitting algorithm is preferred because it is faster than the direct approach of inverting $\mathbf{A}$.

We now turn to the question of convergence of the backfitting algorithm and the uniqueness of the solution. Here, it helps to revisit the analogy to multiple regression. Let $\mathbf{D}$ denote the $n \times p$ design matrix whose ith row is $\mathbf{x}_i^T$ so $\mathbf{D} = (\mathbf{x}_1, \ldots, \mathbf{x}_n)^T$. Consider solving the multiple regression normal equations $\mathbf{D}^T \mathbf{D} \boldsymbol{\beta} = \mathbf{D}^T \mathbf{Y}$ for $\boldsymbol{\beta}$. The elements of $\boldsymbol{\beta}$ are not uniquely determined if any predictors are linearly dependent, or equivalently, if the columns of $\mathbf{D}^T \mathbf{D}$ are linearly dependent. In that case, there would exist a vector $\boldsymbol{\gamma}$ such that $\mathbf{D}^T \mathbf{D} \boldsymbol{\gamma} = \mathbf{0}$. Thus, if $\hat{\boldsymbol{\beta}}$ were a solution to the normal equations, $\hat{\boldsymbol{\beta}} + c\boldsymbol{\gamma}$ would also be a solution for any c.

Analogously, the backfitting estimating equations $\mathbf{A}\hat{\mathbf{s}} = \mathbf{\Lambda}\mathbf{Y}$ will not have a unique solution if there exists any $\boldsymbol{\gamma}$ such that $\mathbf{A}\boldsymbol{\gamma} = \mathbf{0}$. Let $\mathcal{I}_k$ be the space spanned by vectors that pass through the kth smoother unchanged. If these spaces are linearly dependent, then there exist $\boldsymbol{\gamma}_k \in \mathcal{I}_k$ such that $\sum_{k=1}^p \boldsymbol{\gamma}_k = \mathbf{0}$. In this case, $\mathbf{A}\boldsymbol{\gamma} = \mathbf{0}$, where $\boldsymbol{\gamma} = (\boldsymbol{\gamma}_1, \boldsymbol{\gamma}_2, \ldots, \boldsymbol{\gamma}_p)^T$, and therefore there is not a unique solution (see Problem 12.1).

A more complete discussion of these issues is provided by Hastie and Tibshirani [280], from which the following result is derived. Let the p smoothers be linear and each $\mathbf{S}_k$ be symmetric with eigenvalues in $[0, 1]$. Then $\mathbf{A}\boldsymbol{\gamma} = \mathbf{0}$ if and only if there exist linearly dependent $\boldsymbol{\gamma}_k \in \mathcal{I}_k$ that pass through the kth smoother unchanged. In this case, there are many solutions to $\mathbf{A}\hat{\mathbf{s}} = \mathbf{\Lambda}\mathbf{Y}$, and backfitting converges to one of them, depending on the starting values. Otherwise, backfitting converges to the unique solution.

The flexibility of the additive model is further enhanced by allowing the additive components of the model to be multivariate and by allowing different smoothing methods for different components. For example, suppose there are seven predictors, $x_1, \ldots, x_7$, where x_1 is a discrete variable with levels $1, \ldots, c$. Then an additive

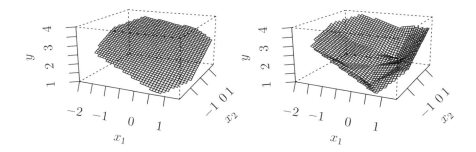

Fig. 12.1 Linear (left) and additive (right) models fitted to the Norwegian paper data in Example 12.1.

model to estimate $E\{Y|\mathbf{x}\}$ might be fitted by backfitting:

$$\hat{\alpha} + \sum_{i=1}^{c-1} \hat{\delta}_i 1_{\{x_1=i\}} + \hat{s}(x_2) + \hat{p}(x_3) + \hat{t}(x_4, x_5) + \hat{f}(x_6, x_7), \qquad (12.6)$$

where the $\hat{\delta}_i$ permit a separate additive effect for each level of X_1, $\hat{s}(x_2)$ is a spline smooth over x_2, $\hat{p}(x_3)$ is a cubic polynomial regression on x_3, $\hat{t}(x_4, x_5)$ is a recursively partitioned regression tree from Section 12.1.4, and $\hat{f}(x_6, x_7)$ is a bivariate kernel smooth. Grouping several predictors in this way provides coarser blocks in the blockwise implementation of the Gauss–Seidel algorithm.

Example 12.1 (Norwegian paper) We consider some data from a paper plant in Halden, Norway [9]. The response is a measure of imperfections in the paper, and there are two predictors. (Our Y, x_1, and x_2 correspond to $16 - Y_5$, X_1, and X_3, respectively, in the author's original notation). The left panel of Figure 12.1 shows the response surface fitted by an ordinary linear model with no interaction. The right panel shows an additive model fitted to the same data. The estimated $\hat{s}_k$ are shown in Figure 12.2. Clearly x_1 has a nonlinear effect on the response; in this sense the additive model is an improvement over the linear regression fit. □

12.1.2 Generalized additive models

Linear regression models can be generalized in several ways. Above, we have replaced linear predictors with smooth nonlinear functions. A different way to generalize linear regression is in the direction of generalized linear models [379].

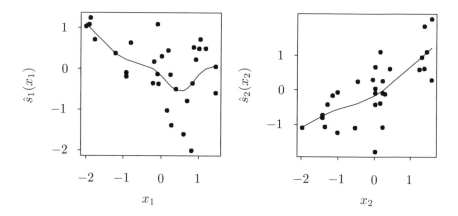

Fig. 12.2 The smooths $\hat{s}_k(x_k)$ fitted with an additive model for the Norwegian paper data in Example 12.1. The points are partial residuals as given on the right hand side of (12.3), namely, $\hat{s}_k(x_{ik})$ plus the overall residual from the final smooth.

Suppose that $Y|\mathbf{x}$ has a distribution in the exponential family. Let $\mu = \mathrm{E}\{Y|\mathbf{x}\}$. A generalized linear model assumes that some function of μ is a linear function of the predictors. In other words, the model is $g(\mu) = \alpha + \sum_{k=1}^{p} \beta_k x_k$, where g is the *link function*. For example, the identity link $g(\mu) = \mu$ is used to model a Gaussian distributed response, $g(\mu) = \log \mu$ is used for log-linear models, and $g(\mu) = \log\left\{\frac{\mu}{1-\mu}\right\}$ is one link used to model Bernoulli data.

Generalized additive models (GAMs) extend the additive models of Section 12.1.1 in a manner analogous to how generalized linear models extend linear models. For response data in an exponential family, a link function g is chosen, and the model is

$$g(\mu) = \alpha + \sum_{k=1}^{p} s_k(x_k), \tag{12.7}$$

where s_k is a smooth function of the kth predictor. The right hand side of (12.7) is denoted η and is called the *additive predictor*. GAMs provide the scope and diversity of generalized linear models, with the additional flexibility of nonlinear smooth effects in the additive predictor.

For generalized linear models, estimation of $\mu = \mathrm{E}\{Y|\mathbf{x}\}$ proceeds via iteratively reweighted least squares. Roughly, the algorithm proceeds by alternating between (i) constructing adjusted response values and corresponding weights, and (ii) fitting a weighted linear regression of the adjusted response on the predictors. These steps are repeated until the fit has converged.

Specifically, we described in Section 2.2.1.1 how the iteratively reweighted least squares approach for fitting generalized linear models in exponential families is in fact the Fisher scoring approach. The Fisher scoring method is ultimately motivated by a linearization of the score function which yields an updating equation for estimating the parameters. The update is achieved by weighted linear regression. Adjusted responses and weights are defined as in (2.41). The updated parameter vector consists of the coefficients resulting from a weighted linear least squares regression for the adjusted responses.

For fitting a GAM, weighted linear regression is replaced by weighted smoothing. The resulting procedure, called *local scoring*, is described below. First, let μ_i be the mean response for observation i, so $\mu_i = \mathrm{E}\{Y_i|\mathbf{x}_i\} = g^{-1}(\eta_i)$, where η_i is called the ith value of the additive predictor; and let $V(\mu_i)$ be the variance function, namely, $\mathrm{var}\{Y_i|\mathbf{x}_i\}$ expressed as a function of μ_i. The algorithm proceeds as follows:

1. Initialize the algorithm at $t = 0$. Set $\hat{\alpha}^{(0)} = g(\bar{Y})$ and $\hat{s}_k^{(0)}(\cdot) = 0$ for $k = 1, \ldots, p$. This also initializes the additive predictor values $\hat{\eta}_i^{(0)} = \hat{\alpha}^{(0)} + \sum_{k=1}^p \hat{s}_k^{(0)}(x_{ik})$ and the fitted values $\hat{\mu}_i^{(0)} = g^{-1}(\hat{\eta}_i^{(0)})$ corresponding to each observation.

2. For $i = 1, \ldots, n$, construct adjusted response values

$$z_i^{(t+1)} = \hat{\eta}_i^{(t)} + \left(Y_i - \hat{\mu}_i^{(t)}\right) \left(\left.\frac{d\mu}{d\eta}\right|_{\eta=\hat{\eta}_i^{(t)}}\right)^{-1}. \qquad (12.8)$$

3. For $i = 1, \ldots, n$, construct the corresponding weights

$$w_i^{(t+1)} = \left(\left.\frac{d\mu}{d\eta}\right|_{\eta=\hat{\eta}_i^{(t)}}\right)^2 \left(V\left(\hat{\mu}_i^{(t)}\right)\right)^{-1}. \qquad (12.9)$$

4. Use a weighted version of the backfitting algorithm from Section 12.1.1 to estimate new additive predictors, $\hat{s}_k^{(t+1)}$. In this step, a weighted additive model of the form (12.7) is fitted to the adjusted response values $z_i^{(t+1)}$ with weights $w_i^{(t+1)}$, yielding $\hat{s}_k^{(t+1)}(x_{ik})$ for $i = 1, \ldots, n$ and $k = 1, \ldots, p$. This step, described further below, also allows calculation of new $\hat{\eta}_i^{(t+1)}$ and $\hat{\mu}_i^{(t+1)}$.

5. Compute a convergence criterion such as

$$\sum_{k=1}^p \sum_{i=1}^n \left(\hat{s}_k^{(t+1)}(x_{ik}) - \hat{s}_k^{(t)}(x_{ik})\right)^2 \bigg/ \sum_{k=1}^p \sum_{i=1}^n \left(\hat{s}_k^{(t)}(x_{ik})\right)^2, \qquad (12.10)$$

and stop when it is small. Otherwise, go to step 2.

To revert to a standard generalized linear model, the only necessary change would be to replace the smoothing in step 4 with weighted least squares.

The fitting of a weighted additive model in step 4 requires weighted smoothing methods. For linear smoothers, one way to introduce weights is to multiply the elements in the ith column of $\mathbf{S}$ by $w_i^{(t+1)}$ for each i, and then standardize each row so it sums to 1. There are other, more natural approaches to weighting some linear

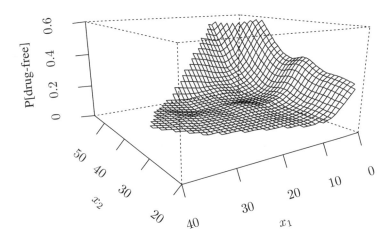

Fig. 12.3 The fit of a generalized additive model to the drug abuse data described in Example 12.2. The vertical axis corresponds to the predicted probability of remaining drug-free for one year.

smoothers (e.g., splines) and nonlinear smoothers. Further details about weighted smooths and local scoring are provided in [280, 485].

As with additive models, the linear predictor in GAMs need not consist solely of univariate smooths of the same type. The ideas in Section 12.1.1 regarding more general and flexible model building apply here too.

Example 12.2 (Drug abuse) The website for this book provides data on 575 patients receiving residential treatment for drug abuse [294]. The response variable is binary, with $Y = 1$ for a patient who remained drug-free for one year and $Y = 0$ otherwise. We will examine two predictors: number of prior drug treatments (x_1) and age of patient (x_2). A simple generalized additive model is given by $Y_i|\mathbf{x}_i \sim \text{Bernoulli}(\pi_i)$ with

$$\log\left\{\frac{\pi_i}{1 - \pi_i}\right\} = \alpha + \beta_1 s_1(x_{i1}) + \beta_2 s_2(x_{i2}). \tag{12.11}$$

Spline smoothing was used in step 4 of the fitting algorithm. Figure 12.3 shows the fitted response surface graphed on the probability scale. Figure 12.4 shows the fitted smooths $\hat{s}_k$ on the logit scale. The raw response data are shown by hash marks along the bottom $(y_i = 0)$ and top $(y_i = 1)$ of each panel. □

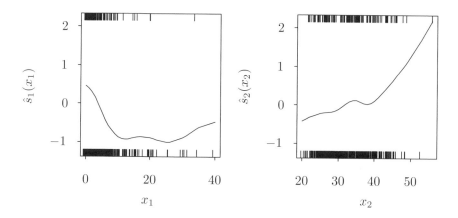

Fig. 12.4 The smooth functions $\hat{s}_k$ fitted with a generalized additive model for the drug-abuse data in Example 12.2. The raw response data are shown via the hash marks along the bottom ($y_i = 0$) and top ($y_i = 1$) of each panel at the locations observed for the corresponding predictor variable.

12.1.3 Other methods related to additive models

Generalized additive models are not the only way to extend the additive model. Several other methods transform the predictors or response in an effort to provide a more effective model for the data. We describe four such approaches below.

12.1.3.1 *Projection pursuit regression* Additive models produce fits composed of p additive surfaces, each of which has a nonlinear profile along one coordinate axis while staying constant in orthogonal directions. This aids interpretation of the model, because each nonlinear smooth reflects the additive effect of one predictor. However, it also limits the ability to fit more general surfaces and interaction effects that are not additively attributable to a single predictor. *Projection pursuit regression* eliminates this constraint by allowing effects to be smooth functions of univariate linear projections of the predictors [184, 331].

Specifically, these models take the form

$$E\{Y|\mathbf{x}\} = \alpha + \sum_{k=1}^{M} s_k(\mathbf{a}_k^T \mathbf{x}), \tag{12.12}$$

where each term $\mathbf{a}_k^T \mathbf{x}$ is a one-dimensional projection of the predictor vector $\mathbf{x} = (x_1, \ldots, x_p)^T$. Thus each s_k has a profile determined by s_k along $\mathbf{a}_k$ and is constant in all orthogonal directions. In the projection pursuit approach, the smooths s_k and the projection vectors $\mathbf{a}_k$ are estimated for $k = 1, \ldots, M$ to obtain the optimal fit. For sufficiently large M, the expression in (12.12) can approximate an arbitrary continuous function of the predictors [140, 331].

To fit such a model, the number of projections, M, must be chosen. When $M > 1$, the model contains several smooth functions of different linear combinations $\mathbf{a}_k^T \mathbf{x}$. The results may therefore be very difficult to interpret, notwithstanding the model's usefulness for prediction. Choosing M is a model selection problem akin to choosing the terms in a multiple regression model, and analogous reasoning should apply. One approach would be to fit a model with small M first, then repeatedly add the most effective next term and refit. A sequence of models is thus produced, until no further additional term substantially improves the fit.

For a given M, fitting (12.12) can be carried out using the following algorithm:

1. Begin with $m = 0$ by setting $\hat{\alpha} = \bar{Y}$.

2. Increment m. Define the current working residual for observation i as

$$r_i^{(m)} = Y_i - \hat{\alpha} - \sum_{k=1}^{m-1} \hat{s}_k(\mathbf{a}_k^T \mathbf{x}_i) \tag{12.13}$$

for $i = 1, \ldots, n$, where the summation vanishes if $m = 1$. These current residuals will be used to fit the mth projection.

3. For any p-vector $\mathbf{a}$ and smoother s_m, define the goodness-of-fit measure

$$Q(\mathbf{a}) = 1 - \frac{\sum_{i=1}^{n} \left(r_i^{(m)} - \hat{s}_m(\mathbf{a}^T \mathbf{x}_i) \right)^2}{\sum_{i=1}^{n} \left(r_i^{(m)} \right)^2}. \tag{12.14}$$

4. For a chosen type of smoother, maximize $Q(\mathbf{a})$ with respect to $\mathbf{a}$. This provides $\mathbf{a}_m$ and $\hat{s}_m$. If $m = M$, stop; otherwise go to step 2.

Example 12.3 (Norwegian paper, continued) We return to the Norwegian paper data of Example 12.1. Figure 12.5 shows the response surface fitted with projection pursuit regression for $M = 2$. A supersmoother (Section 11.4.2) was used for each projection. The fitted surface exhibits some interaction between the predictors that is not captured by either model shown in Figure 12.1. An additive model was not wholly appropriate for these predictors. The heavy lines in Figure 12.5 show the two linear directions onto which the bivariate predictor data were projected. The first projection direction, labeled $\mathbf{a}_1^T \mathbf{x}$, is far from being parallel to either coordinate axis. This allows a better fit of the interaction between the two predictors. The second projection very nearly contributes an additive effect of x_1. To further understand the fitted surface, we can examine the individual $\hat{s}_k$, which are shown in Figure 12.6. These effects along the selected directions provide a more general fit than either the regression or the additive model. □

In addition to predictor–response smoothing, the ideas of projection pursuit have been applied in many other areas, including smoothing for multivariate response data [9] and density estimation [180]. Another approach, known as multivariate

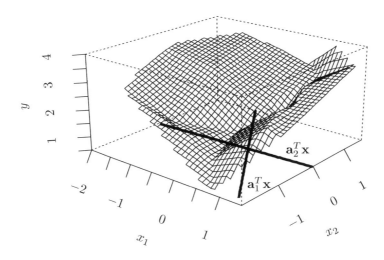

Fig. 12.5 A projection pursuit regression surface fitted to the Norwegian paper data for $M = 2$, as described in Example 12.3.

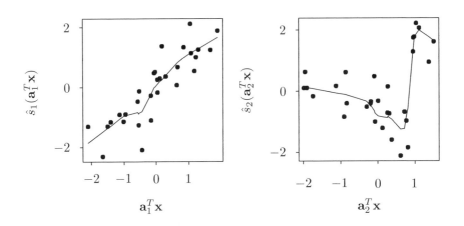

Fig. 12.6 The smooth functions $\hat{s}_k$ fitted with a projection pursuit regression model for the Norwegian paper data. The current residuals, namely the component-fitted smooth plus the overall residual, shown as dots, are plotted against each projection $\mathbf{a}_k^T \mathbf{x}$ for $k = 1, 2$.

adaptive regression splines (MARS), has links to projection pursuit regression, spline smoothing (Section 11.2.5), and regression trees (Section 12.1.4.1) [182]. MARS may perform very well for some datasets, but recent simulations have shown less promising results for high-dimensional data [19].

12.1.3.2   *Neural networks*   Neural networks are a nonlinear modeling method for either continuous or discrete responses, producing a regression or classification model [44, 45, 281, 457]. For a continuous response Y and predictors $\mathbf{x}$, one type of neural network model, called the *feed-forward network*, can be written as

$$g(Y) = \beta_0 + \sum_{m=1}^{M} \beta_m f(\boldsymbol{\alpha}_m^T \mathbf{x} + \gamma_m), \tag{12.15}$$

where β_0, β_m, $\boldsymbol{\alpha}_m$, and γ_m for $m = 1, \ldots, M$ are estimated from the data. We can think of the $f(\boldsymbol{\alpha}_m^T \mathbf{x} + \gamma_m)$ for $m = 1, \ldots, M$ as being analogous to a set of basis functions for the predictor space. These $f(\boldsymbol{\alpha}_m^T \mathbf{x} + \gamma_m)$, whose values are not directly observed, constitute a *hidden layer*, in neural net terminology. Usually the analyst chooses M in advance, but data-driven selection is also possible. In (12.15), the form of the *activation function*, f, is usually chosen to be logistic, namely $f(z) = \frac{1}{1 + \exp\{-z\}}$. We use g as a link function. Parameters are estimated by minimizing the squared error, typically via gradient-based optimization.

Neural networks are related to projection pursuit regression where s_k in (12.12) is replaced by a parametric function f in (12.15), such as the logistic function. Many enhancements to the simple neural network model given above are possible, such as the inclusion of an additional hidden layer using a different activation function, say h. This layer is composed of evaluations of h at a number of linear combinations of the $f(\boldsymbol{\alpha}_m^T \mathbf{x} + \gamma_m)$ for $m = 1, \ldots, M$, roughly serving as a basis for the first hidden layer. Neural networks are very popular in some fields, and a large number of software packages for fitting these models are available.

12.1.3.3   *Alternating conditional expectations*   The alternating conditional expectations (ACE) procedure fits models of the form

$$E\{g(Y)|\mathbf{x}\} = \alpha + \sum_{k=1}^{p} s_k(x_k) \tag{12.16}$$

where g is a smooth function of the response [58]. Unlike most other methods in this chapter, ACE treats the predictors as observations of a random variable $\mathbf{X}$, and model fitting is driven by consideration of the joint distribution of Y and $\mathbf{X}$. Specifically, the idea of ACE is to estimate g and s_k for $k = 1, \ldots, p$ such that the magnitude of the correlation between $g(Y)$ and $\sum_{k=1}^{p} s_k(X_k)$ is maximized subject to the constraint that $\text{var}\{g(Y)\} = 1$. The constant α does not affect this correlation, so it can be ignored.

To fit the ACE model, the following iterative algorithm can be used:

1. Initialize the algorithm by letting $t = 0$ and $\hat{g}^{(0)}(Y_i) = (Y_i - \bar{Y})/\hat{\sigma}_Y$, where $\hat{\sigma}_Y$ is the sample standard deviation of the Y_i values.

2. Generate updated estimates of the additive predictor functions $\hat{s}_k^{(t+1)}$ for $k = 1, \ldots, p$ by fitting an additive model with the $\hat{g}^{(t)}(Y_i)$ values as the response and the $\hat{s}_k^{(t+1)}(X_{ik})$ values as the predictors. The backfitting algorithm from Section 12.1.1 can be used to fit this model.

3. Estimate $\hat{g}^{(t+1)}$ by smoothing the values of $\sum_{k=1}^{p} \hat{s}_k^{(t+1)}(X_{ik})$ (treated as the response) over the Y_i (treated as the predictor).

4. Rescale $\hat{g}^{(t+1)}$ by dividing by the sample standard deviation of the $\hat{g}^{(t+1)}(Y_i)$ values. This step is necessary because otherwise setting both $\hat{g}^{(t+1)}$ and $\sum_{k=1}^{p} \hat{s}_k^{(t+1)}$ to zero functions trivially provides zero residuals regardless of the data.

5. If $\sum_{i=1}^{n} \left[\hat{g}^{(t+1)}(Y_i) - \sum_{k=1}^{p} \hat{s}_k^{(t+1)}(X_{ik}) \right]^2$ has converged according to a relative convergence criterion, stop. Otherwise, increment t and go to step 2.

Maximizing the correlation between $\sum_{k=1}^{p} s_k(X_k)$ and $g(Y)$ is equivalent to minimizing $\mathrm{E} \left\{ [g(Y) - \sum_{k=1}^{p} s_k(X_k)]^2 \right\}$ with respect to g and $\{s_k\}$ subject to the constraint that $\mathrm{var}\{g(Y)\} = 1$. For $p = 1$, this objective is symmetric in X and Y: If the two variables are interchanged, the result remains the same, up to a constant.

ACE provides no fitted model component that directly links $\mathrm{E}\{Y|\mathbf{X}\}$ to the predictors, which impedes prediction. ACE is therefore quite different than the other predictor–response smoothers we have discussed, because it abandons the notion of estimating the regression function; instead it provides a correlational analysis. Consequently, ACE can produce surprising results, especially when there is low correlation between variables. Such problems, and the convergence properties of the fitting algorithm, are discussed in [58, 74, 280].

12.1.3.4 *Additivity and variance stabilization*

Another additive model variant relying on transformation of the response is *additivity and variance stabilization* (AVAS) [535]. The model is the same as (12.16) except that g is constrained to be strictly monotone with

$$\mathrm{var}\left\{ g(Y) \left| \sum_{k=1}^{p} s_k(x_k) \right. \right\} = C \tag{12.17}$$

for some constant C.

To fit the model, the following iterative algorithm can be used:

1. Initialize the algorithm by letting $t = 0$ and $\hat{g}^{(0)}(Y_i) = (Y_i - \bar{Y})/\hat{\sigma}_Y$, where $\hat{\sigma}_Y$ is the sample standard deviation of the Y_i values.

2. Initialize the predictor functions by fitting an additive model to the $\hat{g}^{(0)}(Y_i)$ and predictor data, yielding $\hat{s}_k^{(0)}$ for $k = 1, \ldots, p$, as done for ACE.

3. Denote the current mean response function as $\hat{\mu}^{(t)} = \sum_{k=1}^{p} \hat{s}_k^{(t)}(X_k)$. To estimate the variance-stabilizing transformation, we must first estimate the conditional variance function of $\hat{g}^{(t)}(Y)$ given $\hat{\mu}^{(t)} = u$. This function, $\hat{V}^{(t)}(u)$, is

estimated by smoothing the current log squared residuals against u and exponentiating the result.

4. Given $\hat{V}^{(t)}(u)$, compute the corresponding variance-stabilizing transformation $\psi^{(t)}(z) = \int_0^z \hat{V}^{(t)}(u)^{-1/2} \, du$. This integration can be carried out using a numerical technique from Chapter 5.

5. Update and standardize the response transformation by defining $\hat{g}^{(t+1)}(y) = \left[\psi^{(t)} \left(\hat{g}^{(t)}(y) \right) - \bar{\psi}^{(t)} \right] / \hat{\sigma}_{\psi^{(t)}}$, where $\bar{\psi}^{(t)}$ and $\hat{\sigma}_{\psi^{(t)}}$ denote the sample mean and standard deviation of the $\psi^{(t)} \left(\hat{g}^{(t)}(Y_i) \right)$ values.

6. Update the predictor functions by fitting an additive model to the $\hat{g}^{(t+1)}(Y_i)$ and predictor data, yielding $\hat{s}_k^{(t+1)}$ for $k = 1, \dots, p$, as done for ACE.

7. If $\sum_{i=1}^n \left[\hat{g}^{(t+1)}(Y_i) - \sum_{k=1}^p \hat{s}_k^{(t+1)}(X_{ik}) \right]^2$ has converged according to a relative convergence criterion, stop. Otherwise, increment t and go to step 3.

Unlike ACE, the AVAS procedure is well suited for predictor–response regression problems. Further details of this method are given by [280, 535].

Both ACE and AVAS can be used to suggest parametric transformations for standard multiple regression modeling. In particular, plotting the ACE or AVAS transformed predictor versus the untransformed predictor can sometimes suggest a simple piecewise linear or other transformation for standard regression modeling [136, 290].

12.1.4 Tree-based methods

Tree-based methods recursively partition the predictor space into subregions associated with increasingly homogeneous values of the response variable. An important appeal of such methods is that the fit is often very easy to describe and interpret. For reasons discussed shortly, the summary of the fit is called a *tree*.

The most familiar tree-based method for statisticians is the classification and regression tree (CART) method described by Breiman, Friedman, Olshen, and Stone [59]. Both proprietary and open-source code software to carry out tree-based modeling are widely available [96, 199, 516, 533, 545]. While implementation details vary, all of these methods are fundamentally based on the idea of recursive partitioning.

A tree can be summarized by two sets of information:

- the answers to a series of binary (yes–no) questions, each of which is based on the value of a single predictor; and

- a set of values used to predict the response variable on the basis of answers to these questions.

An example will clarify the nature of a tree.

Example 12.4 (Stream monitoring) Various organisms called macroinvertebrates live in the bed of a stream, called the substrate. To monitor stream health, ecologists

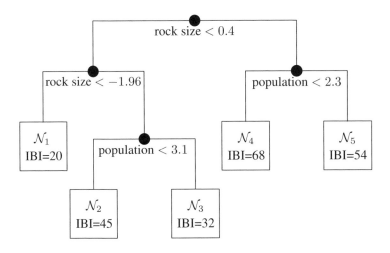

Fig. 12.7 Tree fit to predict IBI for Example 12.4. The root node is the top node of the tree, the parent nodes are the other nodes indicated by the • symbol, and the terminal nodes are $\mathcal{N}_1, \ldots, \mathcal{N}_5$. Follow the left branch from a parent node when the indicated criterion is true, and the right branch when it is false.

use a measure called the index of biotic integrity (IBI), which quantifies the stream's ability to support and maintain a natural biological community. An IBI allows for meaningful measurement of the effect of anthropogenic and other potential stressors on streams [319]. In this example, we consider predicting a macroinvertebrate IBI from two predictors, human population density and rock size in the substrate. The first predictor is the human population density (persons per square kilometer) in the stream's watershed. To improve graphical presentation, the log of population density is used in the analysis below, but the same tree would be selected were the untransformed predictor to be used. The second predictor is the estimated geometric mean of the diameter of rocks collected at the sample location in the substrate, measured in millimeters and transformed logarithmically. These data, considered further in Problem 12.5, were collected by the Environmental Protection Agency as part of a study of 353 sites in the Mid-Atlantic Highlands region of the eastern United States from 1993 to 1998 [161].

Figure 12.7 shows a typical tree. Four binary questions are represented by splits in the tree. Each split is based on the value of one of the predictors. The left branch of a split is taken when the answer is yes, so that the condition labeling that split is met. For example, the top split indicates that the left portion of the tree is for those observations with rock size values of 0.4 or less (sand and smaller). Each position in the tree where a split is made is a *parent node*. The topmost parent node is called the *root node*. All parent nodes except the root node are *internal nodes*. At the bottom of the tree the data have been classified into five *terminal nodes* based on the decisions made at the parent nodes. Associated with each terminal node is the mean value of the IBI for all observations in that node. We would use this value as the prediction for any observation whose predictors lead to this node. For example, we predict IBI $= 20$ for any observation that would be classified in $\mathcal{N}_1$. □

12.1.4.1 *Recursive partitioning regression trees*   Suppose initially that the response variable is continuous. Then tree-based smoothing is often called *recursive partitioning regression*. Section 12.1.4.3 discusses prediction of categorical responses.

Consider predictor–response data where $\mathbf{x}_i$ is a vector of p predictors associated with a response Y_i, for $i = 1, \ldots, n$. For simplicity, assume that all p predictors are continuous. Let q denote the number of terminal nodes in the tree to be fitted.

Tree-based predictions are piecewise constant. If the predictor values for the ith observation place it in the jth terminal node, then the ith predicted response equals a constant, $\hat{a}_j$. Thus, the tree-based smooth is

$$\hat{s}(\mathbf{x}_i) = \sum_{j=1}^{q} \hat{a}_j 1_{\{\mathbf{x}_i \in \mathcal{N}_j\}}. \tag{12.18}$$

This model is fitted using a partitioning process that adaptively partitions predictor space into hyperrectangles, each corresponding to one terminal node. Once the partitioning is complete, $\hat{a}_j$ is set equal to, say, the mean response value of cases falling in the jth terminal node.

Notice that this framework implies that there are a large number of possible trees whenever n and/or p are not trivially small. Any terminal node might be split to create a larger tree. The two branches from any parent node might be collapsed to convert the parent node to a terminal node, forming a *subtree* of the original tree. Any branch itself might be replaced by one based on a different predictor variable and/or criterion. The partitioning process used to fit a tree is described next.

In the simplest case, suppose $q = 2$. Then we seek to split $\Re^p$ into two hyperrectangles using one axis-parallel boundary. The choice can be characterized by a split coordinate, $c \in \{1, \ldots, p\}$, and a split point or threshold, $t \in \Re$. The two terminal nodes are then $\mathcal{N}_1 = \{\mathbf{x}_i : x_{ic} < t\}$ and $\mathcal{N}_2 = \{\mathbf{x}_i : x_{ic} \geq t\}$. Denote the sets of indices of the observations falling in the two nodes as $\mathcal{S}_1$ and $\mathcal{S}_2$, respectively. Using node-specific sample averages yields the fit

$$\hat{s}(\mathbf{x}_i) = 1_{\{i \in \mathcal{S}_1\}} \sum_{j \in \mathcal{S}_1} Y_j / n_1 + 1_{\{i \in \mathcal{S}_2\}} \sum_{j \in \mathcal{S}_2} Y_j / n_2, \tag{12.19}$$

where n_j is the number of observations falling in the jth terminal node.

For continuous predictors and ordered discrete predictors, defining a split in this manner is straightforward. Treatment of an unordered categorical variable is different. Suppose each observation of this variable may take on one of several categories. The set of all such categories must be partitioned into two subsets. Fortunately, we may avoid considering all possible partitions. First, order the categories in order of the average response within each category. Then, treat these ordered categories as if they were observations of an ordered discrete predictor. This strategy permits optimal splits [59]. There are also natural ways to deal with observations having some missing predictor values. Finally, selecting transformations of the predictors is usually not a problem: Tree-based models are invariant to monotone transformations of the predictors, because the split point is determined in terms of the rank of predictor, in most software packages.

To find the best tree with $q = 2$ terminal nodes, we seek to minimize the residual squared error,

$$\text{RSS}(c, t) = \sum_{j=1}^{q} \sum_{i \in \mathcal{S}_j} (Y_i - \hat{a}_j)^2 \tag{12.20}$$

with respect to c and t, where $\hat{a}_j = \sum_{i \in \mathcal{S}_j} Y_i / n_j$. Note that the $\mathcal{S}_j$ are defined using the values of c and t, and that $\text{RSS}(c, t)$ changes only when memberships change in the sets $\mathcal{S}_j$. Minimizing (12.20) is therefore a combinatorial optimization problem. For each coordinate, we need to try at most $n - 1$ splits, and fewer if there are tied predictor values in the coordinate. Therefore, the minimal $\text{RSS}(c, t)$ can be found by searching at most $p(n - 1)$ trees. Exhaustive search to find the best tree is feasible when $q = 2$.

Now suppose $q = 3$. A first split coordinate and split point partition $\Re^p$ into two hyperrectangles. One of these hyperrectangles is then partitioned into two portions using a second split coordinate and split point, applied only within this hyperrectangle. The result is three terminal nodes. There are at most $p(n - 1)$ choices for the first split. For making the second split on any coordinate different from the one used for the first split, there are at most $p(n - 1)$ choices for each possible first split chosen. For a second split on the same coordinate as the first split, there are at most $p(n - 2)$ choices. Carrying this logic on for larger q, we see that there are about $(n - 1)(n - 2) \cdots (n - q + 1)p^{q-1}$ trees to be searched. This enormous number defies exhaustive search.

Instead, a greedy search algorithm is applied (see Section 3.2). Each split is treated sequentially. The best single split is chosen to split the root node. For each child node, a separate split is chosen to split it optimally. Note that the q terminal nodes obtained in this way will usually not minimize the residual squared error over the set of all possible trees having q terminal nodes.

Example 12.5 (Stream monitoring, continued) To understand how terminal nodes in a tree correspond to hyperrectangles in predictor space, recall the stream monitoring data introduced in Example 12.4. Another representation of the tree in Figure 12.7 is given in Figure 12.8. This plot shows the partitioning of the predictor space determined by values of the rock size and population density variables. Each circle is centered at an $\mathbf{x}_i$ observation ($i = 1, \ldots, n$). The area of each circle reflects the magnitude of the IBI value for that observation, with larger circles corresponding to larger IBI values. The rectangular regions labeled $\mathcal{N}_1, \ldots, \mathcal{N}_5$ in this graph correspond to the terminal nodes shown in Figure 12.7. The first split (on the rock size coordinate at the threshold $t = 0.4$) is shown by the vertical line in the middle of the plot. Subsequent splits partition only portions of the predictor space. For example, the region corresponding to rock size exceeding 0.4 is next split into two nodes, $\mathcal{N}_4$ and $\mathcal{N}_5$,

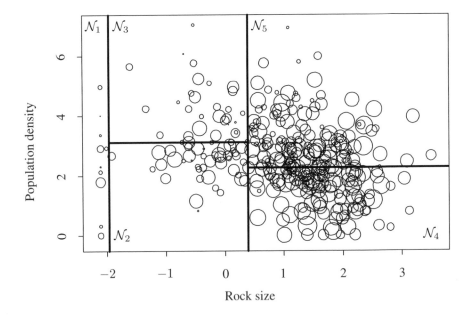

Fig. 12.8 Partitioning of predictor space (rock size and population density variables) for predicting IBI as discussed in Examples 12.4 and 12.5.

based on the value of the population density variable. Note that sequential splitting has drawbacks: an apparently natural division of the data based on whether the population density exceeds about 2.5 is represented by two slightly mismatched splits, because a previous split occurred at 0.4 on the rock size variable. The uncertainty of the tree structure is discussed further in Section 12.1.4.4.

The piecewise constant model for this fitted tree is shown in Figure 12.9, with the IBI on the vertical axis. To best display the surface, the axes have been reversed compared to Figure 12.8. □

12.1.4.2 *Tree pruning*
For a given q, greedy search can be used to fit a tree model. Note that q is, in essence, a smoothing parameter. Large values of q retain high fidelity to the observed data but provide trees with high potential variation in predictions. Such an elaborate model may also sacrifice interpretability. Low values of q provide less predictive variability because there are only a few terminal nodes, but predictive bias may be introduced if responses are not homogeneous within each terminal node. We now discuss how to choose q.

A naive approach for choosing q is to continue splitting terminal nodes until no additional split gives a sufficient reduction in the overall residual sum of squares. This approach may miss important structure in the data, because subsequent splits may be quite valuable even if the current split offers little or no improvement. For example, consider the saddle-shaped response surface obtained when X_1 and X_2 are

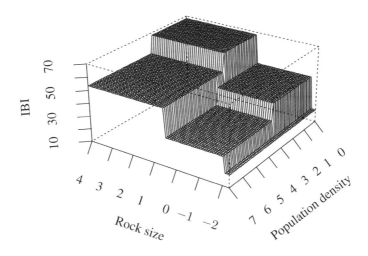

Fig. 12.9 Piecewise constant tree model predictions for IBI as discussed in Example 12.5.

independent predictors distributed uniformly over $[-1, 1]$ and $Y = X_1 X_2$. Then no single split on either predictor variable will be of much benefit, but any first split enables two subsequent splits that will greatly reduce the residual sum of squares.

A more effective strategy for choosing q begins by growing the tree, splitting each terminal node until it has no more than some prespecified minimal number of observations in it or its residual squared error does not exceed some prespecified percentage of the squared error for the root node. The number of terminal nodes in this *full tree* may greatly exceed q. Then, terminal nodes are sequentially recombined from the bottom up a way that doesn't greatly inflate the residual sum of squares. One implementation of this strategy is called *cost–complexity pruning* [59, 457]. The final tree is a subtree of the full tree, selected according to a criterion that balances a penalty for prediction error and a penalty for tree complexity.

Let the full tree be denoted by T_0, and let T denote some subtree of T_0 which can be obtained by pruning everything below some parent nodes in T_0. Let $q(T)$ denote the number of terminal nodes in the tree T. The cost–complexity criterion is given by

$$R_\alpha(T) = r(T) + \alpha q(T), \tag{12.21}$$

where $r(T)$ is the residual sum of squares or some other measure of prediction error for tree T, and α is a user-supplied parameter that penalizes tree complexity. For a given α, the optimal tree is the subtree of T_0 that minimizes $R_\alpha(T)$. When $\alpha = 0$, the full tree, T_0, will be selected as optimal. When $\alpha = \infty$, the tree with only the root node will be selected. If T_0 has $q(T_0)$ terminal nodes, then there are at most $q(T_0)$ subtrees that can be obtained by choosing different values of α.

The best approach for selecting the value for the parameter α in (12.21) relies on cross-validation. The dataset is partitioned into V separate portions of equal size, where V is typically between 3 and 10. For a finite sequence of α values, the algorithm proceeds as follows:

1. Remove one of the V parts of the dataset. This subset is called the validation set.

2. Find the optimal subtree for each value of α in the sequence using the remaining $V - 1$ parts of the data.

3. For each optimal subtree, predict the validation-set responses, and compute the cross-validated sum of squared error based on these validation-set predictions.

Repeat this process for all V parts of the data. For each α, compute the total cross-validated sum of squares over all V data partitions. The value of α that minimizes the cross-validated sum of squares is selected; call it $\hat{\alpha}$. Having estimated the best value for the complexity parameter, we may now prune the full tree for all the data back to the subtree determined by $\hat{\alpha}$.

Efficient algorithms for finding the optimal tree for a sequence of α values (see step 2 above) are available [59, 457]. Indeed, the set of optimal trees for a sequence of α values is nested, with smaller trees corresponding to larger values of α, and all members in the sequence can be visited by sequential recombination of terminal nodes from the bottom up. Various enhancements of this cross-validation strategy have been proposed, including a variant of the above approach that chooses the simplest tree among those trees that nearly achieve the minimum cross-validated sum of squares [533].

Example 12.6 (Stream monitoring, continued) Let us return to the stream ecology example introduced in Example 12.4. A full tree for these data was obtained by splitting until every terminal node has fewer than 10 observations in it or has residual squared error less 1% of the residual squared error for the root node. This process produced a full tree with 53 terminal nodes. Figure 12.10 shows the total cross-validated residual squared error as a function of the number of terminal nodes. This plot was produced using 10-fold cross-validation ($V = 10$). The full tree can be pruned from the bottom up, recombining the least beneficial terminal nodes, until the minimal value of $R_\alpha(T)$ is reached. Note that the correspondence between values of α and tree sizes means that we need only consider a limited collection of α values, and it is therefore more straightforward to plot $R_\alpha(T)$ against $q(T)$ instead of plotting against α. The minimal cross-validated sum of squares is achieved for a tree with five terminal nodes; indeed, this is the tree shown in Figure 12.7.

For this example, the selection of the optimal α, and thus the final tree, varies with different random partitions of the data. The optimal tree typically has between three and thirteen terminal nodes. This uncertainty emphasizes the potential structural

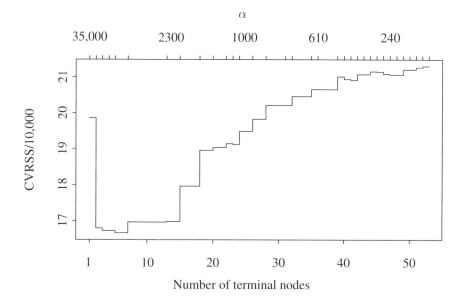

Fig. 12.10 Cross-validation residual sum of squares versus node size for Example 12.6. The top horizonal axis shows the cost–complexity parameter α.

instability of tree-based models, particularly for datasets where the signal is not strong.

□

12.1.4.3 Classification trees

Digressing briefly from this chapter's focus on smoothing, it is worthwhile here to quickly summarize tree-based methods for categorical response variables.

Recursive partitioning models for predicting a categorical response variable are typically called *classification trees* [59, 457]. Let each response observation Y_i take on one of M categories. Let $\hat{p}_{jm}$ denote the proportion of observations in the terminal node $\mathcal{N}_j$ that are of class m (for $m = 1, \ldots, M$). Loosely speaking, all the observations in $\mathcal{N}_j$ are predicted to equal the class that makes up the majority in that node. Such prediction by majority vote within terminal nodes can be modified in two ways. First, votes can be weighted to reflect prior information about the overall prevalence of each class. This permits predictions to be biased toward predominant classes. Second, votes can be weighted to reflect different losses for different types of misclassification [533]. For example, if classes correspond to medical diagnoses, some false positive or false negative diagnoses may be grave errors, while other mistakes may have only minor consequences.

Construction of a classification tree relies on partitioning the predictor space using a greedy strategy similar to the one used for recursive partitioning regression. For a regression tree split, the split coordinate c and split point t are selected by minimizing the total residual sum of squares within the left and right children nodes. For classification trees, a different measure of error is required. The residual squared error is replaced by a measure of node impurity.

There are various approaches to measuring node impurity, but most are based on the following principle. The impurity of node j should be small when observations in that node are concentrated within one class, and the impurity should be large when they are distributed uniformly over all M classes. Two popular measures of impurity are the *entropy*, given for node j by $\sum_{m=1}^{M} \hat{p}_{jm} \log \hat{p}_{jm}$, and the *Gini index*, given by $\sum_{\ell \neq m} \hat{p}_{jl} \hat{p}_{jm}$. These approaches are more effective than simply counting misclassifications, because a split may drastically improve the purity in a node without changing any classifications. This occurs, for example, when majority votes on both sides of a split have the same outcome as the unsplit vote, but the margin of victory in one of the subregions is much narrower than in the other.

Cost–complexity tree pruning can proceed using the same strategies described in Section 12.1.4.2. The entropy or the Gini index can be used for the cost measure $r(T)$ in (12.21). Alternatively, one may let $r(T)$ equal a (possibly weighted) misclassification rate to guide pruning.

12.1.4.4 *Other issues for tree-based methods*   Tree-based methods offer several advantages over other, more traditional modeling approaches. First, tree models fit interactions between predictors and other nonadditive behavior without requiring formal specification of the form of the interaction by the user. Second, there are natural ways to use data with some missing predictor values, both when fitting the model and when using it to make predictions. Some strategies are surveyed in [58, 457].

One disadvantage is that trees can be unstable. Therefore, care must be taken not to overinterpret particular splits. For example, if the two smallest IBI values in $\mathcal{N}_1$ in Figure 12.8 are increased somewhat, then this node is omitted when a new tree is constructed on the revised data. New data can often cause quite different splits to be chosen even if predictions remain relatively unchanged. For example, from Figure 12.8 it is easy to imagine that slightly different data could have led to the root node splitting on population density at the split point 2.5, rather than on rock size at the point 0.4. Trees can also be unstable in that building the full tree to a different size before pruning can cause a different optimal tree to be chosen after pruning.

Another concern is that assessment of uncertainty can be somewhat challenging. There is no simple way to summarize a confidence region for the tree structure itself. Confidence intervals for tree predictions can be obtained using the bootstrap (Chapter 9).

Tree-based methods are popular in computer science, particular for classification [440, 457]. Bayesian alternatives to tree methods have also been proposed [94, 131]. Medical applications of tree-based methods are particularly popular, perhaps because the binary decision tree is simple to explain and to apply as a tool in disease diagnosis [59, 95].

12.2   GENERAL MULTIVARIATE DATA

Finally, we consider high-dimensional data that lie near a low-dimensional manifold such as a curve or surface. For such data, there may be no clear conceptual separation of the variables into predictors and responses. Nevertheless, we may be interested in estimating smooth relationships among variables. In this section we describe one approach, called principal curves, for smoothing multivariate data. Alternative methods for discovering relationships among variables, such as association rules and cluster analysis, are discussed in [281].

12.2.1   Principal curves

A *principal curve* is a special type of one-dimensional nonparametric summary of a p-dimensional general multivariate dataset. Loosely speaking, each point on a principal curve is the average of all data that project onto that point on the curve. We began motivating principal curves in Section 11.6. The data in Figure 11.18 were not suitable for predictor–response smoothing, yet adapting the concept of smoothing to general multivariate data allowed the very good fit shown in the right panel of Figure 11.18. We now describe more precisely the notion of a principal curve and its estimation [279]. Related software includes [277, 323, 546].

12.2.1.1   *Definition and motivation*   General multivariate data may lie near a connected, one-dimensional curve snaking through $\Re^p$. It is this curve we want to estimate. We adopt a time–speed parameterization of curves below to accommodate the most general case.

   We can write a one-dimensional curve in $\Re^p$ as $\mathbf{f}(\tau) = (f_1(\tau), \dots, f_p(\tau))$ for τ between τ_0 and τ_1. Here, τ can be used to indicate distance along the one-dimensional curve in p-dimensional space. The arc length of a curve $\mathbf{f}$ is $\int_{\tau_0}^{\tau_1} \|\mathbf{f}'(\tau)\| \, d\tau$, where

$$\|\mathbf{f}'(\tau)\| = \sqrt{\left(\frac{df_1(\tau)}{d\tau}\right)^2 + \cdots + \left(\frac{df_p(\tau)}{d\tau}\right)^2}.$$

If $\|\mathbf{f}'(\tau)\| = 1$ for all $\tau \in [\tau_0, \tau_1]$, then the arc length between any two points τ_a and τ_b along the curve is $|\tau_a - \tau_b|$. In this case, $\mathbf{f}$ is said to have the *unit-speed parameterization*. It is often helpful to imagine a bug walking forward along the curve at a speed of 1, or backward with a speed of -1 (the designation of forward and backward is arbitrary). Then the amount of time it takes the bug to walk between two points corresponds to the arc length, and the positive or negative sign corresponds to the direction taken. Any smooth curve with $\|\mathbf{f}'(\tau)\| > 0$ for all $\tau \in [\tau_0, \tau_1]$ can be reparameterized to unit speed. If the coordinate functions of a unit-speed curve are smooth, then $\mathbf{f}$ itself is smooth.

   The types of curves we are interested in estimating are smooth, nonintersecting, curves that aren't too wiggly. Specifically, let us assume that $\mathbf{f}$ is a smooth unit-speed curve in $\Re^p$ parameterized over the closed interval $[\tau_0, \tau_1]$ such that $\mathbf{f}(t) \neq \mathbf{f}(r)$ when $r \neq t$ for all $r, t \in [\tau_0, \tau_1]$, and $\mathbf{f}$ has finite length inside any closed ball in $\Re^p$.

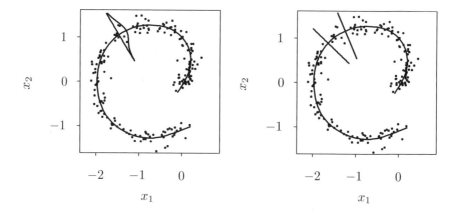

Fig. 12.11 Two panels illustrating the definition of a principal curve and its estimation. In the left panel, the curve $\mathbf{f}$ is intersected by an axis that is orthogonal to $\mathbf{f}$ at a particular τ^*. A conditional density curve is sketched over this axis; if $\mathbf{f}$ is a principal curve, then the mean of this conditional density must equal $\mathbf{f}(\tau^*)$. In the right panel, a neighborhood around τ^* is sketched. Within these boundaries, all points project onto $\mathbf{f}$ near τ^*. The sample mean of these points should be a good approximation to the true mean of the conditional density in the left panel.

For any point $\mathbf{x} \in \Re^p$, define the *projection index* function as $\tau_{\mathbf{f}}(\mathbf{x}) : \Re^p \rightarrow \Re^1$ according to

$$\tau_{\mathbf{f}}(\mathbf{x}) = \sup_\tau \left\{ \tau : \|\mathbf{x} - \mathbf{f}(\tau)\| = \inf_r \|\mathbf{x} - \mathbf{f}(r)\| \right\}. \qquad (12.22)$$

Thus $\tau_{\mathbf{f}}(\mathbf{x})$ is the largest value of τ for which $\mathbf{f}(\tau)$ is closest to $\mathbf{x}$. Points with similar projection indices project orthogonally onto a small portion of the curve $\mathbf{f}$. The projection index will later be used to define neighborhoods.

Suppose that $\mathbf{X}$ is a random vector in $\Re^p$, having a probability density with finite second moment. Unlike in previous sections, we cannot distinguish variables as predictors and response.

We define $\mathbf{f}$ to be a *principal curve* if $\mathbf{f}(\tau^*) = \mathrm{E}\{\mathbf{X}|\tau_{\mathbf{f}}(\mathbf{X}) = \tau^*\}$ for all $\tau^* \in [\tau_0, \tau_1]$. This requirement is sometimes termed *self-consistency*. Figure 12.11 illustrates this notion that the distribution of points orthogonal to the curve at some τ must have mean equal to the curve itself at that point. In the left panel, a distribution is sketched along an axis that is orthogonal to $\mathbf{f}$ at one τ^*. The mean of this density is $\mathbf{f}(\tau^*)$. Note that for ellipsoid distributions, the principal component lines are principal curves. Principal components are reviewed in [402].

Principal curves are motivated by the concept of local averaging: the principal curve connects the means of points in local neighborhoods. For predictor–response smooths, neighborhoods are defined along the predictor coordinate axes. For principal curves, neighborhoods are defined along the curve itself. Points that project nearby on the curve are in the same neighborhood. The right panel of Figure 12.11 illustrates the notion of a local neighborhood along the curve.

12.2.1.2 *Estimation* An iterative algorithm can be used to estimate a principal curve from a sample of p-dimensional data, $\mathbf{X}_1, \ldots, \mathbf{X}_n$. The algorithm is initialized at iteration $t = 0$ by choosing a simple starting curve $\hat{\mathbf{f}}^{(0)}(\tau)$ and setting $\tau^{(0)}(\mathbf{X}) = \tau_{\hat{\mathbf{f}}^{(0)}}(\mathbf{x})$ from (12.22). One reasonable choice would be to set $\hat{\mathbf{f}}^{(0)}(\tau) = \bar{\mathbf{X}} + \mathbf{a}\tau$, where $\mathbf{a}$ is the first linear principal component estimated from the data. The algorithm proceeds as follows:

1. Smooth the kth coordinate of the data. Specifically, for $k = 1, \ldots, p$, smooth X_{ik} against $\tau^{(t)}(\mathbf{X}_i)$ using a standard bivariate predictor–response smoother with span $h^{(t)}$. The projection of the points $\mathbf{X}_i$ onto $\hat{\mathbf{f}}^{(t)}$ for $i = 1, \ldots, n$ provides the predictors $\tau^{(t)}(\mathbf{X}_i)$. The responses are the X_{ik}. The result is $\hat{\mathbf{f}}^{(t+1)}$, which serves as an estimate of $E\{\mathbf{X}|\tau^{(t)}(\mathbf{x})\}$. This implements the scatterplot smoothing strategy of locally averaging the collection of points that nearly project onto the same point on the principal curve.

2. Interpolate between the $\hat{\mathbf{f}}^{(t+1)}(\mathbf{X}_i)$ for $i = 1, \ldots, n$, and compute $\tau_{\hat{\mathbf{f}}^{(t+1)}}(\mathbf{X}_i)$ as the distances along $\hat{\mathbf{f}}^{(t+1)}$. Note that some $\mathbf{X}_i$ may project onto a quite different segment than they did at the previous iteration.

3. Let $\tau^{(t+1)}(\mathbf{X})$ equal $\tau_{\hat{\mathbf{f}}^{(t+1)}}(\mathbf{X})$ transformed to unit speed. This amounts to rescaling the $\tau_{\hat{\mathbf{f}}^{(t+1)}}(\mathbf{X}_i)$ so that each equals the total distance traveled along the polygonal curve to reach it.

4. Evaluate the convergence of $\hat{\mathbf{f}}^{(t+1)}$, and stop if possible; otherwise, increment t and return to step 1. A relative convergence criterion could be constructed based on the total error, $\sum_{i=1}^{n} \|\mathbf{X}_i - \hat{\mathbf{f}}^{(t+1)}(\tau^{(t+1)}(\mathbf{X}_i))\|$.

The result of this algorithm is a piecewise linear polygonal curve that serves as the estimate of the principal curve.

The concept of principal curves can be generalized for multivariate responses. For this purpose, *principal surfaces* are defined analogously to the above. The surface is parameterized by a vector $\boldsymbol{\tau}$, and data points are projected onto the surface. Points that project anywhere on the surface near $\boldsymbol{\tau}^*$ dominate in the local smooth at $\boldsymbol{\tau}^*$.

Example 12.7 (Principal curve for bivariate data) Figure 12.12 illustrates several steps during the iterative process of fitting a principal curve. The sequence of panels should be read across the page from top left to bottom right. In the first panel, the data are plotted. The solid line shaped like a square letter C is $\hat{\mathbf{f}}^{(0)}$. Each data point is connected to $\hat{\mathbf{f}}^{(0)}$ by a line showing its orthogonal projection. As a bug walks along $\hat{\mathbf{f}}^{(0)}(\tau)$ from the top right to the bottom right, $\tau^{(0)}(\mathbf{x})$ increases from 0 to about 7. The second and third panels show each coordinate of the data plotted against the projection index, $\tau^{(0)}(\mathbf{x})$. These coordinatewise smooths correspond to step 1 of the estimation algorithm. A smoothing spline was used in each panel, and the resulting overall estimate, $\hat{\mathbf{f}}^{(1)}$, is shown in the fourth panel. The fifth panel shows $\hat{\mathbf{f}}^{(2)}$. The sixth panel gives the final result when convergence was achieved. □

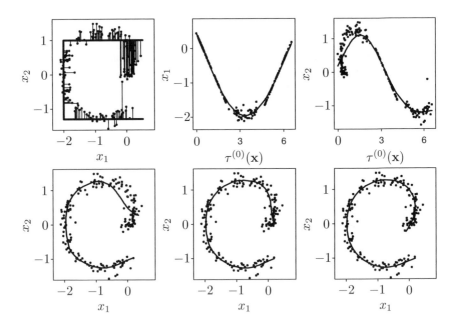

Fig. 12.12 These panels illustrate the progression of the iterative fit of a principal curve. See Example 12.7 for details.

12.2.1.3 Span selection   The principal curve algorithm depends on the selection of a span $h^{(t)}$ at each iteration. Since the smoothing is done coordinatewise, different spans could be used for each coordinate at each iteration, but in practice it is more sensible to standardize the data before analysis and then use a common $h^{(t)}$.

Nevertheless, the selection of $h^{(t)}$ from one iteration to the next remains an issue. The obvious solution is to select $h^{(t)}$ via cross-validation at each iteration. Surprisingly, this doesn't work well. Pervasive undersmoothing arises because the errors in the coordinate functions are autocorrelated. Instead, $h^{(t)} = h$ can be chosen sensibly and remain unchanged until convergence is achieved. Then, one additional iteration of step 1 can be done with a span chosen by cross-validation.

This span selection approach is troubling because the initial span choice clearly can affect the shape of the curve to which the algorithm converges. When the span is then cross-validated after convergence, it is too late to correct $\hat{\mathbf{f}}$ for such an error. Nevertheless, the algorithm seems to work well on a variety of examples where ordinary smoothing techniques would fail catastrophically.

Problems

12.1   For $\mathbf{A}$ defined as in (12.5), smoothing matrices $\mathbf{S}_k$, and n-vectors $\boldsymbol{\gamma}_k$ for $k = 1, \ldots, p$, let $\mathcal{I}_k$ be the space spanned by vectors that pass through $\mathbf{S}_k$ unchanged (i.e.,

Table 12.2 Potential predictors of body fat. Predictors 4–13 are circumference measurements given in centimeters.

1. Age (years)	8. Thigh
2. Weight (pounds)	9. Knee
3. Height (inches)	10. Ankle
4. Neck	11. Extended biceps
5. Chest	12. Forearm
6. Abdomen	13. Wrist
7. Hip	

vectors $\mathbf{v}$ satisfying $\mathbf{S}_k \mathbf{v} = \mathbf{v}$). Prove that $\mathbf{A}\boldsymbol{\gamma} = \mathbf{0}$ (where $\boldsymbol{\gamma} = (\gamma_1 \ \gamma_2 \ \cdots \ \gamma_p)^T$) if and only if $\gamma_k \in \mathcal{I}_k$ for all k and $\sum_{k=1}^{p} \gamma_k = \mathbf{0}$.

12.2 Accurate measurement of body fat can be expensive and time-consuming. Good models to predict body fat accurately using standard measurements are still useful in many contexts. A study was conducted to predict body fat using 13 simple body measurements on 251 men. For each subject, the percentage of body fat as measured using an underwater weighing technique, age, weight, height, and ten body circumference measurements were recorded (Table 12.2). Further details on this study are available in [289, 311]. These data are available from the website for this book. The goal of this problem is to compare and contrast several multivariate smoothing methods applied to these data.

(a) Using a smoother of your own choosing, develop a backfitting algorithm to fit an additive model to these data as described in Section 12.1.1. Compare the results of the additive model with those from a multiple regression model.

(b) Use any available software to estimate models for these data, using five methods: (1) the standard multiple linear regression model (MLR), (2) an additive model (AM), (3) projection pursuit regression (PPR), (4) the alternating conditional expectations procedure (ACE), and (5) the additivity and variance stabilization approach (AVAS).

    i. For MLR, AM, ACE, and AVAS, plot the kth estimated coordinate smooth against the observed values of the kth predictor for $k = 1, \ldots, 13$. In other words, graph the values of $\hat{s}_k(x_{ik})$ versus x_{ik} for $i = 1, \ldots, 251$ as in Figure 12.2. For PPR, imitate Figure 12.6 by plotting each component smooth against the projection coordinate. For all methods, include the observed data points in an appropriate way in each graph. Comment on any differences between the methods.

    ii. Carry out leave-one-out cross-validation analyses where the ith cross-validated residual is computed as the difference between the ith observed response and the ith predicted response obtained when the model is fitted

374 MULTIVARIATE SMOOTHING

omitting the ith data point from the dataset. Use these results to compare the predictive performance of MLR, AM, and PPR using a cross-validated residual sum of squares similar to (11.16).

12.3 For the body fat data of Problem 12.2, compare the performance of at least three different smoothers used within an additive model of the form given in (12.3). Compare the leave-one-out cross-validation mean squared prediction error for the different smoothers. Is one smoother superior to another in the additive model?

12.4 Example 2.5 describes a generalized linear model for data derived from testing an algorithm for human face recognition. The data are available from the book website. The response variable is binary, with $Y_i = 1$ if two images of the same person were correctly matched and $Y_i = 0$ otherwise. There are three predictor variables. The first is the absolute difference in eye-region mean pixel intensity between the two images of the ith person. The second is the absolute difference in nose-cheek-region mean pixel intensity between the two images. The third predictor compares pixel intensity variability between the two images. For each image of the ith person, the median absolute deviation (a robust spread measure) of pixel intensity is computed in two image areas: the forehead and nose-cheek regions. The third predictor is the between-image ratio of these within-image ratios. Fit a generalized additive model to these data. Plot your results and interpret. Compare your results with the fit of an ordinary logistic regression model.

12.5 Consider a larger set of stream monitoring predictors of the index of biotic integrity for macroinvertebrates considered in Example 12.4. The 21 predictors, described in more detail in the website for this book, can be grouped into four categories:

Site chemistry measures: Acid-neutralizing capacity, chloride, specific conductance, total nitrogen, pH, total phosphorus, sulfate

Site habitat measures: Substrate diameter, percent fast water, canopy density above midchannel, channel slope

Site geographic measures: Elevation, longitude, latitude, mean slope above site

Watershed measures: Watershed area above site; human population density; percentages of agricultural, mining, forest, and urban land cover

(a) Construct a regression tree to predict the IBI.

(b) Compare the performance of several strategies for tree pruning. Compare the 10-fold cross-validated mean squared prediction errors for the final trees selected by each strategy.

(c) The variables are categorized above into four groups. Create a regression tree using only the variables from each group in turn. Compare the 10-fold cross-validated mean squared prediction errors for the final trees selected for each group of predictors.

12.6   Discuss how the combinatorial optimization methods from Chapter 3 might be used to improve tree-based methods.

12.7   Find an example for which $\mathbf{X} = \mathbf{f}(\tau) + \epsilon$, where ϵ is a random vector with mean zero, but $\mathbf{f}$ is not a principal curve for $\mathbf{X}$.

12.8   The website for this book provides some artificial data suitable for fitting a principal curve. There are 50 observations of a bivariate variable, and each coordinate has been standardized. Denote these data as $\mathbf{x}_1, \ldots, \mathbf{x}_{50}$.

(a) Plot the data. Let $\hat{\mathbf{f}}^{(0)}$ correspond to the segment of the line through the origin with slope 1 onto which the data project. Superimpose this line on the graph. Imitating the top left panel of Figure 12.12, show how the data project onto $\hat{\mathbf{f}}^{(0)}$.

(b) Compute $\tau^{(0)}(\mathbf{x}_i)$ for each data point $\mathbf{x}_i$. Transform to unit speed. Hint: Show why the transformation $\mathbf{a}^T \mathbf{x}_i$ works, where $\mathbf{a} = (\sqrt{2}/2, \sqrt{2}/2)^T$.

(c) For each coordinate of the data in turn, plot the data values for that coordinate (i.e. the x_{ik} values for $i = 1, \ldots, 50$ and $k = 1$ or $k = 2$) against the projection index values, $\tau^{(0)}(\mathbf{x}_i)$. Smooth the points in each plot, and superimpose the smooth on each graph. This mimics the center and right top panels of Figure 12.12.

(d) Superimpose $\hat{\mathbf{f}}^{(1)}$ over a scatterplot of the data, as in the bottom left panel of Figure 12.12.

(e) Advanced readers may consider automating and extending these steps to produce an iterative algorithm whose iterates converges to the estimated principal curve. Some related software for fitting principal curves in S-Plus is available in [277, 323, 546].

Data Acknowledgments

The datasets used in the examples and exercises discussed in this book are available from the website www.colostate.edu/computationalstatistics/. Many of these were collected by scientific researchers in diverse fields. The remaining data are owned by us or have been simulated for instructional purposes. Further details about ownership of some datasets are given below.

We thank Richard Barker, Department of Statistics, University of Otago, New Zealand, for the fur seal pup data used in Section 7.4 and for hosting us during a great sabbatical.

We thank Ross Beveridge and Bruce Draper, Department of Computer Science, Colorado State University, for the face recognition data used in Example 2.5, and for the opportunity to collaborate with them on this interesting project.

We thank Gordon Reese, Department of Fishery and Wildlife Biology, Colorado State University, for his assistance in extracting the Utah serviceberry data used in Chapter 8.

We thank Alan Herlihy, Department of Fisheries and Wildlife, Oregon State University, for providing the stream monitoring data used in Example 12.4 and for his assistance with the interpretation of the results. These data and the data used in Problem 12.5 were produced by the U.S. Environmental Protection Agency through its Environmental Monitoring and Assessment Program (EMAP) [161, 541].

The leukemia data in Problem 2.3 are used with permission, and are taken from [177]. Copyright American Society of Hematology, 1963.

The oil spill data in Problem 2.5 are derived from data given in [11] and are used with permission from Elsevier. Copyright Elsevier 2000.

The Alzheimer's data used throughout Chapter 5 are reprinted with permission from [134]. Copyright CRC Press, Boca Raton, Florida, 2002.

The pigment moisture data in Problem 7.8 are reprinted from [52] with permission of John Wiley & Sons, Inc. Copyright John Wiley & Sons, Inc. 1978.

The copper-nickel alloy data used throughout Chapter 9 are reprinted from [147] with permission of John Wiley & Sons, Inc. Copyright John Wiley & Sons, Inc. 1966.

The air blast data in Problem 11.6 are reprinted from [299] with permission from Elsevier, Copyright 2001.

The Norwegian paper data in Chapter 12 are reprinted from [9], with permission from Elsevier, Copyright 1996.

Acknowledgments for the remaining datasets from other sources are given at the locations in the text where we first use the data. We thank all these authors and researchers.

References

1. E. H. L. Aarts and P. J. M. van Laarhoven. Statistical cooling: a general approach to combinatorial optimization problems. *Philips Journal of Research*, 40:193–226, 1985.

2. M. Abramowitz and I. A. Stegun, editors. *Handbook of Mathematical Functions,* National Bureau of Standards Applied Mathematics Series, No. 55. US Government Printing Office, Washington, DC, 1964.

3. I. S. Abramson. On bandwidth variation in kernel estimates—a square root law. *Annals of Statistics*, 10:1217–1223, 1982.

4. D. Ackley. *A Connectionist Machine for Genetic Hillclimbing*. Kluwer, Boston, 1987.

5. D. H. Ackley. An empirical study of bit vector function optimization. In L. Davis, editor, *Genetic Algorithms and Simulated Annealing*. Morgan Kauffman, Los Altos, CA, 1987.

6. R. P. Agarwal, M. Meehan, and D. O'Regan. *Fixed Point Theory and Applications*. Cambridge University Press, Cambridge, 2001.

7. H. Akaike. Information theory and an extension of the maximum likelihood principle. In B. N. Petrox and F. Caski, editors, *Proceedings of the Second International Symposium on Information Theory*. Akedemia Kiaodo, Budapest, 1973.

8. J. T. Alander. On optimal population size of genetic algorithms. In *Proceedings of CompEuro 92*, pages 65–70. IEEE Computer Society Press, 1992.

9. M. Aldrin. Moderate projection pursuit regression for multivariate response data. *Computational Statistics and Data Analysis*, 21:501–531, 1996.

10. N. S. Altman. An introduction to kernel and nearest-neighbor nonparametric regression. *The American Statistican*, 46:175–185, 1992.

11. C. M. Anderson and R. P. Labelle. Update of comparative occurrence rates for offshore oil spills. *Spill Science and Technology Bulletin*, 6:303–321, 2000.

12. J. Antonisse. A new interpretation of schema notation that overturns the binary encoding constraint. In J. D. Schaffer, editor, *Proceedings of the 3rd International Conference on Genetic Algorithms*. Morgan Kaufmann, Los Altos, CA, 1989.

13. L. Armijo. Minimization of functions having Lipschitz-continuous first partial derivatives. *Pacific Journal of Mathematics*, 16:1–3, 1966.

14. K. Arms and P. S. Camp. *Biology*. Saunders College Publishing, Fort Worth, TX, 4th edition, 1995.

15. T. Bäck. *Evolutionary Algorithms in Theory and Practice*. Oxford University Press, New York, 1996.

16. J. E. Baker. Adaptive selection methods for genetic algorithms. In J. J. Grefenstette, editor, *Proceedings of an International Conference on Genetic Algorithms and Their Applications*. Lawrence Erlbaum Associates, Hillsdale, NJ, 1985.

17. J. E. Baker. Reducing bias and inefficiency in the selection algorithm. In J. J. Grefenstette, editor, *Proceedings of the 2nd International Conference on Genetic Algorithms and Their Applications*. Lawrence Erlbaum Associates, Hillsdale, NJ, 1987.

18. S. G. Baker. A simple method for computing the observed information matrix when using the EM algorithm with categorical data. *Journal of Computational and Graphical Statistics*, 1:63–76, 1992.

19. D. L. Banks, R. T. Olszewski, and R. Maxion. Comparing methods for multivariate nonparametric regression. *Communications in Statistics—Simulation and Computation*, 32:541–571, 2003.

20. G. A. Barnard. Discussion of paper by M. S. Bartlett. *Journal of the Royal Statistical Society, Series B*, 25:294, 1963.

21. O. E. Barndorff-Nielsen and D. R. Cox. *Inference and Asymptotics*. Chapman & Hall, London, 1994.

22. L. E. Baum, T. Petrie, G. Soules, and N. Weiss. A maximization technique occurring in the statistical analysis of probabilistic functions of Markov chains. *Annals of Mathematical Statistics*, 41:164–171, 1970.

23. R. Beran. Prepivoting to reduce level error of confidence sets. *Biometrika*, 74:457–468, 1987.

24. R. Beran. Prepivoting test statistics: A bootstrap view of asymptotic refinements. *Journal of the American Statistical Association*, 83:687–697, 1988.

25. J. O. Berger. *Statistical Decision Theory: Foundations, Concepts, and Methods*. Springer-Verlag, New York, 1980.

26. J. O. Berger and M.-H. Chen. Predicting retirement patterns: prediction for a multinomial distribution with constrained parameter space. *The Statistician*, 42(4):427–443, 1993.

27. A. Berlinet and L. Devroye. A comparison of kernel density estimates. *Publications de l'Institute de Statistique de l'Université de Paris*, 38:3–59, 1994.

28. D. Bertsimas and J. Tsitsiklis. Simulated annealing. *Statistical Science*, 8:10–15, 1993.

29. J. Besag. Spatial interaction and the statistical analysis of lattice systems (with discussion). *Journal of the Royal Statistical Society, Series B*, 36:192–236, 1974.

30. J. Besag. On the statistical analysis of dirty pictures (with discussion). *Journal of the Royal Statistical Society, Series B*, 48:259–302, 1986.

31. J. Besag. Comment on "Representations of knowledge in complex systems" by Grenander and Miller. *Journal of the Royal Statistical Society, Series B*, 56:591–592, 1994.

32. J. Besag and P. Clifford. Generalized Monte Carlo significance tests. *Biometrika*, 76:633–642, 1989.

33. J. Besag and P. Clifford. Sequential Monte Carlo *p*-values. *Biometrika*, 78:301–304, 1991.

34. J. Besag, P. Green, D. Higdon, and K. Mengersen. Bayesian computation and stochastic systems (with discussion). *Statistical Science*, 10:3–66, 1995.

35. J. Besag and P. J. Green. Spatial statistics and Bayesian computation. *Journal of the Royal Statistical Society, Series B*, 55(1):25–37, 1993.

36. J. Besag and C. Kooperberg. On conditional and intrinsic autoregressions. *Biometrika*, 82:733–746, 1995.

37. J. Besag, J. York, and A. Mollié. Bayesian image restoration, with two applications in spatial statistics (with discussion). *Annals of the Institute of Statistical Mathematics*, 43:1–59, 1991.

38. N. Best, S. Cockings, J. Bennett, J. Wakefield, and P. Elliott. Ecological regression analysis of environmental benzene exposure and childhood leukaemia: sensitivity to data inaccuracies, geographical scale and ecological bias. *Journal of the Royal Statistical Society, Series A*, 164(1):155–174, 2001.

39. N. G. Best, R. A. Arnold, A. Thomas, L. A. Waller, and E. M. Conlon. Bayesian models for spatially correlated disease and exposure data. In J. O. Berger, J. M. Bernardo, A. P. Dawid, D. V. Lindley, and A. F. M. Smith, editors, *Bayesian Statistics 6*, pages 131–156. Oxford University Press, Oxford, 1999.

40. R. J. H. Beverton and S. J. Holt. *On the Dynamics of Exploited Fish Populations*, volume 19 of *Fisheries Investment Series 2*. UK Ministry of Agriculture and Fisheries, London, 1957.

41. P. J. Bickel and D. A. Freedman. Some asymptotics for the bootstrap. *Annals of Statistics*, 9:1196–1217, 1981.

42. C. Biller. Adaptive Bayesian regression splines in semiparametric generalized linear models. *Journal of Computational and Graphical Statistics*, 9(1):122–140, 2000.

43. P. Billingsley. *Probability and Measure*. Wiley, New York, 3rd edition, 1995.

44. C. M. Bishop. *Neural Networks for Pattern Recognition*. Oxford University Press, 1995.

45. C. M. Bishop, editor. *Neural Networks and Machine Learning*. Springer-Verlag, 1998.

46. F. Black and M. Scholes. The pricing of options and corporate liabilities. *Journal of Political Economy*, 81:635–654, 1973.

47. C. L. Blake and C. J. Merz. UCI Repository of Machine Learning Databases, University of California, Irvine, Dept. of Information and Computer Sciences. Available from http://www.ics.uci.edu/~mlearn/MLRepository.html, 1998.

48. L. B. Booker. Improving search in genetic algorithms. In L. Davis, editor, *Genetic Algorithms and Simulated Annealing*. Morgan Kauffman, Los Altos, CA, 1987.

49. D. L. Borchers, S. T. Buckland, and W. Zucchini. *Estimating Animal Abundance*. Springer-Verlag, London, 2002.

50. A. Bowman. An alternative method of cross-validation for the smoothing of density estimates. *Biometrika*, 71:353–360, 1984.

51. G. E. P. Box and D. R. Cox. An analysis of transformations. *Journal of the Royal Statistical Society, Series B*, 26:211–246, 1964.

52. G. E. P. Box, W. G. Hunter, and J. S. Hunter. *Statistics for Experimenters*. Wiley, New York, 1978.

53. P. Boyle, M. Broadie, and P. Glasserman. Monte Carlo methods for security pricing. *Journal of Economic Dynamics and Control*, 21:1267–1321, 1997.

54. R. A. Boyles. On the convergence of the EM algorithm. *Journal of the Royal Statistical Society, Series B*, 45:47–50, 1983.

55. C. J. A. Bradshaw, R. J. Barker, R. G. Harcourt, and L. S. Davis. Estimating survival and capture probability of fur seal pups using multistate mark–recapture models. *Journal of Mammalogy*, 84(1):65–80, 2003.

56. C. J. A. Bradshaw, C. Lalas, and C. M. Thompson. Cluster of colonies in an expanding population of New Zealand fur seals (*Arctocephalus fosteri*). *Journal of Zoology*, 250:41–51, 2000.

57. L. Breiman. Bagging predictors. *Machine Learning*, 24:123–140, 1996.

58. L. Breiman and J. H. Friedman. Estimating optimal transformations for multiple regression and correlation (with discussion). *Journal of the American Statistical Association*, 80:580–619, 1985.

59. L. Breiman, J. H. Friedman, R. A. Olshen, and C. J. Stone. *Classification and Regression Trees*. Wadsworth, 1984.

60. L. Breiman, W. Meisel, and E. Purcell. Variable kernel estimates of multivariate densities. *Technometrics*, 19:135–144, 1977.

61. P. Brémaud. *Markov Chains: Gibbs Fields, Monte Carlo Simulation, and Queues*. Springer-Verlag, New York, 1999.

62. R. P. Brent. *Algorithms for Minimization without Derivatives*. Prentice-Hall, Englewood Cliffs, NJ, 1973.

63. N. E. Breslow and D. G. Clayton. Approximate inference in generalized linear mixed models. *Journal of the American Statistical Association*, 88:9–25, 1993.

64. S. P. Brooks. Markov chain Monte Carlo method and its application. *The Statistician*, 47:69–100, 1998.

65. S. P. Brooks and A. Gelman. General methods for monitoring convergene of iterative simulations. *Journal of Computational and Graphical Statistics*, 7:434–455, 1998.

66. S. P. Brooks and P. Giudici. Markov chain Monte Carlo convergence assessment via two-way analysis of variance. *Journal of Computational and Graphical Statistics*, 9(2):266–285, 2000.

67. S. P. Brooks, P. Giudici, and A. Philippe. Nonparametric convergence assessment for MCMC model selection. *Journal of Computational and Graphical Statistics*, 12(1):1–22, 2003.

68. S. P. Brooks, P. Giudici, and G. O. Roberts. Efficient construction of reversible jump Markov chain Monte Carlo proposal distributions. *Journal of the Royal Statistical Society, Series B*, 65(1):3–39, 2003.

69. S. P. Brooks and B. J. T. Morgan. Optimization using simulated annealing. *The Statistican*, 44:241–257, 1995.

70. S. P. Brooks and G. O. Roberts. Assessing convergence of Markov chain Monte Carlo algorithms. *Statistics and Computing*, 8:319–335, 1999.

71. C. G. Broyden. Quasi-Newton methods and their application to function minimization. *Mathematics of Computation*, 21:368–381, 1967.

72. C. G. Broyden. The convergence of a class of double-rank minimization algorithms. *Journal of the Institute of Mathematics and its Applications*, 6:76–90, 1970.

73. C. G. Broyden. Quasi-Newton methods. In W. Murray, editor, *Numerical Methods for Unconstrained Optimization*, pages 87–106. Academic Press, New York, 1972.

74. A. Buja. Remarks on functional canonical variates, alternating least squares methods, and ACE. *Annals of Statistics*, 18:1032–1069, 1989.

75. K. P. Burnham and D. R. Anderson. *Model Selection and Inference: A Practical Information Theoretic Approach.* Springer-Verlag, New York, 2nd edition, 2002.

76. E. Cameron and L. Pauling. Supplemental ascorbate in the supportive treatment of cancer: re-evaluation of prolongation of survival times in terminal human cancer. *Proceedings of the National Academy of Sciences of the USA*, 75(9):4538–4542, September 1978.

77. R. Cao, A. Cuevas, and W. González-Mantiega. A comparative study of several smoothing methods in density estimation. *Computational Statistics and Data Analysis*, 17:153–176, 1994.

78. O. Cape, C. P. Rober, and T. Ryden. Reversible jump, birth-and-death and more general continuous time Markov chain Monte Carlo sampler. *Journal of the Royal Statistical Society, Series B*, 65(3):679–700, 2003.

79. B. P. Carlin, A. E. Gelfand, and A. F. M. Smith. Hierarchical Bayesian analysis of changepoint problems. *Applied Statistics*, 41:389–405, 1992.

80. B. P. Carlin and T. A. Louis. *Bayes and Empirical Bayes Methods for Data Analysis.* Chapman & Hall, London, 1996.

81. G. Casella and R. L. Berger. *Statistical Inference.* Brooks/Cole, Pacific Grove, CA, 2nd edition, 2001.

82. G. Casella and E. I. George. Explaining the Gibbs sampler. *The American Statistican*, 46(3):167–174, 1992.

83. G. Casella, K. L. Mengersen, C. P. Robert, and D. M. Titterington. Perfect samplers for mixtures of distributions. *Journal of the Royal Statistical Society, Series B*, 64(4):777–790, 2002.

84. G. Casella and C. Robert. Rao–Blackwellization of sampling schemes. *Biometrika*, 83:81–94, 1996.

85. G. Casella and C. P. Robert. Post-processing accept–reject samples: recycling and rescaling. *Journal of Computational and Graphical Statistics*, 7:139–157, 1998.

86. J. M. Chambers and T. J. Hastie, editors. *Statistical Models in S.* Chapman & Hall, New York, 1992.

87. K. S. Chan and J. Ledholter. Monte Carlo EM estimation for time series models involving counts. *Journal of the American Statistical Association*, 90:242–252, 1995.

88. R. N. Chapman. The quantitative analysis of environmental factors. *Ecology*, 9:111–122, 1928.

89. M.-H. Chen and B. W. Schmeiser. Performance of the Gibbs, hit-and-run, and Metropolis samplers. *Journal of Computational and Graphical Statistics*, 2:251–272, 1993.

90. M.-H. Chen and B. W. Schmeiser. General hit-and-run Monte Carlo sampling for evaluating multidimensional integrals. *Operations Research Letters*, 19:161–169, 1996.

91. M.-H. Chen, Q.-M. Shao, and J. G. Ibrahim. *Monte Carlo Methods in Bayesian Computation.* Springer-Verlag, New York, 2000.

92. Y. Chen, P. Diaconis, S. P. Holmes, and J. S. Liu. Sequential Monte Carlo methods for statistical analysis of tables. Working Paper 03-22, Institute of Statistics and Decision Sciences, Duke University, 2004.

93. S. Chib and E. Greenberg. Understanding the Metropolis–Hastings algorithm. *The American Statistican*, 49(4):327–335, 1995.

94. H. A. Chipman, E. I. George, and R. E. McCulloch. Bayesian CART model search (with discussion). *Journal of the American Statistical Association*, 93:935–960, 1998.

95. A. Ciampi, C.-H. Chang, S. Hogg, and S. McKinney. Recursive partitioning: A verstatile method for exploratory data analysis in biostatistics. In I. B. MacNeil and G. J. Umphrey, editors, *Biostatistics*, pages 23–50. Reidel, Dordrecht, Netherlands, 1987.

96. L. A. Clark and D. Pregiborn. Tree-based models. In J. M. Chambers and T. Hastie, editors, *Statistical Models in S*, pages 377–419. Duxbury, New York, 1991.

97. D. Clarkson. S+BEST (S-plus B-spline Statistical Technologies). Available from http://www.insightful.com/downloads/libraries/.

98. W. S. Cleveland. Robust locally weighted regression and smoothing scatter plots. *Journal of the American Statistical Association*, 74:829–836, 1979.

99. W. S. Cleveland, E. Grosse, and W. M. Shyu. Local regression models. In J. M. Chambers and T. J. Hastie, editors, *Statistical Models in S*. Chapman & Hall, New York, 1992.

100. W. S. Cleveland and C. Loader. Smoothing by local regression: principles and methods (with discussion). In W. H. Härdle and M. G. Schimek, editors, *Statistical Theory and Computational Aspects of Smoothing*. Springer-Verlag, New York, 1996.

101. M. Clyde. Discussion of "Bayesian model averaging: a tutorial" by Hoeting, Madigan, Raftery and Volinsky. *Statistical Science*, 14(4):382–417, 1999.

102. A. R. Conn, N. I. M. Gould, and P. L. Toint. Convergence of quasi-Newton matrices generated by the symmetric rank one update. *Mathematical Programming*, 50:177–195, 1991.

103. S. D. Conte and C. de Boor. *Elementary Numerical Analysis: An Algorithmic Approach*. McGraw-Hill, New York, 1980.

104. J. Corander and M. J. Sillanpaa. A unified approach to joint modeling of multiple quantitative and qualitative traits in gene mapping. *Journal of Theoretical Biology*, 218(4):435–446, 2002.

105. J. N. Corcoran and R. L. Tweedie. Perfect sampling from independent Metropolis–Hastings chains. *Journal of Statistical Planning and Inference*, 104(2):297–314, 2002.

106. M. K. Cowles. Efficient model-fitting and model-comparison for high-dimensional Bayesian geostatistical models. *Journal of Statistical Planning and Inference*, 112:221–239, 2003.

107. M. K. Cowles and B. P. Carlin. Markov chain Monte Carlo convergence diagnostics: a comparative review. *Journal of the American Statistical Association*, 91(434):883–904, 1996.

108. M. K. Cowles, G. O. Roberts, and J. S. Rosenthal. Possible biases induced by MCMC convergence diagnostics. *Journal of Statistical Computing and Simulation*, 64(1):87–104, 1999.

109. D. R. Cox and D. V. Hinkley. *Theoretical Statistics*. Chapman & Hall, London, 1974.

110. N. A. C. Cressie. *Statistics for Spatial Data*. Wiley, New York, 1993.

111. V. Černy. A thermodynamical approach to the travelling salesman problem: an efficient simulation algorithm. *Journal of Optimization Theory and Applications*, 45:41–55, 1985.

112. G. Dahlquist and Å. Björck, translated by N. Anderson. *Numerical Methods*. Prentice-Hall, Englewood Cliffs, NJ, 1974.

113. P. Damien, J. Wakefield, and S. Walker. Gibbs sampling for Bayesian non-conjugate and hierarchical models by using auxiliary variables. *Journal of the Royal Statistical Society, Series B*, 61:331–344, 1999.

114. G. B. Dantzig. *Linear Programming and Extensions*. Princeton University Press, Princeton, NJ, 1963.

115. W. C. Davidon. Variable metric methods for minimization. AEC Research and Development Report ANL-5990, Argonne National Laboratory, IL, 1959.

116. L. Davis. Applying adaptive algorithms to epistatic domains. In *Proceedings of the 9th Joint Conference on Artificial Intelligence*, pages 162–164. 1985.

117. L. Davis. Job shop scheduling with genetic algorithms. In *Proceedings of the 1st International Conference on Genetic Algorithms and their Applications*, pages 136–140. 1985.

118. L. Davis. Adapting operator probabilities in genetic algorithms. In J. D. Schaffer, editor, *Proceedings of the 3rd International Conference on Genetic Algorithms*. Morgan Kaufmann, San Mateo, CA, 1989.

119. L. Davis, editor. *Handbook of Genetic Algorithms*. Van Nostrand Reinhold, New York, 1991.

120. P. J. Davis and P. Rabinowitz. *Methods of Numerical Integration*. Academic Press, New York, 1984.

121. A. C. Davison, D. Hinkley, and B. J. Worton. Bootstrap likelihoods. *Biometrika*, 79:113–130, 1992.

122. A. C. Davison and D. V. Hinkley. *Bootstrap Methods and their Applications*. Cambridge University Press, Cambridge, 1997.

123. A. C. Davison, D. V. Hinkley, and E. Schechtman. Efficient bootstrap simulation. *Biometrika*, 73:555–566, 1986.

124. C. de Boor. *A Practical Guide to Splines*. Springer-Verlag, New York, 1978.

125. F. R. de Hoog and M. F. Hutchinson. An efficient method for calculating smoothing splines using orthogonal transformations. *Numerische Mathematik*, 50:311–319, 1987.

126. K. A. DeJong. *An Analysis of the Behavior of a Class of Genetic Adaptive Systems*. Ph.D. thesis, University of Michigan, 1975.

127. P. Dellaportas and J. J. Forster. Markov chain Monte Carlo model determination for hierarchical and graphical log-linear models. *Biometrika*, 86(3):615–633, 1999.

128. P. Dellaportas, J. J. Forster, and I. Ntzoufras. On Bayesian model and variable selection using MCMC. *Statistics and Computing*, 12(2):27–36, 2002.

129. B. Delyon, M. Lavielle, and E. Moulines. Convergence of a stochastic approximation version of the EM algorithm. *Annals of Statistics*, 27:94–128, 1999.

130. A. P. Dempster, N. Laird, and D. B. Rubin. Maximum likelihood from incomplete data via the EM algorithm. *Journal of the Royal Statistical Society, Series B*, 39:1–38, 1977.

131. D. G. T. Denison, B. K. Mallick, and A. F. M. Smith. A Bayesian CART algorithm. *Biometrika*, 85:363–377, 1998.

132. J. E. Dennis, Jr., D. M. Gay, and R. E. Welsch. An adaptive nonlinear least-squares algorithm. *ACM Transactions on Mathematical Software*, 7:369–383, 1981.

133. J. E. Dennis, Jr. and R. B. Schnabel. *Numerical Methods for Unconstrained Optimization and Nonlinear Equations*. Prentice-Hall, Englewood Cliffs, NJ, 1983.

134. G. Der and B. S. Everitt. *A Handbook of Statistical Analyses using SAS*. Chapman & Hall/CRC, Boca Raton, FL, 2nd edition, 2002.

135. E. H. Dereksdóttir and K. G. Magnússon. A strike limit algorithm based on adaptive Kalman filtering with application to aboriginal whaling of bowhead whales. *Journal of Cetacean Research and Management*, 5:29–38, 2003.

136. R. D. Deveaux. Finding transformations for regression using the ACE algorithm. *Sociological Methods & Research*, 18(2–3):327–359, 1989.

137. L. Devroye. *Non-uniform Random Variate Generation.* Springer-Verlag, New York, 1986.

138. L. Devroye. *A Course in Density Estimation.* Birkhäuser, Boston, 1987.

139. L. Devroye and L. Györfi. *Nonparametric Density Estimation: The L_1 View.* Wiley, New York, 1985.

140. P. Diaconis and M. Shahshahani. On non-linear functions of linear combinations. *SIAM Journal of Scientific and Statistical Computing*, 5:175–191, 1984.

141. R. Dias and D. Gamerman. A Bayesian approach to hybrid splines non-parametric regression. *Journal of Statistical Computation and Simulation*, 72(4):285–297, 2002.

142. T. J. DiCiccio and B. Efron. Bootstrap confidence intervals (with discussion). *Statistical Science*, 11:189–228, 1996.

143. P. Dierckx. *Curve and Surface Fitting with Splines.* Clarendon Press, New York, 1993.

144. X. K. Dimakos. A guide to exact simulation. *International Statistical Review*, 69(1):27–48, 2001.

145. P. Djuric, Y. Huang, and T. Ghirmai. Perfect sampling: a review and applications to signal processing. *IEEE Transaction on Signal Processing*, 50(2):345–356, 2002.

146. K. A. Dowsland. Simulated annealing. In C. R. Reeves, editor, *Modern Heuristic Techniques for Combinatorial Problems.* Wiley, New York, 1993.

147. N. R. Draper and H. Smith. *Applied Regression Analysis.* Wiley, New York, 1966.

148. R. P. W. Duin. On the choice of smoothing parameter for Parzen estimators of probability density functions. *IEEE Transactions on Computing*, C-25:1175–1179, 1976.

149. R. Durbin, S. Eddy, A. Krogh, and G. Mitchison. *Biological Sequence Analysis: Probabilistic Models of Proteins and Nucleic Acids.* Cambridge University Press, Cambridge, 1998.

150. E. S. Edgington. *Randomization Tests.* Marcel Dekker, New York, 3rd edition, 1995.

151. R. G. Edwards and A. D. Sokal. Generalization of the Fortuin–Kasteleyn–Swendsen–Wang representation and Monte Carlo algorithm. *Physical Review D*, 38(6):2009–2012, 1988.

152. B. Efron. Bootstrap methods: another look at the jackknife. *Annals of Statistics*, 7:1–26, 1979.

153. B. Efron. Nonparametric standard errors and confidence intervals (with discussion). *Canadian Journal of Statistics*, 9:139–172, 1981.

154. B. Efron. *The Jackknife, the Bootstrap, and Other Resampling Plans.* Number 38 in CBMS–NSF Regional Conference Series in Applied Mathematics. SIAM, Philadelphia, 1982.

155. B. Efron. Better bootstrap confidence intervals (with discussion). *Journal of the American Statistical Association*, 82:171–200, 1987.

156. B. Efron. Computer-intensive methods in statistical regression. *SIAM Review*, 30:421–449, 1988.

157. B. Efron and G. Gong. A leisurely look at the bootstrap, the jackknife, and cross-validation. *The American Statistican*, 37:36–48, 1983.

158. B. Efron and D. V. Hinkley. Assessing the accuracy of the maximum likelihood estimator: observed versus expected Fisher information. *Biometrika*, 65:457–482, 1978.

159. B. Efron and R. J. Tibshirani. *An Introduction to the Bootstrap.* Chapman & Hall, New York, 1993.

160. R. J. Elliott and P. E. Kopp. *Mathematics of Financial Markets.* Springer-Verlag, New York, 1999.

161. Environmental Monitoring and Assessment Program, Mid-Atlantic Highlands Streams Assessment, EPA-903-R-00-015, US Environmental Protection Agency, National Health and Environmental Effects Research Laboratory, Western Ecology Division, Corvallis, OR, 2000.

162. V. A. Epanechnikov. Non-parametric estimation of a multivariate probability density. *Theory of Probability and its Applications,* 14:153–158, 1969.

163. L. J. Eshelman, R. A. Caruana, and J. D. Schaffer. Biases in the crossover landscape. In J. D. Schaffer, editor, *Proceedings of the 3rd International Conference on Genetic Algorithms.* Morgan Kaufmann, Los Altos, CA, 1989.

164. R. L. Eubank. *Spline Smoothing and Nonparametric Regression.* Marcel Dekker, New York, 1988.

165. M. Evans. Adaptive importance sampling and chaining. *Contemporary Mathematics,* 115 *(Statistical Multiple Integration)*:137–143, 1991.

166. M. Evans and T. Swartz. *Approximating Integrals via Monte Carlo and Deterministic Methods.* Oxford University Press, Oxford, 2000.

167. U. Faigle and W. Kern. Some convergence results for probabilistic tabu search. *ORSA Journal on Computing,* 4:32–37, 1992.

168. J. Fan and I. Gijbels. *Local Polynomial Modelling and Its Applications.* Chapman & Hall, New York, 1996.

169. J. A. Fill. An interruptible algorithm for perfect sampling via Markov chains. *The Annals of Applied Probability,* 8(1):131–162, 1998.

170. R. A. Fisher. *Design of Experiments.* Hafner, New York, 1935.

171. G. S. Fishman. *Monte Carlo.* Springer-Verlag, New York, 1996.

172. R. Fletcher. A new approach to variable metric algorithms. *Computer Journal,* 13:317–322, 1970.

173. R. Fletcher. *Practical Methods of Optimization.* Wiley, Chichester, UK, 2nd edition, 1987.

174. R. Fletcher and M. J. D. Powell. A rapidly convergent descent method for minimization. *Computer Journal,* 6:163–168, 1963.

175. D. B. Fogel. *Evolutionary Computation: Toward a New Philosophy of Machine Intelligence.* IEEE Press, Piscataway, NJ, 2nd edition, 2000.

176. B. L. Fox. Simulated annealing: folklore, facts, and directions. In H. Niederreiter and P. J. Shiue, editors, *Monte Carlo and Quasi-Monte-Carlo Methods in Scientific Computing.* Springer-Verlag, New York, 1995.

177. E. J. Freireich, E. Gehan, E. Frei III, L. R. Schroeder, I.J. Wolman, R. Anabari, E. O. Burgert, S. D. Mills, D. Pinkel, O. S. Selawry, J. H. Moon, B. R. Gendel, C. L. Spurr, R. Storrs, F. Haurani, B. Hoogstraten, and S. Lee. The effect of 6-Mercaptopurine on the duraction of steriod-induced remissions in acute leukemia: A model for evaluation fo other potentially useful therapy. *Blood,* 21(6):699–716, June 1963.

178. D. Frenkel and B. Smit. *Understanding Molecular Simulation.* Academic Press, New York, 1996.

179. H. Freund and R. Wolter. Evolution of bit strings II: a simple model of co-evolution. *Complex Systems,* 7:25–42, 1993.

180. J. H. Friedman. A variable span smoother. Technical Report 5, Dept. of Statistics, Stanford University, Palo Alto, CA, 1984.

181. J. H. Friedman. Exploratory projection pursuit. *Journal of the American Statistical Association*, 82:249–266, 1987.

182. J. H. Friedman. Multivariate additive regression splines (with discussion). *Annals of Statistics*, 19(1):1–141, 1991.

183. J. H. Friedman and W. Steutzle. Smoothing of scatterplots. Technical Report ORION-003, Dept. of Statistics, Stanford University, Palo Alto, CA, 1982.

184. J. H. Friedman and W. Stuetzle. Projection pursuit regression. *Journal of the American Statistical Association*, 76:817–823, 1981.

185. J. H. Friedman, W. Stuetzle, and A. Schroeder. Projection pursuit density estimation. *Journal of the American Statistical Association*, 79:599–608, 1984.

186. K. Fukunaga. *Introduction to Statistical Pattern Recognition*. Academic Press, New York, 1972.

187. W. A. Fuller. *Introduction to Statistical Time Series*. Wiley, New York, 1976.

188. G. M. Furnival and R. W. Wilson, Jr. Regressions by leaps and bounds. *Technometrics*, 16:499–511, 1974.

189. M. R. Garey and D. S. Johnson. *Computers and Intractability: A Guide to the Theory of NP-Completeness*. Freeman, San Francisco, 1979.

190. C. Gaspin and T. Schiex. Genetic algorithms for genetic mapping. In J.-K. Hao, E. Lutton, E. Ronald, M. Schoenauer, and D. Snyers, editors, *Artificial Evolution 1997*, pages 145–156. Springer-Verlag, New York, 1997.

191. A. Gelfand and A. F. M. Smith. Sampling based approaches to calculating marginal densities. *Journal of the American Statistical Association*, 85:398–409, 1990.

192. A. E. Gelfand, J. A. Silander, Jr., S. Wu, A. Latimer, P. O. Lewis, A. G. Rebelo, and M. Holder. Explaining species distribution patterns through hierarachical modeling (with discussuion). *Bayesian Analysis*, 2005. To appear.

193. A. Gelman. Iterative and non-iterative simulation algorithms. *Computing Science and Statistics*, 24:433–438, 1992.

194. A. Gelman, J. B. Carlin, H. S. Stern, and D. B. Rubin. *Bayesian Data Analysis*. Chapman & Hall, London, 2nd edition, 2004.

195. A. Gelman and X.-L. Meng. Simulating normalizing constants: from importance sampling to bridge sampling to path sampling. *Statistical Science*, 13:163–185, 1998.

196. A. Gelman and D. B. Rubin. Inference from iterative simulation using multiple sequences (with discussion). *Statistical Science*, 7:457–511, 1992.

197. S. Geman and D. Geman. Stochastic relaxation, Gibbs distributions, and the Bayesian restoration of images. *IEEE Transactions on Pattern Analysis and Machine Intelligence*, PAMI-6(6):721–741, 1984.

198. J. E. Gentle. *Random Number Generation and Monte Carlo Methods*. Springer-Verlag, New York, 1998.

199. R. Gentleman and R. Ihaka. The Comprehensive R Archive Network. Available from http://lib.stat.cmu.edu/R/CRAN/, 2003.

200. E. I. George and R. E. McCulloch. Variable selection via Gibbs sampling. *Journal of the American Statistical Association*, 88:881–889, 1993.

201. E. I. George and C. P. Robert. Capture–recapture estimation via Gibbs sampling. *Biometrika*, 79(4):677–683, 1992.

202. C. J. Geyer. Burn-in is unneccessary. Available from http://www.stat.umn.edu/~charlie/mcmc/burn.html.

203. C. J. Geyer. Markov chain Monte Carlo maximum likelihood. In E. Keramigas, editor, *Computing Science and Statistics: The 23rd Symposium on the Interface*. Interface Foundation, Fairfax Station, VA, 1991.

204. C. J. Geyer. Practical Markov chain Monte Carlo (with discussion). *Statistical Science*, 7:473–511, 1992.

205. C. J. Geyer and E. A. Thompson. Constrained Monte Carlo maximum likelihood for dependent data. *Journal of the Royal Statistical Society, Series B*, 54:657–699, 1992.

206. C. J. Geyer and E. A. Thompson. Annealing Markov chain Monte Carlo with applications to ancestral inference. *Journal of the American Statistical Association*, 90:909–920, 1995.

207. Z. Ghahramani. An introduction to hidden Markov models and Bayesian networks. *International Journal of Pattern Recognition and Artificial Intelligence*, 15:9–42, 2001.

208. W. R. Gilks. Derivative-free adaptive rejection sampling for Gibbs sampling. In J. M. Bernardo, J. O. Berger, A. P. Dawid, and A. F. M. Smith, editors, *Bayesian Statistics 4*. Oxford, 1992. Clarendon.

209. W. R. Gilks. Adaptive rejection sampling, MRC Biostatistics Unit, Software from the BSU. Available from http://www.mrc-bsu.cam.ac.uk/BSUsite/Research/software.shtml, 2004.

210. W. R. Gilks, N. G. Best, and K. K. C. Tan. Adaptive rejection Metropolis sampling within Gibbs sampling. *Applied Statistics*, 44:455–472, 1995.

211. W. R. Gilks, S. Richardson, and D. J. Spiegelhalter. *Markov Chain Monte Carlo Methods in Practice*. Chapman & Hall/CRC, London, 1996.

212. W. R. Gilks and G. O. Roberts. Strategies for improving MCMC. In W. R. Gilks, S. Richardson, and D. J. Spiegelhalter, editors, *Markov Chain Monte Carlo in Practice*, pages 89–114. Chapman & Hall/CRC, London, 1996.

213. W. R. Gilks, A. Thomas, and D. J. Spiegelhalter. A language and program for complex Bayesian modeling. *The Statistician*, 43:169–178, 1994.

214. W. R. Gilks and P. Wild. Adaptive rejection sampling for Gibbs sampling. *Applied Statistics*, 41:337–348, 1992.

215. P. E. Gill, G. H. Golub, W. Murray, and M. A. Saunders. Methods for modifying matrix factorizations. *Mathematics of Computation*, 28:505–535, 1974.

216. P. E. Gill and W. Murray. Newton-type methods for unconstrained and linearly constrained optimization. *Mathematical Programming*, 28:311–350, 1974.

217. P. E. Gill, W. Murray, and M. Wright. *Practical Optimization*. Academic Press, London, 1981.

218. P. Giudici and P. J. Green. Decomposable graphical Gaussian model determination. *Biometrika*, 86(4):785–801, 1999.

219. G. H. Givens. Empirical estimation of safe aboriginal whaling limits for bowhead whales. *Journal of Cetacean Research and Management*, 5:39–44, 2003.

220. G. H. Givens, J. R. Beveridge, and B. A. Draper. How features of the human face affect recognition: a statistical comparison of three face recognition algorithms. In *IEEE Conference on Computer Vision and Pattern Recognition*. 2004. To appear.

221. G. H. Givens, J. R. Beveridge, B. A. Draper, and D. Bolme. A statistical assessment of subject factors in the PCA recognition of human faces. In *IEEE Conference on Computer Vision and Pattern Recognition*. December 2003.

222. G. H. Givens and A. E. Raftery. Local adaptive importance sampling for multivariate densities with strong nonlinear relationships. *Journal of the American Statistical Association*, 91:132–141, 1996.

223. J. R. Gleason. Algorithms for balanced bootstrap simulations. *The American Statistican*, 42:263–266, 1988.

224. F. Glover. Tabu search, Part I. *ORSA Journal on Computing*, 1:190–206, 1989.

225. F. Glover. Tabu search, Part II. *ORSA Journal on Computing*, 2:4–32, 1990.

226. F. Glover and H. J. Greenberg. New approaches for heuristic search: a bilateral link with artificial intelligence. *European Journal of Operational Research*, 39:119–130, 1989.

227. F. Glover and M. Laguna. Tabu search. In C. R. Reeves, editor, *Modern Heuristic Techniques for Combinatorial Problems*. Wiley, New York, 1993.

228. F. Glover and M. Laguna. *Tabu Search*. Kluwer, Boston, 1997.

229. F. Glover, E. Taillard, and D. de Werra. A user's guide to tabu search. *Annals of Operations Research*, 41:3–28, 1993.

230. S. J. Godsill. On the relationship between Markov chain Monte Carlo methods for model uncertainty. *Journal of Computational and Graphical Statistics*, 10(2):230–248, 2001.

231. D. E. Goldberg. *Genetic Algorithms in Search, Optimization, and Machine Learning*. Addison-Wesley, Reading, MA, 1989.

232. D. E. Goldberg. A note on Boltzmann tournament selection for genetic algorithms and population-oriented simulated annealing. *Complex Systems*, 4:445–460, 1990.

233. D. E. Goldberg and K. Deb. A comparative analysis of selection schemes used in genetic algorithms. In G. Rawlins, editor, *Foundations of Genetic Algorithms and Classifier Systems*. Morgan Kaufmann, San Mateo, CA, 1991.

234. D. E. Goldberg, K. Deb, and B. Korb. Messy genetic algorithms revisited: studies in mixed size and scale. *Complex Systems*, 4:415–444, 1990.

235. D. E. Goldberg, K. Deb, and B. Korb. Don't worry, be messy. In R. K. Belew and L. B. Booker, editors, *Proceedings of the 4th International Conference on Genetic Algorithms*. Morgan Kaufmann, San Mateo, CA, 1991.

236. D. E. Goldberg, B. Korb, and K. Deb. Messy genetic algorithms: motivation, analysis, and first results. *Complex Systems*, 3:493–530, 1989.

237. D. E. Goldberg and R. Lingle. Alleles, loci, and the travelling salesman problem. In J. J. Grefenstette, editor, *Proceedings of an International Conference on Genetic Algorithms and their Applications*, pages 154–159. Lawrence Erlbaum Associates, Hillsdale, NJ, 1985.

238. D. Goldfarb. A family of variable metric methods derived by variational means. *Mathematics of Computation*, 24:23–26, 1970.

239. A. A. Goldstein. On steepest descent. *SIAM Journal on Control and Optimization*, 3:147–151, 1965.

240. P. I. Good. *Permutation Tests: A Practical Guide to Resampling Methods for Testing Hypotheses*. Springer-Verlag, New York, 2nd edition, 2000.

241. P. I. Good. *Resampling Methods: A Practical Guide to Data Analysis*. Birkhäuser, Boston, 2nd edition, 2001.

242. B. S. Grant and L. L. Wiseman. Recent history of melanism in American peppered moths. *The Journal of Heredity*, 93:86–90, 2002.

243. P. J. Green. Reversible jump Markov chain Monte Carlo computation and Bayesian model determination. *Biometrika*, 82:711–732, 1995.

244. P. J. Green. Trans-dimensional Markov chain Monte Carlo. In P. J. Green, N. L. Hjort, and S. Richardson, editors, *Highly Structured Stochastic Systems*, pages 179–198. Oxford University Press, Oxford, 2003.

245. P. J. Green and B. W. Silverman. *Nonparametric Regression and Generalized Linear Models*. Chapman & Hall, New York, 1994.

246. J. W. Greene and K. J. Supowit. Simulated annealing without rejected moves. In *Proceedings of the IEEE International Conference on Computer Design*. 1984.

247. J. W. Greene and K. J. Supowit. Simulated annealing without rejected moves. *IEEE Transactions on Computer-Aided Design*, CAD-5:221–228, 1986.

248. U. Grenander and M. Miller. Representations of knowledge in complex systems (with discussion). *Journal of the Royal Statistical Society, Series B*, 56:549–603, 1994.

249. B. Grund, P. Hall, and J. S. Marron. Loss and risk in smoothing parameter selection. *Journal of Nonparametric Statistics*, 4:107–132, 1994.

250. C. Gu. Smoothing spline density estimation: a dimensionless automatic algorithm. *Journal of the American Statistical Association*, 88:495–504, 1993.

251. A. Guisan, T. C. Edwards, Jr., and T. Hastie. Generalized linear and generalized additive models in studies of speices distributions: setting the scene. *Ecological Modelling*, 157:89–100, 2002.

252. J. D. F. Habbema, J. Hermans, and K. Van Der Broek. A stepwise discriminant analysis program using density estimation. In G. Bruckman, editor, *COMPSTAT 1974, Proceedings in Computational Statistics*. Vienna, 1974. Physica-Verlag.

253. S. Haber. Numerical evaluation of multiple integrals. *SIAM Review*, 12:481–526, 1970.

254. R. P. Haining. *Spatial Data Analysis: Theory and Practice*. Cambridge University Press, Cambridge, 2003.

255. B. Hajek. Cooling schedules for optimal annealing. *Mathematics of Operations Research*, 13:311–329, 1988.

256. P. Hall. Large sample optimality of least squares cross-validation in density estimation. *Annals of Statistics*, 11:1156–1174, 1983.

257. P. Hall. Antithetic resampling for the bootstrap. *Biometrika*, 76:713–724, 1989.

258. P. Hall. *The Bootstrap and Edgeworth Expansion*. Springer-Verlag, New York, 1992.

259. P. Hall and J. S. Marron. Extent to which least squares cross-validation minimises integrated squared error in nonparametric density estimation. *Probability Theory and Related Fields*, 74:567–581, 1987.

260. P. Hall and J. S. Marron. Lower bounds for bandwidth selection in density estimation. *Probability Theory and Related Fields*, 90:149–173, 1991.

261. P. Hall, J. S. Marron, and B. U. Park. Smoothed cross-validation. *Probability Theory and Related Fields*, 92:1–20, 1992.

262. P. Hall, S. J. Sheather, M. C. Jones, and J. S. Marron. On optimal data-based bandwidth selection in kernel density estimation. *Biometrika*, 78:263–269, 1991.

263. P. Hall and S. R. Wilson. Two guidelines for bootstrap hypothesis testing. *Biometrics*, 47:757–762, 1991.

264. J. M. Hammersley and K. W. Morton. A new Monte Carlo technique: antithetic variates. *Proceedings of the Cambridge Philosophical Society*, 52:449–475, 1956.

265. M. H. Hansen and C. Kooperberg. Spline adaptation in extended linear models (with discussion). *Statistical Science*, 17:2–51, 2002.

266. P. Hansen and B. Jaumard. Algorithms for the maximum satisfiability problem. *Computing*, 44:279–303, 1990.

267. W. Härdle. Resistant smoothing using the fast Fourier transform. *Applied Statistics*, 36:104–111, 1986.

268. W. Härdle. *Applied Nonparametric Regression.* Cambridge University Press, Cambridge, 1990.

269. W. Härdle. *Smoothing Techniques: With Implementation in S.* Springer-Verlag, New York, 1991.

270. W. Härdle, P. Hall, and J. S. Marron. How far are automatically chosen regression smoothing parameters from their optimum? (with discussion). *Journal of the American Statistical Association*, 83:86–99, 1988.

271. W. Härdle, P. Hall, and J. S. Marron. Regression smoothing parameters that are not far from their optimum. *Journal of the American Statistical Association*, 87:227–233, 1992.

272. W. Härdle and J. S. Marron. Random approximations to an error criterion of nonparametric statistics. *Journal of Multivariate Analysis*, 20:91–113, 1986.

273. W. Härdle and M. G. Schimek, editors. *Statistical Theory and Computational Aspects of Smoothing*. Physica-Verlag, Heidelberg, 1996.

274. W. Härdle and D. Scott. Smoothing by weighted averaging using rounded points. *Computational Statistics*, 7:97–128, 1992.

275. G. H. Hardy. Mendelian proportions in a mixed population. *Science*, 28:49–50, 1908.

276. J. A. Hartigan and M. A. Wong. A k-means clustering algorithm. *Applied Statistics*, 28:100–108, 1979.

277. T. J. Hastie. Principal curve library for S. Available from http://lib.stat.cmu.edu/, 2004.

278. T. J. Hastie and D. Pregibon. Generalized linear models. In J. M. Chambers and T. J. Hastie, editors, *Statistical Models in S*. Chapman & Hall, New York, 1993.

279. T. J. Hastie and W. Steutzle. Principal curves. *Journal of the American Statistical Association*, 84:502–516, 1989.

280. T. J. Hastie and R. J. Tibshirani. *Generalized Additive Models.* Chapman & Hall, New York, 1990.

281. T. J. Hastie, R. J. Tibshirani, and J. Friedman. *The Elements of Statistical Learning: Data Mining, Inference, and Prediction*. Springer-Verlag, New York, 2001.

282. W. K. Hastings. Monte Carlo sampling methods using Markov chains and their applications. *Biometrika*, 57:97–109, 1970.

283. P. Henrici. *Elements of Numerical Analysis*. Wiley, New York, 1964.

284. T. Hesterberg. Weighted average importance sampling and defensive mixture distributions. *Technometrics*, 37:185–194, 1995.

285. D. Higdon. Comment on "Spatial statistics and Bayesian computation" by Besag and Green. *Journal of the Royal Statistical Society, Series B*, 55(1):78, 1993.

286. D. M. Higdon. Auxiliary variable methods for Markov chain Monte Carlo with applications. *Journal of the American Statistical Association*, 93:585–595, 1998.

287. S. E. Hills and A. F. M. Smith. Parameterization issues in Bayesian inference. In J. M. Bernardo, J. O. Berger, A. P. Dawid, and A. F. M. Smith, editors, *Bayesian Statistics 4*, pages 227–246. Oxford University Press, Oxford, 1992.

288. J. S. U. Hjorth. *Computer Intensive Statistical Methods: Validation, Model Selection, and Bootstrap*. Chapman & Hall, New York, 1994.

289. J. A. Hoeting, D. Madigan, A. E. Raftery, and C. T. Volinsky. Bayesian model averaging: a tutorial (with discussion). *Statistical Science*, 14:382–417, 1999.

290. J. A. Hoeting, A. E. Raftery, and D. Madigan. Bayesian variable and transformation selection in linear regression. *Journal of Computational and Graphical Statistics*, 11(3):485–507, 2002.

291. J. H. Holland. *Adaptation in Natural and Artificial Systems*. University of Michigan Press, Ann Arbor, 1975.

292. C. C. Holmes and B. K. Mallick. Generalized nonlinear modeling with multivariate free-knot regression splines. *Journal of the American Statistical Association*, 98(462):352–368, 2003.

293. A. Homaifar, S. Guan, and G. E. Liepins. Schema analysis of the traveling salesman problem using genetic algorithms. *Complex Systems*, 6:533–552, 1992.

294. D. W. Hosmer and S. Lemeshow. *Applied Logistic Regression*. Wiley, New York, 2000.

295. Y. F. Huang and P. M. Djuric. Variable selection by perfect sampling. *EURASIP Journal on Applied Signal Processing*, pages 38–45, 2002.

296. P. J. Huber. Projection pursuit. *Annals of Statistics*, 13:435–475, 1985.

297. K. Hukushima and K. Nemoto. Exchange Monte Carlo method and application to spin glass simulations. *Journal of the Physical Society of Japan*, 64:1604–1608, 1996.

298. J. N. Hwang, S. R. Lay, and A. Lippman. Nonparametric multivariate density estimation: a comparative study. *IEEE Transactions on Signal Processing*, 42:2795–2810, 1994.

299. A. C. Jacinto, R. D. Ambrosini, and R. F. Danesi. Experimental and computational analysis of plates under air blast loading. *International Journal of Impact Engineering*, 25:927–947, 2001.

300. M. Jamshidian and R. I. Jennrich. Conjugate gradient acceleration of the EM algorithm. *Journal of the American Statistical Association*, 88:221–228, 1993.

301. M. Jamshidian and R. I. Jennrich. Acceleration of the EM algorithm by using quasi-Newton methods. *Journal of the Royal Statistical Society, Series B*, 59:569–587, 1997.

302. M. Jamshidian and R. I. Jennrich. Standard errors for EM estimation. *Journal of the Royal Statistical Society, Series B*, 62:257–270, 2000.

303. C. Z. Janikow and Z. Michalewicz. An experimental comparison of binary and floating point representations in genetic algorithms. In R. K. Belew and L. B. Booker, editors, *Proceedings of the 4th International Conference on Genetic Algorithms*. Morgan Kaufmann, San Mateo, CA, 1991.

304. B. Jansen. *Interior Point Techniques in Optimization: Complementarity, Sensitivity and Algorithms*. Kluwer, Boston, 1997.

305. P. Jarratt. A review of methods for solving nonlinear algebraic equations in one variable. In P. Rabinowitz, editor, *Numerical Methods for Nonlinear Algebraic Equations*. Gordon and Breach, London, 1970.

306. R. G. Jarrett. A note on the intervals between coal-mining disasters. *Biometrika*, 66:191–193, 1979.

307. H. Jeffreys. *Theory of Probability*. Oxford University Press, New York, 3rd edition, 1961.

308. D. S. Johnson. *Bayesian Analysis of State-Space Models for Discrete Response Compositions*. Ph.D. thesis, Colorado State University, 2003.

309. D. S. Johnson, C. R. Aragon, L. A. McGeoch, and C. Schevon. Optimization by simulated annealing: an experimental evaluation; part I, graph partitioning. *Operations Research*, 37:865–892, 1989.

310. L. W. Johnson and R. D. Riess. *Numerical Analysis*. Addison-Wesley, Reading, MA, 1982.

311. R. W. Johnson. Fitting percentage of body fat to simple body measurements. *Journal of Statistics Education*, 4(1), 1996.

312. M. C. Jones. Variable kernel density estimates. *Australian Journal of Statistics*, 32:361–371, 1990.

313. M. C. Jones. The roles of ISE and MISE in density estimation. *Statistics and Probability Letters*, 12:51–56, 1991.

314. M. C. Jones, J. S. Marron, and B. U. Park. A simple root n bandwidth selector. *Annals of Statistics*, 19:1919–1932, 1991.

315. M. C. Jones, J. S. Marron, and S. J. Sheather. A brief survey of bandwidth selection for density estimation. *Journal of the American Statistical Association*, 91:401–407, 1996.

316. M. C. Jones, J. S. Marron, and S. J. Sheather. Progress in data-based bandwidth selection for kernel density estimation. *Computational Statistics*, 11:337–381, 1996.

317. B. H. Juang and L. R. Rabiner. Hidden Markov models for speech recognition. *Technometrics*, 33:251–272, 1991.

318. N. Karmarkar. A new polynomial-time algorithm for linear programming. *Combinatorica*, 4:373–395, 1984.

319. J. R. Karr and D. R. Dudley. Ecological perspectives on water quality goals. *Environmental Management*, 5(1):55–68, 1981.

320. R. E. Kass, B. P. Carlin, A. Gelman, and R. M. Neal. Markov chain Monte Carlo in practice: a roundtable discussion. *The American Statistican*, 52:93–100, 1998.

321. R. E. Kass and A. E. Raftery. Bayes factors. *Journal of the American Statistical Association*, 90:773–795, 1995.

322. D. E. Kaufman and R. L. Smith. Direction choice for accelerated convergence in hit-and-run sampling. *Operations Research*, 46:84–95, 1998.

323. B. Kégl. Principal curve webpage. Available from http://www.iro.umontreal.ca/~kegl/research/pcurves/.

324. A. G. Z. Kemna and A. C. F. Vorst. A pricing method for options based on average asset values. *Journal of Banking and Finance*, 14:113–129, 1990.

325. M. Kendall and A. Stuart. *The Advanced Theory of Statistics*, volume 1. Macmillan, New York, 4th edition, 1977.

326. W. J. Kennedy, Jr. and J. E. Gentle. *Statistical Computing*. Marcel Dekker, New York, 1980.

327. H. F. Khalfan, R. H. Byrd, and R. B. Schnabel. A theoretical and experimental study of the symmetric rank-one update. *SIAM Journal of Optimization*, 3:1–24, 1993.

328. D. R. Kincaid and E. W. Cheney. *Numerical Analysis*. Wadsworth, Belmont, CA, 1991.

329. R. Kindermann and J. L. Snell. *Markov Random Fields and their Applications*, volume 1 of Contemporary Mathematics. American Mathematical Society, Providence, 1980.

330. S. Kirkpatrick, C. D. Gellat, and M. P. Vecchi. Optimization by simulated annealing. *Science*, 220:671–680, 1983.

331. S. Klinke and J. Grassmann. Projection pursuit regression. In M. G. Schimek, editor, *Smoothing and Regression: Approaches, Computation, and Application*, pages 277–327. Wiley, New York, 2000.

332. T. Kloek and H. K. Van Dijk. Bayesian estimates of equation system parameters: an application of integration by Monte Carlo. *Econometrica*, 46:1–20, 1978.

333. L. Knorr-Held and H. Rue. On block updating in Markov random field models for disease mapping. *Scandinavian Journal of Statistics*, 29(4):567–614, 2002.

334. D. Knuth. *The Art of Computer Programming 2: Seminumerical Algorithms*. Addison-Wesley, Reading, MA, 3rd edition, 1997.

335. M. Kofler. *Maple: An Introduction and Reference*. Addison-Wesley, Reading, MA, 1997.

336. A. Kong, J. S. Liu, and W. H. Wong. Sequential imputations and Bayesian missing data problems. *Journal of the American Statistical Association*, 89:278–288, 1994.

337. A. S. Konrod. *Nodes and Weights of Quadrature Formulas*. Consultants Bureau Enterprises, Inc., New York, 1966.

338. C. Kooperberg. Polspline. Available from http://cran.r-project.org/src/contrib/Descriptions/polspline, 2004.

339. C. Kooperberg and C. J. Stone. Logspline density estimation. *Computational Statistics and Data Analysis*, 12:327–347, 1991.

340. C. Kooperberg and C. J. Stone. Logspline density estimation for censored data. *Journal of Computational and Graphical Statistics*, 1:301–328, 1992.

341. C. Kooperberg, C. J. Stone, and Y. K. Truong. Hazard regression. *Journal of the American Statistical Association*, 90:78–94, 1995.

342. T. Koski. *Hidden Markov Models of Bioinformatics*. Kluwer, Dordrecht, Netherlands, 2001.

343. V. I. Krylov, translated by A. H. Stroud. *Approximate Calculation of Integrals*. Macmillan, New York, 1962.

344. H. R. Künsch. The jackknife and the bootstrap for general stationary observations. *Annals of Statistics*, 17:1217–1241, 1989.

345. C. Lalas and B. Murphy. Increase in the abundance of New Zealand fur seals at the Catlins, South Island, New Zealand. *Journal of the Royal Society of New Zealand*, 28:287–294, 1998.

346. D. Lamberton and B. Lapeyre. *Introduction to Stochastic Calculus Applied to Finance*. Chapman & Hall, London, 1996.

347. K. Lange. A gradient algorithm locally equivalent to the EM algorithm. *Journal of the Royal Statistical Society, Series B*, 57:425–437, 1995.

348. K. Lange. A quasi-Newton acceleration of the EM algorithm. *Statistica Sinica*, 5:1–18, 1995.

349. K. Lange. *Numerical Analysis for Statisticians*. Springer-Verlag, New York, 1999.

350. K. Lange, D. R. Hunter, and I. Yang. Optimization transfer using surrogate objective functions (with discussion). *Journal of Computational and Graphical Statistics*, 9:1–59, 2000.

351. A. B. Lawson. *Statistical Methods in Spatial Epidemiology*. Wiley, New York, 2001.

352. H. Li and G. S. Maddala. Bootstrapping time series models (with discussion). *Econometric Reviews*, 15:115–195, 1996.

353. S. Z. Li. *Markov Random Field Modeling in Image Analysis*. Springer-Verlag, Tokoyo, 2001.

354. R. J. A. Little and D. B. Rubin. *Statistical Analysis with Missing Data*. Wiley, Hoboken, NJ, 2nd edition, 2002.

355. E. L. Little, Jr. *Atlas of United States Trees, Minor Western Hardwoods*, volume 3 of Miscellaneous Publication 1314. US Department of Agriculture, 1976.

356. C. Liu and D. B. Rubin. The ECME algorithm: a simple extension of EM and ECM with faster monotone convergence. *Biometrika*, 81:633–648, 1994.

357. J. S. Liu. *Monte Carlo Strategies in Scientific Computing*. Springer-Verlag, New York, 2001.

358. J. S. Liu and R. Chen. Sequential Monte Carlo methods for dynamical systems. *Journal of the American Statistical Association*, 93:1032–1044, 1998.

359. J. S. Liu, F. Liang, and W. H. Wong. The multiple-try method and local optimization in Metropolis sampling. *Journal of the American Statistical Association*, 95:121–134, 2000.

360. J. S. Liu, D. B. Rubin, and Y. Wu. Parameter expansion to accelerate EM: the PX-EM algorithm. *Biometrika*, 85:755–770, 1998.

361. C. R. Loader. Bandwidth selection: classical or plug-in? *Annals of Statistics*, 27:415–438, 1999.

362. P. O. Loftsgaarden and C. P. Quesenberry. A nonparametric estimate of a multivariate probability density function. *Annals of Mathematical Statistics*, 28:1049–1051, 1965.

363. T. A. Louis. Finding the observed information matrix when using the EM algorithm. *Journal of the Royal Statistical Society, Series B*, 44:226–233, 1982.

364. M. Lundy and A. Mees. Convergence of an annealing algorithm. *Mathematical Programming*, 34:111–124, 1986.

365. S. N. MacEachern and L. M. Berliner. Subsampling the Gibbs sampler. *The American Statistican*, 48(3):188–190, 1994.

366. N. Madras. *Lecture Notes on Monte Carlo Methods*. American Mathematical Society, Providence, RI, 2002.

367. N. Madras and M. Piccioni. Importance sampling for families of distributions. *The Annals of Applied Probability*, 9:1202–1225, 1999.

368. B. A. Maguire, E. S. Pearson, and A. H. A. Wynn. The time intervals between industrial accidents. *Biometrika*, 39:168–180, 1952.

369. C. L. Mallows. Some comments on C_p. *Technometrics*, 15:661–675, 1973.

370. E. Mammen. Resampling methods for nonparametric regression. In M. G. Schimek, editor, *Smoothing and Regression: Approaches, Computation, and Application*. Wiley, New York, 2000.

371. E. Marinari and G. Parisi. Simulated tempering: a new Monte Carlo scheme. *Europhysics Letters*, 19:451–458, 1992.

372. J. S. Maritz. *Distribution Free Statistical Methods*. Chapman & Hall, London, 2nd edition, 1996.

373. J. S. Marron and D. Nolan. Canonical kernels for density estimation. *Statistics and Probability Letters*, 7:195–199, 1988.

374. G. Marsaglia. Random variables and computers. In *Transactions of the Third Prague Conference on Information Theory, Statistical Decision Functions and Random Processes*. Czechoslovak Academy of Sciences, Prague, 1964.

375. G. Marsaglia. The squeeze method for generating gamma variates. *Computers and Mathematics with Applications*, 3:321–325, 1977.

376. G. Marsaglia. The exact-approximation method for generating random variables in a computer. *Journal of the American Statistical Association*, 79:218–221, 1984.

377. G. Marsaglia and W. W. Tsang. A simple method for generating gamma variables. *ACM Transactions on Mathematical Software*, 26:363–372, 2000.

378. W. L. Martinez and A. R. Martinez. *Computational Statistics Handbook with MATLAB.* Chapman & Hall/CRC, Boca Raton, FL, 2002.

379. P. McCullagh and J. A. Nelder. *Generalized Linear Models.* Chapman & Hall, New York, 1989.

380. G. J. McLachlan and T. Krishnan. *The EM Algorithm and Extensions.* Wiley, New York, 1997.

381. I. Meilijson. A fast improvement to the EM algorithm on its own terms. *Journal of the Royal Statistical Society, Series B*, 51:127–138, 1989.

382. J. Meinguet. Multivariate interpolation at arbitrary points made simple. *Journal of Applied Mathematics and Physics*, 30:292–304, 1979.

383. X.-L. Meng. On the rate of convergence of the ECM algorithm. *Annals of Statistics*, 22:326–339, 1994.

384. X.-L. Meng and D. B. Rubin. Using EM to obtain asymptotic variance–covariance matrices: the SEM algorithm. *Journal of the American Statistical Association*, 86:899–909, 1991.

385. X.-L. Meng and D. B. Rubin. Maximum likelihood estimation via the ECM algorithm: a general framework. *Biometrika*, 80:267–278, 1993.

386. X.-L. Meng and D. B. Rubin. On the global and componentwise rates of convergence of the EM algorithm. *Linear Algebra and its Applications*, 199:413–425, 1994.

387. X.-L. Meng and D. van Dyk. The EM algorithm—an old folk-song sung to a fast new tune. *Journal of the Royal Statistical Society, Series B*, 59:511–567, 1997.

388. X.-L. Meng and W. H. Wong. Simulating ratios of normalizing constants via a simple identity: a theoretical exploration. *Statistica Sinica*, 6:831–860, 1996.

389. K. L. Mengersen, C. P. Robert, and C. Guihenneuc-Jouyaux. MCMC convergence diagnostics: a "reviewwww" (with discussion). In J. O. Berger, J. M. Bernardo, A. P. Dawid, D. V. Lindley, and A. F. M. Smith, editors, *Bayesian Statistics 6*, pages 415–440. Oxford University Press, Oxford, 1999.

390. R. C. Merton. Theory of rational option pricing. *Bell Journal of Economics and Management Science*, 4:141–183, 1973.

391. N. Metropolis, A. W. Rosenbluth, M. N. Rosenbluth, A. H. Teller, and E. Teller. Equation of state calculation by fast computing machines. *Journal of Chemical Physics*, 21:1087–1091, 1953.

392. N. Metropolis and S. Ulam. The Monte Carlo method. *Journal of the American Statistical Association*, 44:335–341, 1949.

393. S. P. Meyn and R. L. Tweedie. *Markov Chains and Stochastic Stability.* Springer-Verlag, New York, 1993.

394. Z. Michalewicz. *Genetic Algorithms + Data Structures = Evolution Programs.* Springer-Verlag, New York, 1992.

395. Z. Michalewicz and D. B. Fogel. *How to Solve It: Modern Heuristics.* Springer-Verlag, New York, 2000.

396. A. J. Miller. *Subset Selection in Regression.* Chapman & Hall/CRC, Boca Raton, FL, 2nd edition, 2002.

397. A. Mira, J. Møller, and G. O. Roberts. Perfect slice samplers. *Journal of the Royal Statistical Society, Series B*, 63(3):593–606, 2001.

398. A. Mira and L. Tierney. Efficiency and convergence properties of slice samplers. *Scandinavian Journal of Statistics*, 29(1):1–12, 2002.

399. J. Møller. Perfect simulation of conditionally specified models. *Journal of the Royal Statistical Society, Series B*, 61(1):251–264, 1999.

400. J. F. Monahan. *Numerical Methods of Statistics*. Cambridge University Press, Cambridge, 2001.

401. A. M. Mood, F. A. Graybill, and D. C. Boes. *Introduction to the Theory of Statistics*. McGraw-Hill, New York, 3rd edition, 1974.

402. R. J. Muirhead. *Aspects of Multivariate Statistical Theory*. Wiley, New York, 1982.

403. D. J. Murdoch and P. J. Green. Exact sampling from a continuous state space. *Scandinavian Journal of Statistics*, 25(3):483–502, 1998.

404. D. J. Murdoch and J. S. Rosenthal. Efficient use of exact samples. *Statistics and Computing*, 10:237–243, 2000.

405. W. Murray, editor. *Numerical Methods for Unconstrained Optimization*. Academic Press, New York, 1972.

406. E. A. Nadaraya. On estimating regression. *Theory of Probability and Its Applications*, 10:186–190, 1964.

407. Y. Nagata and S. Kobayashi. Edge assembly crossover: a high-power genetic algorithm for the traveling salesman problem. In T. Bäck, editor, *Proceedings of the 7th International Conference on Genetic Algorithms*. Morgan Kaufmann, Los Altos, CA, 1997.

408. J. C. Naylor and A. F. M. Smith. Applications of a method for the efficient computation of posterior distributions. *Applied Statistics*, 31:214–225, 1982.

409. R. Neal. Sampling from multimodal distributions using tempered transitions. *Statistics and Computing*, 6:353–366, 1996.

410. R. M. Neal. Slice sampling. *The Annals of Statistics*, 31(3):705–767, 1999.

411. J. A. Nelder and R. Mead. A simplex method for function minimization. *Computer Journal*, 7:308–313, 1965.

412. J. Neter, M. H. Kutner, C. J. Nachtsheim, and W. Wasserman. *Applied Linear Statistical Models*. Irwin, Chicago, 1996.

413. M. A. Newton and C. J. Geyer. Bootstrap recycling: a Monte Carlo alternative to the nested bootstrap. *Journal of the American Statistical Association*, 89:905–912, 1994.

414. M. A. Newton and A. E. Raftery. Approximate Bayesian inference with the weighted likelihood bootstrap (with discussion). *Journal of the Royal Statistical Society, Series B*, 56:3–48, 1994.

415. J. Nocedal and S. J. Wright. *Numerical Optimization*. Springer-Verlag, New York, 1999.

416. I. Ntzoufras, P. Dellaportas, and J. J. Forster. Bayesian variable and link determination for generalised linear models. *Journal of Statistical Planning and Inference*, 111(1-2):165–180, 2003.

417. J. Null. Golden Gate Weather Services, Climate of San Francisco. Available from http://ggweather.com/sf/climate.html.

418. Numerical Recipes Home Page. Available from http://www.nr.com, 2003.

419. M. S. Oh and J. O. Berger. Adaptive importance sampling in Monte Carlo integration. *Journal of Statistical Computation and Simulation*, 41:143–168, 1992.

420. M. S. Oh and J. O. Berger. Integration of multimodal functions by Monte Carlo importance sampling. *Journal of the American Statistical Association*, 88:450–456, 1993.

421. I. Oliver, D. Smith, and J. R. Holland. A study of permutation crossover operators on the traveling salesman problem. In J. J. Grefenstette, editor, *Proceedings of the 2nd*

International Conference on Genetic Algorithms, pages 224–230. Lawrence Erlbaum Associates, Hillsdale, NJ, 1987.

422. J. M. Ortega, W. C. Rheinboldt, and J. M. Orrega. *Iterative Solution of Nonlinear Equations in Several Variables*. SIAM, Philadelphia, 2000.

423. A. M. Ostrowski. *Solution of Equations and Systems of Equations*. Academic Press, New York, 2nd edition, 1966.

424. F. O'Sullivan. Discussion of "Some aspects of the spline smoothing approach to nonparametric regression curve fitting" by Silverman. *Journal of the Royal Statistical Society, Series B*, 47:39–40, 1985.

425. C. H. Papadimitriou and K. Steiglitz. *Combinatorial Optimization: Algorithms and Complexity*. Prentice-Hall, Englewood Cliffs, NJ, 1982.

426. B. U. Park and J. S. Marron. Comparison of data-driven bandwidth selectors. *Journal of the American Statistical Association*, 85:66–72, 1990.

427. B. U. Park and B. A. Turlach. Practical performance of several data driven bandwidth selectors. *Computational Statistics*, 7:251–270, 1992.

428. C. Pascutto, J. C. Wakefield, N. G. Best, S. Richardson, L. Bernardinelli, A. Staines, and P. Elliott. Statistical issues in the analysis of disease mapping data. *Statistics in Medicine*, 19:2493–2519, 2000.

429. A. Penttinen. *Modelling Interaction in Spatial Point Patterns: Parameter Estimation by the Maximum Likelihood Method*. Ph.D. thesis, University of Jyväskylä, 1984.

430. A. Philippe. Processing simulation output by Riemann sums. *Journal of Statistical Computation and Simulation*, 59:295–314, 1997.

431. A. Philippe and C. P. Robert. Riemann sums for MCMC estimation and convergence monitoring. *Statistics and Computing*, 11:103–115, 2001.

432. D. B. Phillips and A. F. M. Smith. Bayesian model comparison via jump diffusions. In S. T. Richardson W. R. Gilks and D. J. Spiegelhalter, editors, *Markov Chain Monte Carlo in Practice*, pages 215–240. Chapman & Hall/CRC, London, 1996.

433. E. J. G. Pitman. Significance tests which may be applied to samples from any population. *Royal Statistical Society Supplement*, 4:119–130, 225–232, 1937.

434. E. J. G. Pitman. Significance tests which may be applied to samples from any population. Part iii. The analysis of variance test. *Biometrika*, 29:322–335, 1938.

435. M. J. D. Powell. A view of unconstrained optimization. In L. C. W. Dixon, editor, *Optimization in Action*, pages 53–72. Academic Press, London, 1976.

436. G. Pozrikidis. *Numerical Computation in Science and Engineering*. Oxford University Press, New York, 1998.

437. J. Propp and D. Wilson. Coupling from the past: a user's guide. In D. Aldous and J. Propp, editors, *Microsurveys in Discrete Probability*, volume 41 of DIMACS Series in Discrete Mathematics and Theoretical Computer Science, pages 181–192. American Mathematical Society, 1998.

438. J. G. Propp and D. B. Wilson. Exact sampling with coupled Markov chains and applications to statistical mechanics. *Random Structures and Algorithms*, 9:223–252, 1996.

439. M. H. Protter and C. B. Morrey. *A First Course in Real Analysis*. Springer-Verlag, New York, 1977.

440. J. R. Quinlan. *C4.5 : Programs for Machine Learning*. Morgan Kaufmann, San Mateo, CA, 1993.

441. L. R. Rabiner and B. H. Juang. An introduction to hidden Markov models. *IEEE Acoustics, Speech, and Signal Processing Magazine*, 3:4–16, 1986.

442. N. J. Radcliffe. Equivalence class analysis of genetic algorithms. *Complex Systems*, 5:183–205, 1991.

443. A. E. Raftery and V. E. Akman. Bayesian analysis of a Poisson process with a change point. *Biometrika*, 73:85–89, 1986.

444. A. E. Raftery and S. M. Lewis. How many iterations in the Gibbs sampler? In J. M. Bernardo, J. O. Berger, A. P. Dawid, and A. F. M. Smith, editors, *Bayesian Statistics 4*, pages 763–773. Oxford University Press, Oxford, 1992.

445. A. E. Raftery, D. Madigan, and J. A. Hoeting. Bayesian model averaging for linear regression models. *Journal of the American Statistical Association*, 92:179–191, 1997.

446. A. E. Raftery and J. E. Zeh. Estimating bowhead whale, *Balaena mysticetus*, population size and rate of increase from the 1993 census. *Journal of the American Statistical Association*, 93:451–463, 1998.

447. R. A. Redner and H. F. Walker. Mixture densities, maximum likelihood and the EM algorithm. *SIAM Review*, 26:195–239, 1984.

448. C. R. Reeves. Genetic algorithms. In C. R. Reeves, editor, *Modern Heuristic Techniques for Combinatorial Problems*. Wiley, New York, 1993.

449. C. R. Reeves. A genetic algorithm for flowshop sequencing. *Computers and Operations Research*, 22(1):5–13, 1995.

450. C. R. Reeves and J. E. Rowe. *Genetic Algorithms—Principles and Perspectives*. Kluwer, Norwell, MA, 2003.

451. C. R. Reeves and N. C. Steele. A genetic algorithm approach to designing neural network architecture. In *Proceedings of the 8th International Conference on Systems Engineering*. 1991.

452. J. R. Rice. *Numerical Methods, Software, and Analysis*. McGraw-Hill, New York, 1983.

453. S. Richardson and P. J. Green. On Bayesian analysis of mixtures with an unknown number of components (with discussion). *Journal of the Royal Statistical Society, Series B*, 59:731–792, 1997. Correction, 1998, p. 661.

454. C. J. F. Ridders. 3-point iterations derived from exponential curve fitting. *IEEE Transactions on Circuits and Systems*, 26:669–670, 1979.

455. B. Ripley. Computer generation of random variables. *International Statistical Review*, 51:301–319, 1983.

456. B. Ripley. *Stochastic Simulation*. Wiley, New York, 1987.

457. B. D. Ripley. *Pattern Recognition and Neural Networks*. Cambridge University Press, 1996.

458. C. Ritter and M. A. Tanner. Facilitating the Gibbs sampler: the Gibbs stopper and the griddy-Gibbs sampler. *Journal of the American Statistical Association*, 87(419):861–868, 1992.

459. C. P. Robert. *Discretization and MCMC Convergence Assessment*, volume 135 of Lecture Notes in Statistics. Springer-Verlag, New York, 1998.

460. C. P. Robert and G. Casella. *Monte Carlo Statistical Methods*. Springer-Verlag, New York, 1999.

461. G. O. Roberts, A. Gelman, and W. R. Gilks. Weak convergence and optimal scaling or random walk Metropolis algorithms. *The Annals of Probability*, 7(1):110–120, 1997.

462. G. O. Roberts and J. S. Rosenthal. Convergence of slice sampler Markov chains. *Journal of the Royal Statistical Society, Series B*, 61:643–660, 1999.

463. G. O. Roberts and S. K. Sahu. Updating schemes, correlation structure, blocking and parameterization for the Gibbs sampler. *Journal of the Royal Statistical Society, Series B*, 59(2):291–317, 1997.

464. G. O. Roberts and R. L. Tweedie. Exponential convergence of Langevin diffusions and their discrete approximations. *Bernoulli*, 2:344–364, 1996.

465. C. Roos, T. Terlaky, and J. P. Vial. *Theory and Algorithms for Linear Optimization: An Interior Point Approach*. Wiley, Chichester, UK, 1997.

466. S. M. Ross. *Simulation*. Academic Press, San Diego, CA, 2nd edition, 1997.

467. S. M. Ross. *Introduction to Probability Models*. Academic Press, 7th edition, 2000.

468. R. Y. Rubenstein. *Simulation and the Monte Carlo Method*. Wiley, New York, 1981.

469. D. B. Rubin. The Bayesian bootstrap. *Annals of Statistics*, 9:130–134, 1981.

470. D. B. Rubin. A noniterative sampling/importance resampling alternative to the data augmentation algorithm for creating a few imputations when fractions of missing information are modest: the SIR algorihm. Discussion of M. A. Tanner and W. H. Wong. *Journal of the American Statistical Association*, 82:543–546, 1987.

471. D. B. Rubin. Using the SIR algorithm to simulate posterior distributions. In J. M. Bernardo, M. H. DeGroot, D. V. Lindley, and A. F. Smith, editors, *Bayesian Statistics 3*, pages 395–402. Clarendon Press, Oxford, 1988.

472. M. Rudemo. Empirical choice of histograms and kernel density estimators. *Scandinavian Journal of Statistics*, 9:65–78, 1982.

473. W. Rudin. *Principles of Mathematical Analysis*. McGraw-Hill, New York, 3rd edition, 1976.

474. H. Rue. Fast sampling of Gaussian Markov random fields. *Journal of the Royal Statistical Society, Series B*, 63:325–338, 2001.

475. D. Ruppert, S. J. Sheather, and M. P. Wand. An effective bandwidth selector for local least squares regression. *Journal of the American Statistical Association*, 90:1257–1270, 1995.

476. *S-Plus, Version 6.1*. Copyright 1998, 2002, Insightful Corporation. Available from http://www.insightful.com.

477. S. M. Sait and H. Youssef. *Iterative Computer Algorithms with Applications to Engineering: Solving Combinatorial Optimization Problems*. IEEE Computer Society Press, Los Alamitos, CA, 1999.

478. D. B. Sanders, J. M. Mazzarella, D. C. Kim, J. A. Surace, and B. T. Soifer. The IRAS revised bright galaxy sample (RGBS). *The Astronomical Journal*, 126:1607–1664, 2003.

479. G. Sansone. *Orthogonal Functions*. Interscience Publishers, New York, 1959.

480. D. J. Sargent, J. S. Hodges, and B. P. Carlin. Structured Markov chain Monte Carlo. *Journal of Computational and Graphical Statistics*, 9(2):217–234, 2000.

481. L. Scaccia and P. J. Green. Bayesian growth curves using normal mixtures with nonparametric weights. *Journal of Computational and Graphical Statistics*, 12(2):308–331, 2003.

482. J. D. Schaffer, R. A. Caruana, L. J. Eshelman, and R. Das. A study of control parameters affecting online performance of genetic algorithms for function optimization. In J. D. Schaffer, editor, *Proceedings of the 3rd International Conference on Genetic Algorithms*. Morgan Kaufmann, Los Altos, CA, 1989.

483. T. Schiex and C. Gaspin. CARTHAGENE: constructing and joining maximum likelihood genetic maps. In T. Gaasterland, P. D. Karp, K. Karplus, C. Ouzounis, C. Sander, and

A. Valencia, editors, *Proceedings of the 5th International Conference on Intelligent Systems for Molecular Biology*, pages 258–267. Menlo Park, CA, 1997. Association for Artificial Intelligence (AAAI).

484. M. G. Schimek, editor. *Smoothing and Regression: Approaches, Computation, and Application*. Wiley, New York, 2000.

485. M. G. Schimek and B. A. Turlach. Additive and generalized additive models. In M. G. Schimek, editor, *Smoothing and Regression: Approaches, Computation, and Application*, pages 277–327. Wiley, New York, 2000.

486. U. Schneider and J. N. Corcoran. Perfect simulation for Bayesian model selection in a linear regression model. *Journal of Statistical Planning and Inference*, 126(1):153–171, 2004.

487. C. Schumacher, D. Whitley, and M. Vose. The no free lunch and problem description length. In *Genetic and Evolutionary Computation Conference, GECCO-2001*, pages 565–570. Morgan Kaufmann, San Mateo, CA, 2001.

488. L. L. Schumaker. *Spline Functions: Basic Theory*. Wiley, New York, 1993.

489. E. F. Schuster and G. G. Gregory. On the nonconsistency of maximum likelihood density estimators. In W. G. Eddy, editor, *Proceedings of the Thirteenth Interface of Computer Science and Statistics*, pages 295–298. Springer-Verlag, New York, 1981.

490. G. Schwartz. Estimating the dimension of a model. *Annals of Statistics*, 6:497–511, 1978.

491. D. W. Scott. Average shifted histograms: effective nonparametric estimators in several dimensions. *Annals of Statistics*, 13:1024–1040, 1985.

492. D. W. Scott. *Multivariate Density Estimation: Theory, Practice, and Visualization*. Wiley, New York, 1992.

493. D. W. Scott and L. E. Factor. Monte Carlo study of three data-based nonparametric density estimators. *Journal of the American Statistical Association*, 76:9–15, 1981.

494. D. W. Scott and G. R. Terrell. Biased and unbiased cross-validation in density estimation. *Journal of the American Statistical Association*, 82:1131–1146, 1987.

495. J. M. Scott, P. J. Heglund, M. L. Morrison, J. B. Haufler, M. G. Raphael, W. Q. Wall, and F. B. Samson, editors. *Predicting Species Occurrences—Issues of Accuracy and Scale*. Island Press, Washington, DC, 2002.

496. G. A. F. Seber. *The Estimation of Animal Abundance and Related Parameters*. Charles Griffin, London, 2nd edition, 1982.

497. R. J. Serfling. *Approximation Theorems of Mathematical Statistics*. Wiley, New York, 1980.

498. R. Seydel. *Tools for Computational Finance*. Springer-Verlag, Berlin, 2002.

499. K. Shahookar and P. Mazumder. VLSI cell placement techniques. *ACM Computing Surveys*, 23:143–220, 1991.

500. D. F. Shanno. Conditioning of quasi-Newton methods for function minimization. *Mathematics of Computation*, 24:647–657, 1970.

501. J. Shao and D. Tu. *The Jackknife and Bootstrap*. Springer-Verlag, New York, 1995.

502. S. J. Sheather. The performance of six popular bandwidth selection methods on some real data sets. *Computational Statistics*, 7:225–250, 1992.

503. S. J. Sheather and M. C. Jones. A reliable data-based bandwidth selection method for kernel density estimation. *Journal of the Royal Statistical Society, Series B*, 53:683–690, 1991.

504. G. R. Shorack. *Probability for Statisticians*. Springer-Verlag, New York, 2000.

505. B. W. Silverman. Kernel density estimation using the fast Fourier transform. *Applied Statistics*, 31:93–99, 1982.

506. B. W. Silverman. Some aspects of the spline smoothing approach to non-parametric regression curve fitting (with discussion). *Journal of the Royal Statistical Society, Series B*, 47:1–52, 1985.

507. B. W. Silverman. *Density Estimation for Statistics and Data Analysis*. Chapman & Hall, London, 1986.

508. J. S. Simonoff. *Smoothing Methods in Statistics*. Springer-Verlag, New York, 1996.

509. D. J. Sirag and P. T. Weisser. Towards a unified thermodynamic genetic operator. In J. J. Grefenstette, editor, *Proceedings of the 2nd International Conference on Genetic Algorithms and Their Applications*. Lawrence Erlbaum Associates, Hillsdale, NJ, 1987.

510. A. F. M. Smith and G. O. Roberts. Bayesian computation via the Gibbs sampler and related Markov chain Monte Carlo methods (with discussion). *Journal of the Royal Statistical Society, Series B*, 55:3–23, 1993.

511. A. F. M. Smith, A. M. Skene, J. E. H. Shaw, and J. C. Naylor. Progress with numerical and graphical methods for practical Bayesian statistics. *The Statistician*, 36:75–82, 1987.

512. B. Smith. *Bayesian Output Analysis Program (BOA) User's Manual, Version 1.0*. Dept. of Biostatistics, College of Public Health, University of Iowa, 2003. Available from http://www.public-health.uiowa.edu/boa.

513. P. J. Smith, M. Shafi, and H. Gao. Quick simulation: a review of importance sampling techniques in communications systems. *IEEE Journal on Selected Areas in Communications*, 15:597–613, 1997.

514. D. Sorenson and D. Gianola. *Likelihood, Bayesian and MCMC Methods in Quantitative Genetics*. Springer-Verlag, New York, 2002.

515. D. Spiegelhalter, D. Thomas, N. Best, and D. Lunn. *WinBUGS User Manual, Version 1.4*. MRC Biostatistics Unit, Institute of Public Health, Cambridge, 2003. Available from http://www.mrc-bsu.cam.ac.uk/bugs.

516. D. Steinberg. Salford Systems. Available from http://www.salford-systems.com, 2003.

517. M. Stephens. Bayesian analysis of mixture models with an unknown number of components—an alternative to reversible jump methods. *The Annals of Statistics*, 28(1):40–74, 2000.

518. D. S. Stoffer and K. D. Wall. Bootstrapping state-space models: Gaussian maximum likelihood estimation and the Kalman filter. *Journal of the American Statistical Association*, 86:1024–1033, 1991.

519. C. J. Stone. An asymptotically optimal window selection rule for kernel density estimation. *Annals of Statistics*, 12:1285–1297, 1984.

520. C. J. Stone, M. Hansen, C. Kooperberg, and Y. K. Truong. Polynomial splines and their tensor products in extended linear modeling (with discussion). *Annals of Statistics*, 25:1371–1470, 1997.

521. M. Stone. Cross-validatory choice and assessment of statistical predictions. *Journal of the Royal Statistical Society, Series B*, 36:111–147, 1974.

522. O. Stramer and R. L. Tweedie. Langevin-type models I: diffusions with given stationary distributions, and their discretizations. *Methodology and Computing in Applied Probability*, 1:283–306, 1999.

523. O. Stramer and R. L. Tweedie. Langevin-type models II: self-targeting candidates for MCMC algorithms. *Methodology and Computing in Applied Probability*, 1:307–328, 1999.

524. A. H. Stroud. *Approximate Calculation of Multiple Integrals*. Prentice-Hall, Englewood Cliffs, NJ, 1971.

525. A. H. Stroud and D. Secrest. *Gaussian Quadrature Formulas*. Prentice-Hall, Englewood Cliffs, NJ, 1966.

526. R. H. Swendsen and J.-S. Wang. Nonuniversal critical dynamics in Monte Carlo simulations. *Physical Review Letters*, 58(2):86–88, 1987.

527. G. Syswerda. Uniform crossover in genetic algorithms. In J. D. Schaffer, editor, *Proceedings of the 3rd International Conference on Genetic Algorithms*, pages 2–9. Morgan Kaufmann, Los Altos, CA, 1989.

528. G. Syswerda. Schedule optimization using genetic algorithms. In L. Davis, editor, *Handbook of Genetic Algorithms*, pages 332–349. Van Nostrand Reinhold, New York, 1991.

529. M. A. Tanner. *Tools for Statistical Inference: Methods for the Exploration of Posterior Distributions and Likelihood Functions*. Springer-Verlag, New York, 2nd edition, 1993.

530. M. A. Tanner. *Tools for Statistical Inference: Methods for the Exploration of Posterior Distributions and Likelihood Functions*. Springer-Verlag, New York, 3rd edition, 1996.

531. G. R. Terrell. The maximal smoothing principle in density estimation. *Journal of the American Statistical Association*, 85:470–477, 1990.

532. G. R. Terrell and D. W. Scott. Variable kernel density estimation. *Annals of Statistics*, 20:1236–1265, 1992.

533. T. Therneau and B. Atkinson. An introduction to recursive partitioning using the RPART routines. Technical Report, Mayo Clinic. Available from http://lib.stat.cmu.edu, 1997.

534. R. A. Thisted. *Elements of Statistical Computing: Numerical Computation*. Chapman & Hall, New York, 1988.

535. R. Tibshirani. Estimating optimal transformations for regression via additivity and variance stabilization. *Journal of the American Statistical Association*, 82:559–568, 1988.

536. R. Tibshirani and K. Knight. Model search by bootstrap "bumping". *Journal of Computational and Graphical Statistics*, 8:671–686, 1999.

537. L. Tierney. Markov chains for exploring posterior distributions (with discussion). *Annals of Statistics*, 22:1701–1786, 1994.

538. D. M. Titterington. Recursive parameter estimation using incomplete data. *Journal of the Royal Statistical Society, Series B*, 46:257–267, 1984.

539. H. Tjelmeland and J. Besag. Markov random fields with higher-order interactions. *Scandinavian Journal of Statistics*, 25:415–433, 1998.

540. G. L. Tyler, G. Balmino, D. P. Hinson, W. L. Sjogren, D. E. Smith, R. Woo, J. W. Armstrong, F. M. Flasar, R. A. Simpson, S. Asmar, A. Anabtawi, and P. Priest. Mars Global Surveyor Radio Science Data Products. Data can be obtained from the website http://www-star.stanford.edu/projects/mgs/public.html, 2004.

541. U.S. Environmental Protection Agency, Environmental Monitoring and Assessment Program (EMAP). Available from http://www.epa.gov/emap.

542. D. A. van Dyk and X.-L. Meng. The art of data augmentation (with discussion). *Journal of Computational and Graphical Statistics*, 10(1):1–111, 2001.

543. P. J. M. van Laarhoven and E. H. L. Aarts. *Simulated Annealing: Theory and Applications*. Kluwer, Boston, 1987.

544. W. N. Venables and B. D. Ripley. *Modern Applied Statistics with S-Plus.* Springer-Verlag, New York, 1994.

545. W. N. Venables and B. D. Ripley. *Modern Applied Statistics with S-Plus.* Springer-Verlag, New York, 3rd edition, 2002.

546. J. J. Verbeek. Principal curve webpage. Available from http://carol.wins.uva.nl/~jverbeek/pc/index_en.html.

547. C. Vogl and S. Xu. QTL analysis in arbitrary pedigrees with incomplete marker information. *Heredity*, 89(5):339–345, 2002.

548. M. D. Vose. *The Simple Genetic Algorithm: Foundations and Theory.* MIT Press, Cambridge, MA, 1999.

549. M. D. Vose. Form invariance and implicit parallelism. *Evolutionary Computation*, 9:355–370, 2001.

550. R. Waagepetersen and D. Sorensen. A tutorial on reversible jump MCMC with a view toward applications in QTL-mapping. *International Statistical Review*, 69(1):49–61, 2001.

551. G. Wahba. *Spline Models for Observational Data.* SIAM, Philadelphia, 1990.

552. F. H. Walters, L. R. Parker, S. L. Morgan, and S. N. Deming. *Sequential Simplex Optimization.* CRC Press, Boca Raton, FL, 1991.

553. M. P. Wand and M. C. Jones. *Kernel Smoothing.* Chapman & Hall, New York, 1995.

554. M. P. Wand, J. S. Marron, and D. Ruppert. Transformations in density estimation. *Journal of the American Statistical Association*, 86:343–353, 1991.

555. M. R. Watnik. Pay for play: are baseball salaries based on performance? *Journal of Statistics Education*, 6(2), 1998.

556. G. S. Watson. Smooth regression analysis. *Sankhyā, Series A*, 26:359–372, 1964.

557. G. C. G. Wei and M. A. Tanner. A Monte Carlo implementation of the EM algorithm and the poor man's data augmentation algorithms. *Journal of the American Statistical Association*, 85:699–704, 1990.

558. M. West. Modelling with mixtures. In J. M. Bernardo, M. H. DeGroot, and D. V. Lindley, editors, *Bayesian Statistics 2*, pages 503–524. Oxford, 1992. Oxford University Press.

559. M. West. Approximating posterior distributions by mixtures. *Journal of the Royal Statistical Society, Series B*, 55:409–422, 1993.

560. S. R. White. Concepts of scale in simulated annealing. In *Proceedings of the IEEE International Conference on Computer Design*. 1984.

561. D. Whitley. The GENITOR algorithm and selection pressure: shy rank-based allocation of reproductive trials is best. In J. D. Schaffer, editor, *Proceedings of the 3rd International Conference on Genetic Algorithms*. Morgan Kaufmann, Los Altos, CA, 1989.

562. D. Whitley. A genetic algorithm tutorial. *Statistics and Computing*, 4:65–85, 1994.

563. D. Whitley. An overview of evolutionary algorithms. *Journal of Information and Software Technology*, 43:817–831, 2001.

564. D. Whitley, T. Starkweather, and D. Fuquay. Scheduling problems and traveling salesman: the genetic edge recombination operator. In J. D. Schaffer, editor, *Proceedings of the 3rd International Conference on Genetic Algorithms*, pages 133–140. Morgan Kaufmann, Los Altos, CA, 1989.

565. D. Whitley, T. Starkweather, and D. Shaner. The traveling salesman and sequence scheduling: quality solutions using genetic edge recombination. In L. Davis, editor,

Handbook of Genetic Algorithms, pages 350–372. Von Nostrand Reinhold, New York, 1991.

566. P. Wilmott, J. Dewynne, and S. Howison. *Option Pricing: Mathmatical Models and Computation*. Oxford Financial Press, Oxford, 1997.

567. D. B. Wilson. How to couple from the past using a read-once source of randomness. *Random Structures and Algorithms*, 16(1):85–113, 2000.

568. D. B. Wilson. Web site for perfectly random sampling with Markov chains. Available from http://dbwilson.com/exact, August 2002.

569. G. Winkler. *Image Analysis, Random Fields and Markov Chain Monte Carlo Methods*. Springer-Verlag, Berlin, 2nd edition, 2003.

570. P. Wolfe. Convergence conditions for ascent methods. *SIAM Review*, 11:226–235, 1969.

571. R. Wolfinger and M. O'Connell. Generalized linear models: a pseudo-likelihood approach. *Journal of Computational and Graphical Statistics*, 48:233–243, 1993.

572. S. Wolfram. *Mathematica: A System for Doing Mathematics by Computer*. Addison-Wesley, Redwood City, CA, 1988.

573. D. H. Wolpert and W. G. Macready. No free lunch theorems for search. Technical Report SFI-TR-95-02-010, Santa Fe Institute, NM, 1995.

574. M. A. Woodbury. Discussion of "The analysis of incomplete data" by Hartley and Hocking. *Biometrics*, 27:808–813, 1971.

575. B. J. Worton. Optimal smoothing parameters for multivariate fixed and adaptive kernel methods. *Journal of Statistical Computation and Simulation*, 32:45–57, 1989.

576. C. F. J. Wu. On the convergence properties of the EM algorithm. *Annals of Statistics*, 11:95–103, 1983.

577. H. Youssef, S. M. Sait, K. Nassar, and M. S. T. Benton. Performance driven standard-cell placement using genetic algorithm. In *GLSVLSI'95: Fifth Great Lakes Symposium on VLSI*. 1995.

578. B. Yu and P. Mykland. Looking at Markov samplers through cusum plots: a simple diagnostic idea. *Statistics and Computing*, 8:275–286, 1998.

579. P. Zhang. Nonparametric importance sampling. *Journal of the American Statistical Association*, 91:1245–1253, 1996.

580. W. Zhao, A. Krishnaswamy, R. Chellappa, D. L. Swets, and J. Weng. Discriminant analysis of principal components for face recognition. In H. Wechsler, P. J. Phillips, V. Bruce, F. F. Soulie, and T. S. Huang, editors, *Face Recognition: From Theory to Applications*, pages 73–85. Springer-Verlag, Berlin, 1998.

581. Z. Zheng. On swapping and simulated tempering algorithms. *Stochastic Processes and Their Applications*, 104:131–154, 2003.

Index

A

absolute convergence criterion, 23, 32
accelerated bias-corrected bootstrap, *see* bootstrap, BC_a
accelerated EM methods, 110–113
ACE, *see* alternating conditional expectations
activation function, 358
adaptive importance sampling, 159–160
adaptive kernel, 301
adaptive quadrature, 140
adaptive rejection sampling, 151–155
additive model, 348–351
additive predictor, 352
additivity and variance stabilization, 359–360
AIC, *see* Akaike information criterion
AIDS, 114–115
air blast pressure, 344–345
Aitken acceleration
    for EM algorithm, 110–111
Akaike information criterion, 54
allele, 52
almost everywhere, 13
almost sure convergence, 13
alternating conditional expectations, 358–359
Alzheimer's disease, 124–126, 128–129, 133–135, 138

AMISE, *see* asymptotic mean integrated squared error
annealing, 67
antithetic bootstrap, 269
antithetic sampling, 169–171
aperiodic Markov chain, 16, 184–185
ascent algorithm, 37–39, 56
    backtracking, 37–38
    random ascent, 56
    steepest ascent, 37, 56
    step length, 37
Asian option, 174
aspiration criteria, 63
asymptotic mean integrated squared error, 284, 292, 300
asymptotically unbiased estimator, 14
auxiliary variable methods, 219–223, 225, 241–244
    for Markov random fields, 241–244
AVAS, *see* additivity and variance stabilization

B

backfitting, 43, 349
backtracking, 37–38, 40
bagging, 270
balanced bootstrap, 268–269
balloon estimator, 302
bandwidth, 280–292, 300–301, 303–304

WILEY SERIES IN PROBABILITY AND STATISTICS

ESTABLISHED BY WALTER A. SHEWHART AND SAMUEL S. WILKS

The **Wiley Series in Probability and Statistics** is well established and authoritative. It covers many topics of current research interest in both pure and applied statistics and probability theory. Written by leading statisticians and institutions, the titles span both state-of-the-art developments in the field and classical methods.

Reflecting the wide range of current research in statistics, the series encompasses applied, methodological and theoretical statistics, ranging from applications and new techniques made possible by advances in computerized practice to rigorous treatment of theoretical approaches.

This series provides essential and invaluable reading for all statisticians, whether in academia, industry, government, or research.

† BELSLEY, KUH, and WELSCH · Regression Diagnostics: Identifying Influential Data and Sources of Collinearity

BENDAT and PIERSOL · Random Data: Analysis and Measurement Procedures, *Third Edition*

BERRY, CHALONER, and GEWEKE · Bayesian Analysis in Statistics and Econometrics: Essays in Honor of Arnold Zellner

BERNARDO and SMITH · Bayesian Theory

BHAT and MILLER · Elements of Applied Stochastic Processes, *Third Edition*

BHATTACHARYA and WAYMIRE · Stochastic Processes with Applications

† BIEMER, GROVES, LYBERG, MATHIOWETZ, and SUDMAN · Measurement Errors in Surveys

BILLINGSLEY · Convergence of Probability Measures, *Second Edition*

BILLINGSLEY · Probability and Measure, *Third Edition*

BIRKES and DODGE · Alternative Methods of Regression

BLISCHKE AND MURTHY (editors) · Case Studies in Reliability and Maintenance

BLISCHKE AND MURTHY · Reliability: Modeling, Prediction, and Optimization

BLOOMFIELD · Fourier Analysis of Time Series: An Introduction, *Second Edition*

BOLLEN · Structural Equations with Latent Variables

BOROVKOV · Ergodicity and Stability of Stochastic Processes

BOULEAU · Numerical Methods for Stochastic Processes

BOX · Bayesian Inference in Statistical Analysis

BOX · R. A. Fisher, the Life of a Scientist

BOX and DRAPER · Empirical Model-Building and Response Surfaces

* BOX and DRAPER · Evolutionary Operation: A Statistical Method for Process Improvement

BOX, HUNTER, and HUNTER · Statistics for Experimenters: An Introduction to Design, Data Analysis, and Model Building

BOX and LUCEÑO · Statistical Control by Monitoring and Feedback Adjustment

BRANDIMARTE · Numerical Methods in Finance: A MATLAB-Based Introduction

BROWN and HOLLANDER · Statistics: A Biomedical Introduction

BRUNNER, DOMHOF, and LANGER · Nonparametric Analysis of Longitudinal Data in Factorial Experiments

BUCKLEW · Large Deviation Techniques in Decision, Simulation, and Estimation

CAIROLI and DALANG · Sequential Stochastic Optimization

CASTILLO, HADI, BALAKRISHNAN, and SARABIA · Extreme Value and Related Models with Applications in Engineering and Science

CHAN · Time Series: Applications to Finance

CHATTERJEE and HADI · Sensitivity Analysis in Linear Regression

CHATTERJEE and PRICE · Regression Analysis by Example, *Third Edition*

CHERNICK · Bootstrap Methods: A Practitioner's Guide

CHERNICK and FRIIS · Introductory Biostatistics for the Health Sciences

CHILÈS and DELFINER · Geostatistics: Modeling Spatial Uncertainty

CHOW and LIU · Design and Analysis of Clinical Trials: Concepts and Methodologies, *Second Edition*

CLARKE and DISNEY · Probability and Random Processes: A First Course with Applications, *Second Edition*

* COCHRAN and COX · Experimental Designs, *Second Edition*

CONGDON · Applied Bayesian Modelling

CONGDON · Bayesian Statistical Modelling

CONOVER · Practical Nonparametric Statistics, *Third Edition*

COOK · Regression Graphics

COOK and WEISBERG · Applied Regression Including Computing and Graphics

COOK and WEISBERG · An Introduction to Regression Graphics

*Now available in a lower priced paperback edition in the Wiley Classics Library.
†Now available in a lower priced paperback edition in the Wiley–Interscience Paperback Series.

CORNELL · Experiments with Mixtures, Designs, Models, and the Analysis of Mixture Data, *Third Edition*

COVER and THOMAS · Elements of Information Theory

COX · A Handbook of Introductory Statistical Methods

* COX · Planning of Experiments

CRESSIE · Statistics for Spatial Data, *Revised Edition*

CSÖRGŐ and HORVÁTH · Limit Theorems in Change Point Analysis

DANIEL · Applications of Statistics to Industrial Experimentation

DANIEL · Biostatistics: A Foundation for Analysis in the Health Sciences, *Eighth Edition*

* DANIEL · Fitting Equations to Data: Computer Analysis of Multifactor Data, *Second Edition*

DASU and JOHNSON · Exploratory Data Mining and Data Cleaning

DAVID and NAGARAJA · Order Statistics, *Third Edition*

* DEGROOT, FIENBERG, and KADANE · Statistics and the Law

DEL CASTILLO · Statistical Process Adjustment for Quality Control

DeMARIS · Regression with Social Data: Modeling Continuous and Limited Response Variables

DEMIDENKO · Mixed Models: Theory and Applications

DENISON, HOLMES, MALLICK and SMITH · Bayesian Methods for Nonlinear Classification and Regression

DETTE and STUDDEN · The Theory of Canonical Moments with Applications in Statistics, Probability, and Analysis

DEY and MUKERJEE · Fractional Factorial Plans

DILLON and GOLDSTEIN · Multivariate Analysis: Methods and Applications

DODGE · Alternative Methods of Regression

* DODGE and ROMIG · Sampling Inspection Tables, *Second Edition*

* DOOB · Stochastic Processes

DOWDY, WEARDEN, and CHILKO · Statistics for Research, *Third Edition*

DRAPER and SMITH · Applied Regression Analysis, *Third Edition*

DRYDEN and MARDIA · Statistical Shape Analysis

DUDEWICZ and MISHRA · Modern Mathematical Statistics

DUNN and CLARK · Basic Statistics: A Primer for the Biomedical Sciences, *Third Edition*

DUPUIS and ELLIS · A Weak Convergence Approach to the Theory of Large Deviations

* ELANDT-JOHNSON and JOHNSON · Survival Models and Data Analysis

ENDERS · Applied Econometric Time Series

ETHIER and KURTZ · Markov Processes: Characterization and Convergence

EVANS, HASTINGS, and PEACOCK · Statistical Distributions, *Third Edition*

FELLER · An Introduction to Probability Theory and Its Applications, Volume I, *Third Edition, Revised;* Volume II, *Second Edition*

FISHER and VAN BELLE · Biostatistics: A Methodology for the Health Sciences

FITZMAURICE, LAIRD, and WARE · Applied Longitudinal Analysis

* FLEISS · The Design and Analysis of Clinical Experiments

FLEISS · Statistical Methods for Rates and Proportions, *Third Edition*

FLEMING and HARRINGTON · Counting Processes and Survival Analysis

FULLER · Introduction to Statistical Time Series, *Second Edition*

FULLER · Measurement Error Models

GALLANT · Nonlinear Statistical Models

GHOSH, MUKHOPADHYAY, and SEN · Sequential Estimation

GIESBRECHT and GUMPERTZ · Planning, Construction, and Statistical Analysis of Comparative Experiments

GIFI · Nonlinear Multivariate Analysis

GIVENS and HOETING · Computational Statistics

GLASSERMAN and YAO · Monotone Structure in Discrete-Event Systems
GNANADESIKAN · Methods for Statistical Data Analysis of Multivariate Observations, *Second Edition*
GOLDSTEIN and LEWIS · Assessment: Problems, Development, and Statistical Issues
GREENWOOD and NIKULIN · A Guide to Chi-Squared Testing
GROSS and HARRIS · Fundamentals of Queueing Theory, *Third Edition*
† GROVES · Survey Errors and Survey Costs
* HAHN and SHAPIRO · Statistical Models in Engineering
HAHN and MEEKER · Statistical Intervals: A Guide for Practitioners
HALD · A History of Probability and Statistics and their Applications Before 1750
HALD · A History of Mathematical Statistics from 1750 to 1930
HAMPEL · Robust Statistics: The Approach Based on Influence Functions
HANNAN and DEISTLER · The Statistical Theory of Linear Systems
HEIBERGER · Computation for the Analysis of Designed Experiments
HEDAYAT and SINHA · Design and Inference in Finite Population Sampling
HELLER · MACSYMA for Statisticians
HINKELMAN and KEMPTHORNE: · Design and Analysis of Experiments, Volume 1: Introduction to Experimental Design
HOAGLIN, MOSTELLER, and TUKEY · Exploratory Approach to Analysis of Variance
HOAGLIN, MOSTELLER, and TUKEY · Exploring Data Tables, Trends and Shapes
* HOAGLIN, MOSTELLER, and TUKEY · Understanding Robust and Exploratory Data Analysis
HOCHBERG and TAMHANE · Multiple Comparison Procedures
HOCKING · Methods and Applications of Linear Models: Regression and the Analysis of Variance, *Second Edition*
HOEL · Introduction to Mathematical Statistics, *Fifth Edition*
HOGG and KLUGMAN · Loss Distributions
HOLLANDER and WOLFE · Nonparametric Statistical Methods, *Second Edition*
HOSMER and LEMESHOW · Applied Logistic Regression, *Second Edition*
HOSMER and LEMESHOW · Applied Survival Analysis: Regression Modeling of Time to Event Data
† HUBER · Robust Statistics
HUBERTY · Applied Discriminant Analysis
HUNT and KENNEDY · Financial Derivatives in Theory and Practice
HUSKOVA, BERAN, and DUPAC · Collected Works of Jaroslav Hajek— with Commentary
HUZURBAZAR · Flowgraph Models for Multistate Time-to-Event Data
IMAN and CONOVER · A Modern Approach to Statistics
† JACKSON · A User's Guide to Principle Components
JOHN · Statistical Methods in Engineering and Quality Assurance
JOHNSON · Multivariate Statistical Simulation
JOHNSON and BALAKRISHNAN · Advances in the Theory and Practice of Statistics: A Volume in Honor of Samuel Kotz
JOHNSON and BHATTACHARYYA · Statistics: Principles and Methods, *Fifth Edition*
JOHNSON and KOTZ · Distributions in Statistics
JOHNSON and KOTZ (editors) · Leading Personalities in Statistical Sciences: From the Seventeenth Century to the Present
JOHNSON, KOTZ, and BALAKRISHNAN · Continuous Univariate Distributions, Volume 1, *Second Edition*
JOHNSON, KOTZ, and BALAKRISHNAN · Continuous Univariate Distributions, Volume 2, *Second Edition*
JOHNSON, KOTZ, and BALAKRISHNAN · Discrete Multivariate Distributions

*Now available in a lower priced paperback edition in the Wiley Classics Library.
†Now available in a lower priced paperback edition in the Wiley–Interscience Paperback Series.

JOHNSON, KOTZ, and KEMP · Univariate Discrete Distributions, *Second Edition*
JUDGE, GRIFFITHS, HILL, LÜTKEPOHL, and LEE · The Theory and Practice of
Econometrics, *Second Edition*
JUREČKOVÁ and SEN · Robust Statistical Procedures: Aymptotics and Interrelations
JUREK and MASON · Operator-Limit Distributions in Probability Theory
KADANE · Bayesian Methods and Ethics in a Clinical Trial Design
KADANE AND SCHUM · A Probabilistic Analysis of the Sacco and Vanzetti Evidence
KALBFLEISCH and PRENTICE · The Statistical Analysis of Failure Time Data, *Second
Edition*
KASS and VOS · Geometrical Foundations of Asymptotic Inference
KAUFMAN and ROUSSEEUW · Finding Groups in Data: An Introduction to Cluster
Analysis
KEDEM and FOKIANOS · Regression Models for Time Series Analysis
KENDALL, BARDEN, CARNE, and LE · Shape and Shape Theory
KHURI · Advanced Calculus with Applications in Statistics, *Second Edition*
KHURI, MATHEW, and SINHA · Statistical Tests for Mixed Linear Models
* KISH · Statistical Design for Research
KLEIBER and KOTZ · Statistical Size Distributions in Economics and Actuarial Sciences
KLUGMAN, PANJER, and WILLMOT · Loss Models: From Data to Decisions,
Second Edition
KLUGMAN, PANJER, and WILLMOT · Solutions Manual to Accompany Loss Models:
From Data to Decisions, *Second Edition*
KOTZ, BALAKRISHNAN, and JOHNSON · Continuous Multivariate Distributions,
Volume 1, *Second Edition*
KOTZ and JOHNSON (editors) · Encyclopedia of Statistical Sciences: Volumes 1 to 9
with Index
KOTZ and JOHNSON (editors) · Encyclopedia of Statistical Sciences: Supplement
Volume
KOTZ, READ, and BANKS (editors) · Encyclopedia of Statistical Sciences: Update
Volume 1
KOTZ, READ, and BANKS (editors) · Encyclopedia of Statistical Sciences: Update
Volume 2
KOVALENKO, KUZNETZOV, and PEGG · Mathematical Theory of Reliability of
Time-Dependent Systems with Practical Applications
LACHIN · Biostatistical Methods: The Assessment of Relative Risks
LAD · Operational Subjective Statistical Methods: A Mathematical, Philosophical, and
Historical Introduction
LAMPERTI · Probability: A Survey of the Mathematical Theory, *Second Edition*
LANGE, RYAN, BILLARD, BRILLINGER, CONQUEST, and GREENHOUSE ·
Case Studies in Biometry
LARSON · Introduction to Probability Theory and Statistical Inference, *Third Edition*
LAWLESS · Statistical Models and Methods for Lifetime Data, *Second Edition*
LAWSON · Statistical Methods in Spatial Epidemiology
LE · Applied Categorical Data Analysis
LE · Applied Survival Analysis
LEE and WANG · Statistical Methods for Survival Data Analysis, *Third Edition*
LePAGE and BILLARD · Exploring the Limits of Bootstrap
LEYLAND and GOLDSTEIN (editors) · Multilevel Modelling of Health Statistics
LIAO · Statistical Group Comparison
LINDVALL · Lectures on the Coupling Method
LINHART and ZUCCHINI · Model Selection
LITTLE and RUBIN · Statistical Analysis with Missing Data, *Second Edition*
LLOYD · The Statistical Analysis of Categorical Data

---

*Now available in a lower priced paperback edition in the Wiley Classics Library.
†Now available in a lower priced paperback edition in the Wiley–Interscience Paperback Series.

*Now available in a lower priced paperback edition in the Wiley Classics Library.

†Now available in a lower priced paperback edition in the Wiley–Interscience Paperback Series.

PRESS · Subjective and Objective Bayesian Statistics, *Second Edition*

PRESS and TANUR · The Subjectivity of Scientists and the Bayesian Approach

PUKELSHEIM · Optimal Experimental Design

PURI, VILAPLANA, and WERTZ · New Perspectives in Theoretical and Applied Statistics

PUTERMAN · Markov Decision Processes: Discrete Stochastic Dynamic Programming

* RAO · Linear Statistical Inference and Its Applications, *Second Edition*

RAUSAND and HØYLAND · System Reliability Theory: Models, Statistical Methods, and Applications, *Second Edition*

RENCHER · Linear Models in Statistics

RENCHER · Methods of Multivariate Analysis, *Second Edition*

RENCHER · Multivariate Statistical Inference with Applications

* RIPLEY · Spatial Statistics

RIPLEY · Stochastic Simulation

ROBINSON · Practical Strategies for Experimenting

ROHATGI and SALEH · An Introduction to Probability and Statistics, *Second Edition*

ROLSKI, SCHMIDLI, SCHMIDT, and TEUGELS · Stochastic Processes for Insurance and Finance

ROSENBERGER and LACHIN · Randomization in Clinical Trials: Theory and Practice

ROSS · Introduction to Probability and Statistics for Engineers and Scientists

† ROUSSEEUW and LEROY · Robust Regression and Outlier Detection

* RUBIN · Multiple Imputation for Nonresponse in Surveys

RUBINSTEIN · Simulation and the Monte Carlo Method

RUBINSTEIN and MELAMED · Modern Simulation and Modeling

RYAN · Modern Regression Methods

RYAN · Statistical Methods for Quality Improvement, *Second Edition*

SALTELLI, CHAN, and SCOTT (editors) · Sensitivity Analysis

* SCHEFFE · The Analysis of Variance

SCHIMEK · Smoothing and Regression: Approaches, Computation, and Application

SCHOTT · Matrix Analysis for Statistics, *Second Edition*

SCHOUTENS · Levy Processes in Finance: Pricing Financial Derivatives

SCHUSS · Theory and Applications of Stochastic Differential Equations

SCOTT · Multivariate Density Estimation: Theory, Practice, and Visualization

* SEARLE · Linear Models

SEARLE · Linear Models for Unbalanced Data

SEARLE · Matrix Algebra Useful for Statistics

SEARLE, CASELLA, and McCULLOCH · Variance Components

SEARLE and WILLETT · Matrix Algebra for Applied Economics

SEBER and LEE · Linear Regression Analysis, *Second Edition*

† SEBER · Multivariate Observations

† SEBER and WILD · Nonlinear Regression

SENNOTT · Stochastic Dynamic Programming and the Control of Queueing Systems

* SERFLING · Approximation Theorems of Mathematical Statistics

SHAFER and VOVK · Probability and Finance: It's Only a Game!

SILVAPULLE and SEN · Constrained Statistical Inference: Inequality, Order, and Shape Restrictions

SMALL and McLEISH · Hilbert Space Methods in Probability and Statistical Inference

SRIVASTAVA · Methods of Multivariate Statistics

STAPLETON · Linear Statistical Models

STAUDTE and SHEATHER · Robust Estimation and Testing

STOYAN, KENDALL, and MECKE · Stochastic Geometry and Its Applications, *Second Edition*

STOYAN and STOYAN · Fractals, Random Shapes and Point Fields: Methods of Geometrical Statistics

STYAN · The Collected Papers of T. W. Anderson: 1943–1985
SUTTON, ABRAMS, JONES, SHELDON, and SONG · Methods for Meta-Analysis in Medical Research
TANAKA · Time Series Analysis: Nonstationary and Noninvertible Distribution Theory
THOMPSON · Empirical Model Building
THOMPSON · Sampling, *Second Edition*
THOMPSON · Simulation: A Modeler's Approach
THOMPSON and SEBER · Adaptive Sampling
THOMPSON, WILLIAMS, and FINDLAY · Models for Investors in Real World Markets
TIAO, BISGAARD, HILL, PEÑA, and STIGLER (editors) · Box on Quality and Discovery: with Design, Control, and Robustness
TIERNEY · LISP-STAT: An Object-Oriented Environment for Statistical Computing and Dynamic Graphics
TSAY · Analysis of Financial Time Series
UPTON and FINGLETON · Spatial Data Analysis by Example, Volume II: Categorical and Directional Data
VAN BELLE · Statistical Rules of Thumb
VAN BELLE, FISHER, HEAGERTY, and LUMLEY · Biostatistics: A Methodology for the Health Sciences, *Second Edition*
VESTRUP · The Theory of Measures and Integration
VIDAKOVIC · Statistical Modeling by Wavelets
VINOD and REAGLE · Preparing for the Worst: Incorporating Downside Risk in Stock Market Investments
WALLER and GOTWAY · Applied Spatial Statistics for Public Health Data
WEERAHANDI · Generalized Inference in Repeated Measures: Exact Methods in MANOVA and Mixed Models
WEISBERG · Applied Linear Regression, *Third Edition*
WELSH · Aspects of Statistical Inference
WESTFALL and YOUNG · Resampling-Based Multiple Testing: Examples and Methods for *p*-Value Adjustment
WHITTAKER · Graphical Models in Applied Multivariate Statistics
WINKER · Optimization Heuristics in Economics: Applications of Threshold Accepting
WONNACOTT and WONNACOTT · Econometrics, *Second Edition*
WOODING · Planning Pharmaceutical Clinical Trials: Basic Statistical Principles
WOODWORTH · Biostatistics: A Bayesian Introduction
WOOLSON and CLARKE · Statistical Methods for the Analysis of Biomedical Data, *Second Edition*
WU and HAMADA · Experiments: Planning, Analysis, and Parameter Design Optimization
YANG · The Construction Theory of Denumerable Markov Processes
* ZELLNER · An Introduction to Bayesian Inference in Econometrics
ZHOU, OBUCHOWSKI, and McCLISH · Statistical Methods in Diagnostic Medicine